Springer Series in Operations Research and Financial Engineering

Series Editors

Thomas V. Mikosch, Køebenhavns Universitet, Copenhagen, Denmark

Sidney I. Resnick, Cornell University, Ithaca, USA

Bert Zwart, Centrum Wiskunde & Informatica, Amsterdam, The Netherlands

Editorial Board

Torben G. Andersen, Northwestern University, Evanston, USA

Dmitriy Drusvyatskiy, University of Washington, Seattle, USA

Avishai Mandelbaum, Technion - Israel Institute of Technology, Haifa, Israel

Jack Muckstadt, Cornell University, Ithaca, USA

Per Mykland, University of Chicago, Chicago, USA

Philip E. Protter, Columbia University, New York, USA

Claudia Sagastizabal, IMPA – Instituto Nacional de Matemáti, Rio de Janeiro, Brazil

David B. Shmoys, Cornell University, Ithaca, USA

David Glavind Skovmand, Køebenhavns Universitet, Copenhagen, Denmark

Josef Teichmann, ETH Zürich, Zürich, Switzerland

The Springer Series in Operations Research and Financial Engineering publishes monographs and textbooks on important topics in theory and practice of Operations Research, Management Science, and Financial Engineering. The Series is distinguished by high standards in content and exposition, and special attention to timely or emerging practice in industry, business, and government. Subject areas include:

Linear, integer and non-linear programming including applications; dynamic programming and stochastic control; interior point methods; multi-objective optimization; Supply chain management, including inventory control, logistics, planning and scheduling; Game theory Risk management and risk analysis, including actuarial science and insurance mathematics; Queuing models, point processes, extreme value theory, and heavy-tailed phenomena; Networked systems, including telecommunication, transportation, and many others; Quantitative finance: portfolio modeling, options, and derivative securities; Revenue management and quantitative marketing Innovative statistical applications such as detection and inference in very large and/or high dimensional data streams; Computational economics

Rafael Correa · Abderrahim Hantoute ·
Marco A. López

Fundamentals of Convex Analysis and Optimization

A Supremum Function Approach

Rafael Correa
Department of Mathematical Engineering
(DIM), Center for Mathematical
Modeling (CMM)
University of Chile
Santiago, Chile

Universidad de O'Higgins
Rancagua, Chile

Marco A. López
Department of Mathematics
University of Alicante
Alicante, Spain

CIAO, Federation University
Ballarat, Australia

Abderrahim Hantoute
Department of Mathematics
University of Alicante
Alicante, Spain

ISSN 1431-8598   ISSN 2197-1773 (electronic)
Springer Series in Operations Research and Financial Engineering
ISBN 978-3-031-29550-8   ISBN 978-3-031-29551-5 (eBook)
https://doi.org/10.1007/978-3-031-29551-5

Mathematics Subject Classification: 90C25, 90C46, 90C48, 49J53, 46N10, 49K27, 52A41, 90C31, 45A03, 49K40, 49N15

© The Editor(s) (if applicable) and The Author(s), under exclusive license to Springer Nature Switzerland AG 2023

This work is subject to copyright. All rights are solely and exclusively licensed by the Publisher, whether the whole or part of the material is concerned, specifically the rights of translation, reprinting, reuse of illustrations, recitation, broadcasting, reproduction on microfilms or in any other physical way, and transmission or information storage and retrieval, electronic adaptation, computer software, or by similar or dissimilar methodology now known or hereafter developed.

The use of general descriptive names, registered names, trademarks, service marks, etc. in this publication does not imply, even in the absence of a specific statement, that such names are exempt from the relevant protective laws and regulations and therefore free for general use.

The publisher, the authors, and the editors are safe to assume that the advice and information in this book are believed to be true and accurate at the date of publication. Neither the publisher nor the authors or the editors give a warranty, expressed or implied, with respect to the material contained herein or for any errors or omissions that may have been made. The publisher remains neutral with regard to jurisdictional claims in published maps and institutional affiliations.

This Springer imprint is published by the registered company Springer Nature Switzerland AG
The registered company address is: Gewerbestrasse 11, 6330 Cham, Switzerland

To our wives, Isabel M., Perla, and María Pilar
To Halima (in memoriam) and Mohammed

Preface

This book provides a novel approach to convex analysis and convex optimization, based on subdifferential calculus of pointwise suprema of convex functions. The main goal in writing this book consists of analyzing the subdifferential of the supremum of an arbitrary collection of convex functions, defined on a separated locally convex space, in terms of the subdifferentials of the data functions. We provide a series of results in this line, but in different settings and under different assumptions. Since many convex functions such as the Fenchel conjugate, the sum, the composition with affine applications, etc. can be written as a supremum, the formulas provided in the book lead to calculus rules unifying many results in the literature.

We present in this book a rather original approach to convex analysis, which we hope will be appreciated by the community of optimizers, mathematical analysts, operations researchers, etc. The contents and style are appropriate for graduate and doctoral students in mathematics, economics, physics, engineering, etc., and also for specialists and practitioners. The book offers a source of alternative perspectives to developments supplied in other convex analysis and optimization texts, and it provides challenging and motivating material for experts in functional analysis and convex geometry who may be interested in the applications of their work. A considerable part of the book, more precisely chapters 2, 3, and 4, constitutes a source of valuable didactic material for an advanced course on convex analysis and optimization.

Deriving calculus rules for subdifferentials is one of the main issues raised in convex analysis. The study of the subdifferential of the supremum function has attracted the attention of specialists in convex analysis; in fact, many earlier contributions dealing with pointwise supremum functions can be found in the literature, starting in the 1960s. Let us quote a paragraph from the second page of J.-B. Hiriart-Urruty, in his article. "Convex analysis and optimization in the past 50 years: some snapshots" [107]: "One of the most specific constructions in convex or nonsmooth analysis is certainly taking the supremum of a (possibly infinite) collection of functions". In the years 1965–1970, various calculus rules concerning the subdifferential of supremum functions started to emerge; working in that direction and using various assumptions, several authors contributed to these calculus rules. Among them we can cite B. N. Pshenichnyi, A. D. Ioffe, V. L. Levin, R. T. Rockafellar, A. Sotskov, etc. The mathematical interest in the main subject of the book has been widely recognized by prestigious authors since the very beginning of convex analysis history. A sample of remarkable contributions to this topic are attributed to A. Brøndsted [29], F. H. Clarke [40], J. M. Danskin [67], J.-B. Hiriart-Urruty and R. R. Phelps [111], A. D. Ioffe and V. H. Tikhomirov [115], V. L. Levin [131], O. Lopez and L. Thibault [139], B. S. Mordukhovich and T. Nghia [158], B. N. Pshenichnyi [169], R. T. Rockafellar [177], M. Valadier [191], M. Volle [195], etc. See, for instance, V. M. Tikhomirov [190] to trace out the historical origins of the subject. A short historical review of some of these results appears in the introduction and bibliographic notes of the corresponding chapters.

Furthermore, any formula for the subdifferential of the supremum function can be seen as a useful tool in deriving Karush–Kuhn–Tucker optimality conditions for a convex optimization problem. This is due to the fact that any family of convex constraints, even an infinite one, can be replaced with a unique convex constraint via the supremum function. An alternative approach consists of replacing the constraints with the indicator function of the feasible set. It turns out that, under certain constraint qualifications, its subdifferential (i.e., the normal cone to the feasible set) appears in the so-called Fermat optimality principle, and its relation to the subdifferential of the supremum function can then be exploited.

The context of locally convex spaces has been chosen with the aim of proposing formulations of our results at their maximum level of generality in order to compare them with the pre-existing results in the literature. Around 40 examples have been included to clarify the meaning of concepts and results. Each chapter concludes with a list of exercises which are strongly related to its contents. There are in total 131 exercises, and their solutions, detailed or schematic, are given in chapter 9 ("Exercises - Solutions"). Some of them are used inside the proofs of some results to shorten them by keeping exclusively the core part of the arguments. The last section of each chapter, entitled "Bibliographical notes", is devoted to supplying historical notes, comments on related results, etc. The following paragraphs summarize the contents of the book.

- Introductory chapter 1 intends to motivate the reader and provide a detailed account of the objectives and contents of the book, and their relation to antecedents in the literature.

- Section 2.1 in chapter 2 contains background material, including a brief introduction to locally convex spaces, duality pairs, weak topologies, separation theorems, and the Banach–Alaoglu–Bourbaki theorem, among other basic results. The relation between convexity and continuity is reviewed in section 2.2, whereas the last section of this chapter, section 2.3, presents typical examples of convex functions.

- Chapter 3 provides an extensive background on convex analysis, including preliminary results and the notation used in the book. Special emphasis is put on the Fenchel–Moreau–Rockafellar theorem (Theorem 3.2.2) and its consequences, including dual representations of the support function of sublevel sets and extensions of the minimax theorem. The purpose of this chapter is to show that such a crucial theorem constitutes the main tool for deriving many other fundamental results of convex analysis.

- Chapter 4 complements chapter 3 by presenting an in-depth study of key concepts and results involving the subdifferential and the Fenchel conjugate, such as the duality theory, integration of the subdifferential, convexity in Banach spaces, subdifferential calculus rules, etc. Some results and/or their proofs are new, justifying their inclusion in a book that aspires to become a reference text.

- Chapter 5 is a key part of the book, with characterizations of the subdifferential of the supremum function provided. It starts by presenting some particular formulas for the approximate subdifferential of the supremum function, which are proved using classic tools of convex analysis presented in chapter 4. Next, the main formula of the subdifferential, involving exclusively the ε-subdifferential of the data functions, is given in Theorem 5.2.2. Other simpler formulas are derived under additional continuity assumptions.

- The main purpose of chapter 6 is to specify the fundamental results of the previous chapter in the presence of specific structures such as the so-called compact-continuous setting. This allows simpler characterizations which appeal to the subgradients of active (instead of ε-active) functions. In order to unify the last setting and the general framework, different compactification procedures are proposed in section 6.2. All previous characterizations include, aside from the subdifferentials of ε-active (or active) functions, additional terms relying on the normal cone to the effective domain of the supremum function (or finite-dimensional sections of such a domain). Alternatively, homogeneous formulas where this normal cone is dropped out are also derived in section 6.4. The last section of this chapter presents a family of different qualification conditions under which the above formulas become more manageable.

- Chapter 7 emphasizes the unifying character of the supremum function in modeling most operations in convex analysis. Indeed, the primary purpose of this chapter is to characterize the subdifferential of the sum and the composition with linear mappings. To this aim, general formulas involving the ε-subgradients are established for convex functions with a mere lower semicontinuity-like property (see section 7.1). Further symmetric and asymmetric qualifications in section 7.2 allow replacing the ε-subgradients with (exact) subgradients, and both settings are combined in section 7.3. Particular instances are examined in relation to the family of infinite-dimensional polyhedral functions.

- Chapter 8 presents some selected topics which are closely related to the material in the previous chapters. These topics can be seen, to a certain extent, as theoretical applications of the main results in the book. The first part contains extensions of the Farkas lemma for systems of convex inequalities, and provides constraint qualifications, such as the Farkas–Minkowski property in infinite convex optimization. Section 8.2 is intended to establish different Karush–Kuhn–Tucker and Fritz–John optimality conditions for infinite convex optimization. The approach adopted consists of replacing the set of constraints with a single one associated with a supremum function, and appeals to properties of its subdifferentials, exhaustively studied in the previous chapters. The chapter also gives an account of other applications devoted to convexification processes in optimization, integration in locally convex spaces, variational characterizations of convexity, and the theory of Chebychev sets.

Much of the book has come out of work done at the Universidad de Alicante (UA), Universidad de Chile (UCH), and Universidad de O'Higgins (UOH), over the last ten years. We would like to acknowledge the support provided by the Center for Mathematical Modeling (CMM-UCH) and the Department of Mathematics of UA. We also would like to express our gratitude to our colleagues and friends of the Optimization Laboratory of UA, the Optimization and Equilibrium Group of CMM, and the optimization group of the Engineering Institute of UOH, who helped to create the ideal scientific atmosphere for the production of this book. We are particularly grateful to all our collaborators of the papers and projects that are used in the book. Our thanks also go to Pedro Pérez-Aros and Anton Svensson for their help in the revision of the manuscript, to Donna Chernyk, Editor in Mathematics, Casey Russell, Editorial Assistant, and Boopalan Renu and Velmurugan Vidyalakshmi, responsible for the production team, for their assistance during the publication process, as well as to Lisa Lupiani and Verónica Ojeda for their technical support. Funding support for the authors' research has been provided by the following projects: Contract Beatriz Galindo BEA-GAL 18/00205 of MICIU (Spain) and Universidad de Alicante, AICO/2021/165 of Generalitat Valenciana, PGC2018-097960-B-C21 of MICINN (Spain), Fondecyt 1190012 and 1190110 of ANID (Chile), and Basal FB210005 (Chile). The research of the third author was also supported by the Australian ARC—Discovery Projects DP 180100602.

Santiago, Chile	Rafael Correa
Alicante, Spain	Abderrahim Hantoute
Alicante, Spain	Marco A. López

Contents

Contents xi

1 Introduction 1
    1.1 Motivation 2
    1.2 Historical antecedents 5
    1.3 Working framework and objectives 7

2 Preliminaries 15
    2.1 Functional analysis background 15
    2.2 Convexity and continuity 36
    2.3 Examples of convex functions 46
    2.4 Exercises 52
    2.5 Bibliographical notes 56

3 Fenchel–Moreau–Rockafellar theory 59
    3.1 Conjugation theory 60
    3.2 Fenchel–Moreau–Rockafellar theorem 68
    3.3 Dual representations of support functions 75
    3.4 Minimax theory 83
    3.5 Exercises 91
    3.6 Bibliographical notes 94

4 Fundamental topics in convex analysis 95

- 4.1 Subdifferential theory 95
- 4.2 Convex duality 131
- 4.3 Convexity in Banach spaces 142
- 4.4 Subdifferential integration 150
- 4.5 Exercises 165
- 4.6 Bibliographical notes 171

5 Supremum of convex functions 173

- 5.1 Conjugacy-based approach 174
- 5.2 Main subdifferential formulas 187
- 5.3 The role of continuity assumptions 207
- 5.4 Exercises 217
- 5.5 Bibliographical notes 223

6 The supremum in specific contexts 227

- 6.1 The compact-continuous setting 227
- 6.2 Compactification approach 241
- 6.3 Main subdifferential formula revisited 259
- 6.4 Homogeneous formulas 267
- 6.5 Qualification conditions 274
- 6.6 Exercises 279
- 6.7 Bibliographical notes 282

7 Other subdifferential calculus rules 283

- 7.1 Subdifferential of the sum 283
- 7.2 Symmetric versus asymmetric conditions 287
- 7.3 Supremum-sum subdifferential calculus 296
- 7.4 Exercises 303
- 7.5 Bibliographical notes 304

8 Miscellaneous 307

- 8.1 Convex systems and Farkas-type qualifications 308
- 8.2 Optimality and duality in (semi)infinite convex optimization 317
- 8.3 Convexification processes in optimization 330
- 8.4 Non-convex integration 342
- 8.5 Variational characterization of convexity 350

	8.6	Chebychev sets and convexity	354
	8.7	Exercises	358
	8.8	Bibliographical notes	361
9	**Exercises - Solutions**		365
	9.1	Exercises of chapter 2	365
	9.2	Exercises of chapter 3	373
	9.3	Exercises of chapter 4	380
	9.4	Exercises of chapter 5	394
	9.5	Exercises of chapter 6	407
	9.6	Exercises of chapter 7	417
	9.7	Exercises of chapter 8	420

Glossary of notations 427

Bibliography 431

Index 441

Chapter 1

Introduction

Whenever an operation on a family of convex functions preserves convexity, one naturally wonders whether the subdifferential of this new function can be written in terms of the subdifferentials of the original data functions.

It is well-known that many operations with convex functions preserve convexity. Specific to convex analysis is the classical operation which consists of taking the pointwise supremum of an arbitrarily indexed family of convex functions. It has no equivalence in the classical theory of differential analysis, and constitutes a largely used tool in convex optimization, in theory as well as in practice (see, for instance, [8], [108], and references therein). Another motivation for developing calculus rules for the supremum function is the need for handling convex analysis tools in the framework of stability and well/ill-posedness in linear semi-infinite optimization (see [33–35, 100] and others).

The main goal of this book is to present explicit characterizations for the subdifferential mapping of the supremum function of an arbitrarily indexed family of convex functions, exclusively in terms of data functions. The convex functions we consider in the book are general, defined on a separated locally convex space, and not necessarily lower semicontinuous or even proper. In the most general context, the so-called non-continuous setting, the index set over which the supremum is taken is arbitrary, without any algebraic or topological structure. Moreover, generally we do not assume regularity conditions such as continuity of the supremum function, continuity of the data functions, conditions on their domains, and the like.

Since many convex functions can be written as the supremum of particular families of convex functions (perhaps, affine), the second main goal of the book is to develop a unified approach for the framework of calculus rules in convex analysis. In fact, our characterizations of the subdifferential of the supremum function also allow us to obtain formulas for the subdifferential of the resulting function in many operations such as the sum of convex functions, the composition of an affine continuous mapping with a convex function, and conjugation. In this way, we provide more direct and easier proofs for the basic chain rules when some supplementary qualification conditions are assumed, and our approach gives rise to a unifying view of many well-known calculus rules in convex analysis.

1.1 Motivation

In order to provide motivation by showing the practical interest of the supremum function, the main subject of this book, we present in this section some examples developed in the Euclidean space $\mathbb{R}^n$ for the sake of simplicity, although the framework of the book is that of locally convex spaces. The first one constitutes an application to the following optimization problem:

$$(\text{P}) \quad \text{Min } g(x)$$
$$\text{s.t. } f_t(x) \leq 0, \ t \in T,$$

where $g, f_t : \mathbb{R}^n \to \mathbb{R}, \ t \in T$, with T being an arbitrary index set. The supremum function can be used here to transform (P) into an unconstrained problem

$$(\widetilde{\text{P}}) \quad \text{Min } f(x)$$
$$\text{s.t. } x \in \mathbb{R}^n.$$

In fact, it is easy to see that every optimal solution $\bar{x}$ of (P) is also optimal for the unconstrained problem $(\widetilde{\text{P}})$, taking as the objective

1.1. MOTIVATION

function the supremum function $f : \mathbb{R}^n \to \mathbb{R} \cup \{+\infty\}$ given by

$$f(x) := \sup\{g(x) - g(\bar{x});\ f_t(x),\ t \in T\}.$$

Actually, for any other feasible point x, one has $g(x) - g(\bar{x}) \geq 0$, and so $f(x) \geq f(\bar{x}) = 0$. If x is not feasible, one has $\sup_t f_t(x) > 0$. In this way, one could derive optimality conditions for (P) just by dealing with its unconstrained representation involving the supremum function f.

The supremum function also allows unifying many other operations in convex analysis. For instance, given a function $g : \mathbb{R}^n \to \mathbb{R}$, a convex function $h : \mathbb{R}^m \to \mathbb{R}$, and a linear mapping $A : \mathbb{R}^n \to \mathbb{R}^m$ with transpose A', the function $f := g + h \circ A$ can be written as

$$\begin{aligned} f(x) &= g(x) + h(Ax) \\ &= g(x) + \sup\{\langle y, Ax \rangle - h^*(y) : y \in \mathbb{R}^m\} \\ &= \sup\{g(x) + \langle A'y, x \rangle - h^*(y) : y \in \mathbb{R}^m\}, \end{aligned}$$

where $h^* : \mathbb{R}^m \to \mathbb{R} \cup \{+\infty\}$ is the conjugate of h; i.e., $h^*(y) := \sup\{\langle y, x \rangle - h(x) : x \in X\}$. The linearization of h above is made possible thanks to the Moreau theorem (see Theorem 3.2.2), which establishes in this case that h and its biconjugate h^{**} coincide. In other words, f is the supremum of the family of functions $g(\cdot) + \langle A'y, \cdot \rangle - h^*(y)$, $y \in \mathbb{R}^m$, which are sums of g and affine functions. Moreover, provided that g is also convex, f becomes the supremum of affine functions

$$f(x) = \sup\{\langle z + A'y, x \rangle - g^*(z) - h^*(y) : y, z \in \mathbb{R}^m\}.$$

The Moreau theorem is a key tool in convex analysis and its consequences are conveniently exploited throughout the book. In fact, it is equivalent to the property that the conjugate of $f := \sup_{t \in T} f_t$, where the functions $f_t : \mathbb{R}^n \to \mathbb{R}$, $t \in T$, are convex, coincides with the closed convex hull of the function $\inf_{t \in T} f_t^*$ (see Corollary 3.2.7).

Supremum functions can also be used to study geometrical properties of sublevel sets of a function $h : \mathbb{R}^n \to \mathbb{R}$; i.e.,

$$[h \leq 0] := \{x \in \mathbb{R}^n : h(x) \leq 0\}.$$

This analysis can be done by considering, for instance, the indicator function $\mathrm{I}_{[h \leq 0]} : \mathbb{R}^n \to \mathbb{R} \cup \{+\infty\}$, which is equal to 0 if $h(x) \leq 0$ and $+\infty$ otherwise. We observe that

$$\mathrm{I}_{[h\leq 0]}(x) = \sup\{(\alpha h)(x) : \alpha > 0\}$$

and, consequently, normal vectors to the set $[h \leq 0]$ can be seen as subgradients of the supremum function $f(\cdot) := \sup\{(\alpha h)(\cdot) : \alpha > 0\}$. This information is used in section 8.2 to derive optimality conditions for problem (P) by writing it as

$$(\text{P}) \quad \text{Min } g(x) + \mathrm{I}_{[h\leq 0]}(x)$$
$$\text{s.t. } x \in \mathbb{R}^n,$$

where $h := \sup_{t \in T} f_t$. Additional qualification conditions would then lead to Karush–Kuhn–Tucker optimality conditions involving subgradients of both g and a finite number of functions f_t. Results related to the Farkas lemma are also obtained in section 8.1 by appealing to the supremum function.

The supremum function is useful in convexification processes as well. Let us consider the optimization problem $\text{Min}_{x \in X} f(x)$, where $f : \mathbb{R}^n \to \mathbb{R}$ is a general function, possibly non-convex, and its associated convex relaxation given by $\text{Min}_{x \in X}(\overline{co} f)(x)$, where $(\overline{co} f) : \mathbb{R}^n \to \mathbb{R}$ is the closed convex hull of f, assumed here to be proper for simplicity. Then one can see that both problems have the same optimal value, and that every optimal solution of the original problem is also optimal for its convex relaxation. We wonder how we can obtain optimal solutions of the original problem from the optimal solutions of the relaxed one. At this moment, as it is detailed in section 8.3, the supremum function comes into play since the optimal solutions of the relaxed problem turn out to be subgradients of f^* at the origin (remember that f^* is itself a supremum function).

Many specific functions in convex and functional analysis can be expressed as supremum functions. For instance, in a normed space $(X, \|\cdot\|)$, the dual norm is defined as $\|x^*\|_* := \sup\{\langle x^*, x\rangle : \|x\| \leq 1\}$; that is, $\|\cdot\|_*$ is the supremum of continuous linear functions. Similarly, the primal dual norm is also the supremum of continuous linear functions on X^*, as we can show that $\|x\| = \sup\{\langle x^*, x\rangle : \|x^*\|_* \leq 1\}$. More generally, the support and gauge functions of a set in X or X^* are examples of supremum functions.

The supremum function is also a useful device in robust optimization. Actually, in the case that problem (P) includes some uncertainty, say (P) is written as

$$(\text{P}) \quad \text{Min } f(x, u)$$
$$\text{s.t. } x \in \mathbb{R}^n, \ u \in \mathcal{U},$$

for some given set of parameters $\mathcal{U}$, the so-called pessimistic robust counterpart is the ordinary optimization problem with objective function $f := \sup_{u \in \mathcal{U}} f(\cdot, u)$.

1.2 Historical antecedents

In what follows, we will summarize the contents of the book, and relate them to their antecedents by means of a brief presentation of the state of the art. The common thread throughout the book is the supremum of arbitrary families of convex functions. More precisely, we consider a family of convex functions $\{f_t,\ t \in T\}$, where T is an arbitrary index set, defined on a (real) separated locally convex space X with values in the extended real line $\overline{\mathbb{R}}$, and its associated pointwise supremum function

$$f := \sup_{t \in T} f_t.$$

This new function is convex and our aim is to express its subdifferential by means of the subdifferential of the f_t's.

We start in chapters 3 and 4 by reviewing the main and fundamental results of convex analysis, via the crucial Fenchel–Moreau–Rockafellar theorem (Theorem 3.2.2). This result actually constitutes a conjugation rule for the (proper) supremum function f as (see Proposition 3.2.6 and Corollary 3.2.7)

$$f^* := \overline{\mathrm{co}}\left(\inf_{t \in T} f_t^*\right),$$

where the star in the superscript represents the Fenchel conjugate and $\overline{\mathrm{co}}$ is used for the closed convex hull. Chapter 3 also includes dual representations of the support function of sublevel sets and extensions of the minimax theorem, all presented as consequences of the above-mentioned Fenchel–Moreau–Rockafellar theorem. Along the same line, we develop in chapter 4 an in-depth study of key concepts and results involving the subdifferential and the Fenchel conjugate, covering the convex duality theory, integration of the subdifferential, convexity in Banach spaces, standard subdifferential calculus rules, etc. Some results of these two chapters and/or their proofs are new.

Concerning the subdifferential of the supremum function f, when the index set T is finite and the data functions f_t's are all continuous, a basic result due to Dubovitskii and Milyutin (see, e.g., [115]) asserts that, at every point $x \in X$, the subdifferential of f is completely characterized by means of the subdifferential of active data functions at the reference point x. More explicitly, we have

$$\partial f(x) = \mathrm{co}\left(\bigcup_{t \in T(x)} \partial f_t(x)\right),$$

where

$$T(x) := \{t \in T : f_t(x) = f(x)\}$$

is the set of active indices, and co stands for the convex hull. The last formula extends to the more general setting, where T is a separated compact topological space and the function $(t, z) \mapsto f_t(z)$ is upper semicontinuous (usc, for short) with respect to t and continuous with respect to z. In this case, the subdifferential of f is given by

$$\partial f(x) = \overline{\mathrm{co}}\left(\bigcup_{t \in T(x)} \partial f_t(x)\right), \tag{1.1}$$

where $\overline{\mathrm{co}}$ stands for the closed convex hull. The closure in the last formula can be removed in the finite-dimensional setting. Formula (1.1) goes back to [191], whereas the case of finite-valued functions defined on $\mathbb{R}^n$ was established in [131] (see, also, [115, Section 4.2]). The continuity assumption on the data functions was weakened in [177, Theorem 4] and [196] in the case of finitely many functions. The case of the maximum of a finite family of Fréchet-differentiable functions was studied in [67] and [169], and extended to Lipschitz functions in [40] using the notion of generalized gradient (see, also, [130]). A recent development for non-convex Lipschitz functions is due to [157] using different concepts of nonsmooth subdifferentials.

Even in simple situations, involving finitely many convex functions, the problem is not easy, and simple examples in the Euclidean space show that these nice formulas do not hold in general (e.g., Example 5.2.1). Nevertheless, in order to overcome this difficulty, Brøndsted [29] used the concept of ε-subdifferential to establish the following formula, which is valid when $T = \{1, 2, ..., k\}$ and all functions f_i, $i \in T$, agree at x and belong to $\Gamma_0(X)$, the family of proper convex and lower semicontinuous (lsc, in brief) functions:

$$\partial f(x) = \bigcap_{\varepsilon>0} \overline{\mathrm{co}} \left(\bigcup_{i=1,\ldots,k} \partial_\varepsilon f_i(x) \right).$$

In regard to infinite collections of convex functions (T is infinite), a remarkable result was established by Valadier in [191] where, assuming that the supremum function f is finite and continuous at x, the subdifferential $\partial f(x)$ is expressed by considering not only x but all nearby points around it. More precisely, in the particular context of normed spaces, and denoting by $\|\cdot\|$ the norm in X, the following formula is given in [191]:

$$\partial f(x) = \bigcap_{\varepsilon>0} \overline{\mathrm{co}} \{\bigcup \partial f_t(y) : \ t \in T_\varepsilon(x) \text{ and } \|y - x\| \leq \varepsilon\},$$

where $T_\varepsilon(x)$ is the set of ε-active indices,

$$T_\varepsilon(x) := \{t \in T : \ f_t(x) \geq f(x) - \varepsilon\}.$$

By using the concept of ε-subdifferential, Volle [195] obtained in normed spaces another characterization of $\partial f(x)$ where only the nominal point x appears:

$$\partial f(x) = \bigcap_{\varepsilon>0} \overline{\mathrm{co}} \left(\bigcup_{t \in T_\varepsilon(x)} \partial_\varepsilon f_t(x) \right). \tag{1.2}$$

It is straightforward that the two formulas above are identical if the functions f_t are affine. Moreover, when the space X is Banach, each of the two formulas can be obtained from the other one. This equivalence is proved by using the Brøndsted–Rockafellar theorem, expressing the ε-subdifferential by means of exact subdifferentials at nearby points. Actually, the advantage of using enlargements of the subdifferential, such as the ε-subdifferential, is to avoid qualification-type conditions. This idea is exploited in the survey paper [110] to provide many calculus rules without regularity conditions.

1.3 Working framework and objectives

Throughout this book we work in a (real) separated locally convex space (lcs) X. By $\mathcal{N}_X$ and $\mathcal{P}$ we represent a neighborhood base of closed convex balanced neighborhoods of θ (θ-neighborhoods) and a saturated family of seminorms generating the topology in X, respectively. The topological dual space of X is X^*, and the associated duality pairing is represented by $\langle \cdot, \cdot \rangle$. Unless otherwise stated, X^* is endowed with a locally convex topology which makes (X, X^*) a compatible dual pair. When another lcs Y is given, the lcs $X \times Y$ is endowed with the locally convex topology of the Cartesian product associated with the bilinear form $\langle (x^*, y^*), (x, y) \rangle = \langle x^*, x \rangle + \langle y^*, y \rangle_Y$, where $\langle \cdot, \cdot \rangle_Y$ is the bilinear form in Y, and $x^* \in X^*$, $y^* \in Y^*$, $x \in X$, $y \in Y$. All these concepts and related results will be defined in section 2.1.

Given a set $C \subset X^*$, by $\operatorname{cl} C$ or, indistinctly $\overline{C}$, we usually refer to the closure of C with respect to such a compatible topology. One of these topologies is the w^*-topology; however, sometimes we write $\operatorname{cl}^{w^*} C$ for those results which are specifically valid for the w^*-topology (as in the Alaoglu–Banach–Bourbaki theorem). Of course, if C is convex, then $\operatorname{cl}^{w^*} C$ coincides with the closure of C for any other compatible topology.

The following key formula will constitute the most general representation of the subdifferential of the supremum function $f = \sup_{t \in T} f_t$ when f_t, $t \in T$, are given convex functions defined on the lcs X. It was established in [103, Theorem 4] (see [141, Theorem 4] and [134] for related formulas) when $f_t \in \Gamma_0(X)$, $t \in T$:

$$\partial f(x) = \bigcap_{L \in \mathcal{F}(x),\, \varepsilon > 0} \overline{\operatorname{co}} \left(\bigcup_{t \in T_\varepsilon(x)} \partial_\varepsilon f_t(x) + \mathrm{N}_{L \cap \operatorname{dom} f}(x) \right), \quad (1.3)$$

where $\operatorname{dom} f$ is the (effective) domain of f,

$$\mathcal{F}(x) := \{ L \subset X : L \text{ is a finite-dimensional linear subspace such that } x \in L \}, \quad (1.4)$$

and $\mathrm{N}_{L \cap \operatorname{dom} f}(x)$ is the normal cone to $L \cap \operatorname{dom} f$ at the point x. In fact, (1.3) is also valid for general convex functions $f_t : X \to \overline{\mathbb{R}}$, $t \in T$, under the following closedness condition, which is held in various situations (e.g., Proposition 5.2.4),

$$\operatorname{cl} f = \sup_{t \in T}(\operatorname{cl} f_t), \quad (1.5)$$

where cl also stands for the closed hull.

A variant of formula (1.3), which does not assume condition (1.5), involves the augmented functions $f_t + \mathrm{I}_{L \cap \operatorname{dom} f}$ instead of the original f_t's:

1.3. WORKING FRAMEWORK AND OBJECTIVES 9

$$\partial f(x) = \bigcap_{L \in \mathcal{F}(x), \varepsilon > 0} \overline{co} \left(\bigcup_{t \in T_\varepsilon(x)} \partial_\varepsilon (f_t + I_{L \cap \mathrm{dom}\, f})(x) \right), \qquad (1.6)$$

where $I_{L \cap \mathrm{dom}\, f}$ is the indicator function of the set $L \cap \mathrm{dom}\, f$. If, additionally, we are in the so-called continuous-compact setting, which stands for the compactness of the index set T and the upper semi-continuity of the mappings $t \in T \mapsto f_t(z)$, $z \in X$, we get

$$\partial f(x) = \bigcap_{L \in \mathcal{F}(x)} co \left\{ \bigcup_{t \in T(x)} \partial (f_t + I_{L \cap \mathrm{dom}\, f})(x) \right\},$$

where now, instead of $T_\varepsilon(x)$, we use $T(x) := T_0(x)$ the set of active indices at x. The last formula has the advantage that the closure and the intersection over ε in (1.6) are dropped out. We can always appeal to a compactification strategy consisting of compactifying T (Stone–Čech or one-point extensions), and enlarging the original set of functions by taking upper limits of the mappings $t \mapsto f_t(z)$, $z \in X$. The main results in this line are Theorems 6.2.5 and 6.2.8.

Under additional continuity assumptions, we can remove the finite-dimensional sections $L \cap \mathrm{dom}\, f$ in (1.3). More specifically, if the closedness criterion (1.5) holds and, for a given $x \in \mathrm{dom}\, f$, we assume that either $\mathrm{ri}(\mathrm{cone}(\mathrm{dom}\, f - x)) \neq \emptyset$ or $\mathrm{cone}(\mathrm{dom}\, f - x)$ is closed, where $\mathrm{ri}\, C$ and $\mathrm{cone}\, C$ refer to the relative interior of C and the cone generated by $C \cup \{\theta\}$, respectively, then

$$\partial f(x) = \bigcap_{\varepsilon > 0} \overline{co} \left(\bigcup_{t \in T_\varepsilon(x)} \partial_\varepsilon f_t(x) + \mathrm{N}_{\mathrm{dom}\, f}(x) \right). \qquad (1.7)$$

In addition, if the interior of $\mathrm{cone}(\mathrm{dom}\, f - x)$ is non-empty, then (1.7) gives rise to

$$\partial f(x) = \mathrm{N}_{\mathrm{dom}\, f}(x) + \bigcap_{\varepsilon > 0} \overline{co} \left(\bigcup_{t \in T_\varepsilon(x)} \partial_\varepsilon f_t(x) \right). \qquad (1.8)$$

In particular, if f is finite and continuous at the nominal point x, then $\mathrm{N}_{\mathrm{dom}\, f}(x) = \{\theta\}$ and (1.8) yields (1.2).

If $f_t \in \Gamma_0(X)$, $t \in T$, and we are in the continuous-compact setting introduced above, the following result, involving only the active functions at the reference point x, is established in [53, Theorem 3.8],

$$\partial f(x) = \bigcap_{L \in \mathcal{F}(x),\, \varepsilon > 0} \overline{\operatorname{co}} \left(\bigcup_{t \in T(x)} \partial_\varepsilon f_t(x) + \mathrm{N}_{L \cap \operatorname{dom} f}(x) \right). \tag{1.9}$$

In order to avoid the presence of normal cones, and without requiring the continuity of f at the nominal point x, we provide in the continuous-compact setting in section 6.4 the following formula, for a family $\{f_t,\, t \in T\} \subset \Gamma_0(X)$ satisfying the condition $\inf_{t \in T} f_t(x) > -\infty$,

$$\partial f(x) = \bigcap_{\varepsilon > 0} \overline{\operatorname{co}} \left(\left(\bigcup_{t \in T(x)} \partial_\varepsilon f_t(x) \right) + \left(\bigcup_{t \in T \setminus T(x)} \{0, \varepsilon\} \partial_\varepsilon f_t(x) \right) \right). \tag{1.10}$$

The proof of (1.10) uses the following expression for the normal cone $\mathrm{N}_{\operatorname{dom} f}(x)$, with $x \in \operatorname{dom} f$ and assuming that $\inf_{t \in T} f_t(x) > -\infty$:

$$\mathrm{N}_{\operatorname{dom} f}(x) = \left[\overline{\operatorname{co}} \left(\bigcup_{t \in T} \partial_\varepsilon f_t(x) \right) \right]_\infty \quad \text{for every } \varepsilon > 0,$$

where C_∞ stands for the recession cone of C. If we remove the continuous-compact assumption, we get in section 6.4 a refinement of (1.10) involving finite subfamilies of indices.

In chapter 7, and based on our main formula (1.3), we obtain the Hiriart-Urruty and Phelps formula ([111])

$$\partial(g + f \circ A)(x) = \bigcap_{\varepsilon > 0} \operatorname{cl}\left(\partial_\varepsilon g(x) + A^* \partial_\varepsilon f(Ax) \right),$$

involving two convex functions $f : Y \to \overline{\mathbb{R}}$ and $g : X \to \overline{\mathbb{R}}$, where Y is also a (real) separated locally convex space, and $A : X \to Y$ is a continuous linear mapping with continuous adjoint A^*. The last formula assumes that $\operatorname{cl}(g + f \circ A) = (\operatorname{cl} g) + (\operatorname{cl} f) \circ A$, instead of the slightly stronger assumption that $f \in \Gamma_0(Y)$, $g \in \Gamma_0(X)$. In fact, that assumption constitutes a counterpart to the closedness condition (1.5) for the sum operation.

In this chapter we also deal with asymmetric subdifferential sum rules, which are given in terms of the exact subdifferential of one function (the most qualified one), and the approximate subdifferential of the other. They require that the domains of f and g overlap sufficiently, or that the epigraphs enjoy certain closedness-type properties. More precisely, we prove that for $f, g \in \Gamma_0(X)$ and $x \in \operatorname{dom} f \cap \operatorname{dom} g$, if at least one of the following conditions, involving the effective

1.3. WORKING FRAMEWORK AND OBJECTIVES

domains and the epigraphs, holds: (i) $\mathbb{R}_+(\operatorname{epi} g - (x, g(x)))$ is closed, (ii) $\operatorname{dom} f \cap \operatorname{ri}(\operatorname{dom} g) \neq \emptyset$ and $g_{|\operatorname{aff}(\operatorname{dom} g)}$ is continuous on $\operatorname{ri}(\operatorname{dom} g)$, then

$$\partial (f+g)(x) = \bigcap_{\varepsilon > 0} \operatorname{cl}(\partial_\varepsilon f(x) + \partial g(x)).$$

Observe that if the set $\operatorname{epi} g$ is polyhedral, then condition (i) holds. We even get the quasi-exact subdifferential rule

$$\partial (f+g)(x) = \operatorname{cl}(\partial f(x) + \partial g(x)),$$

under any one of the following assumptions: (iii) $\mathbb{R}_+(\operatorname{epi} f - (x, f(x)))$ is closed, $\operatorname{dom} f \cap \operatorname{ri}(\operatorname{dom} g) \neq \emptyset$, and $g_{|\operatorname{aff}(\operatorname{dom} g)}$ is continuous on $\operatorname{ri}(\operatorname{dom} g)$, (iv) $\mathbb{R}_+(\operatorname{epi} f - (x, f(x)))$ and $\mathbb{R}_+(\operatorname{epi} g - (x, g(x)))$ are closed, and (v) $\operatorname{ri}(\operatorname{dom} f) \cap \operatorname{ri}(\operatorname{dom} g) \neq \emptyset$ and $f_{|\operatorname{aff}(\operatorname{dom} f)}$ and $g_{|\operatorname{aff}(\operatorname{dom} g)}$ are continuous on $\operatorname{ri}(\operatorname{dom} f)$ and $\operatorname{ri}(\operatorname{dom} g)$, respectively. Moreover, under (v), if $\partial f(x)$ or $\partial g(x)$ is locally compact, then $\partial (f+g)(x) = \partial f(x) + \partial g(x)$.

We also characterize in chapter 7 the subdifferential of the sum $f + g$, when $f := \sup_{t \in T} f_t$, and g, $f_t : X \to \mathbb{R}_\infty$, $t \in T$ ($\neq \emptyset$), are all proper convex functions. In particular, under the assumption

$$\operatorname{cl}(f+g)(x) = \sup_{t \in T}(\operatorname{cl} f_t)(x) + g(x) \text{ for all } x \in \operatorname{dom} f \cap \operatorname{dom} g,$$

we derive the following formula, which constitutes an extension of (1.3),

$$\partial(f+g)(x) = \bigcap_{L \in \mathcal{F}(x),\, \varepsilon > 0} \overline{\operatorname{co}} \left\{ \bigcup_{t \in T_\varepsilon(x)} \partial_\varepsilon f_t(x) + \partial(g + \mathrm{I}_{L \cap \operatorname{dom} f})(x) \right\}.$$

Miscellaneous chapter 8 starts by considering the convex infinite optimization problem

$$\begin{aligned} \text{(P)} \quad &\text{Min } g(x) \\ &\text{s.t. } f_t(x) \leq 0,\ t \in T, \\ &\qquad x \in C, \end{aligned} \qquad (1.11)$$

where $\{g,\, f_t,\, t \in T\} \subset \Gamma_0(X)$ and C is a non-empty closed convex subset of X. We assume that the constraint system

$$\mathcal{S} := \{f_t(x) \leq 0, \ t \in T; \ x \in C\},$$

is consistent; i.e., the set of feasible solutions, represented by F, is assumed to be non-empty. An important particular case is that in which $f_t(x) = \langle a_t^*, x \rangle - b_t$, with $a_t^* \in X^*$, $t \in T$. When T is infinite, the objective function g is linear and continuous, and X is the Euclidean space, (P) becomes the so-called linear semi-infinite optimization problem. The purpose of the first two sections of this chapter is twofold. The first is to study some constraint qualifications, in particular the Farkas–Minkowski property and the local Farkas–Minkowski property; and the second is to provide optimality conditions for problem (P) by appealing to the properties of the supremum function $\tilde{f} = \sup_{t \in T} f_t$. Here, in this introduction, and for the sake of simplicity, we exclusively focus on the Farkas–Minkowski property.

A key tool in our approach is the so-called characteristic cone of system $\mathcal{S}$ defined as follows:

$$K := \text{cone co} \left\{ \bigcup_{t \in T} \text{epi } f_t^* \cup \text{epi } \sigma_C \right\} = \text{cone co} \left\{ \bigcup_{t \in T} \text{epi } f_t^* \right\} + \text{epi } \sigma_C.$$

It can be easily proved that, if $F \neq \emptyset$, then $\text{epi } \sigma_F = \text{cl } K$, where σ_F is the support function of the feasible set. In fact, a kind of generalized Farkas lemma can be formulated in terms of the characteristic cone: given $\varphi, \psi \in \Gamma_0(X)$, we have $\varphi(x) \leq \psi(x)$ for all $x \in F$, assumed non-empty, if and only if $\text{epi } \varphi^* \subset \text{cl } (\text{epi } \psi^* + K)$. A straightforward consequence of the last result is the following characterization of continuous linear consequences of $\mathcal{S}$: given $(a^*, \alpha) \in X^* \times \mathbb{R}$, the inequality $\langle a^*, x \rangle \leq \alpha$ holds for all $x \in F$, assumed non-empty, if and only if $(a^*, \alpha) \in \text{cl } K$.

The system $\mathcal{S} = \{f_t(x) \leq 0, \ t \in T; \ x \in C\}$, assumed consistent, is said to enjoy the Farkas–Minkowski (FM) property if K is w^*-closed. This property has a clear geometrical meaning, namely if $\mathcal{S}$ has the FM property, then every continuous linear consequence $\langle a^*, x \rangle \leq \alpha$ of $\mathcal{S}$ is also a consequence of a finite subsystem

$$\widehat{\mathcal{S}} := \{f_t(x) \leq 0, \ t \in \widehat{T}; \ x \in C\}, \text{ with } \widehat{T} \subset T \text{ and } |\widehat{T}| < \infty,$$

and the converse statement holds if $\mathcal{S}$ is linear; i.e., $\mathcal{S} = \{\langle a_t^*, x \rangle \leq b_t, \ t \in T\}$.

Moreover, the Farkas–Minkowski property is crucial in formulating Karush–Kuhn–Tucker and Fritz–John optimality conditions for prob-

1.3. WORKING FRAMEWORK AND OBJECTIVES

lem (P). In fact, let us suppose that $\mathcal{S}$ is Farkas–Minkowski and that g is continuous at some point of F, and let $\bar{x} \in F \cap \mathrm{dom}\, g$. Then $\bar{x}$ is a (global) minimum of (P) if and only if there exists $\lambda \in \mathbb{R}_+^{(T)}$ such that $\partial f_t(\bar{x}) \neq \emptyset$, for all $t \in \mathrm{supp}\,\lambda := \{t \in T : \lambda_t > 0\}$, and the Karush–Kuhn–Tucker conditions hold; that is,

$$\theta \in \partial g(\bar{x}) + \sum_{t \in \mathrm{supp}\,\lambda} \lambda_t \partial f_t(\bar{x}) + \mathrm{N}_C(\bar{x}) \text{ and } \lambda_t f_t(\bar{x}) = 0,\ \forall t \in T, \quad \text{(KKT'1)}$$

where
$$\mathbb{R}_+^{(T)} := \{\lambda : T \to \mathbb{R}_+ : \mathrm{supp}\,\lambda \text{ finite}\}.$$

We are also interested in comparing the set of optimal solutions of a given possibly non-convex optimization problem

$$\text{(P)} \quad \begin{array}{l} \text{Min } g(x) \\ \text{s.t. } x \in X, \end{array}$$

and its convex relaxation

$$\text{(P}_r\text{)} \quad \begin{array}{l} \text{Min } (\overline{\mathrm{co}}g)(x) \\ \text{s.t. } x \in X. \end{array}$$

This analysis reveals that the solution set of (P$_r$) can be written by means of either the global approximate solutions of (P) or the global exact ones, plus a term which reflects the asymptotic behavior of the function g (see Theorem 8.3.2 and its consequences).

We make use of subdifferential calculus rules of the supremum function given in the book to establish an integration theory for the Fenchel subdifferential (Theorem 8.4.3) and the ε-subdifferential (Theorem 8.4.7) of non-convex functions, both in Banach and locally convex spaces. This theory aims to extend the classical result of the integration formula of convex functions by Moreau and Rockafellar, stating that for every $f_1, f_2 \in \Gamma_0(X)$

$$\partial f_1 \subset \partial f_2 \Rightarrow f_1 \text{ and } f_2 \text{ coincide up to some constant.}$$

The last two sections are addressed to establish some variational characterizations for convexity of functions and sets. While the case of functions relies on the convexity of the solution sets of linear perturbations (see Theorem 8.5.2 and its consequences), the one establishing the convexity of sets comes as a consequence of the convexity/uniqueness of projections (see Theorem 8.6.8). Sets with unique projections are called Chebychev sets.

Chapter 2

Preliminaries

2.1 Functional analysis background

In this section, we present some basic concepts and results of functional analysis that will be used throughout the book.

Nets

Given a *binary relation* $\preccurlyeq$ in a non-empty set I, we say that the pair $(I, \preccurlyeq)$ is a *directed set* if the following properties hold:

(a) $i \preccurlyeq i$ for all $i \in I$ (*reflexivity*).
(b) $i \preccurlyeq j$ and $j \preccurlyeq k$ imply that $i \preccurlyeq k$ (*transitivity*).
(c) Any pair $i, j \in I$ has an *upper bound* $k \in I$; i.e., $i \preccurlyeq k$ and $j \preccurlyeq k$.

A *net* in a non-empty set S is a function $x : I \to S$, where $(I, \preccurlyeq)$ is a directed set. It is customary to denote a net x by $(x_i)_{i \in I}$, where $x_i = x(i) \in S$. The set I is called the *index set* of the net $(x_i)_{i \in I}$.

Given a subset $V \subset S$, we say that a net $(x_i)_{i \in I}$ in S is *eventually* in V if there exists $i_0 \in I$ such that $x_i \in V$ for all $i \in I$ such that $i_0 \preccurlyeq i$. The net $(x_i)_{i \in I}$ is *frequently* in V if, for each $i \in I$, there exists $j \in I$, $i \preccurlyeq j$, such that $x_j \in V$.

A net $(y_j)_{j \in J}$ is a *subnet* of the net $(x_i)_{i \in I}$, if there is a function $\phi : J \to I$ satisfying

(a) $y_j = x_{\phi(j)}$ for all $j \in J$.
(b) For all $i_0 \in I$, there exists $j_0 \in J$ such that $i_0 \preccurlyeq \phi(j)$ when $j_0 \preccurlyeq j$.

It is clear that the order in I is not necessarily the same as in J, although we have used the same symbol.

A binary relation $\preccurlyeq$ in the set I is a *partial order* if it is reflexive, transitive, and *antisymmetric*; that is, $i \preccurlyeq j$ and $j \preccurlyeq i$ imply $i = j$. A binary relation $\sim$ in I is an *equivalence relation* if it satisfies the reflexivity and the transitivity properties together with the symmetry property; that is, $i \sim j$ if and only if $j \sim i$ for all $i, j \in I$. The *equivalence class* of $i \in I$ is the set $\langle i \rangle := \{j \in I : i \sim j\}$, and the *quotient set* of I by $\sim$ is $I / \sim := \{\langle i \rangle : i \in I\}$.

Topological spaces

A family $\mathfrak{T}_X$ of subsets of a set X is called a *topology* in X if it contains the empty set, $\emptyset$, the whole set X, as well as all arbitrary unions and finite intersections of its elements. We say that the pair $(X, \mathfrak{T}_X)$ is a *topological space*. The elements of $\mathfrak{T}_X$ are called *open sets* and their complements are the *closed sets*. A *neighborhood* of $x \in X$ is a set $V \subset X$ such that $x \in W \subset V$ for some $W \in \mathfrak{T}_X$. We denote by $\mathcal{V}_X(x)$ the family of all neighborhoods of x. A subfamily of $\mathcal{V}_X(x)$ is a *neighborhood base* of x if every $V \in \mathcal{V}_X(x)$ contains an element of this subfamily. Given two topologies $\mathfrak{T}_1$ and $\mathfrak{T}_2$ on X, we say that $\mathfrak{T}_2$ is *finer* (*coarser*) than $\mathfrak{T}_1$ if $\mathfrak{T}_1 \subset \mathfrak{T}_2$ ($\mathfrak{T}_2 \subset \mathfrak{T}_1$, respectively). A topological subspace of $(X, \mathfrak{T}_X)$ is a pair $(Y, \mathfrak{T}_Y)$ such that $Y \subset X$ and

$$\mathfrak{T}_Y := \{W \cap Y : W \in \mathfrak{T}_X\}.$$

The topology $\mathfrak{T}_Y$ is called the *relative* (or *induced*) *topology* of $\mathfrak{T}_X$ in Y.

Given a mapping f between two non-empty sets X, Y, the sets $f(A) := \{f(a) : a \in A\}$ and $f^{-1}(B) := \{a \in X : f(a) \in B\}$ are the *image* of $A \subset X$ and *pre-image* of $B \subset Y$, respectively. The *range* and the *domain* of f are the sets $\operatorname{Im} f := f(X)$ and $f^{-1}(Y)$, respectively. The *graph* of f is the set $\operatorname{gph} f := \{(x, y) \in X \times Y : y = f(x)\}$. The mapping f is said to be *injective* if $x = y$ whenever $f(x) = f(y)$, *surjective* if $\operatorname{Im} f = Y$, and *bijective* if it enjoys both properties. These definitions are also valid for a set-valued mapping $F : X \rightrightarrows Y$ as it can be seen as a usual mapping between X and the *power set* of Y, denoted by 2^Y.

Given a topological space $(X, \mathfrak{T}_X)$ and an equivalence relation $\sim$ in X, the *quotient canonical projection* associated with $\sim$ is the surjective mapping $q : X \to X/\sim$ defined as

$$q(x) := \langle x \rangle. \qquad (2.1)$$

2.1. FUNCTIONAL ANALYSIS BACKGROUND

The *quotient topological space* is the topological space $(X/\sim, \mathfrak{T}_\sim)$, where
$$\mathfrak{T}_\sim := \{U \subset X/\sim : q^{-1}(U) \in \mathfrak{T}_X\}.$$

Equivalently, $\mathfrak{T}_\sim$ is the finest topology that makes q continuous.

The *interior* of $C \subset X$ is the largest open set contained in C and is denoted by $\mathfrak{T}_X\text{-int}\, C$ (or just $\text{int}\, C$ when no confusion is possible); that is, $\text{int}\, C = \cup_{W \in \mathfrak{T}_X, W \subset C} W$. The *closure* of C is the smallest closed set that contains C, and it is denoted by $\text{cl}\, C$ (or $\overline{C}$); that is, the intersection of all closed sets containing C. If $\text{cl}\, C = X$ we say that C is *dense* in X. The *boundary* of C is $\text{bd}\, C := \overline{C} \setminus (\text{int}\, C)$.

A *distance* (or *metric*) on a set X is a nonnegative function $d : X \times X \to [0, +\infty[$, defined on the *Cartesian product* $X \times X$ that satisfies the following three conditions, for all $x, y, z \in X$,

(a) $d(x,y) = 0$ if and only if $x = y$,
(b) $d(x,y) = d(y,x)$, and
(c) $d(x,y) \le d(x,z) + d(z,y)$.

Let $\mathfrak{T}_d$ be the topology formed by $\emptyset$, X, and the sets U satisfying the following property: for each $x \in U$ there exists $r > 0$ such that

$$\{y \in X : d(x,y) < r\} \subset U.$$

The pair $(X, d) := (X, \mathfrak{T}_d)$ is called a *metric space*, and a sequence $(x_k)_k \subset X$ is called a *Cauchy sequence* in (X, d) if, for every $\varepsilon > 0$, there exists $k_\varepsilon \ge 1$ such that $d(x_p, x_q) \le \varepsilon$ for all $p, q \ge k_\varepsilon$. A metric space is *complete* if every Cauchy sequence in X is convergent. By the *Baire category theorem*, a complete metric space is a *Baire space*; that is, for every *countable* collection $\{C_n \subset X, n \ge 1\}$ of closed subsets of a complete metric space X such that $\text{int}\,(\cup_{n \ge 1} C_n) \ne \emptyset$, there exists some $n_0 \ge 1$ satisfying $\text{int}(C_{n_0}) \ne \emptyset$.

A function $f : (X, d) \to \mathbb{R}$ is said to be *Lipschitz continuous* (or, just, *Lipschitz*) if there exists some $l \ge 0$ such that

$$|f(x_1) - f(x_2)| \le l\, d(x_1, x_2) \text{ for all } x_1, x_2 \in X.$$

It is *locally Lipschitz* at $x \in X$ if the last inequality holds for all x_1, x_2 in some neighborhood of x.

A topological space X is *separated* (or *Hausdorff*) if for all $x_1, x_2 \in X$ ($x_1 \ne x_2$) there exist $V_1 \in \mathcal{V}_X(x_1)$ and $V_2 \in \mathcal{V}_X(x_2)$ such that $V_1 \cap V_2 = \emptyset$. A topological space X is *separable* if $X = \text{cl}(C)$ for a countable subset $C \subset X$. A subset $C \subset X$ is *compact* if, whenever C is contained in the union of a family of open sets, it is also contained in the union of a finite subfamily. In this section the given topological spaces are assumed to be Hausdorff. We say that C is *locally compact* if each point $x \in C$ has a compact neighborhood for the induced topology in C by $\mathfrak{T}_X$.

A net $(x_i)_i$ in X *converges* to $x \in X$ if it is eventually in every neighborhood of x; we write $x_i \to x$ or $\lim_i x_i = x$. A point x is said to be a *cluster point* of a net $(x_i)_{i \in I}$ in X if there exists a subnet $(y_j)_{j \in J}$ of $(x_i)_{i \in I}$ that converges to x. For any set $C \subset X$, $x \in \operatorname{cl} C$ if and only if there exists a net $(x_i)_i$ in C such that $x_i \to x$. It follows that C is closed if and only if the limit of every convergent net in C belongs to C.

The limit of a convergent net is unique, and every subnet of a convergent net converges to the same point. Moreover, it turns out that a set $C \subset X$ is compact if and only if every net in C has a subnet that converges to a point in C. Consequently, a compact set in X is closed, and every closed subset of it is also compact. We say that $C \subset X$ is *sequentially closed* if the limit of every convergent sequence in C belongs to C. Moreover, the set C is *sequentially compact* if every sequence in C has a subsequence that converges to a point in C.

A mapping $f : X \to Y$ between two topological spaces $(X, \mathfrak{T}_X)$ and $(Y, \mathfrak{T}_Y)$ is *continuous* at $x \in X$ if, for all $W \in \mathcal{V}_Y(f(x))$, there exists $V \in \mathcal{V}_X(x)$ such that $f(V) \subset W$. Equivalently, f is continuous at x if and only if $f(x_i) \to f(x)$ for every net $(x_i)_i \subset X$ such that $x_i \to x$. Similarly, f is sequentially continuous at x if the last condition holds when we replace nets by sequences. We say that f is continuous (sequentially continuous) on X if it is continuous (sequentially continuous, respectively) at every point in X. If f is continuous, then the image $f(C)$ of a compact set $C \subset X$ is also compact in Y. A function $f : X \to Y$ is said to be a *homeomorphism* if it is a continuous *bijection* with a continuous inverse.

We extend the usual order in $\mathbb{R}$ to $\overline{\mathbb{R}} = \mathbb{R} \cup \{+\infty, -\infty\}$ by setting $-\infty < r < +\infty$ for all $r \in \mathbb{R}$. We also denote

$$\mathbb{R}_\infty := \mathbb{R} \cup \{+\infty\}, \ \mathbb{R}_+ := [0, +\infty[, \ \mathbb{R}_-^* :=]-\infty, 0[, \ \mathbb{R}_+^* :=]0, +\infty[,$$

and adopt the convention

$$(+\infty) + (-\infty) = (-\infty) + (+\infty) = +\infty, \ 0(+\infty) = +\infty, \ 0(-\infty) = 0. \tag{2.2}$$

By $\operatorname{sign}(\cdot) : \mathbb{R} \setminus \{0\} \to \{-1, 1\}$, we refer to the *sign function*; i.e., $\operatorname{sign}(\alpha) = -1$ if $\alpha < 0$, and $\operatorname{sign}(\alpha) = +1$ if $\alpha > 0$.

2.1. FUNCTIONAL ANALYSIS BACKGROUND

The *upper and lower limits* of a net $(r_i)_{i \in I}$ in $\overline{\mathbb{R}}$ are, respectively,

$$\limsup_i r_i = \inf_{i \in I} \left(\sup_{i \preccurlyeq j} r_j \right) \quad \text{and} \quad \liminf_i r_i = \sup_{i \in I} \left(\inf_{i \preccurlyeq j} r_j \right).$$

It is clear that $\liminf_i r_i \leq \limsup_i r_i$ and when the upper and lower limits coincide in $\overline{\mathbb{R}}$, the common value is called the *limit* of $(r_i)_{i \in I}$; it is written as

$$\lim_i r_i := \liminf_i r_i = \limsup_i r_i.$$

If the net $(r_i)_{i \in I} \subset \mathbb{R}$ is non-increasing; that is, $i_1 \preccurlyeq i_2$, $i_1, i_2 \in I \Rightarrow r_{i_2} \leq r_{i_1}$, then the limit of $(r_i)_{i \in I}$ exists in $\overline{\mathbb{R}}$ and is given by

$$\lim_i r_i = \inf_i r_i.$$

Similarly, if $(r_i)_{i \in I}$ is non-decreasing, then the limit exists in $\overline{\mathbb{R}}$ and is given by

$$\lim_i r_i = \sup_i r_i.$$

Given a topological space $(X, \mathfrak{T}_X)$, a function $f : X \to \overline{\mathbb{R}}$ is *lower semicontinuous* (or *lsc*, for short) at $x \in X$ if, for every net $x_i \to x$,

$$f(x) \leq \liminf_i f(x_i).$$

The function f is lsc if it is lsc at every point of X. The function f is *sequentially lsc* at $x \in X$ (sequentially lsc, respectively) if these last conditions hold with sequences instead of nets.

The function f is *inf-compact* if all its *sublevel sets*,

$$[f \leq \lambda] := \{x \in X : f(x) \leq \lambda\}, \quad \lambda \in \mathbb{R},$$

are compact sets.

The most fundamental theorem in optimization theory is the *Weierstrass theorem*.

Theorem 2.1.1 *Let $f : X \to \overline{\mathbb{R}}$ be an lsc function and let $C \subset X$ be a non-empty compact set. Then f achieves its infimum over the set C.*

Given a family $(X_t, \mathfrak{T}_t)_{t \in T}$ of topological spaces, the product topology $\mathfrak{T}$ in the *Cartesian product*

$$X := \prod_{t \in T} X_t$$

is the weakest (i.e., the smallest) topology on X for which all the *canonical projections* $\pi_t : X \to X_t$ are continuous. In this way, a net $(x_i)_{i \in I}$ in X converges to $x \in X$ if and only if $\pi_t(x_i) \to \pi_t(x)$ for all $t \in T$. The space $(X, \mathfrak{T})$ is called the *product topological space* of the X_t's.

We give now the *Tychonoff theorem*, which is considered the most important theorem in topology.

Theorem 2.1.2 *The product topological space $(X, \mathfrak{T})$ is compact if and only if each factor $(X_t, \mathfrak{T}_t)$ is a compact topological space.*

Let $\mathbb{R}^T$ denote the real linear space of functions from a given set T to $\mathbb{R}$. Again, the *support* of a function $\lambda \in \mathbb{R}^T$ is the set

$$\operatorname{supp} \lambda := \{t \in T : \lambda_t \neq 0\},$$

where $\lambda_t := \lambda(t)$. We denote

$$\mathbb{R}^{(T)} := \{\lambda \in \mathbb{R}^T : \quad \operatorname{supp} \lambda \text{ is finite}\}. \tag{2.3}$$

In particular, if the *cardinal* of T, denoted by $|T|$, is finite, say $|T| = k$, then we set $\mathbb{R}^{(T)} \equiv \mathbb{R}^T \equiv \mathbb{R}^k$ and, similarly, $\mathbb{R}_+^{(T)} \equiv \mathbb{R}_+^T \equiv \mathbb{R}_+^k$. For $u \in \mathbb{R}^T$ and $\lambda \in \mathbb{R}^{(T)}$, we write

$$\lambda(u) := \sum_{t \in T} \lambda_t u_t := \begin{cases} \sum_{t \in \operatorname{supp} \lambda} \lambda_t u_t, & \text{if } \operatorname{supp} \lambda \neq \emptyset, \\ 0, & \text{otherwise,} \end{cases} \tag{2.4}$$

The nonnegative cone of $\mathbb{R}^{(T)}$ is

$$\mathbb{R}_+^{(T)} := \left\{\lambda \in \mathbb{R}^{(T)} : \lambda_t \geq 0 \text{ for all } t \in T\right\},$$

and we denote

$$\Delta(T) := \left\{\lambda \in \mathbb{R}_+^{(T)} : \sum_{t \in T} \lambda_t = 1\right\}; \tag{2.5}$$

in particular, the *canonical simplex* in $\mathbb{R}^k$, $k \in \mathbb{N}$, is given by

2.1. FUNCTIONAL ANALYSIS BACKGROUND

$$\Delta_k := \left\{ \lambda \in \mathbb{R}_+^k : \sum_{1 \leq i \leq k} \lambda_i = 1 \right\}.$$

Sometimes, we also use the notation

$$\Delta_k^* = \{\lambda \in \Delta_k : \lambda_i > 0 \text{ for all } 1 \leq i \leq k\}.$$

Operations and concepts in vector spaces

Next we define standard operations and concepts in a (real) vector space X. We denote by θ the zero vector in X. Given two non-empty sets C and D in X, we define the *algebraic* (or *Minkowski*) *sum* of C and D by

$$C + D := \{c + d : c \in C, d \in D\}, \quad C + \emptyset = \emptyset + D = \emptyset, \qquad (2.6)$$

with $C + d, d \in X$, ($c + D, c \in X$), representing the set $C + \{d\}$ ($\{c\} + D$, respectively). If $\emptyset \neq \Lambda \subset \mathbb{R}$, the *scalar product* of Λ and C is the set

$$\Lambda C := \{\lambda c : \lambda \in \Lambda, c \in C\}, \quad \Lambda \emptyset = \emptyset, \qquad (2.7)$$

and, in particular, for $x \in X$ and $\lambda \in \mathbb{R}$ we write $\Lambda x := \Lambda\{x\}$ and $\lambda C := \{\lambda\} C$. This last set is a *scalar multiple* of C.

A set $C \subset X$ is *convex* if, for every $x, y \in C$,

$$[x, y] := \{\lambda x + (1 - \lambda) y : \lambda \in [0, 1]\} \subset C,$$

and *affine* if, for every $x, y \in C$,

$$\{\lambda x + (1 - \lambda) y : \lambda \in \mathbb{R}\} \subset C.$$

We say that $C \subset X$ is a *hyperplane* (*affine hyperplane*) if it is a proper linear subspace (affine set, respectively), which is maximal for the inclusion.

The set C is a *cone* if it contains θ and $\mathbb{R}_+ C \subset C$; *balanced* if $[-1, 1]C \subset C$; and *absorbing* if, for every $x \in X$, there exists $\alpha > 0$ such that $\lambda x \in C$ when $|\lambda| < \alpha$. The intersection of a family of convex sets is convex; the scalar multiple of a convex set is convex; and the sum of two convex sets is convex. The *convex*, the *affine*, the *linear* (or *span*), and the *conic hulls* of a set C are defined as

$$\operatorname{co} C = \left\{ \sum_{c \in C} \lambda_c c : \lambda \in \Delta(C) \right\},$$

$$\operatorname{aff} C = \left\{ \sum_{c \in C} \lambda_c c : \lambda \in \mathbb{R}^{(C)},\ \sum_{c \in C} \lambda_c = 1 \right\},$$

$$\operatorname{span} C = \left\{ \sum_{c \in C} \lambda_c c : \lambda \in \mathbb{R}^{(C)} \right\},\ \text{and}$$

$$\operatorname{cone} C = \mathbb{R}_+ C \cup \{\theta\}.$$

Observe that C is convex if and only if $C = \operatorname{co} C$, and that C is a cone if and only if $C \cup \{\theta\} = \operatorname{cone} C$; in particular, $\operatorname{co} \emptyset = \emptyset$ and $\operatorname{cone} \emptyset = \{\theta\}$. Equivalently, $\operatorname{co} C$ is the intersection of all convex sets containing C. By $\dim C$ we represent the *dimension* of the vector space $\operatorname{aff} C$, and we say that C is finite-dimensional if $\dim C$ is finite. In an n-dimensional vector space, the convex hull is characterized by the celebrated *Carathéodory theorem*, stating that every $x \in \operatorname{co} C$ is a convex combination of no more than $n + 1$ elements of C.

Topological vector spaces

Given a real vector space X and a topology $\mathfrak{T}_X$ in X, we say that $(X, \mathfrak{T}_X)$ is a *topological vector space* (tvs, for short) if the two algebraic operations

$$(x, y) \in X \times X \mapsto x + y \in X \text{ and } (\alpha, x) \in \mathbb{R} \times X \mapsto \alpha x \in X \quad (2.8)$$

are continuous. We denote by X^* the *topological dual* of X; that is, the real vector space of *continuous linear* functions from X to $\mathbb{R}$. Associated with X and its topological dual X^*, the *bilinear function* $(x^*, x) \in X^* \times X \mapsto \langle x^*, x \rangle := x^*(x) \in \mathbb{R}$ is called a *dual pairing*, while the pair (X, X^*) is referred to as a *dual pair*.

We have the following basic properties in a tvs: The set ΛC is open if C is open and $0 \notin \Lambda$; the sum of an open set and an arbitrary set is open; if D is open, then $\overline{C} + D = \overline{C + D}$ for any set C; the sum of a compact set and a closed set is closed; the sum of two compact sets is compact; and the scalar multiple of a compact set is compact. A linear function from X to $\mathbb{R}$ is continuous if and only if it is continuous at θ. A set $C \subset X$ is a hyperplane (affine hyperplane) if and only if $C = \{x \in X : \ell(x) = 0\}$ ($C = \{x \in X : \ell(x) = \alpha\}$, $\alpha \in \mathbb{R}$, respectively) for some nonzero linear function $\ell : X \to \mathbb{R}$. The hyperplane C is closed if and only if ℓ is continuous, and it is dense if and only if ℓ is not continuous. A (closed) *half-space* of X is a set of the form $\{x \in X : \ell(x) \le \alpha\}$, $\alpha \in \mathbb{R}$, where $\ell : X \to \mathbb{R}$ is a nonzero continuous linear function.

2.1. FUNCTIONAL ANALYSIS BACKGROUND

In addition, the following relations hold for every pair of sets C, $D \subset X$:
$$(\operatorname{cl} C) + (\operatorname{cl} D) \subset \operatorname{cl}(C + D) = \operatorname{cl}(C + \operatorname{cl} D), \tag{2.9}$$

$$\operatorname{cl}(C + D) = \operatorname{cl}(C) + D, \text{ when } D \text{ is compact}, \tag{2.10}$$

and
$$\bigcap_{U \in \mathcal{V}_X(\theta)} \operatorname{cl}(C + U) = \bigcap_{U \in \mathcal{V}_X(\theta)} (C + U) = \operatorname{cl}(C). \tag{2.11}$$

Thanks to the continuity of the operations in (2.8), every element $V \in \mathcal{V}_X(\theta)$ is absorbing and satisfies the following properties:
(i) $x + V \in \mathcal{V}_X(x)$ for all $x \in X$.
(ii) There exists $W \in \mathcal{V}_X(\theta)$ such that $W + W \subset V$.
(iii) $\lambda V \in \mathcal{V}_X(\theta)$ for all $\lambda \neq 0$.

If X is a finite-dimensional tvs, then there is a unique (Hausdorff) topology $\mathfrak{T}_X$ such that $(X, \mathfrak{T}_X)$ is a tvs; it is the Euclidean topology. Every finite-dimensional vector subspace of a tvs is closed. A tvs X is finite-dimensional if and only if θ has a compact neighborhood.

We denote
$$\overline{\operatorname{co}} C := \operatorname{cl}(\operatorname{co} C) \text{ and } \overline{\operatorname{cone}} C := \operatorname{cl}(\operatorname{cone} C).$$

Correspondingly, $\overline{\operatorname{co}} C$ is the intersection of all closed convex sets containing C. If C is convex, then the interior and the closure of C are also convex; in fact, we have the following property:
$$[x, y[\subset \operatorname{int} C \text{ for every } x \in \operatorname{int} C \text{ and } y \in \operatorname{cl} C. \tag{2.12}$$

Moreover, if $\operatorname{int} C \neq \emptyset$, then
$$\operatorname{cl}(\operatorname{int} C) = \operatorname{cl} C \text{ and } \operatorname{int}(\operatorname{cl} C) = \operatorname{int} C.$$

Relation (2.12) can be relaxed using the (topological) *relative interior* of C, denoted by $\operatorname{ri} C$, which is the interior of C in the topology relative to $\operatorname{aff} C$ if $\operatorname{aff} C$ is closed, and the empty set otherwise. One of the main features of the relative interior is that it is non-empty for every non-empty finite-dimensional convex set. In addition, the relative interior enjoys nice properties in finite dimensions, including the following relations which hold for convex sets $C \subset \mathbb{R}^n$, $D \subset \mathbb{R}^m$, and a linear mapping $T : \mathbb{R}^n \to \mathbb{R}^m$,

$$\mathrm{ri}(TC) = T(\mathrm{ri}\, C) \tag{2.13}$$

and, provided that the pre-image $T^{-1}(\mathrm{ri}\, D) \neq \emptyset$,

$$\mathrm{ri}(T^{-1}D) = T^{-1}(\mathrm{ri}\, D). \tag{2.14}$$

In fact, for every non-empty convex set $C \subset X$, we have the following extension of (2.12):

$$[x,y[\, \subset \mathrm{ri}\, C, \text{ for every } x \in \mathrm{ri}\, C \text{ and } y \in \mathrm{cl}\, C; \tag{2.15}$$

this result is sometimes referred to as the *accessibility lemma*. Therefore, when $\mathrm{ri}(C) \neq \emptyset$, we also have

$$\overline{C} = \overline{\mathrm{ri}(C)} \text{ and } \mathrm{ri}(\overline{C}) = \mathrm{ri}(C). \tag{2.16}$$

In particular, this last relation holds for every non-empty finite-dimensional convex set C.

Other useful consequences of (2.15) come next: If C and D are convex sets in X such that $\mathrm{ri}(C) \cap \mathrm{ri}(D) \neq \emptyset$, then we have

$$\mathrm{ri}(C) \cap \mathrm{ri}(D) = \mathrm{ri}(C \cap D) \tag{2.17}$$

and

$$\overline{C \cap D} = \overline{C} \cap \overline{D} = \overline{\mathrm{ri}(C) \cap \mathrm{ri}(D)} = \overline{\mathrm{ri}(C) \cap \mathrm{ri}(D)}. \tag{2.18}$$

Also, if $\mathrm{int}(D) \neq \emptyset$, so that $\mathrm{int}(D) = \mathrm{ri}(D)$, then

$$\mathrm{int}(C + D) = C + \mathrm{int}(D), \tag{2.19}$$

and if, in addition, $C \cap \mathrm{int}(D) \neq \emptyset$, then

$$\overline{C \cap D} = \overline{C} \cap \overline{D}. \tag{2.20}$$

All these results, involving the interior and the relative interior of convex sets, are based on the separation theorems that come below.

We also consider the *algebraic interior* of a set C contained in a vector space X, denoted by C^i. It is the set of points $c \in C$ such that

$$\mathbb{R}_+(C - c) = X;$$

2.1. FUNCTIONAL ANALYSIS BACKGROUND

that is, the set $C - c$ is absorbing. In a tvs we have $\operatorname{int} C \subset C^i$, while the equality $C^i = \operatorname{int} C$ holds for convex sets C such that, for instance, $\operatorname{int} C \neq \emptyset$ or C is finite-dimensional (another important case is mentioned below).

Given a non-empty closed convex set $C \neq \emptyset$, the *recession cone* of C is defined as

$$C_\infty := \{u \in X : x + \mathbb{R}_+ u \subset C\},$$

where x is an arbitrary point of C. The recession cone is closed and convex, and collapses to θ when C is *bounded*; that is, for any $V \in \mathcal{V}_X(\theta)$, there exists $\lambda > 0$ such that $C \subset \lambda V$. The *lineality space* of a non-empty closed convex set C is

$$\operatorname{lin} C := C_\infty \cap (-C_\infty). \qquad (2.21)$$

If $C, D \subset X$ are non-empty closed convex sets such that $C \subset D$, then

$$C_\infty \subset D_\infty. \qquad (2.22)$$

If $C_i \subset X$, $i \in I$, is a family of non-empty closed convex sets such that $\bigcap_{i \in I} C_i \neq \emptyset$, then

$$\left(\bigcap_{i \in I} C_i\right)_\infty = \bigcap_{i \in I} (C_i)_\infty. \qquad (2.23)$$

One of the most important and far-reaching results of functional analysis is the *Hahn–Banach extension theorem* (see (2.32) for the convexity of functions).

Theorem 2.1.3 *Let X be a vector space, and $\ell : L \to \mathbb{R}$ be a linear function defined on a linear subspace $L \subset X$, which is dominated by a convex function $g : X \to \mathbb{R}$ (that is, $\ell(x) \leq g(x)$ for all $x \in L$). Then there is a linear extension $\tilde{\ell}$ of ℓ to X (that is, $\tilde{\ell}(x) = \ell(x)$ for all $x \in L$), which is also dominated by g.*

We give now the geometric version of the last theorem, which is a cornerstone in convex analysis.

Theorem 2.1.4 *Two non-empty disjoint convex subsets, C and D, in a vector space X such that one of them has a non-empty algebraic interior, can be properly separated by a nonzero linear function $\ell : X \to \mathbb{R}$; that is, there exists $\alpha \in \mathbb{R}$ such that $\ell(x) \leq \alpha \leq \ell(y)$ for all $x \in C$ and $y \in D$, and there exists a point $z \in C \cup D$ with $\ell(z) \neq \alpha$.*

The two previous theorems do not mention any topology, whereas the next result ensures the separation by means of a *continuous* linear function.

Theorem 2.1.5 *If X is a tvs and $\operatorname{int} C \neq \emptyset$ in the previous theorem, then $\operatorname{cl} C$ and $\operatorname{cl} D$ can be properly separated by a nonzero continuous linear function.*

To obtain a stronger separation property than the one given above, we need to assume that the tvs X is locally convex. As we said before, this is the framework of the book. A tvs is said to be a *locally convex space* (lcs, for short) if every neighborhood of θ includes a convex neighborhood. The associated topology of an lcs is called a locally convex topology. For example, observing that $\mathbb{R}^T \equiv \Pi_{t \in T} X_t$ with $X_t := \mathbb{R}$ for all $t \in T$, it follows that $\mathbb{R}^T$ endowed with the product topology is an lcs and that $\mathbb{R}^{(T)}$ is its topological dual space with the pairing defined in (2.4).

In an lcs every neighborhood of θ includes a closed convex balanced neighborhood, so that the family

$$\mathcal{N}_X := \{V \in \mathcal{V}_X(\theta) : V \text{ is convex, closed, and balanced}\}$$

is a neighborhood base of θ. The elements of $\mathcal{N}_X$ are called *θ-neighborhoods* in this book.

Theorem 2.1.6 *Given two non-empty disjoint convex sets in an lcs, C and D, if one of them is compact and the other one is closed, then there is an $\ell \in X^* \setminus \{\theta\}$ that strongly separates them; that is, there exist $\alpha, \beta \in \mathbb{R}$ such that $\ell(x) \leq \alpha < \beta \leq \ell(y)$ for all $x \in C$ and $y \in D$.*

The typical example of convex sets in an lcs that cannot be properly separated from any point of its complement are the dense subspaces (this is the case of the hyperplane $\{x \in X : \ell(x) = 0\}$, where $\ell : X \to \mathbb{R}$ is a non-continuous linear function).

Theorem 2.1.6 easily leads us to the following result, which is frequently used throughout the book.

Corollary 2.1.7 *Given a non-empty closed convex set C in an lcs, for every point $x \notin C$ there are $\ell \in X^* \setminus \{\theta\}$ and $\alpha \in \mathbb{R}$ such that $\ell(x) < \alpha \leq \ell(y)$ for all $y \in C$.*

Next, we see the most important consequences of the strong separation property in Theorem 2.1.6, which is stated in an lcs X. The first one is the *Dieudonné theorem*, exploiting the notion of the recession cone for checking the closedness of the sum of two convex sets.

2.1. FUNCTIONAL ANALYSIS BACKGROUND

Theorem 2.1.8 *Given two non-empty closed convex sets in an lcs, C and D, we assume that one of them is locally compact and $C_\infty \cap (-D_\infty)$ is a linear subspace. Then $C + D$ is closed.*

Other consequences of Theorem 2.1.6 follow:

(*i*) A linear subspace $L \subset X$ fails to be dense in X if and only if there exists $\ell \in X^* \setminus \{\theta\}$ such that $L = \{x \in X : \ell(x) = 0\}$.

(*ii*) The elements of X^* separate points in X; that is, given x, y in X such that $x \neq y$, there exists $\ell \in X^*$ such that $\ell(x) \neq \ell(y)$.

(*iii*) If $C \subset X$ is closed and convex, then it is the intersection of all closed half-spaces that contain it. The analytic counterpart to this result constitutes the subject of section 3.2.

(*iv*) If C is a convex cone, then for all $x \in X$ we have that either $x \in \operatorname{cl} C$ or there exists some $\ell \in X^*$ such that $\ell(x) > 0$ and $\ell(z) \leq 0$ for all $z \in C$.

(*v*) If Y is a linear subspace of X and $\ell \in Y^*$, then ℓ can be extended to a linear function $\tilde{\ell} \in X^*$.

Associated with a subset C in an lcs X, we consider the sets

$$C^\circ := \{x^* \in X^* : \langle x^*, x \rangle \leq 1 \text{ for all } x \in C\},$$
$$C^- := (\operatorname{cone} C)^\circ = \{x^* \in X^* : \langle x^*, x \rangle \leq 0 \text{ for all } x \in C\}, \text{ and}$$
$$C^\perp := (-C^-) \cap C^- = \{x^* \in X^* : \langle x^*, x \rangle = 0 \text{ for all } x \in C\},$$

i.e., the (one-sided) *polar*, the *negative dual cone*, and the *orthogonal subspace* (or *annihilator*) of C, respectively. Observe that C° is a closed convex set containing θ, C^- is a closed convex cone, and $C^\perp = (\operatorname{aff} C)^\perp$ is a closed linear subspace. Additionally, given $\varepsilon \geq 0$, we define the ε-*normal set* to C at $x \in C$ by

$$\operatorname{N}_C^\varepsilon(x) := \{x^* \in X^* : \langle x^*, y - x \rangle \leq \varepsilon \text{ for all } y \in C\}, \tag{2.24}$$

with $\operatorname{N}_C(x) := \operatorname{N}_C^0(x)$ being the *normal cone* of C at x. Observe that $\operatorname{N}_C(x) = (C - x)^-$.

Given an lcs $(X, \mathfrak{T}_X)$ with the topological dual X^* endowed with a locally convex topology $\mathfrak{T}_{X^*}$, we say that $((X, \mathfrak{T}_X), (X^*, \mathfrak{T}_{X^*}))$ is a *compatible dual pair* if the dual of $(X^*, \mathfrak{T}_{X^*})$ is identified with X. In such a case, the topologies $\mathfrak{T}_X$ and $\mathfrak{T}_{X^*}$ are said to be *compatible* (or *consistent*) *topologies* for the dual pair (X, X^*).

Observe that an lcs X can be regarded as a linear subspace of $\mathbb{R}^{X^*}$, using the identification $x \in X \mapsto \langle \cdot, x \rangle \in \mathbb{R}^{X^*}$, where $\langle x^*, x \rangle := x^*(x)$. In this way, the space X inherits the product topology of $\mathbb{R}^{X^*}$, which gives rise to the *weak-topology* $\sigma(X, X^*)$ on X, also denoted by w. Since $\mathbb{R}^{X^*}$ is an lcs, so is the topological space (X, w). We use the symbol $\to^w$ (or $\to$ when no confusion is possible) to represent the convergence in the topology $\sigma(X, X^*)$, so that a net $x_i \to^w x$ in (X, w) if and only if $\langle x_i, x^* \rangle \to \langle x, x^* \rangle$ in $\mathbb{R}$ for all $x^* \in X^*$.

A *seminorm* on a linear space X is a function $p: X \to \mathbb{R}$ that satisfies $p(x + y) \leq p(x) + p(y)$ and $p(\lambda x) = |\lambda| p(x)$ for all $x, y \in X$ and all $\lambda \in \mathbb{R}$. Note that a seminorm is nonnegative and satisfies $p(\theta) = 0$ together with $-p(x) \leq p(-x)$. A seminorm p that satisfies the implication $p(x) = 0 \Rightarrow x = \theta$ is called a *norm*. An important family of seminorms in an lcs X is

$$\{p_{x^*} := |\langle x^*, \cdot \rangle| : x^* \in X^*\}.$$

This family generates the weak topology on X; in fact, the *semiballs*

$$B_{x^*}(\theta, r) := \{x \in X : p_{x^*}(x) \leq r\}, x^* \in X^* \text{ and } r > 0,$$

are θ-neighborhoods. In addition, for any $U \in \mathcal{N}_X$ in the weak topology, there exists a finite number of elements $x_1^*, \ldots, x_n^* \in X^*$ and $r > 0$ such that

$$\bigcap_{1 \leq i \leq n} B_{x_i^*}(\theta, r) = \{x \in X : p_{x_i^*}(x) \leq r, \ i = 1, \ldots, n\} \subset U.$$

The *weak*-topology* defined on X^*, denoted by $\sigma(X^*, X)$ (also by w^*-topology or, simply, w^*), is the locally convex topology on X^* generated by the family of seminorms

$$\{p_x := |\langle \cdot, x \rangle|, \ x \in X\}.$$

Similarly as above, given any $V \in \mathcal{N}_{X^*}$ in the w^*-topology, there exist $x_1, \ldots, x_m \in X$ and $r > 0$ such that

$$\bigcap_{1 \leq i \leq m} B_{x_i}(\theta, r) = \{x^* \in X^* : p_{x_i}(x^*) \leq r, \ i = 1, \ldots, m\} \subset V;$$

that is, in particular, for $L := \text{span}\{x_1, \ldots, x_m\}$ we have that $L^\perp \subset V$. Observe that a net $(x_i^*)_i$ converges to x^* in (X^*, w^*), also written as $x_i^* \to^{w^*} x^*$ (or $x_i^* \to x^*$ when no confusion is possible), if and only if $\langle x_i^*, x \rangle \to \langle x^*, x \rangle$ for all $x \in X$. A net $(x_i)_i$ is said to be $\mathfrak{T}_{X^*}$-boundedly w^*-convergent to x^*, where $\mathfrak{T}_{X^*}$ is a given locally convex topology on X^*, if it is $\mathfrak{T}_{X^*}$-bounded and satisfies $x_i^* \to^{w^*} x^*$.

2.1. FUNCTIONAL ANALYSIS BACKGROUND

The weak and weak* topologies, among other important ones, are compatible with the dual pair (X, X^*). In fact, we have the identification $(X, w)^* = X^*$, and the w-topology is the coarsest of the locally convex topologies preserving the family of continuous linear functions on X. Similarly, we have the identification $(X^*, w^*)^* = X$; that is, for every w^*-continuous linear function $\ell : X^* \to \mathbb{R}$, there exists a unique $x \in X$ such that

$$\ell(x^*) = \langle x^*, x \rangle \text{ for all } x^* \in X^*.$$

Therefore, $((X, \mathfrak{T}_X), (X^*, w^*))$ and $((X, w), (X^*, w^*))$ are compatible dual pairs for the bilinear function $(x^*, x) \in X^* \times X \mapsto \langle x^*, x \rangle \in \mathbb{R}$. This fundamental property of the weak topology shows that closed convex subsets of X and lsc convex functions defined on X are the same if we consider in X the weak topology instead of the original one.

Another important topology in X for which the dual of X remains X^* is the one generated by the family of seminorms

$$\rho_A(x) = \sup\{|\langle x^*, x \rangle| : x^* \in A\},$$

where A is a non-empty w^*-compact subset in X^* (implying that, for all $x \in X$, the set $\{\langle x^*, x \rangle : x^* \in A\}$ is bounded in $\mathbb{R}$). This topology is called the *Mackey-topology* and is denoted by $\tau(X, X^*)$ (also written as τ). This is the finest of the locally convex topologies $\mathfrak{T}$ such that the dual of $(X, \mathfrak{T})$ is X^*. In parallel, the Mackey-topology defined on X^*, $\tau(X^*, X)$, is the one generated by the family of seminorms

$$\rho_A(x^*) = \sup\{|\langle x^*, x \rangle| : x \in A\},$$

where A is any non-empty w-compact subset in X. Hence, it is the finest of the locally convex topologies $\mathfrak{T}$ such that the dual of $(X^*, \mathfrak{T})$ is X. Any locally convex topology $\mathfrak{T}$ on X (on X^*) that satisfies $w \subset \mathfrak{T} \subset \tau$ ($w^* \subset \mathfrak{T} \subset \tau(X^*, X)$), respectively) satisfies $(X, \mathfrak{T})^* = X^*$ $((X^*, \mathfrak{T})^* = X$, respectively). We also consider the *strong topology* $\beta(X^*, X)$, also written as β, which is generated by the family of seminorms

$$\rho_A(x^*) = \sup\{|\langle x^*, x \rangle| : x \in A\},$$

where A is any non-empty w-bounded subset in X. Hence, we have $\tau(X^*, X) \subset \beta(X^*, X)$ with a possibly strict inclusion, and so

$(X^*, \beta(X^*, X))^*$ may not coincide with X. Observe that one can also define the strong topology $\beta(X, X^*)$, generated by the family of seminorms

$$\rho_A(x) = \sup\{|\langle x^*, x \rangle| : x^* \in A\},$$

for A being a non-empty w^*-bounded subset in X^*. However, it turns out that the topology $\beta(X, X^*)$ coincides with the Mackey-topology $\tau(X, X^*)$ as a consequence of the Alaoglu–Banach–Bourbaki theorem given below (Theorem 2.1.9). In other words, the difference between the Mackey and the strong topologies only occurs in the dual space X^*.

To show that every locally convex topology on a vector space X is generated by the family of seminorms, we use the *Minkowski function* (also called *Minkowski gauge*), defined for a non-empty set $C \subset X$ as

$$p_C(x) := \inf\{\lambda \geq 0 : x \in \lambda C\}, \tag{2.25}$$

with the convention $\inf \emptyset = +\infty$. Then the collection of functions

$$p_U(x) := \inf\{\lambda \geq 0 : x \in \lambda U\},$$

where U runs the family $\mathcal{N}_X$ gives the desired family of seminorms that defines the locally convex topology of X. Conversely, we consider a *saturated family* $\mathcal{P}$ of seminorms; that is, for every $p_1, p_2 \in \mathcal{P}$ there exists a $p \in \mathcal{P}$ such that $\max\{p_1, p_2\} \leq p$. Then $\mathcal{P}$ defines a locally convex topology on X for which the semiballs

$$B_p(\theta, r) := \{y \in X : p(y) < r\}, \ r > 0, \ p \in \mathcal{P},$$

constitute a neighborhood base of θ. If Y is a linear subspace, then the locally convex topology induced on Y by the one of X is generated by the family of seminorms $\{p_{|Y} : p \in \mathcal{P}\}$. The associated locally convex topology is separated if and only if, for every $x \neq \theta$, there exists some $p \in \mathcal{P}$ such that $p(x) \neq 0$.

Now, we come to the *Alaoglu–Banach–Bourbaki theorem*, which is one of the most useful theorems in functional analysis.

Theorem 2.1.9 *If X is an lcs and $V \in \mathcal{N}_X$, then the polar V° is w^*-compact in X^*.*

Given a linear subspace Y of an lcs X, we consider the following equivalence relation in the topological dual space X^*:

$$x_1^* \sim x_2^* \Leftrightarrow x_1^* - x_2^* \in Y^\perp,$$

2.1. FUNCTIONAL ANALYSIS BACKGROUND

and denote $X^*/Y^\perp := X^*/\sim$. Let $h : X^*/Y^\perp \to Y^*$ be the mapping defined by

$$h(\langle x^* \rangle) := z^*_{|Y}, \text{ for any } z^* \in \langle x^* \rangle.$$

Then h defines a *linear isomorphism* from $X^*/Y^\perp$, endowed with the quotient topology, onto the dual space Y^* of Y; in other words, h is a continuous linear bijection with a continuous inverse. Then, since the canonical quotient mapping q (see (2.1)) is *open* (the image of every open set is open), for every $V \in \mathcal{N}_{X^*}$ we have $q(V) \in \mathcal{N}_{X^*/Y^\perp}$, and so

$$V_{|Y} := h(\{\langle x^* \rangle : x^* \in V\}) = \left\{x^*_{|Y} : x^* \in V\right\} \in \mathcal{N}_{Y^*}.$$

To finish this summary of locally convex spaces, we focus on the case when the associated topology is defined by a norm $\|\cdot\|$. In such a case, we say that X is a *normed space* and represent it by $(X, \|\cdot\|)$ (or, simply, by X when no confusion is possible). The closure of a set $A \subset X$ with respect to the *norm*-topology is denoted by $\text{cl}^{\|\cdot\|}(A)$ (or simply $\text{cl}(A)$ if no confusion is possible). A normed space is *Banach* if it is complete (as a metric space). The *dual norm* on X^* is denoted similarly (when no confusion is possible) and is defined by

$$\|x^*\| := \sup\{|\langle x^*, x \rangle| : \|x\| \le 1\}, \ x^* \in X^*;$$

equivalently, we have $\|x^*\| = \sup\{|\langle x^*, x \rangle| : \|x\| = 1\}$. In particular, for all $x \in X$ and $x^* \in X^*$, we get

$$|\langle x^*, x \rangle| \le \|x^*\| \|x\|,$$

known as the *Cauchy–Schwarz inequality*. It is easy to see that $(B_X)^\circ = B_{X^*}$. The space X^* endowed with this dual norm is always Banach (even if the normed space X is not). It is worth observing that the strong topology $\beta(X^*, X)$ coincides with the norm topology. The dual norm in a general Banach space is a typical example of a w^*-lsc convex function which is β-continuous but not necessarily w^*-continuous. By $B_X(x, r)$ (or $B(x, r)$ if no confusion is possible) we represent the closed *ball* in X with center x and radius $r > 0$. In particular, $B_X := B_X(\theta, 1)$ is the *closed unit ball* in X. It is known that in any infinite-dimensional normed space the *unit sphere* $\{x \in X : \|x\| = 1\}$ is weakly dense in B_X.

An *inner product* defined in a linear space X is a bilinear form $\langle \cdot, \cdot \rangle : X \times X \to \mathbb{R}$ (that is, the mappings $\langle \cdot, y \rangle$ and $\langle x, \cdot \rangle$, $x, y \in X$, are both linear), which is *symmetric* and satisfies $\langle x, x\rangle \geq 0$ for all $x \in X$, whereas $\langle x, x\rangle = 0$ if and only if $x = \theta$. An inner product $\langle \cdot, \cdot \rangle$ induces a norm which is defined by $\|x\| := \sqrt{\langle x, x\rangle}$. A *Hilbert space* is a normed space whose norm is induced by an inner product. The *Kadec–Klee property* holds in Hilbert spaces; that is, every sequence $(x_k)_k$, which is weakly convergent to x and satisfies $\|x_k\| \to \|x\|$, is *norm*-convergent to x.

If X and Y are Banach, then the product space $X \times Y$ is also Banach when endowed with the sum (or the maximum) of the norms. In this normed setting, the *Mazur theorem* asserts that every closed convex subset of X is closed for the topology $\sigma(X, X^*)$.

The convex hull of a compact set in a finite-dimensional tvs is compact, as a consequence of the Carathéodory theorem, but this property may fail in infinite dimensions. However, the convex hull of a finite union of convex compact sets is always compact. In addition, if the space is Banach, then the closed convex hull of a compact set is compact, and the closed convex hull of a weakly compact set is weakly compact (this fact corresponds to the *Krein–Šmulian theorem*).

The following result gives a topological counterpart to the Hahn–Banach extension theorem above.

Theorem 2.1.10 *A continuous linear function $\ell \in Y^*$, where Y is a linear subspace of a normed space X, admits a continuous linear extension $\tilde{\ell} \in X^*$ which preserves the norm; i.e., $\|\tilde{\ell}\| = \|\ell\|$.*

We also recall the celebrated *Eberlein–Šmulian theorem*, which establishes that the compactness and the sequential compactness in $(X, \sigma(X, X^*))$ coincide in every normed space X.

The dual of $(X^*, \|\cdot\|)$ is called the *bidual* of X and is denoted by X^{**}. It is also a Banach space for the dual norm

$$\|z\| := \sup\{|\langle z, x^*\rangle| : \|x^*\| \leq 1\}, \; z \in X^{**}.$$

The normed space X is *embedded* in X^{**} in a natural way by means of the *injection mapping*

$$x \in X \mapsto \hat{x} := \langle \cdot, x\rangle \in X^{**}. \qquad (2.26)$$

It can be shown that

$$\|\hat{x}\| = \|x\| = \max\{\langle x^*, x\rangle : \|x^*\| \leq 1\} \text{ for all } x \in X. \qquad (2.27)$$

2.1. FUNCTIONAL ANALYSIS BACKGROUND

We see that the above injection mapping is a *linear isometry* (i.e., an injective linear mapping from X to X^{**} such that $\|x\| = \|\hat{x}\|$ for all $x \in X$). The bidual space X^{**} can also be endowed with the corresponding w^*-topology $\sigma(X^{**}, X^*)$, also denoted by w^{**}. Hence, we write $\to^{w^{**}}$ for the convergence with respect to the topology w^{**} in X^{**} and represent the associated closure of a set $C \subset X^{**}$ by $\mathrm{cl}^{w^{**}} C$.

A Banach space is called *reflexive* if

$$\hat{X} := \{\hat{x} : x \in X\} \equiv X^{**};$$

that is, when the isometry $x \in X \mapsto \hat{x} \in X^{**}$ is surjective. In this case, according to the Mazur theorem above, every *norm*-closed convex subset of X^* is closed for the topology $\sigma(X^*, X^{**}) \equiv \sigma(X^*, X)$.

In a dual pair (X, X^*), where X is a normed space, all compatible topologies on X have the same bounded sets (this is the *Mackey theorem*).

We have the following important special case of Theorem 2.1.9, which supports the proof of (2.27).

Theorem 2.1.11 *Given a normed space X, B_{X^*} is w^*-compact in X^*. Consequently, a subset $C \subset X^*$ is w^*-compact if and only if it is w^*-closed and norm-bounded.*

More properties of normed spaces come next:
(i) $(X, \tau(X, X^*)) = (X, \|\cdot\|)$, and $\sigma(X^{**}, X^*)$ induces $\sigma(X, X^*)$ on $X \subset X^{**}$.
(ii) B_X is $\sigma(X^{**}, X^*)$-dense in $B_{X^{**}}$; consequently, X is $\sigma(X^{**}, X^*)$-dense in X^{**} (*Goldstein theorem*).

In addition, for a Banach space X, the following statements are equivalent:
(i) X is reflexive.
(ii) B_X is weakly compact.
(iii) X^* is reflexive.

In a Banach space X, every convex subset C which is a *countable union* of closed convex sets satisfies $C^i = \mathrm{int}\, C$ (see Exercise 3(iii) in chapter 3).

The following result is the fundamental *James theorem*.

Theorem 2.1.12 *A non-empty weakly closed bounded set C in a Banach space X is weakly compact if and only if every continuous linear function on X attains its maximum on C.*

A mapping $A : X \to 2^{X^*}$ is a *monotone operator* if

$$\langle x^* - y^*, x - y \rangle \geq 0 \text{ for all } x, y \in X, x^* \in Ax, \text{ and } y^* \in Ay. \quad (2.28)$$

We say that A is *cyclically monotone* if

$$\sum_{i=0}^{n} \langle x_i^*, x_{i+1} - x_i \rangle + \langle x_{n+1}^*, x_0 - x_{n+1} \rangle \leq 0, \quad (2.29)$$

for all $x_i \in X$, $x_i^* \in Ax_i$, $i = 0, 1, \ldots, n+1$, $n \geq 1$. It is clear that every cyclically monotone operator is monotone. The converse also holds when $X = \mathbb{R}$.

A *maximally monotone* (*maximally cyclically monotone*) operator $A : X \to 2^{X^*}$ is a monotone (cyclically monotone, respectively) operator such that, for any monotone (cyclically monotone, respectively) operator $B : X \to 2^{X^*}$ such that $\text{gph } B \supset \text{gph } A$, we have $A = B$.

Differentiability

Given an extended real-valued function $f : X \to \overline{\mathbb{R}}$, defined on a linear space X, the *directional derivative* of f at the point $x \in f^{-1}(\mathbb{R})$ in the *direction* $u \in X$ is defined by

$$f'(x; u) := \lim_{t \downarrow 0} \frac{f(x + tu) - f(x)}{t}. \quad (2.30)$$

In particular, when $f : \mathbb{R} \to \mathbb{R}_\infty$, the functions

$$s \mapsto f'_+(s) := f'(s; 1) = \lim_{t \downarrow 0} \frac{f(s+t) - f(s)}{t}$$

and

$$s \mapsto f'_-(s) := -f'(s; -1) = \lim_{t \uparrow 0} \frac{f(s+t) - f(s)}{t}$$

are called the *right* and *left derivatives* of f, respectively.

If X is an lcs, $f : X \to \mathbb{R}$, $f'(x; u)$ exists for all $u \in X$, and $f'(x; \cdot) \in X^*$, then we say that f is *Gâteaux-differentiable* at x. The mapping $f'_G(x) := f'(x; \cdot)$ is called the *Gâteaux-derivative* of f at x. If X is normed, then we say that f is *Fréchet-differentiable* at x if there exists $f'(x) \in X^*$ such that

$$\lim_{u \to 0} \frac{f(x+u) - f(x) - \langle f'(x), u \rangle}{\|u\|} = 0;$$

2.1. FUNCTIONAL ANALYSIS BACKGROUND

we call $f'(x)$ the *Fréchet-derivative* of f at x. If X is a Hilbert space with an inner product $\langle \cdot, \cdot \rangle$, then, by the *Riesz representation theorem*, the dual X^* is identified with X, and the function f is Fréchet-differentiable at $x \in X$ if and only if there exists a vector $\nabla f(x) \in X$, called *gradient* of f at x, such that

$$f'(x)(v) = \langle \nabla f(x), v \rangle \text{ for all } v \in X.$$

If f is Gâteaux-differentiable at x and $f'_G(\cdot)$ is continuous at x, then f is Fréchet-differentiable at x with $f'(x) = f'_G(x)$. Moreover, the Fréchet and Gâteaux-differentiability coincide for convex functions defined on the Euclidean space. For example, if X is Hilbert, then the function $f := \|\cdot\|^2$ is $\mathcal{C}^1$ (in fact, $f'(x)(v) = 2\langle x, v \rangle$ for all $x, v \in X$).

A Banach space X is said to be an *Asplund space* (or just *Asplund*) if every continuous convex function defined on X is Fréchet-differentiable in a dense G_δ-*set*; that is, a set which is the intersection of countably many open sets. If X is separable, then X is Asplund if and only if X^* is separable. All reflexive Banach spaces are Asplund.

Given a convex subset C of a Banach space X, a point $x \in C$ is said to be an *exposed point* of C if there exists $x^* \in X^* \setminus \{\theta\}$ such that x is the unique point that satisfies

$$\sup_{c \in C} \langle c, x^* \rangle = \langle x, x^* \rangle.$$

More restrictively, the point x is said to be a *strongly exposed point* of C if there exists some $x^* \in X^* \setminus \{\theta\}$ such that every sequence $(x_n)_n \subset C$, satisfying

$$\langle x_n, x^* \rangle \to \sup_{c \in C} \langle c, x^* \rangle,$$

converges to x. Equivalently, $x \in C$ is a strongly exposed point of C if and only if the (*support*) function $z^* \in X^* \mapsto \sup_{c \in C} \langle c, z^* \rangle$ is Fréchet-differentiable at x^* with Fréchet-derivative x.

The Banach space X is said to have the *Radon–Nikodym property* (*RNP*, for short) if every non-empty closed bounded convex set C can be written as the closed convex hull of its strongly exposed points. It is known that X has the RNP if and only if X^* is a w^*-*Asplund*; that is, when every (*norm-*)continuous and w^*-lsc convex function defined on X^* is Fréchet-differentiable in a dense G_δ-subset of X^*. In addition, the space X is Asplund if and only if X^* has the RNP.

2.2 Convexity and continuity

In this section we review the main algebraic and topological properties of the extended real-valued convex functions. We work in a (real) separated locally convex space (lcs) X as described in section 1.3; that is, X and X^* form a compatible dual pair via the duality pairing $\langle x^*, x \rangle := x^*(x)$, $x^* \in X^*$, $x \in X$; by $\mathcal{N}_X$ we represent a neighborhood base of closed convex balanced neighborhoods of θ (θ-neighborhoods), and $\mathcal{P}$ is a saturated family of seminorms generating the topology in X.

Given an extended real-valued function $f : X \to \overline{\mathbb{R}}$, also written as $f \in \overline{\mathbb{R}}^X$, we introduce some geometric objects associated with it. First, since f is allowed to take the value $+\infty$, we define the *effective domain* or (simply, *domain*) of f by

$$\operatorname{dom} f := \{x \in X : f(x) < +\infty\} = f^{-1}(\mathbb{R} \cup \{-\infty\}).$$

This definition does not exclude those points where f takes the value $-\infty$. It is clear that the domain of the sum or maximum of two functions is the intersection of the associated domains, and that this is not true for the supremum of an arbitrary family of functions. Sometimes, functions are defined only on subsets of X where they are finite, by writing $f : C \subset X \to \mathbb{R}$ for some given set $C \subset X$.

An intrinsic geometric object in convex analysis, which further captures the properties of a function f, is the *epigraph* of f defined by

$$\operatorname{epi} f := \{(x, \lambda) \in X \times \mathbb{R} : f(x) \leq \lambda\}.$$

This set provides an exact identification of the function f as it allows its recovery by writing

$$f(x) = \inf\{\lambda \in \mathbb{R} : (x, \lambda) \in \operatorname{epi} f\}. \tag{2.31}$$

The epigraph is then placed above the *graph* of f, which is the set

$$\operatorname{gph} f := \{(x, \lambda) \in X \times \mathbb{R} : f(x) = \lambda\}.$$

A related concept is the *strict epigraph* given by

$$\operatorname{epi}_s f := \{(x, \lambda) \in X \times \mathbb{R} : f(x) < \lambda\}.$$

We can check that both the epigraph and the strict epigraph have the same closure.

2.2. CONVEXITY AND CONTINUITY

Now we give the definition of convex functions, taking into account the current convention (2.2).

Definition 2.2.1 *A function* $f : X \to \overline{\mathbb{R}}$ *is* convex *if the following inequality, called* Jensen inequality, *holds for all* $x_1, x_2 \in X$ *and* $\lambda \in [0, 1]$:

$$f(\lambda x_1 + (1 - \lambda)x_2) \leq \lambda f(x_1) + (1 - \lambda)f(x_2). \qquad (2.32)$$

Observe that, due to our conventions (2.2) on the sum and product operations in $\overline{\mathbb{R}}$, we can restrict the inequality in (2.32) to $x_1, x_2 \in \text{dom } f$ and $\lambda \in \,]0, 1[$. The geometric meaning of convexity is made clear, thanks to the concepts of epigraph and strict epigraph. The following proposition shows this fact.

Proposition 2.2.2 *For every function* $f : X \to \overline{\mathbb{R}}$, *the following assertions are equivalent:*
 (i) f *is convex.*
 (ii) $\text{epi } f$ *is a convex set in* $X \times \mathbb{R}$.
 (iii) $\text{epi}_s f$ *is a convex set in* $X \times \mathbb{R}$.

Proof. $(i) \Rightarrow (iii)$ Take $(x_i, \lambda_i) \in \text{epi}_s f$, $i = 1, 2$, and $\alpha \in \,]0, 1[$. Then, by the convexity of f,

$$f(\alpha x_1 + (1-\alpha)x_2) \leq \alpha f(x_1) + (1-\alpha)f(x_2) < \alpha \lambda_1 + (1-\alpha)\lambda_2,$$

and $\alpha(x_1, \lambda_1) + (1-\alpha)(x_2, \lambda_2) \in \text{epi}_s f$.

$(iii) \Rightarrow (ii)$ Take $(x_i, \lambda_i) \in \text{epi } f$, $i = 1, 2$, and $\alpha \in \,]0, 1[$. So, for each fixed $\varepsilon > 0$, $(x_i, \lambda_i + \varepsilon) \in \text{epi}_s f$ and (iii) implies that $\alpha(x_1, \lambda_1 + \varepsilon) + (1-\alpha)(x_2, \lambda_2 + \varepsilon) \in \text{epi}_s f$. Thus,

$$f(\alpha x_1 + (1-\alpha)x_2) < \alpha \lambda_1 + (1-\alpha)\lambda_2 + \varepsilon,$$

and (ii) follows when $\varepsilon \downarrow 0$.

$(ii) \Rightarrow (i)$ Take $x_1, x_2 \in \text{dom } f$ and $\lambda \in \,]0, 1[$. Given $k \geq 1$, we denote $\alpha_{k,i} := \max\{f(x_i), -k\}$, $i = 1, 2$. So, $(x_i, \alpha_{k,i}) \in \text{epi } f$, $i = 1, 2$, and (ii) yields

$$f(\lambda x_1 + (1-\lambda)x_2) \leq \lambda \alpha_{k,1} + (1-\lambda)\alpha_{k,2} \text{ for all } k \geq 1.$$

Hence, the Jensen inequality follows when $k \uparrow +\infty$. ∎

A function $f : X \to \overline{\mathbb{R}}$ is said to be *proper* if it has a non-empty (effective) domain and never takes the value $-\infty$. This is a natural assumption in extended real-valued functions. However, we often allow non-proper functions, not for the sake of generality, but because our theory will sometimes lead to new functions that may not be proper (see Exercise 39, for an example illustrating this situation). Obviously, a function $f : X \to \mathbb{R}_\infty$ is proper if and only if it has a non-empty domain. Clearly, the functions which are identically equal to $+\infty$ or to $-\infty$ are not proper. In addition, the sum of two proper functions may not be proper, unless their effective domains intersect.

We say that $f : X \to \overline{\mathbb{R}}$ is *positively homogeneous* if $f(\theta) = 0$ and

$$f(\lambda x) = \lambda f(x) \text{ for all } x \in \text{dom } f \text{ and all } \lambda > 0.$$

Observe that the condition $f(\theta) = 0$ above can be replaced with the less restrictive relation $f(\theta) \in \mathbb{R}$ (see Exercise 8). As for convexity, the positive homogeneity is also characterized by means of the epigraph. Indeed, a proper function $f : X \to \mathbb{R}_\infty$ is positively homogeneous if and only if its epigraph is a cone in $X \times \mathbb{R}$.

The function $f : X \to \overline{\mathbb{R}}$ is *subadditive* if for all $x_1, x_2 \in X$:

$$f(x_1 + x_2) \leq f(x_1) + f(x_2),$$

and *sublinear* if it is subadditive and positively homogeneous. Equivalently, a sublinear function is a positive homogeneous convex function.

We now turn to the topological side, showing that the epigraph also captures the lower semicontinuity of the associated function. In this case, the sublevel sets come into play since their closure also characterizes the lower semicontinuity property. There is a subtle difference here with convexity, the (non-empty) sublevel sets of convex functions are convex, but we can have non-convex functions with convex sublevel sets. The property of having convex sublevel sets in fact characterizes the so-called *quasi-convex functions*.

Proposition 2.2.3 *For every function $f : X \to \overline{\mathbb{R}}$, the following assertions are equivalent:*
  (i) f *is lsc.*
  (ii) epi f *is a closed set in $X \times \mathbb{R}$.*
  (iii) $[f \leq \lambda]$ *is closed in X for all $\lambda \in \mathbb{R}$.*

Due to the equivalence of (i) and (ii), lsc functions are also called *closed*. Note that the sequential lower semicontinuity of f is also characterized by statements (ii)–(iii), provided that the condition on the closure is replaced with the sequential closure. Moreover, due to the Mazur theorem, every lsc convex function on X is weakly lsc.

2.2. CONVEXITY AND CONTINUITY

When a given function $f : X \to \overline{\mathbb{R}}$ is not lsc (or convex), we sometimes proceed by replacing it with an appropriate lsc (or convex, respectively) approximation. The *convex hull*, *closed hull*, and *closed convex hull* of f are respectively defined by

$$\operatorname{co} f := \sup\{g : X \to \overline{\mathbb{R}} : g \text{ is convex and } g \leq f\},$$
$$\operatorname{cl} f := \sup\{g : X \to \overline{\mathbb{R}} : g \text{ is lsc and } g \leq f\},$$
$$\overline{\operatorname{co}} f := \sup\{g : X \to \overline{\mathbb{R}} : g \text{ is convex, lsc, and } g \leq f\};$$

the closed hull is also represented by $\bar{f}$.

Closed and closed convex hulls can also be obtained by operating topologically and algebraically on the epigraph.

Proposition 2.2.4 *For every function $f : X \to \overline{\mathbb{R}}$, we have*

$$\operatorname{epi}(\operatorname{cl} f) = \operatorname{cl}(\operatorname{epi} f) \quad \text{and} \quad \operatorname{epi}(\overline{\operatorname{co}} f) = \overline{\operatorname{co}}(\operatorname{epi} f). \tag{2.33}$$

Note that there is no equivalent relation to (2.33) for the convex hull; in fact, co (epi f) need not be an epigraph. We also can write cl f and co f in terms of lower limits and convex combinations, respectively: for every $x \in X$,

$$(\operatorname{cl} f)(x) = \liminf_{y \to x} f(y) = \sup_{U \in \mathcal{N}_x} \inf_{y \in U} f(x+y), \tag{2.34}$$

and (see the definition of $\Delta(\operatorname{epi} f)$ in (2.5))

$$(\operatorname{co} f)(x) = \inf \left\{ \sum_{(z,s) \in \operatorname{epi} f} \lambda_{(z,s)} s : \sum_{(z,s) \in \operatorname{epi} f} \lambda_{(z,s)} z = x, \ \lambda \in \Delta(\operatorname{epi} f) \right\}$$

$$= \inf \left\{ \sum_{z \in \operatorname{dom} f} \lambda_z f(z) : \sum_{z \in \operatorname{dom} f} \lambda_z z = x, \ \lambda \in \Delta(\operatorname{dom} f) \right\}.$$

Then we can easily check that

$$\operatorname{cl}(\operatorname{dom}(\operatorname{cl} f)) = \operatorname{cl}(\operatorname{dom} f), \quad \operatorname{dom}(\operatorname{co} f) = \operatorname{co}(\operatorname{dom} f), \tag{2.35}$$

and, as a consequence, we get

$$\operatorname{cl}(\operatorname{dom}(\overline{\operatorname{co}} f)) = \overline{\operatorname{co}}(\operatorname{dom} f). \tag{2.36}$$

Moreover, we have the following equalities:

$$\inf f = \inf(\operatorname{cl} f) = \inf(\operatorname{co} f) = \inf(\overline{\operatorname{co}} f), \qquad (2.37)$$

showing that the above hulls do not change the value of the infimum of the original function.

To continue the previous discussion on the properness assumption, note that when X is infinite-dimensional the closed hull $\operatorname{cl} f$ may be non-proper, even when f is proper and convex (Exercise 39). In addition, if f is convex and lsc but not proper, then $f(x) = -\infty$ for all $x \in \operatorname{dom} f$; i.e., f cannot take finite values. In other words, if f is an lsc convex function which is finite at some point, then it must be proper.

Now we introduce the most important family of functions in convex analysis, which is

$$\Gamma_0(X) := \{f : X \to \mathbb{R}_\infty : f \text{ is proper, lsc, and convex}\}.$$

The goal of the next paragraph is to review the strong link between convexity and continuity, extending the well-known continuity properties of linear functions. This connection confirms that convex functions can be regarded as natural generalizations of linear functions. The key to this connection lies in the fact that, for every convex function $f : X \to \overline{\mathbb{R}}$, and every $x \in \operatorname{dom} f$ and $u \in X$, the quotient

$$s \mapsto \frac{f(x + su) - f(x)}{s}$$

is non-decreasing with $s > 0$. Indeed, for every $0 < s_1 < s_2$, the convexity of f gives rise to

$$\begin{aligned} f(x + s_1 u) &= f\left(\frac{s_1}{s_2}(x + s_2 u) + \left(1 - \frac{s_1}{s_2}\right)x\right) \\ &\leq \frac{s_1}{s_2} f(x + s_2 u) + \left(1 - \frac{s_1}{s_2}\right) f(x), \qquad (2.38) \end{aligned}$$

which easily leads us to the desired non-decreasingness property. Therefore, the directional derivative of f at x in the direction u turns out to be (see (2.30))

$$f'(x; u) = \lim_{s \downarrow 0} \frac{f(x + su) - f(x)}{s} = \inf_{s > 0} \frac{f(x + su) - f(x)}{s}; \qquad (2.39)$$

2.2. CONVEXITY AND CONTINUITY

that is, the last limits exist in $\overline{\mathbb{R}}$. For the same reason we have the equality

$$\lim_{s\uparrow+\infty} \frac{f(x+su)-f(x)}{s} = \sup_{s>0} \frac{f(x+su)-f(x)}{s},$$

describing the behavior of f at $+\infty$ in the direction u.

Relation (2.39) constitutes the key property relating convexity to continuity. Moreover, when $f : \mathbb{R} \to \mathbb{R}_\infty$ is a convex function and $t_1, t_2, t_3 \in \text{dom } f$ are such that $t_1 < t_2 < t_3$, we have that (see Exercise 16)

$$\frac{f(t_2)-f(t_1)}{t_2-t_1} \leq \frac{f(t_3)-f(t_1)}{t_3-t_1} \leq \frac{f(t_3)-f(t_2)}{t_3-t_2}. \tag{2.40}$$

As a consequence of (2.40) we obtain the following result.

Proposition 2.2.5 *Given a convex function* $f : X \to \mathbb{R}_\infty$, $x \in \text{dom } f$, *and* $u \in X$, *for every*

$$\alpha \in \left[\sup_{t<0} t^{-1}(f(x+tu) - f(x)), \inf_{t>0} t^{-1}(f(x+tu) - f(x))\right],$$

the function $\varphi_\alpha : \mathbb{R} \to \overline{\mathbb{R}}$ *defined by*

$$\varphi_\alpha(s) := \begin{cases} s^{-1}(f(x+su)-f(x)), & \text{if } s \neq 0, \\ \alpha, & \text{if } s = 0, \end{cases} \tag{2.41}$$

is non-decreasing on $\mathbb{R}$.

In other words, Proposition 2.2.5 says that the multifunction which assigns to $s \in \mathbb{R}$ the value

$$\begin{cases} \{s^{-1}(f(x+su)-f(x))\}, & \text{if } s \neq 0, \\ \left[\sup_{t<0} t^{-1}(f(x+tu)-f(x)), \inf_{t>0} t^{-1}(f(x+tu)-f(x))\right], & \text{if } s = 0, \end{cases}$$

is monotone.

The following proposition constitutes a remarkable topological property of convex functions. More precisely, the statement of the proposition shows a Lipschitz-like behavior of convex functions.

Proposition 2.2.6 *Given a convex function* $f : X \to \mathbb{R}_\infty$ *and* $x \in \text{dom } f$, *we suppose the existence of some* $m \geq 0$ *and* $U \in \mathcal{N}_X$ *such*

that
$$f(x+y) - f(x) \le m \text{ for all } y \in U.$$

Then
$$|f(x+y) - f(x)| \le m p_U(y) \text{ for all } y \in U,$$

and, consequently, for all $\rho \in \,]0,1[$ we have that

$$|f(y) - f(z)| \le m \frac{1+\rho}{1-\rho} p_U(y-z) \text{ for all } y,\, z \in x + \rho U.$$

Proof. We only give the proof of the first statement. We may suppose, without loss of generality, that $x = \theta$ and $f(\theta) = 0$, so that the current assumption reads
$$f(y) \le m \text{ for all } y \in U.$$

Fix $y \in U$. If $p_U(y) > 0$, then $y/p_U(y) \in U$ and the convexity of f yields

$$f(y) = f\left(p_U(y) \frac{y}{p_U(y)}\right) \le p_U(y) f\left(\frac{y}{p_U(y)}\right) \le m p_U(y).$$

If $p_U(y) = 0$, for all $0 < \varepsilon < 1$, we have $y/\varepsilon \in U$ and, again by the convexity of f,

$$f(y) = f(\varepsilon y/\varepsilon) \le \varepsilon f(y/\varepsilon) \le \varepsilon m,$$

implying that $f(y) \le 0$ as $\varepsilon \downarrow 0$. Hence,

$$f(y) \le m p_U(y) \text{ for all } y \in U.$$

Moreover, as $-y \in U$, we also have

$$0 = f(\theta) = f(y/2 + (1/2)(-y))$$
$$\le (1/2) f(y) + (1/2) f(-y) \le (1/2) f(y) + (1/2) m p_U(y),$$

showing that $-f(y) \le m p_U(y)$. The first statement is proved. ∎

As the first consequence of the last proposition, we conclude the following important result which shows the main inheritance of continuity of convex functions from continuity of linear functions.

Corollary 2.2.7 *The following three statements are equivalent, for every proper convex function $f : X \to \mathbb{R}_\infty$:*

2.2. CONVEXITY AND CONTINUITY

(i) f is bounded above around a point of its domain.
(ii) f is continuous at some point in $\operatorname{dom} f$.
(iii) f is continuous on $\operatorname{int}(\operatorname{dom} f)$.

Proof. The equivalence of assertions (i) and (ii) is Proposition 2.2.6. To see that assertion (ii) also implies the stronger statement in (iii), let $x_0 \in \operatorname{dom} f$ be a continuity point of f and take any $x \in \operatorname{int}(\operatorname{dom} f)$ different from x_0. Let $U \in \mathcal{N}_X$ such that

$$f(x_0 + u) \leq f(x_0) + 1 \text{ for all } u \in U,$$

and choose $\rho > 0$ such that

$$z_0 := x + \rho(x - x_0) \in \operatorname{int}(\operatorname{dom} f);$$

hence, $z_0 \neq x$, and there exists some $\lambda_0 \in \,]0, 1[$ such that

$$x = \lambda_0 z_0 + (1 - \lambda_0) x_0.$$

Then, for every $u \in U$,

$$f(x + (1 - \lambda_0)u) = f(\lambda_0 z_0 + (1 - \lambda_0)x_0 + (1 - \lambda_0)u)$$
$$\leq \lambda_0 f(z_0) + (1 - \lambda_0) f(x_0 + u) \leq \lambda_0 f(z_0) + (1 - \lambda_0)m,$$

and the desired equivalence comes from Proposition 2.2.6 when applied taking $U_{\lambda_0} := (1 - \lambda_0)U \in \mathcal{N}_X$. ■

Furthermore, in a normed space the equivalences in Corollary 2.2.7 remain true if the continuity assumption in (ii) and (iii) is replaced with the local Lipschitzianity property. The following property is also useful.

Corollary 2.2.8 *Assume that X is Banach. If $f \in \Gamma_0(X)$, then f is locally Lipschitz on $\operatorname{int}(\operatorname{dom} f)$ whenever the last set is non-empty.*

Proof. We write $\operatorname{dom} f = \cup_{n \geq 1} [f \leq n]$, where each set $[f \leq n]$ is a closed subset of X. Since X is a Baire space and $\operatorname{int}(\operatorname{dom} f) \neq \emptyset$, by the Baire category theorem there exists some $n_0 \geq 1$ such that the set $\operatorname{int}([f \leq n_0]) \neq \emptyset$. It follows that f is bounded above around a point in $[f \leq n_0] \subset \operatorname{dom} f$, and the equivalences above imply the desired result. ■

The lower semicontinuity condition required in the previous corollary is removed in finite dimensions.

Corollary 2.2.9 *Assume that $f : \mathbb{R}^n \to \mathbb{R}_\infty$ is a proper convex function. Then $\operatorname{ri}(\operatorname{dom} f)$ is non-empty and the function $f_{|\operatorname{aff}(\operatorname{dom} f)}$ is finite and continuous on $\operatorname{ri}(\operatorname{dom} f)$. More generally, f is lsc on $\operatorname{ri}(\operatorname{dom} f)$.*

We deduce from the previous corollary that every function $f \in \Gamma_0(X)$ is continuous relative to each closed segment within its domain:

Corollary 2.2.10 *Given a convex function $f : X \to \overline{\mathbb{R}}$ and $x, y \in \operatorname{dom} f$, the function $\varphi : \mathbb{R} \to \overline{\mathbb{R}}$ defined by*

$$\varphi(t) := f(tx + (1-t)y)$$

is continuous on $]0,1[$. It is continuous relative to $[0,1[$ $(]0,1])$ when, additionally, f is lsc at y (x, respectively).

Proof. The function φ is continuous on $]0,1[$ by Corollary 2.2.9. Moreover, by convexity of f, we have $\varphi(t) \leq t\varphi(1) + (1-t)\varphi(0)$ and, using the lower semicontinuity of f at y, by taking limits we obtain

$$\varphi(0) = f(y) \leq \liminf_{t \downarrow 0} \varphi(t) \leq \limsup_{t \downarrow 0} \varphi(t) \leq \limsup_{t \downarrow 0} (tf(x) + (1-t)f(y)) = \varphi(0).$$

Thus, $\lim_{t \downarrow 0} \varphi(t) = \varphi(0)$ and φ is continuous at 0 relative to $[0,1[$. The same reasoning shows that φ is continuous at 1 relative to $]0,1]$ when f is lsc at x. ∎

Convexity of functions is preserved under many operations. For example, the convexity of the sum of two convex functions follows easily by the Jensen inequality. The convexity of the supremum of a family of convex functions comes from the fact that the epigraph of the supremum is the intersection of their epigraphs (see (2.47)). Further properties of the supremum are given in the forthcoming section.

We close this section with a property of the closed hull operation that will be used several times in the book. This result will help us to develop calculus rules in section 4.1 for the subdifferential of the function $f + g \circ A$ (allowing us to reduce the problem to that of a problem involving lsc functions).

Proposition 2.2.11 *Let X, Y be two lcs, $f : Y \to \overline{\mathbb{R}}$ and $g : X \to \overline{\mathbb{R}}$ convex functions, and $A : X \to Y$ a continuous affine mapping. Assume that f is finite and continuous at Ax_0 for some $x_0 \in \operatorname{dom} g$. Then we have that*

$$\operatorname{cl}(f \circ A + g) = (\operatorname{cl} f) \circ A + (\operatorname{cl} g). \tag{2.42}$$

2.2. CONVEXITY AND CONTINUITY

Proof. For the sake of brevity, we suppose that $X = Y$ and A is the identity mapping. First, we observe that the inequality $\operatorname{cl} f + \operatorname{cl} g \leq \operatorname{cl}(f + g)$ always holds because $\operatorname{cl} f + \operatorname{cl} g \leq f + g$ and $\operatorname{cl} f + \operatorname{cl} g$ is obviously lsc. To prove the converse inequality, we pick $x \in (\operatorname{dom}(\operatorname{cl} f)) \cap (\operatorname{dom}(\operatorname{cl} g))$; that is, due to (2.35),

$$x \in (\operatorname{dom}(\operatorname{cl} f)) \cap \operatorname{cl}(\operatorname{dom}(\operatorname{cl} g)) = (\operatorname{dom}(\operatorname{cl} f)) \cap \operatorname{cl}(\operatorname{dom} g).$$

Let us fix $\lambda \in]0,1[$ and denote $x_\lambda := \lambda x_0 + (1 - \lambda)x$. Since $x_0 \in \operatorname{int}(\operatorname{dom} f)$ and $x \in \operatorname{dom}(\operatorname{cl} f) \subset \operatorname{cl}(\operatorname{dom}(\operatorname{cl} f)) = \operatorname{cl}(\operatorname{dom} f)$, again by (2.35), (2.15) yields $x_\lambda \in \operatorname{int}(\operatorname{dom} f)$, and Proposition 2.2.6 entails the continuity of f at x_λ.

Now, by (2.34), we choose a net $z_i \to_i x_\lambda$ such that $(\operatorname{cl} g)(x_\lambda) = \lim_i g(z_i)$. Observe that $x_i := (1 - \lambda)^{-1}(z_i - \lambda x_0) \to_i x$. Since $\lim_i f(z_i) = f(x_\lambda) = (\operatorname{cl} f)(x_\lambda)$, by the continuity of f at x_λ, we obtain

$$(\operatorname{cl}(f + g))(x_\lambda) \leq \liminf_i (f(z_i) + g(z_i))$$
$$= \lim_i (f(z_i) + g(z_i)) = (\operatorname{cl} f)(x_\lambda) + (\operatorname{cl} g)(x_\lambda),$$

and the convexity assumption gives rise to

$$(\operatorname{cl}(f + g))(x_\lambda) \leq \lambda((\operatorname{cl} f)(x_0) + (\operatorname{cl} g)(x_0)) + (1 - \lambda)((\operatorname{cl} f)(x) + (\operatorname{cl} g)(x))$$
$$\leq \lambda(f(x_0) + g(x_0)) + (1 - \lambda)((\operatorname{cl} f)(x) + (\operatorname{cl} g)(x)).$$

Whence, as $\lambda \downarrow 0$ and $g(x_0) \in \mathbb{R}$ we get

$$\liminf_{\lambda \downarrow 0} (\operatorname{cl}(f + g))(x_\lambda) \leq (\operatorname{cl} f)(x) + (\operatorname{cl} g)(x),$$

and this yields $(\operatorname{cl}(f + g))(x) \leq (\operatorname{cl} f)(x) + (\operatorname{cl} g)(x)$. If $g(x_0) = -\infty$, then $(\operatorname{cl}(f + g))(x_\lambda) = -\infty$ and we deduce that $(\operatorname{cl}(f + g))(x) \leq \liminf_{\lambda \downarrow 0}(\operatorname{cl}(f + g))(x_\lambda) = -\infty \leq (\operatorname{cl} f)(x) + (\operatorname{cl} g)(x)$. The proof is complete. ∎

Observe that, due to (2.37), relation (2.42) implies

$$\inf_{x \in X}(f + g \circ A)(x) = \inf_{x \in X} \operatorname{cl}(f + g \circ A)(x) = \inf_{x \in X}((\operatorname{cl} f) + (\operatorname{cl} g) \circ A)(x), \quad (2.43)$$

resulting in an optimization problem involving only lsc functions. It is this precise property that will allow us in chapter 4 to establish different subdifferential calculus rules and duality results for convex functions not necessarily lsc. A counterpart to (2.43) for *perturbation functions* will be given in Proposition 4.1.23.

2.3 Examples of convex functions

In this section, we present some particular convex functions, which are crucial in convex analysis and optimization.

The *indicator* function of a set $C \subset X$ is defined as

$$I_C(x) := \begin{cases} 0, & \text{if } x \in C, \\ +\infty, & \text{if } x \in X \setminus C. \end{cases}$$

We have $\operatorname{dom} I_C = C$ and $\operatorname{epi} I_C = C \times \mathbb{R}_+$, so that I_C is a convex (lsc) function if and only if C is a convex (closed, respectively) set. The function I_C is proper if and only if the set C is non-empty. More generally, the closed and closed convex hulls of I_C are the indicator functions

$$\operatorname{cl} I_C = I_{\operatorname{cl} C} \text{ and } \overline{\operatorname{co}} I_C = I_{\overline{\operatorname{co}} C}.$$

The indicator function provides a good device for penalizing constrained optimization problems. In fact, if $f : X \to \mathbb{R}_\infty$ is a given function, then for every non-empty set $C \subset X$ we have

$$\inf_{x \in C} f(x) = \inf_{x \in X} \{f(x) + I_C(x)\}.$$

We also have the following properties, when $C := \cap_i C_i$ and $D := \cup_i C_i$ for some arbitrary family of sets $C_i \subset X$,

$$I_C = \sup_i I_{C_i} \text{ and } I_D = \inf_i I_{C_i}.$$

The operation of taking the supremum of an arbitrary family of convex functions is a usual operation preserving convexity. Given an arbitrary family of convex functions $f_t : X \to \overline{\mathbb{R}}$, $t \in T$, where T is a fixed index set, the associated *pointwise supremum function* is defined as

$$f := \sup_{t \in T} f_t. \qquad (2.44)$$

Observe that f can be expressed as

$$f = \sup_{\lambda \in \Delta(T)} \sum_{t \in T} \lambda_t f_t, \qquad (2.45)$$

2.3. EXAMPLES OF CONVEX FUNCTIONS

where $\Delta(T)$ is defined as in (2.5); that is,

$$\Delta(T) = \left\{ \lambda \in \mathbb{R}_+^{(T)} : \sum_{t \in T} \lambda_t = 1 \right\}. \tag{2.46}$$

The function f is convex, and lsc when all the f_t's are so. These assertions are easily seen in the relation

$$\operatorname{epi} f = \bigcap_{t \in T} \operatorname{epi} f_t. \tag{2.47}$$

Generally, we cannot replace the epigraph by the strict epigraph in (2.47), nor can we express the domain of f as the intersection of the domains of the f_t's. However, when T is Hausdorff compact and the mappings $t \mapsto f_t(x)$ are upper semicontinuous (usc, for short) for all $x \in X$, then (see Exercise 9)

$$\operatorname{dom} f = \bigcap_{t \in T} \operatorname{dom} f_t.$$

An interesting example of supremum functions is the support function. Given a set $C \subset X^*$, the *support function* of C is the function $\sigma_C : X \to \overline{\mathbb{R}}$ defined as

$$\sigma_C(x) := \sup\{\langle x^*, x \rangle : x^* \in C\},$$

with the convention that $\sigma_\emptyset \equiv -\infty$. Similarly, we can define σ_C on X^* when $C \subset X$. The function σ_C is always convex and lsc as it is the pointwise supremum of continuous linear functions. Here the convexity of C is superfluous; in fact, we have the following relation resembling (2.45),

$$\sigma_C = \sigma_{\operatorname{cl} C} = \sigma_{\overline{\operatorname{co}} C}. \tag{2.48}$$

Also, provided that $C \neq \emptyset$, we have

$$\operatorname{dom} \sigma_C = \mathbb{R}_+ C^\circ, \tag{2.49}$$

and so σ_C is proper (when $C \neq \emptyset$). Also, given two non-empty sets C, $D \subset X^*$, we verify that

$$\sigma_{C+D} = \sigma_C + \sigma_D \text{ and } \sigma_{C \cup D} = \max\{\sigma_C, \sigma_C\}. \tag{2.50}$$

An operation related to the supremum, which also preserves the convexity, is the *pointwise upper limit* of convex functions. If $(f_i)_i$ is a net of convex functions, with $(I, \succeq)$ being a directed set, then the pointwise upper limit function $f : X \to \overline{\mathbb{R}}$ defined as $f := \limsup_i f_i$ satisfies for all $x, y \in X$ and $\lambda \in {]}0, 1{[}$:

$$\begin{aligned}
f(\lambda x + (1-\lambda)y) &= \limsup_i f_i(\lambda x + (1-\lambda)y) \\
&\leq \limsup_i (\lambda f_i(x) + (1-\lambda) f_i(y)) \\
&\leq \lambda \limsup_i f_i(x) + (1-\lambda) \limsup_i f_i(y) \\
&= \lambda f(x) + (1-\lambda) f(y).
\end{aligned} \quad (2.51)$$

Equivalently, the convexity of the upper limit function f also follows from the expression of its strict epigraph,

$$\operatorname{epi}_s f = \bigcup_i \bigcap_{j \succeq i} \operatorname{epi}_s f_j.$$

To see that the supremum function $f = \sup_{t \in T} f_t$ is a particular instance of the upper limit function, we endow the family of finite subsets of T,

$$\mathcal{F}(T) := \{ S \subset T : T \text{ is finite} \}$$

with the partial order "$\preccurlyeq$" of ascending inclusions:

$$S_1 \preccurlyeq S_2 \iff S_1 \subset S_2, \text{ for all } S_1, S_2 \in \mathcal{F}(T).$$

Then we verify that

$$\sup_{t \in T} f_t = \inf_{S_0 \in \mathcal{F}(T)} \sup_{\substack{S \supset S_0 \\ S \in \mathcal{F}(T)}} \left(\max_{t \in S} f_t \right) = \limsup_{S \in \mathcal{F}(T)} \left(\max_{t \in S} f_t \right).$$

The convexity is also preserved by the *pointwise lower limit function*, $\liminf_i f_i$, of any non-increasing net $(f_i)_i$ of convex functions, because

$$\liminf_i f_i = \inf_i f_i = \limsup_i f_i. \quad (2.52)$$

We saw in section 2.1 that the Minkowski gauge is a fundamental device for characterizing locally convex topologies by means of seminorms. Remember that the Minkowski gauge (see (2.25)) of a given non-empty set $C \subset X$ is given by

$$p_C(x) = \inf\{ \lambda \geq 0 : x \in \lambda C \},$$

with $\inf \emptyset = +\infty$. If C is a closed set containing θ, then

2.3. EXAMPLES OF CONVEX FUNCTIONS

$$[p_C \leq \alpha] = \alpha C \text{ for all } \alpha > 0 \tag{2.53}$$

and so p_C is lsc. In addition, due to our convention $0(+\infty) = +\infty$, the positive homogeneity of p_C is established below only on the set $\operatorname{dom} p_C$; that is,

$$p_C(\alpha x) = \alpha p_C(x) \text{ for all } x \in \operatorname{dom} p_C \text{ and } \alpha \geq 0.$$

The proof of the following proposition is postponed to Exercise 1.

Proposition 2.3.1 *The Minkowski gauge of a non-empty set $C \subset X$ is positively homogeneous in its domain. Moreover, we have*

$$\operatorname{epi}_s p_C = \mathbb{R}_+^* \left((C \cup \{\theta\}) \times \ ]1, +\infty[\right), \tag{2.54}$$

and p_C is convex whenever C is convex.

In the case when $U \in \mathcal{N}_X$, the function p_U is a continuous seminorm and satisfies

$$U = \{x \in X : \ p_U(x) \leq 1\} \text{ and } \operatorname{int} U = \{x \in X : \ p_U(x) < 1\}. \tag{2.55}$$

The inf-convolution operation is very useful in convex analysis, especially in the context of regularization processes. Given two functions $f, g : X \to \overline{\mathbb{R}}$, the *inf-convolution* of f and g is the function $f \square g : X \to \overline{\mathbb{R}}$ defined by

$$(f \square g)(x) := \inf\{f(y) + g(x - y) : y \in X\}.$$

We say that the inf-convolution is *exact at x* when the infimum above is attained. We have

$$\operatorname{dom}(f \square g) = \operatorname{dom} f + \operatorname{dom} g, \tag{2.56}$$

because $x \in \operatorname{dom}(f \square g)$ if and only if there exists some $y \in X$ such that $f(y) + g(x - y) < +\infty$. And thanks to our sign rules, if and only if $x = y + (x - y) \in \operatorname{dom} f + \operatorname{dom} g$. The inf-convolution is also called the *epigraphical sum*, due to the following relations:

$$\operatorname{epi}_s(f \square g) = \operatorname{epi}_s f + \operatorname{epi}_s g, \tag{2.57}$$

and

$$\operatorname{epi}(\operatorname{cl}(f \square g)) = \operatorname{cl}(\operatorname{epi} f + \operatorname{epi} g). \tag{2.58}$$

Hence, it is immediate from (2.57) that $f \square g$ is convex when f and g are convex. The inf-convolution has a regularizing effect. For instance, if the function g is continuous at some point $y_0 \in \operatorname{dom} g$, then we can find some $U \in \mathcal{N}_X$ such that, for all $x_0 \in \operatorname{dom} f$ and all $u \in U$,

$$(f \square g)(x_0 + y_0 + u) \leq f(x_0) + g(y_0 + u) \leq f(x_0) + g(y_0) + 1,$$

and $f \square g$ is continuous at $x_0 + y_0$, due to Proposition 2.2.6.

A typical example of functions that can be expressed as an inf-convolution is the *distance function* to a non-empty subset C of a normed space X, $d_C : X \to \mathbb{R}_+$, defined as

$$d_C(x) := \inf_{y \in C} \|x - y\|.$$

In fact,

$$d_C = \|\cdot\| \square I_C,$$

and so d_C is convex whenever C is convex. A point $c \in C$ is said to be a *projection* of $x \in X$ on C if $\|x - c\| = d_C(x)$. The set of projections of the point x on C is denoted by $\pi_C(x)$.

Given a function $f : X \to \mathbb{R}_\infty$, defined on the normed space X, the *Moreau–Yosida approximation* of f with a parameter $\gamma > 0$ is the function $f^\gamma : X \to \overline{\mathbb{R}}$ defined as

$$f^\gamma(x) := f \square \left(\frac{1}{2\gamma} \|\cdot\|^2 \right).$$

Then f^γ is convex whenever f is convex. In particular, we verify that $(I_C)^\gamma = \frac{1}{2\gamma} d_C^2$.

More general than the inf-convolution operation is the *post-composition with linear mappings*. Given two locally convex spaces X and Y, a function $f : X \to \overline{\mathbb{R}}$, and a linear mapping $A : X \to Y$, the post-composition of f with A is the function $Af : Y \to \overline{\mathbb{R}}$ defined by

$$(Af)(y) := \inf \{ f(x) : Ax = y \}.$$

We say that Af is *exact at* y when the infimum is attained. Observe that, for every pair of functions $f, g : X \to \overline{\mathbb{R}}$, we have

$$(f \square g)(x) = \inf\{f(x_1) + g(x_2) : x_1 + x_2 = x\} = \inf\{h(x_1, x_2) : A(x_1, x_2) = x\},$$

2.3. EXAMPLES OF CONVEX FUNCTIONS

where $h : X \times X \to \overline{\mathbb{R}}$ and $A : X \times X \to X$ are defined by

$$h(x_1, x_2) := f(x_1) + g(x_2) \text{ and } A(x_1, x_2) = x_1 + x_2,$$

showing that $f \square g = Ah$.

Given a continuous linear mapping $A : X \to Y$, we consider the continuous linear mapping $\hat{A} : X \times \mathbb{R} \to Y \times \mathbb{R}$ defined by

$$\hat{A}(x, \alpha) := (Ax, \alpha), \tag{2.59}$$

which satisfies

$$\mathrm{epi}_s(Af) = \hat{A}(\mathrm{epi}_s f). \tag{2.60}$$

Actually, $(y, \alpha) \in \mathrm{epi}_s(Af)$ if and only if there exists some $x \in A^{-1}(y)$ such that $f(x) < \alpha$; equivalently, there exists some $x \in X$ such that $(y, \alpha) = (Ax, \alpha) = \hat{A}(x, \alpha)$ and $(x, \alpha) \in \mathrm{epi}_s f$. Consequently, Af is convex whenever f is convex. Also, we have that

$$\mathrm{dom}(Af) = A(\mathrm{dom}\, f). \tag{2.61}$$

Indeed, $y \in \mathrm{dom}(Af)$ if and only if there exists $x \in X$ such that $Ax = y$ and $f(x) < +\infty$; that is, if and only if $y \in A(\mathrm{dom}\, f)$.

Given a function $f \in \Gamma_0(X)$, the *recession function* of f is the function $f^\infty : X \to \mathbb{R}_\infty$ defined by

$$f^\infty(u) := \sup_{s > 0} \frac{f(x + su) - f(x)}{s},$$

for any $x \in \mathrm{dom}\, f$. Since the quotient mapping

$$s \mapsto s^{-1}(f(x + su) - f(x))$$

is non-decreasing with $s > 0$, we also have that

$$f^\infty(u) = \lim_{s \uparrow \infty} \frac{f(x + su) - f(x)}{s}.$$

Moreover, it readily follows from the definition of f^∞ that

$$\mathrm{epi}\, f^\infty = [\mathrm{epi}\, f]_\infty,$$

so that f^∞ is a proper closed sublinear function. In particular, for every non-empty closed and convex set $C \subset X$ we have

$$(I_C)^\infty = I_{C_\infty}. \tag{2.62}$$

Let X and Y be two lcs. Given a function $F : X \times Y \to \mathbb{R}_\infty$, the *marginal function* of F is the function $f : X \to \overline{\mathbb{R}}$ defined by

$$f(x) := \inf_{y \in Y} F(x, y).$$

It is worth noting that the inf-convolution of any two functions f_1, f_2 can also be viewed as a marginal function, taking

$$F(x, y) := f_1(y) + f_2(x - y).$$

The strict epigraph of f is the *projection* of $\text{epi}_s F$ onto $X \times \mathbb{R}$; that is,

$$\text{epi}_s f = \pi_{X \times \mathbb{R}}(\text{epi}_s F) := \{(x, \lambda) \in X \times \mathbb{R} : \exists y \in Y \text{ such that } (x, y, \lambda) \in \text{epi}_s F\}.$$

Certainly, $(x, \alpha) \in \text{epi}_s f$ if and only if there exists some $y \in Y$ such that $F(x, y) < \alpha$; hence, if and only if there exists some $y \in Y$ such that $(x, y, \alpha) \in \text{epi}_s F$. Thus, provided that F is convex, the convexity of the set $\pi_{X \times \mathbb{R}}(\text{epi}_s F)$ entails the convexity of the marginal value function f.

2.4 Exercises

Exercise 1 *Given a non-empty set $C \subset X$, prove that:*
(i) The Minkowski gauge of C, p_C, is positively homogeneous in its domain.
(ii) The strict epigraph of p_C is expressed as

$$\text{epi}_s p_C = \mathbb{R}_+^* \left((C \cup \{\theta\}) \times \;]1, +\infty[\right). \tag{2.63}$$

(iii) The function p_C is convex whenever C is convex.

Exercise 2 *Given a polyhedral set*

$$C := \{z \in X : \langle a_i^*, z \rangle \leq b_i, i \in 1, ..., m\}, m \geq 1,$$

prove that for every $x \in C$ the set $\mathbb{R}_+(C - x)$ is closed.

Exercise 3 *Let A be a convex subset of X. Prove that $A^i = \text{int } A$ in each one of the following situations: (i) $\text{int } A \neq \emptyset$, (ii) X is finite-dimensional, and (iii) A is a countable union of closed convex subsets*

2.4. EXERCISES

and X is Banach. Apply (iii) to show that $\text{int}(\text{dom } f) = (\text{dom } f)^i$ when $f \in \Gamma_0(X)$.

Exercise 4 Consider the following sets in ℓ_2,

$$A_k := \left\{ \sum_{i=1}^k \alpha_i e_i : \alpha_k > 0 \right\}, \quad k \geq 1,$$

together with $A := \cup_{k \geq 1} A_k$. Prove that $A^i = \text{int } A = \emptyset$.

Exercise 5 (i) Given a non-empty convex set $C \subset X$ such that either $\text{ri}(C) \neq \emptyset$ or C is closed, prove that $\cup_{L \in \mathcal{F}_X} \text{cl}(L \cap C) = \text{cl}(C)$, where $\mathcal{F}_X$ denotes the family of all finite-dimensional linear subspaces of X.

(ii) Give an example of a non-empty convex set $C \subset X$ such that $\cup_{L \in \mathcal{F}_X} \text{cl}(L \cap C) \subsetneq \text{cl}(C)$.

(iii) Give an example of a function $f \in \Gamma_0(\ell_2)$ such that $\cup_{L \in \mathcal{F}_X} \text{cl}(L \cap \text{dom } f) \subsetneq \text{cl}(\text{dom } f)$.

Exercise 6 Given a set $A \subset X$ such that $\theta \in A$, prove that

$$\inf_{\alpha > 0} (I_{\alpha A^\circ} + \alpha) = \sigma_A. \tag{2.64}$$

Exercise 7 Let us consider a family of sets $\{A_{i,p} \subset X, i \in J, p \in \mathcal{P}\}$, where $(J, \preccurlyeq)$ is a directed set and $\mathcal{P}$ is the family of seminorms generating the topology of X. We assume that

$$i_1 \preccurlyeq i_2 \text{ and } p_1 \leq p_2, \ i_1, i_2 \in J, \ p_1, p_2 \in \mathcal{P} \Rightarrow A_{i_1, p_1} \supset A_{i_2, p_2}.$$

Prove that the function $h := \inf_{i \in J, p \in \mathcal{P}} \sigma_{A_{i,p}}$ is sublinear.

Exercise 8 Given a proper function $f : X \to \mathbb{R}_\infty$, prove that f is positively homogeneous if and only if $f(\theta) = 0$ and $f(\lambda x) = \lambda f(x)$ for all $x \in \text{dom } f$ and $\lambda > 0$, if and only if $\theta \in \text{dom } f$ and $f(\lambda x) = \lambda f(x)$ for all $x \in \text{dom } f$ and $\lambda > 0$.

Exercise 9 Given proper convex functions f_t, $t \in T$, such that T is compact and the mappings $t \mapsto f_t(z)$, $z \in X$, are usc, prove that

$$\text{dom } f = \bigcap_{t \in T} \text{dom } f_t, \tag{2.65}$$

and, for every $x \in \text{dom } f$,

$$\mathbb{R}_+(\text{dom } f - x) = \bigcap_{t \in T} \mathbb{R}_+(\text{dom } f_t - x). \tag{2.66}$$

Exercise 10 Given non-empty sets $A, B \subset X$, $C \subset X^*$, prove the following statements:

(i) If $\mathcal{F}$ is the family of finite-dimensional linear subspaces of X, then

$$\bigcap_{U \in \mathcal{N}_{X^*}} (A + U) = \mathrm{cl}\,(A), \qquad (2.67)$$

and, for every $x \in X$,

$$\bigcap_{L \in \mathcal{F}} \mathrm{cl}\,\left(C + L^\perp\right) = \bigcap_{L \in \mathcal{F}(x)} \mathrm{cl}\,\left(C + L^\perp\right) = \mathrm{cl}\,(C),$$

where the closure is taken with respect to the w^*-topology, and $\mathcal{F}(x) = \{L \in \mathcal{F} : x \in L\}$.

(ii) If A is closed and B is convex compact with $\theta \in B$, then

$$\bigcap_{\varepsilon > 0} (A + \varepsilon B) = A. \qquad (2.68)$$

More generally, given closed sets $A_\varepsilon \subset X$, $\varepsilon > 0$, non-decreasing with respect to ε; that is, if $0 < \varepsilon_1 \leq \varepsilon_2$, then $A_{\varepsilon_1} \subset A_{\varepsilon_2}$, and B is convex compact with $\theta \in B$, we have that

$$\bigcap_{\varepsilon > 0} (A_\varepsilon + \varepsilon B) = \bigcap_{\varepsilon > 0} A_\varepsilon.$$

(iii) Let $C_L \subset X^*$, $L \in \mathcal{F}$, be non-increasing with respect to the L's in $\mathcal{F}$; that is, if $L_1 \subset L_2$, $L_1, L_2 \in \mathcal{F}$, then $C_{L_1} \supset C_{L_2}$. Then, we have that

$$\bigcap_{L \in \mathcal{F}} \mathrm{cl}\,\left(C_L + L^\perp\right) = \bigcap_{L \in \mathcal{F}} \mathrm{cl}\,(C_L).$$

Moreover, for every $x \in X$ we can replace $\mathcal{F}$ with $\mathcal{F}(x)$ in the equality above.

(iv) Let $(A_\varepsilon)_{\varepsilon > 0} \subset X^*$ be a family of non-decreasing sets with respect to ε. We have that

$$\bigcap_{\varepsilon > 0,\, L \in \mathcal{F}} \overline{\mathrm{co}}\,\left(A_\varepsilon + L^\perp\right) = \bigcap_{\varepsilon > 0,\, U \in \mathcal{N}_{X^*}} \overline{\mathrm{co}}\,(A_\varepsilon + U) = \bigcap_{\varepsilon > 0} \overline{\mathrm{co}}\,(A_\varepsilon),$$

and, provided that X is a normed space,

$$\bigcap_{\varepsilon > 0} \overline{\mathrm{co}}(A_\varepsilon + \varepsilon B_{X^*}) = \bigcap_{\varepsilon > 0} \overline{\mathrm{co}}(A_\varepsilon).$$

2.4. EXERCISES

Exercise 11 *Given a non-empty set $A \subset X$ and a compact interval $\Lambda \subset \mathbb{R}$ such that $0 \notin \Lambda$, prove that*

$$\Lambda(\overline{co}A) = \overline{co}(\Lambda A).$$

Exercise 12 *Suppose that $(\Lambda_\varepsilon)_{\varepsilon>0}$ is a non-decreasing family of closed sets in $\mathbb{R}$ (i.e., if $\varepsilon' < \varepsilon$, then $\Lambda_{\varepsilon'} \subset \Lambda_\varepsilon$) such that $\bigcap_{\varepsilon>0} \Lambda_\varepsilon = \{1\}$. Let $(A_\varepsilon)_{\varepsilon>0}$ be another non-decreasing family of closed sets in X (or in X^*). Prove that*

$$\bigcap_{\varepsilon>0} \Lambda_\varepsilon A_\varepsilon = \bigcap_{\varepsilon>0} A_\varepsilon.$$

Exercise 13 *Consider a finite family $\{C_i, i = 1, \ldots, m\}$, $m \geq 1$, of non-empty convex subsets of $\mathbb{R}^n$, and denote $C := \bigcap_{i=1,\ldots,m} C_i$. Prove that*

$$\bigcap_{i=1,\ldots,m} \mathrm{ri}(C_i) \neq \emptyset \tag{2.69}$$

if and only if

$$\mathrm{ri}(C_i) \cap C \neq \emptyset \text{ for } i = 1, \ldots, m. \tag{2.70}$$

Exercise 14 *Let $f : X \to \overline{\mathbb{R}}$ be a convex function, and $A \subset X$ be a convex set such that $(\mathrm{ri}\, A) \cap \mathrm{dom}\, f \neq \emptyset$. Prove that $\inf_A f = \inf_{\mathrm{ri}\, A} f = \inf_{\mathrm{cl}\, A} f$.*

Exercise 15 *Despite the fact that $\overline{co}\, f \leq f$ on X, it happens that $\overline{co}\, f$ ends by "behaving like f at infinity", as we have*

$$\liminf_{\|x\| \to \infty} \frac{f(x) - (\overline{co}\, f)(x)}{\|x\|} = 0. \tag{2.71}$$

Prove (2.71), and give an example of a function $f : \mathbb{R} \to \mathbb{R}$ satisfying

$$\lim_{|x| \to \infty} \{f(x) - (\overline{co}\, f)(x)\} = +\infty.$$

Exercise 16 *Given a convex function $f : \mathbb{R} \to \mathbb{R}_\infty$, prove that*

$$\frac{f(t_2) - f(t_1)}{t_2 - t_1} \leq \frac{f(t_3) - f(t_1)}{t_3 - t_1} \leq \frac{f(t_3) - f(t_2)}{t_3 - t_2},$$

for every $t_1, t_2, t_3 \in \mathrm{dom}\, f$ such that $t_1 < t_2 < t_3$.

Exercise 17 Given a function $f \in \Gamma_0(X)$, prove that

$$\mathrm{cl}(f\square f) = f\square f = 2f\left(\frac{1}{2}\cdot\right). \qquad (2.72)$$

Exercise 18 Let f and g be two proper convex functions defined on X. Suppose that $\mathrm{ri}(\mathrm{dom}\, f) \cap \mathrm{ri}(\mathrm{dom}\, g) \neq \emptyset$ and that the functions $f_{|\mathrm{aff}(\mathrm{dom}\, f)}$ and $g_{|\mathrm{aff}(\mathrm{dom}\, g)}$ are continuous on $\mathrm{ri}(\mathrm{dom}\, f)$ and $\mathrm{ri}(\mathrm{dom}\, g)$, respectively. Prove that

$$\mathrm{ri}(\mathrm{dom}(f+g)) = \mathrm{ri}(\mathrm{dom}\, f) \cap \mathrm{ri}(\mathrm{dom}\, g), \qquad (2.73)$$

and

$$(f+g)_{|\mathrm{aff}(\mathrm{dom}(f+g))} \text{ is continuous on } \mathrm{ri}(\mathrm{dom}(f+g)).$$

Exercise 19 Prove that the following statements are equivalent, for any proper convex functions f, g defined on X,
  (i) $\mathrm{dom}\, f \cap \mathrm{ri}(\mathrm{dom}\, g) \neq \emptyset$ and $g_{|\mathrm{aff}(\mathrm{dom}\, g)}$ is continuous on $\mathrm{ri}(\mathrm{dom}\, g)$.
  (ii) $((\mathrm{dom}\, f - x) \times \mathbb{R}) \cap (\mathrm{ri}(\mathrm{epi}\, g - (x, g(x)))) \neq \emptyset$ for all $x \in \mathrm{dom}\, g$.
  (iii) $((\mathrm{dom}\, f - x) \times \mathbb{R}) \cap (\mathrm{ri}(\mathrm{epi}\, g - (x, g(x)))) \neq \emptyset$ for $x \in \mathrm{dom}\, g$.

Exercise 20 Let f, g be two proper convex functions defined on X. Prove that $\overline{f+g} = \overline{f} + \overline{g}$ in each one of the following cases:
  (i) $\mathrm{ri}(\mathrm{dom}\, f) \cap \mathrm{ri}(\mathrm{dom}\, g) \neq \emptyset$ and the functions $f_{|\mathrm{aff}(\mathrm{dom}\, f)}$ and $g_{|\mathrm{aff}(\mathrm{dom}\, g)}$ are continuous on $\mathrm{ri}(\mathrm{dom}\, f)$ and $\mathrm{ri}(\mathrm{dom}\, g)$, respectively.
  (ii) $\mathrm{dom}\, f \cap \mathrm{ri}(\mathrm{dom}\, g) \neq \emptyset$, the function $g_{|\mathrm{aff}(\mathrm{dom}\, g)}$ is continuous on $\mathrm{ri}(\mathrm{dom}\, g)$, and $\overline{f+g} = \overline{f} + g$.

2.5 Bibliographical notes

J.-B. Hiriart-Urruty, in his article "Convex Analysis and Optimization in the Past 50 Years: Some Snapshots"([107]) claims that the development of convex analysis in its first 50 years owes much to W. Fenchel (1905–1988), J. J. Moreau (1923-2014), and R. T. Rockafellar. He also says that the years 1962–1963 should be considered the date of birth of modern convex analysis with applications to optimization, and crucial concepts such as the subdifferential, the Fenchel conjugate, the proximal mapping, the inf-convolution, etc. date back to this period.

2.5. BIBLIOGRAPHICAL NOTES

A selection of books on variational and convex analysis, where the reader will find complementary information on the topics included in this book, are A. Auslender and M. Teboulle [8], D. Azé [9], V. Barbu and Th. Precupanu [11], H. H. Bauschke and P. L. Combettes [12], D. P. Bertsekas *et al.* [15], F. Bonnans and A. Shapiro [16], J. M. Borwein and A. Lewis [20], J. M. Borwein and J. D. Vanderwerff [23], J. M. Borwein and Q. J. Zhu [24], R. I. Boţ [26], C. Castaing and M. Valadier [36], A. L. Dontchev [75], I. Ekeland and R. Temam [80], W. Fenchel [86], J. R. Giles [88], J.-B. Hiriart-Urruty and C. Lemaréchal [108], A. D. Ioffe [112], A. D. Ioffe and V. M. Tikhomirov [115], P.-J. Laurent [129], R. Lucchetti [146], B. S. Mordukhovich [154], B. S. Mordukhovich and N. M. Nam [156], Pallaschke and Rolewicz [164], J.-P. Penot [166], R. R. Phelps [168], B. N. Pshenichnyi [170], R. T. Rockafellar [174, 176], R. T. Rockafellar and R. Wets [178], J. Stoer and C. Witzgall [186], J. van Tiel [189], L. Thibault [187], C. Zălinescu [201], among others. General references for functional analysis and topology are [1], [77], [81], [163], [182], etc.

Theorem 2.1.8 is the so-called Dieudonné theorem (see, e.g., [201, Theorem 1.1.8]). The continuity properties of convex functions given in Proposition 2.2.6 and its consequences are well-known and can be consulted in the references above. The first statement in Exercise 3 is a slight extension of Lemma 2.71 in [16]. A finite-dimensional version of Exercise 9 has been given in [100]. Exercise 15 can be found in [28] (see also [109]).

Chapter 3

Fenchel–Moreau–Rockafellar theory

This chapter and the following one offer a crash course in convex analysis, including the fundamental results in the theory of convex functions which are used throughout this book. In the present chapter we review the Fenchel–Moreau–Rockafellar theory, giving new proofs while highlighting the role of separation theorems. These results are then applied to provide dual representations of support functions, which are used in section 4.2 to develop a general duality theory. We also apply the Fenchel–Moreau–Rockafellar theorem to give non-convex slight extensions of the classical minimax theorem.

As in the previous chapter, X is a (real) separated lcs with $\mathcal{N}_X$ being a neighborhood base of θ-neighborhoods. We denote by $\mathcal{P}$ the family of continuous seminorms on X. The topological dual space X^* of X is, unless otherwise stated, endowed with a locally convex topology making (X, X^*) a compatible dual pair. The associated duality pairing is represented by $\langle \cdot, \cdot \rangle$. By $\operatorname{cl} C$ or, interchangeably $\overline{C}$, we represent the closure of $C \subset X^*$ with respect to such a compatible topology. However, we also sometimes write $\operatorname{cl}^{w^*} C$ when such a specification is needed.

3.1 Conjugation theory

In the present section, we study the main features of the conjugation theory, which is considered the cornerstone of convex analysis.

Definition 3.1.1 *Given a function $f : X \to \overline{\mathbb{R}}$, the* Fenchel conjugate *of f is the function $f^* : X^* \to \overline{\mathbb{R}}$ defined by*

$$f^*(x^*) := \sup\{\langle x^*, x\rangle - f(x) : x \in X\}.$$

Equivalently, we have

$$f^*(x^*) = \sup\{\langle x^*, x\rangle - f(x) : x \in \operatorname{dom} f\}.$$

Notice that if $\operatorname{dom} f = \emptyset$, then $f^* \equiv -\infty$, and if f takes somewhere the value $-\infty$, then $f^* \equiv +\infty$. Sometimes, f^* is called the *dual* function of f, and we say that f *dualizes* to f^*. The function f^* is a w^*-lsc convex function for being the pointwise supremum of (w^*-) continuous affine functions.

If $f = I_C$ for a non-empty set $C \subset X$, then f^* reduces to the support function of C

$$I_C^* = \sigma_C. \tag{3.1}$$

The following relation is also easy to prove: For every $x^* \in X^*$, $\alpha > 0$, and $\beta \in \mathbb{R}$, we have

$$(\alpha f - \langle x^*, \cdot\rangle + \beta)^* = \alpha f^*(\alpha^{-1}(\cdot + x^*)) - \beta. \tag{3.2}$$

As we said before, f^* is w^*-lsc and convex regardless of what the original function f is like, but possibly lacking properness. Furthermore, as follows from the definition of f^*, we have the equality

$$\inf_X f = -f^*(\theta). \tag{3.3}$$

Below are some other simple facts related to the conjugation operation. The primal and dual norms in a normed space are typical examples of conjugate functions.

Example 3.1.2 *Consider a normed space $(X, \|\cdot\|)$ and $f := \|\cdot\|$. Then*

$$f^* = I_{B_{X^*}} \quad \text{and} \quad \|\cdot\|_* = \sigma_{B_X} = (I_{B_X})^*.$$

3.1. CONJUGATION THEORY

Indeed, the equalities $\|\cdot\|_* = \sigma_{B_X} = (I_{B_X})^*$ come from the own definition of the dual norm and (3.1). To show the equality $f^* = I_{B_{X^*}}$, we take $x^* \in X^*$. Then

$$f^*(x^*) = \sup_{x \in X}\{\langle x^*, x\rangle - \|x\|\} = \sup_{\alpha > 0} \alpha(\sup_{z \in B_X}\{\langle x^*, z\rangle - \|z\|\}).$$

If $x^* \notin B_{X^*}$, then $\|x^*\|_* = \sigma_{B_X}(x^*) > 1$ and there exists $z_0 \in B_X$ such that $\langle x^*, z_0\rangle > 1 \geq \|z_0\|$, implying that $\sup_{z \in B_X}\{\langle x^*, z\rangle - \|z\|\} > 0$. Therefore, $f^*(x^*) = +\infty$. Otherwise, if $x^* \in B_{X^*}$, then the Cauchy–Schwarz inequality yields

$$\sup_{z \in B_X}\{\langle x^*, z\rangle - \|z\|\} \leq \sup_{z \in B_X}\{\|z\|(\|x^*\| - 1)\} \leq 0, \qquad (3.4)$$

and we deduce that $f^*(x^*) = 0$; that is, $f^*(x^*) = I_{B_{X^*}}$.

A primal condition satisfied by f is said to be *dualized* into a dual property satisfied by f^* if the first property implies the second. Both properties are dual to each other if they are equivalent. The following statement gives an example of these dualized properties, others will be given in Proposition 3.3.7.

Proposition 3.1.3 *Consider a convex function $f : X \to \mathbb{R}_\infty$, which is finite and continuous at $x \in X$. Then the function $f^*(\cdot) - \langle \cdot, x\rangle$ is inf-compact with respect to the w^*-topology.*

Proof. We may assume that $x = \theta$, due to the relation $(f(\cdot + x_0))^* = f^*(\cdot) - \langle \cdot, x_0\rangle$. Given $\alpha \in \mathbb{R}$, we take $m > -\alpha$ and $U \in \mathcal{N}_X$ such that $f(u) \leq m$ for all $u \in U$ (by the continuity assumption). Then, for all $x^* \in [f^* \leq \alpha]$ and $u \in U$, we have

$$\langle x^*, u\rangle - m \leq \langle x^*, u\rangle - f(u) \leq f^*(x^*) \leq \alpha;$$

that is, $x^* \in (m + \alpha)U^\circ$. Thus, $[f^* \leq \alpha] \subset (m + \alpha)U^\circ$ and Theorem 2.1.9 implies that the (w^*-closed) set $[f^* \leq \alpha]$ is w^*-compact. ∎

A proper conjugate requires that the original function f be minorized by a continuous affine mapping as stated in the following proposition.

Proposition 3.1.4 *The following statements are equivalent, for every lsc convex function $f : X \to \overline{\mathbb{R}}$:*
(i) $f \in \Gamma_0(X)$.
(ii) $\mathrm{dom}\, f \neq \emptyset$ and f is minorized by a continuous affine mapping.
(iii) $f^* \in \Gamma_0(X^*)$.

Proof. We are going to prove the equivalences $(i) \Leftrightarrow (ii)$ and $(ii) \Leftrightarrow (iii)$.

$(ii) \Rightarrow (i)$ It is obvious, since f is convex and lsc by assumption.

$(i) \Rightarrow (ii)$ Suppose that $f \in \Gamma_0(X)$. Then $\operatorname{dom} f \neq \emptyset$ and there exists $x_0 \in \operatorname{dom} f$ such that $f(x_0) > -\infty$, and so $(x_0, f(x_0) - 1) \notin \operatorname{epi} f$. Next, Corollary 2.1.7 yields a nonzero vector $(z^*, \alpha) \in X^* \times \mathbb{R}$ such that
$$\sigma_{\operatorname{epi} f}(z^*, \alpha) < \langle z^*, x_0 \rangle + (f(x_0) - 1)\alpha =: \beta; \qquad (3.5)$$
that is, $\langle z^*, x \rangle + \alpha(f(x) + t) < \beta$ for all $x \in \operatorname{dom} f$ and $t \geq 0$. Then, letting $t \to +\infty$, we deduce that $\alpha \leq 0$. More precisely, we have $\alpha < 0$ (taking $x = x_0$ in the last inequality), and so $\langle -\alpha^{-1} z^*, x \rangle + \alpha^{-1} \beta \leq f(x)$ for all $x \in X$; that is, (ii) follows and the equivalence $(i) \Leftrightarrow (ii)$ holds.

$(ii) \Rightarrow (iii)$ Suppose that $\operatorname{dom} f \neq \emptyset$ and $f \geq \langle z^*, \cdot \rangle + \alpha$ for some $z^* \in X^*$ and $\alpha \in \mathbb{R}$. Hence, $f^*(z^*) \leq -\alpha < +\infty$ and $\operatorname{dom} f^* \neq \emptyset$. At the same time, for any $x_0 \in \operatorname{dom} f$, we have
$$f^*(x^*) \geq \langle x^*, x_0 \rangle - f(x_0) > -\infty \text{ for all } x^* \in X^*,$$
and $f^* \in \Gamma_0(X^*)$.

$(iii) \Rightarrow (ii)$ Suppose that $f^* \in \Gamma_0(X^*)$. If $f(y_0) = -\infty$ for some $y_0 \in X$, then $f^* \equiv +\infty$ and f^* would be non-proper. If $f \equiv +\infty$, then $f^* \equiv -\infty$ and again f^* would be non-proper. Therefore, f is proper and minorized by any of the continuous affine mappings $x \mapsto \langle z^*, x \rangle - f^*(z^*)$, $z^* \in \operatorname{dom} f^*$. Thus, the equivalence $(ii) \Leftrightarrow (iii)$ also holds. ∎

The following inequality, called *Fenchel inequality*, is also a simple consequence of the definition of the conjugate,
$$\langle x^*, x \rangle \leq f(x) + f^*(x^*) \text{ for all } x \in X, \ x^* \in X^*. \qquad (3.6)$$

In addition, as a consequence of (2.37), the conjugation operation does not distinguish between a given function and its closed and closed convex hulls. We have the following proposition.

Proposition 3.1.5 *For every function* $f : X \to \overline{\mathbb{R}}$, *we have*
$$f^* = (\operatorname{cl} f)^* = (\operatorname{co} f)^* = (\overline{\operatorname{co}} f)^*. \qquad (3.7)$$

Proof. First, since $\overline{\operatorname{co}} f \leq \operatorname{cl} f \leq f$ and $\overline{\operatorname{co}} f \leq \operatorname{co} f \leq f$, we get $f^* \leq (\operatorname{cl} f)^* \leq (\overline{\operatorname{co}} f)^*$ and $f^* \leq (\operatorname{co} f)^* \leq (\overline{\operatorname{co}} f)^*$. So, we only need to prove that $(\overline{\operatorname{co}} f)^* = f^*$. In fact, for each $x^* \in X^*$

3.1. CONJUGATION THEORY

$$(\overline{\mathrm{co}}f)^*(x^*) = \sup_{x \in X}\{\langle x^*, x\rangle - (\overline{\mathrm{co}}f)(x)\} = -\inf_{x \in X}\{(\overline{\mathrm{co}}f)(x) - \langle x^*, x\rangle\}.$$
(3.8)

At the same time, we have

$$\begin{aligned}\overline{\mathrm{co}}(f - x^*) &= \sup\{g : g \text{ convex, lsc, and } g \leq f - x^*\}\\ &= \sup\{(g + x^*) - x^* : g \text{ convex, lsc, and } g + x^* \leq f\}\\ &= \sup\{h - x^* : h \text{ convex, lsc, and } h \leq f\} = (\overline{\mathrm{co}}f) - x^*,\end{aligned}$$

and (2.37) together with (3.8) leads us to

$$(\overline{\mathrm{co}}f)^*(x^*) = -\inf\{\overline{\mathrm{co}}(f - x^*)\} = -\inf\{f - x^*\} = f^*(x^*).$$

■

Conjugation can also be used for a function $g : X^* \to \overline{\mathbb{R}}$; namely,

$$g^*(x) := \sup\{\langle x^*, x\rangle - g(x^*) : x^* \in \mathrm{dom}\, g\}.$$

So, in particular, we come to the concept of biconjugate functions.

Definition 3.1.6 *Given a function $f : X \to \overline{\mathbb{R}}$, the* biconjugate *of f is the function $f^{**} : X \to \overline{\mathbb{R}}$ defined as*

$$f^{**}(x) := \sup\{\langle x^*, x\rangle - f^*(x^*) : x^* \in \mathrm{dom}\, f^*\}.$$

Notice that, for every $x \in X$,

$$\begin{aligned}f^{**}(x) &= \sup_{x^* \in X^*}\{\langle x^*, x\rangle - \sup_{z \in X}\{\langle x^*, z\rangle - f(z)\}\}\\ &\leq \sup_{x^* \in X^*}\{\langle x^*, x\rangle - (\langle x^*, x\rangle - f(x))\} = f(x).\end{aligned}$$

Thus, since f^{**} is clearly convex and lsc, we get

$$f^{**} \leq \overline{\mathrm{co}}f \leq \mathrm{cl}\, f \leq f,$$
(3.9)

and the closed convex function f^{**} gives a lower lsc convex estimate to f. The purpose of the Fenchel–Moreau–Rockafellar theorem, which is the subject of next section 3.2, is to see that f^{**} is nothing else but $\overline{\mathrm{co}}f$ (when the latter function is proper). In the same way as with the conjugate, we can define the function $f^{***} : X^* \to \overline{\mathbb{R}}$ as

$$f^{***} := (f^{**})^*.$$

However, this new function does not provide additional information, since it generally coincides with the conjugate of the initial function. This fact is at the heart of section 3.2.

Next, we see how the conjugation operation behaves with respect to some operations on convex functions. In the first place, it is easy to see that the infimum dualizes to the supremum; that is, for a family of functions $f_i : X \to \overline{\mathbb{R}}$, we always have

$$\left(\inf_i f_i\right)^* = \sup_i f_i^*. \qquad (3.10)$$

The converse, which goes from the supremum to the infimum, is less direct as we see in Proposition 3.2.6 below.

The second result connects the conjugate of a function to the support of its epigraph.

Proposition 3.1.7 *Given a proper function $f : X \to \mathbb{R}_\infty$, for all $x^* \in X^*$ and $\alpha \in \mathbb{R}$, we have*

$$\sigma_{\mathrm{epi} f}(x^*, -\alpha) = \begin{cases} \alpha f^*(\alpha^{-1} x^*), & \text{if } \alpha > 0, \\ \sigma_{\mathrm{dom} f}(x^*), & \text{if } \alpha = 0, \\ +\infty, & \text{if } \alpha < 0. \end{cases} \qquad (3.11)$$

Proof. Using (2.31), for every $x \in X$ we write

$$f(x) = \inf\{\lambda \in \mathbb{R} : (x, \lambda) \in \mathrm{epi}\, f\} = \inf\{\lambda + \mathrm{I}_{\mathrm{epi}\, f}(x, \lambda) : \lambda \in \mathbb{R}\},$$

and, thanks to (3.10), the conjugate of f is given by

$$f^*(x^*) = \sup_{x \in X,\ \lambda \in \mathbb{R}} \{\langle x^*, x\rangle - \lambda - \mathrm{I}_{\mathrm{epi}\, f}(x, \lambda)\}$$
$$= \mathrm{I}^*_{\mathrm{epi}\, f}(x^*, -1) = \sigma_{\mathrm{epi} f}(x^*, -1),$$

where the last equality comes from (3.1). Therefore, for every $\alpha > 0$,

$$\sigma_{\mathrm{epi} f}(x^*, -\alpha) = \alpha \sigma_{\mathrm{epi} f}(\alpha^{-1} x^*, -1) = \alpha f^*(\alpha^{-1} x^*),$$

and we conclude as we can easily verify that $\sigma_{\mathrm{epi} f}(x^*, 0) = \sigma_{\mathrm{dom} f}(x^*)$ and $\sigma_{\mathrm{epi} f}(x^*, -\alpha) = +\infty$ for all $\alpha < 0$. ∎

The following proposition compares the support function of the domains of f and f^{**}. The proof of this property, which will be part of the proof of the aforementioned Fenchel–Moreau–Rockafellar theorem, is based on Proposition 3.1.7.

3.1. CONJUGATION THEORY

Proposition 3.1.8 *For every function $f : X \to \mathbb{R}_\infty$ having a proper conjugate, we have*
$$\sigma_{\mathrm{dom} f^{**}} = \sigma_{\mathrm{dom} f}.$$

Proof. First, without loss of generality, we may suppose that $f \in \Gamma_0(X)$ and $f^*(\theta) = 0$ (see Exercise 21). Furthermore, by (3.9), we have the inequality $\sigma_{\mathrm{dom} f} \leq \sigma_{\mathrm{dom} f^{**}}$, and so it suffices to show that

$$\sigma_{\mathrm{dom} f^{**}}(x^*) \leq \sigma_{\mathrm{dom} f}(x^*), \tag{3.12}$$

for any given $x^* \in \mathrm{dom}\, \sigma_{\mathrm{dom} f}$. We consider the lsc convex function $\varphi : \mathbb{R} \to \mathbb{R}_\infty$ defined as

$$\varphi(\alpha) := \sigma_{\mathrm{epi} f}(x^*, -\alpha),$$

so that $\varphi(0) = \sigma_{\mathrm{dom} f}(x^*) < +\infty$ and $\varphi(\alpha) = +\infty$ for all $\alpha < 0$. Moreover, $\mathrm{dom}\, \varphi$ cannot be reduced to $\{0\}$. Indeed, otherwise, if $\mathrm{dom}\, \varphi = \{0\}$, then Proposition 3.1.7 would imply that

$$\varphi(\alpha) = \sigma_{\mathrm{epi} f}(x^*, -\alpha) = \alpha f^*(\alpha^{-1} x^*) = +\infty \text{ for all } \alpha > 0.$$

In particular, taking $\alpha = 1$, we obtain $f^*(x^*) = +\infty$, and so

$$+\infty = \sup_{x \in \mathrm{dom}\, f} \{\langle x^*, x \rangle - f(x)\} \leq \sigma_{\mathrm{dom} f}(x^*) - \inf_{\mathrm{dom}\, f} f(x)$$
$$= \sigma_{\mathrm{dom} f}(x^*) + f^*(\theta) = \sigma_{\mathrm{dom} f}(x^*) = \varphi(0),$$

producing a contradiction with $\varphi(0) < +\infty$. Now, since $\{0\} \neq \mathrm{dom}\, \varphi \subset \mathbb{R}_+$, Corollary 2.2.9 and Proposition 3.1.7 entail

$$\varphi(0) = \lim_{\alpha \downarrow 0} \varphi(\alpha) = \lim_{\alpha \downarrow 0} \sigma_{\mathrm{epi} f}(x^*, -\alpha) = \lim_{\alpha \downarrow 0} \alpha f^*(\alpha^{-1} x^*).$$

Thus, since f^{**} is proper by Proposition 3.1.4, for all $x \in \mathrm{dom}\, f^{**}$, we have $f^{**}(x) \in \mathbb{R}$, and (3.6) yields

$$\varphi(0) = \lim_{\alpha \downarrow 0} \alpha(f^*(\alpha^{-1} x^*) + f^{**}(x)) \geq \liminf_{\alpha \downarrow 0} \alpha(\langle \alpha^{-1} x^*, x \rangle) = \langle x^*, x \rangle.$$

So,
$$\sigma_{\mathrm{dom} f}(x^*) = \varphi(0) \geq \sup_{x \in \mathrm{dom}\, f^{**}} \langle x^*, x \rangle = \sigma_{\mathrm{dom} f^{**}}(x^*),$$

and (3.12) follows. ∎

We introduce the concept of star product function.

Definition 3.1.9 *The* star product *of a function $f : X \to \overline{\mathbb{R}}$ by $\alpha \in \mathbb{R}$ is the function $\alpha * f : X \to \overline{\mathbb{R}}$ defined by*

$$(\alpha * f)(x) := \begin{cases} \alpha f(\alpha^{-1}x), & \text{if } \alpha > 0, \\ \sigma_{\text{dom} f^*}(x), & \text{if } \alpha = 0, \\ +\infty, & \text{if } \alpha < 0. \end{cases} \quad (3.13)$$

It is clear that the star product function $\alpha * f$ is convex and lsc whenever $\alpha \leq 0$. Furthermore, when $\alpha > 0$, one has

$$\text{epi}(\alpha * f) = \alpha \, \text{epi} \, f,$$

because $(x, \lambda) \in \text{epi}(\alpha * f)$ if and only if $f(\alpha^{-1}x) \leq \alpha^{-1}\lambda$, if and only if $(x, \lambda) \in \alpha \, \text{epi} \, f$. Hence, the star product function $\alpha * f$ is convex if and only if f is convex.

We also introduce the concept of perspective function.

Definition 3.1.10 *The* perspective function *of $f : X \to \overline{\mathbb{R}}$ is the function $P_f : \mathbb{R} \times X \to \overline{\mathbb{R}}$ defined by*

$$P_f(\alpha, x) := (\alpha * f)(x).$$

The following proposition establishes some properties of the perspective function. More details are given in Proposition 3.2.5.

Proposition 3.1.11 *For every function $f : X \to \overline{\mathbb{R}}$, we have*

$$\text{epi} \, P_f = \mathbb{R}_+^*(\{1\} \times \text{epi} \, f\}) \cup (\{0\} \times \text{epi} \, \sigma_{\text{dom} f^*}\})$$

and, whenever f has a proper conjugate,

$$P_{f^*}(\alpha, x^*) = \sigma_{\text{epi} f}(x^*, -\alpha) \text{ for all } x^* \in X^* \text{ and } \alpha \in \mathbb{R}. \quad (3.14)$$

Proof. The first statement follows from the definition of P_f. To prove the other statement we observe, thanks to Proposition 3.1.8 and the assumption that f^* is proper, that

$$P_{f^*}(\alpha, x^*) = (\alpha * f^*)(x^*) = \begin{cases} \alpha f^*(\alpha^{-1}x^*), & \text{if } \alpha > 0, \\ \sigma_{\text{dom} f}(x^*), & \text{if } \alpha = 0, \\ +\infty, & \text{if } \alpha < 0. \end{cases}$$

3.1. CONJUGATION THEORY

Thus, since f is also proper, Proposition 3.1.7 yields $P_{f^*}(\alpha, x^*) = \sigma_{\text{epi} f}(x^*, -\alpha)$. ∎

The following proposition shows the relationship of "duality" between the star product function $\alpha * f^*$ and the usual product αf, explaining the meaning of the term "star product".

Proposition 3.1.12 *For every function $f : X \to \mathbb{R}_\infty$, with proper conjugate, and every $\alpha \geq 0$, we have*

$$(\alpha f)^* = \alpha * f^*.$$

Proof. First assume that $\alpha > 0$. Then, for every $x^* \in X^*$,

$$\begin{aligned}(\alpha f)^*(x^*) &= \sup_{x \in X}(\langle x^*, x\rangle - (\alpha f)(x)) \\ &= \alpha \sup_{x \in X}(\langle \alpha^{-1} x^*, x\rangle - f(x)) \\ &= \alpha f^*(\alpha^{-1} x^*) = (\alpha * f^*)(x^*).\end{aligned}$$

Second, if $\alpha = 0$, then $\alpha f = I_{\text{dom} f}$ by (2.2), and (3.1) together with Proposition 3.1.8 entails $(0f)^* = \sigma_{\text{dom} f} = \sigma_{\text{dom} f^{**}} = 0 * f^*$. Here the last equality comes from the definition of $0 * f^*$ in (3.13). ∎

Next, we study the effect of conjugation on inf-convolution, which in fact dualizes to post-composition with linear mappings. To do this, we recall that the continuous *adjoint mapping* of a continuous linear mapping $A : X \to Y$, given between two lcs X and Y, is the continuous linear mapping $A^* : Y^* \to X^*$ defined by

$$\langle A^* y^*, x\rangle = \langle y^*, Ax\rangle \text{ for all } x \in X, \ y^* \in Y^*. \tag{3.15}$$

Similarly, we can define the *second adjoint* of A by $A^{**} := (A^*)^*$. Observe that $A^{**} : X^{**} \to Y^{**}$, so that

$$\langle A^{**} z, y^*\rangle = \langle z, A^* y^*\rangle \text{ for all } z \in X^{**} \text{ and } y^* \in Y^*. \tag{3.16}$$

Therefore, provided that X^* and Y^* are endowed with compatible topologies for the pairs (X, X^*) and (Y, Y^*), it turns out that $A^{**} = A$.

Proposition 3.1.13 *Assume that $A : X \to Y$ is a continuous linear mapping with continuous adjoint A^*. Then, for every functions $f : Y \to \overline{\mathbb{R}}$ and $g : X \to \overline{\mathbb{R}}$, we have*

$$(f \square (Ag))^* = f^* + g^* \circ A^*.$$

Proof. Fix $y^* \in Y^*$. Then, since $(Ag)(y) = \inf_{Ax=y} g(x)$ by definition, we have

$$(f\square(Ag))^*(y^*) = \sup_{y \in Y}(\langle y^*, y \rangle - (f\square(Ag))(y))$$

$$= \sup_{y,z \in Y} \left(\langle y^*, y+z \rangle - f(z) - \inf_{x \in X,\ Ax=y} g(x) \right)$$

$$= \sup_{x \in X,\ y,z \in Y,\ Ax=y} (\langle y^*, y+z \rangle - f(z) - g(x)).$$

Hence, using the definition of A^*,

$$(f\square(Ag))^*(y^*) = \sup_{x \in X,\ z \in Y} (\langle y^*, Ax+z \rangle - f(z) - g(x))$$

$$= \sup_{x \in X} (\langle A^*y^*, x \rangle - g(x)) + f^*(y^*)$$

$$= f^*(y^*) + g^*(A^*y^*),$$

and we are done. ∎

3.2 Fenchel–Moreau–Rockafellar theorem

The *Fenchel–Moreau–Rockafellar theorem*, presented in Theorem 3.2.2 below, constitutes the main tool for deriving many other fundamental results of convex analysis. In this section, we provide a new proof of this result based on Lemma 3.2.1 below, which itself is a particular instance of the Fenchel–Moreau–Rockafellar theorem. As the proposed approach confirms, the keystone in this development is essentially the separation theorem.

Lemma 3.2.1 *For every non-empty closed convex set $A \subset X$, we have*

$$(\sigma_A)^* = I_A. \qquad (3.17)$$

Consequently, $I_A^{**} = I_A$ and $(\sigma_A)^{**} = \sigma_A$.

Proof. We denote $f := I_A$, so that $f^* = \sigma_A$ and $(\sigma_A)^* = f^{**} \leq f$, due to (3.9). So, using the Fenchel inequality (3.6),

$$0 = \langle \theta, x \rangle - \sigma_A(\theta) \leq (\sigma_A)^*(x) = f^{**}(x) \leq f(x) \text{ for all } x \in X,$$

3.2. FENCHEL–MOREAU–ROCKAFELLAR THEOREM

and $(\sigma_A)^*(x) = 0 = f(x)$ if $x \in A$. Otherwise, for $x \notin A$, the separation theorem (Corollary 2.1.7) yields $x_0^* \in X^*$ such that $\beta_0 := \langle x_0^*, x \rangle - \sigma_A(x_0^*) > 0$. So,

$$(\sigma_A)^*(x) \geq \langle \lambda x_0^*, x \rangle - \sigma_A(\lambda x_0^*) = \lambda \beta_0 \text{ for all } \lambda > 0,$$

and we deduce that $(\sigma_A)^*(x) = +\infty = f(x)$. In other words, $(\sigma_A)^*(x) = f(x)$ for all $x \in X$, and (3.17) follows. The last conclusion of the lemma also holds because (3.1) and (3.17) together give rise to $I_A^{**} = \sigma_A^* = I_A$, which leads us to $(\sigma_A)^{**} = (I_A^{**})^* = I_A^* = \sigma_A$. ∎

We now give the Fenchel–Moreau–Rockafellar Theorem.

Theorem 3.2.2 (Fenchel–Moreau–Rockafellar Theorem) *For every function $f : X \to \mathbb{R}_\infty$, the following assertions are true:*
(i) As long as f is proper, we have

$$f = f^{**} \text{ if and only if } f \in \Gamma_0(X). \tag{3.18}$$

(ii) Whenever f admits a continuous affine minorant, we have

$$f^{**} = \overline{\text{co}} f. \tag{3.19}$$

Proof. (i) Suppose that f is proper. If $f = f^{**}$, then f is convex and lsc, so $f \in \Gamma_0(X)$. To prove the converse statement, we assume that $f \in \Gamma_0(X)$, so that epi f is a non-empty closed convex subset of $X \times \mathbb{R}$. Then, by Lemma 3.2.1, for any $(x, \lambda) \in X \times \mathbb{R}$, we have

$$I_{\text{epi }f}(x, \lambda) = I_{\text{epi }f}^{**}(x, \lambda) = (\sigma_{\text{epi }f})^*(x, \lambda)$$
$$= \sup_{x^* \in X^*, \, \alpha \in \mathbb{R}} \{\langle x, x^* \rangle + \alpha \lambda - \sigma_{\text{epi} f}(x^*, \alpha)\},$$
$$= \sup_{x^* \in X^*, \, \alpha \in \mathbb{R}} \{\langle x, x^* \rangle - \alpha \lambda - \sigma_{\text{epi} f}(x^*, -\alpha)\},$$

which reads, applying Propositions 3.1.7 and 3.1.8 (the latter proposition ensures that $\sigma_{\text{dom} f} = \sigma_{\text{dom} f^{**}}$),

$$I_{\text{epi }f}(x, \lambda) = \max \left\{ \begin{array}{l} \sup\limits_{\substack{x^* \in X^* \\ \alpha > 0}} \left(\langle x, x^* \rangle - \alpha \lambda - \alpha f^*(\alpha^{-1} x^*) \right), \\ \sup\limits_{x^* \in X^*} \{\langle x, x^* \rangle - \sigma_{\text{dom} f^{**}}(x^*)\} \end{array} \right\}.$$

Moreover, again by Lemma 3.2.1, we have

$$\sup_{x^*\in X^*} \{\langle x, x^*\rangle - \sigma_{\mathrm{dom} f^{**}}(x^*)\} = (\sigma_{\mathrm{cl}(\mathrm{dom}\, f^{**})})^*(x) = \mathrm{I}_{\mathrm{cl}(\mathrm{dom}\, f^{**})}(x),$$

and, therefore, the above relation simplifies to

$$\mathrm{I}_{\mathrm{epi}\, f}(x,\lambda) = \max\left\{\sup_{\substack{x^*\in X^*\\ \alpha>0}}\{\alpha(\langle x,x^*\rangle - \lambda - f^*(x^*))\},\ \mathrm{I}_{\mathrm{cl}(\mathrm{dom}\, f^{**})}(x)\right\}$$

$$= \max\left\{\sup_{\alpha>0}\{\alpha(f^{**}(x) - \lambda)\}\},\ \mathrm{I}_{\mathrm{cl}(\mathrm{dom}\, f^{**})}(x)\right\}. \qquad (3.20)$$

Consequently, for each $x \in X$ such that $f^{**}(x) > -\infty$ we get

$$\mathrm{I}_{\mathrm{epi}\, f}(x,\lambda) = \max\{\mathrm{I}_{\mathrm{epi}\, f^{**}}(x,\lambda),\ \mathrm{I}_{\mathrm{cl}(\mathrm{dom}\, f^{**})}(x)\} = \mathrm{I}_{\mathrm{epi}\, f^{**}}(x,\lambda), \qquad (3.21)$$

where the last equality holds due to the implication $(x,\lambda) \in \mathrm{epi}\, f^{**} \Rightarrow x \in \mathrm{dom}\, f^{**} \subset \mathrm{cl}(\mathrm{dom}\, f^{**})$. Furthermore, when $f^{**}(x) = -\infty$, (3.20) entails

$$\mathrm{I}_{\mathrm{epi}\, f}(x,\lambda) = \max\{-\infty,\ \mathrm{I}_{\mathrm{cl}(\mathrm{dom}\, f^{**})}(x)\} = \mathrm{I}_{\mathrm{cl}(\mathrm{dom}\, f^{**})}(x) = 0 = \mathrm{I}_{\mathrm{epi}\, f^{**}}(x,\lambda).$$

In other words, (3.21) holds for all $(x,\lambda) \in X \times \mathbb{R}$, and we deduce that $f = f^{**}$.

(ii) Suppose now that f admits a continuous affine minorant. If $f \equiv +\infty$, then direct calculations produce $f^{**} = \overline{\mathrm{co}} f \equiv +\infty$, and (3.19) holds trivially. Otherwise, since f admits a continuous affine minorant, the functions f and $\overline{\mathrm{co}} f$ are proper and, thanks to (3.7), relation (3.18) entails $\overline{\mathrm{co}} f = (\overline{\mathrm{co}} f)^{**} = f^{**}$. ∎

The following corollary, which is essentially the Hahn–Banach separation theorem, can be regarded as a geometric version of Theorem 3.2.2.

Corollary 3.2.3 *Let $C \subset X$ be a non-empty closed convex set. Then C is the intersection of all the closed half-spaces that contain it.*

Proof. It is obvious that C is included in the intersection of all the closed half-spaces that contain it. Conversely, let x be in such an intersection and denote

$$H_{x^*} := \{z \in X : \langle z, x^*\rangle \leq \sigma_C(x^*)\},\ x^* \in \mathrm{dom}\, \sigma_C.$$

Then, since each H_{x^*} is a closed half-space containing the set C, $x \in H_{x^*}$ and we obtain $\langle x, x^*\rangle \leq \sigma_C(x^*)$ for all $x^* \in \mathrm{dom}\, \sigma_C$. Consequently,

3.2. FENCHEL–MOREAU–ROCKAFELLAR THEOREM

since $I_C = (I_C)^{**} = (\sigma_C)^*$ by Theorem 3.2.2, we deduce that

$$I_C(x) = \sup_{x^* \in \operatorname{dom}\sigma_C} \{\langle x, x^* \rangle - \sigma_C(x^*)\} \leq 0;$$

that is, $x \in C$ as required. ∎

Next, we continue with Example 3.1.2 which shows the equivalent representation of the norm using the closed dual unit ball.

Example 3.2.4 *If we consider a normed space $(X, \|\cdot\|)$, then*

$$\|\cdot\| = \sigma_{B_{X^*}}. \tag{3.22}$$

In fact, denoting $f := \|\cdot\|$, by Example 3.1.2, we have $f^* = I_{B_{X^*}}$. Thus, since $f \in \Gamma_0(X)$, Theorem 3.2.2 and (3.1) yield $\|\cdot\| = f = f^{**} = (I_{B_{X^*}})^* = \sigma_{B_{X^*}}$.

As a first consequence of Theorem 3.2.2, we deduce the convexity and the lower semicontinuity of perspective functions.

Proposition 3.2.5 *For any function $f \in \Gamma_0(X)$, we have*

$$P_f(\alpha, x) = \sigma_{\operatorname{epi} f^*}(x, -\alpha) \text{ for all } x \in X \text{ and } \alpha \in \mathbb{R}.$$

Consequently, P_f is proper, lsc, and convex.

Proof. Applying Theorem 3.2.2, by (3.14) we have

$$P_f(\alpha, x) = P_{f^{**}}(\alpha, x) = \sigma_{\operatorname{epi} f^*}(x, -\alpha),$$

for every $x \in X$ and $\alpha \in \mathbb{R}$. As $f^* \in \Gamma_0(X^*)$, due to Proposition 3.1.4, we have $\operatorname{epi} f^* \neq \emptyset$ and $P_f \in \Gamma_0(\mathbb{R} \times X)$. ∎

The second application of Theorem 3.2.2 allows a useful expression of the conjugate of pointwise suprema.

Proposition 3.2.6 *Given a family $\{f_t, t \in T\} \subset \Gamma_0(X)$, we assume that $f := \sup_{t \in T} f_t$ is proper. Then we have*

$$f^* = \overline{\operatorname{co}}\left(\inf_{t \in T} f_t^*\right), \tag{3.23}$$

and, consequently,

$$\operatorname{epi} f^* = \overline{\operatorname{co}}\left(\bigcup_{t \in T} \operatorname{epi} f_t^*\right). \tag{3.24}$$

Proof. Using (3.7) and (3.10), Theorem 3.2.2 implies
$$\left(\overline{\mathrm{co}}\left(\inf_{t\in T} f_t^*\right)\right)^* = \left(\inf_{t\in T} f_t^*\right)^* = \sup_{t\in T} f_t^{**} = \sup_{t\in T} f_t = f.$$

In particular, since f is proper, the function $\overline{\mathrm{co}}\left(\inf_{t\in T} f_t^*\right)$ must also be proper. Therefore, applying Theorem 3.2.2 once again, the last relation entails
$$f^* = \left(\overline{\mathrm{co}}\left(\inf_{t\in T} f_t^*\right)\right)^{**} = \overline{\mathrm{co}}\left(\inf_{t\in T} f_t^*\right),$$
and (3.23) follows. Finally, statement (3.24) is the geometrical counterpart to (3.23). ∎

It is worth noting that Proposition 3.2.6 also entails Theorem 3.2.2, which shows that these two results are indeed equivalent.

Corollary 3.2.7 *The following assertions are equivalent:*

(i) For every function $f : X \to \mathbb{R}_\infty$ with a proper conjugate, we have $f^{**} = \overline{\mathrm{co}} f$.

(ii) For every family $\{f_t, t \in T\} \subset \Gamma_0(X)$ with $f := \sup_{t\in T} f_t$ being proper, we have $f^* = \overline{\mathrm{co}}\left(\inf_{t\in T} f_t^*\right)$.

Proof. The implication $(i) \Rightarrow (ii)$ is the statement of Proposition 3.2.6. To prove that $(ii) \Rightarrow (i)$, we choose a function $f : X \to \mathbb{R}_\infty$ that has a proper conjugate. So, f is proper and we can write
$$f^* = \sup_{z\in \mathrm{dom}\, f} f_z, \text{ where } f_z := \langle \cdot, z\rangle - f(z).$$

Given that $\{f_z, z \in \mathrm{dom}\, f\} \subset \Gamma_0(X^*)$ and $f_z^* = \mathrm{I}_{\{z\}}(\cdot) + f(z)$, assertion (ii) implies that
$$f^{**} = \left(\sup_{z\in X} f_z\right)^* = \overline{\mathrm{co}}\left(\inf_{z\in X} f_z^*\right) = \overline{\mathrm{co}}\left(\inf_{z\in X}\left(\mathrm{I}_{\{z\}}(\cdot) + f(z)\right)\right).$$

But we have $\inf_{z\in X}(\mathrm{I}_{\{z\}}(\cdot) + f(z)) = f$, then the last relation reads $f^{**} = \overline{\mathrm{co}} f$. ∎

The following result specifies Proposition 3.2.6 to monotone families of functions.

Proposition 3.2.8 *The following assertions are true:*

(i) Given a non-decreasing net $(f_i)_i \subset \Gamma_0(X)$ such that $f := \sup_i f_i$ is proper, we have $f^* = \mathrm{cl}\left(\inf_i f_i^*\right)$.

3.2. FENCHEL–MOREAU–ROCKAFELLAR THEOREM

(ii) Let $(A_i)_i \subset X$ be a non-increasing net of non-empty closed convex sets. Then $\cap_i A_i \neq \emptyset$ if and only if the function $\mathrm{cl}\,(\inf_i \sigma_{A_i})$ is proper. Furthermore, under each one of these two equivalent properties, we have $\sigma_{\cap_i A_i} = \mathrm{cl}\,(\inf_i \sigma_{A_i})$.

Proof. (i) Proposition 3.2.6 entails $f^* = (\sup_i f_i)^* = \overline{\mathrm{co}}\,(\inf_i f_i^*)$. But the net $(f_i^*)_i$ is non-increasing, so the function $\inf_i f_i^*$ is convex (by (2.52)). Thus, the last relation reads

$$f^* = \mathrm{cl}\left(\mathrm{co}\left(\inf_i f_i^*\right)\right) = \mathrm{cl}\left(\inf_i f_i^*\right).$$

(ii) Take $f_i := \mathrm{I}_{A_i}$, so that $f_i^* = \sigma_{A_i}$. Then, using (3.7) and (3.10), Lemma 3.2.1 gives rise to

$$\left(\mathrm{cl}\left(\inf_i \sigma_{A_i}\right)\right)^* = \left(\inf_i \sigma_{A_i}\right)^* = \sup_i(\sigma_{A_i})^* = \sup_i \mathrm{I}_{A_i} = \mathrm{I}_{\cap_i A_i}.$$

Thus, using Proposition 3.1.4, $\cap_i A_i \neq \emptyset$ if and only if the function $(\mathrm{cl}\,(\inf_i \sigma_{A_i}))^*$ is proper, if and only if the closed convex function $\mathrm{cl}\,(\inf_i \sigma_{A_i})$ is proper.

Finally, if $\cap_i A_i \neq \emptyset$, then, since the net $(f_i)_i \subset \Gamma_0(X)$ is non-increasing, assertion (i) gives rise to

$$\sigma_{\cap_i A_i} = (\mathrm{I}_{\cap_i A_i})^* = \left(\sup_i \mathrm{I}_{A_i}\right)^* = \mathrm{cl}\left(\inf_i (\mathrm{I}_{A_i})^*\right) = \mathrm{cl}\left(\inf_i \sigma_{A_i}\right).$$

∎

The following corollary is another consequence of Theorem 3.2.2.

Corollary 3.2.9 Let A, B be non-empty sets. Then $\overline{\mathrm{co}}A = \overline{\mathrm{co}}B$ if and only if $\sigma_A = \sigma_B$.

Proof. If $\overline{\mathrm{co}}A = \overline{\mathrm{co}}B$, then $\mathrm{I}_{\overline{\mathrm{co}}A} = \mathrm{I}_{\overline{\mathrm{co}}B}$ and, taking the conjugates,

$$\sigma_A = \sigma_{\overline{\mathrm{co}}A} = \mathrm{I}^*_{\overline{\mathrm{co}}A} = \mathrm{I}^*_{\overline{\mathrm{co}}B} = \sigma_{\overline{\mathrm{co}}B} = \sigma_B.$$

Conversely, if $\sigma_A = \sigma_B$, then, again taking the conjugates, Theorem 3.2.2 entails

$$\mathrm{I}_{\overline{\mathrm{co}}A} = (\mathrm{I}_{\overline{\mathrm{co}}A})^{**} = (\sigma_{\overline{\mathrm{co}}A})^* = (\sigma_A)^* = (\sigma_B)^* = (\sigma_{\overline{\mathrm{co}}B})^* = \mathrm{I}_{\overline{\mathrm{co}}B},$$

and the equality $\overline{\mathrm{co}}A = \overline{\mathrm{co}}B$ follows. ∎

The following result is an important fact that characterizes the recession function in terms of the associated conjugate. Our proof is also based on Theorem 3.2.2.

Proposition 3.2.10 *For every function $f \in \Gamma_0(X)$, we have*

$$f^\infty = \sigma_{\mathrm{dom} f^*}. \tag{3.25}$$

As a consequence, we also have that

$$(f^*)^\infty = \sigma_{\mathrm{dom} f}. \tag{3.26}$$

Proof. First, suppose that $f(\theta) = 0$. We introduce the functions

$$\varphi_s := f(s \cdot) \in \Gamma_0(X), \ s > 0,$$

so that $(s^{-1}\varphi_s)_{s>0}$ is non-decreasing and $\sup_{s>0} s^{-1}\varphi_s = f^\infty$; therefore, $f^\infty \in \Gamma_0(X)$. Next, for each fixed $x^* \in X^*$, Proposition 3.2.8(i) leads us for all $x^* \in X^*$ to

$$(f^\infty)^*(x^*) = \left(\sup_{s>0} s^{-1}\varphi_s\right)^*(x^*) = \mathrm{cl}\left(\inf_{s>0}(s^{-1}\varphi_s)^*\right)(x^*). \tag{3.27}$$

Notice that, due to Theorem 3.2.2,

$$\inf_X f^* = -f^{**}(\theta) = -f(\theta) = 0, \tag{3.28}$$

and, for every $s > 0$ and $z^* \in X^*$,

$$(s^{-1}\varphi_s)^*(z^*) = \sup_{x \in X}\{\langle z^*, x\rangle - s^{-1}\varphi_s(x)\} = \sup_{x \in X}\{\langle z^*, x\rangle - s^{-1}f(sx)\}$$
$$= s^{-1}\sup_{x \in X}\{\langle z^*, sx\rangle - f(sx)\} = s^{-1}f^*(z^*).$$

Thus, using (3.28),

$$\inf_{s>0}(s^{-1}\varphi_s)^*(z^*) = \inf_{s>0} s^{-1}f^*(z^*) = I_{\mathrm{dom} f^*}(z^*),$$

and (3.27) leads us to

$$(f^\infty)^*(x^*) = \mathrm{cl}\,(I_{\mathrm{dom} f^*})(x^*) = I_{\mathrm{cl}(\mathrm{dom} f^*)}(x^*). \tag{3.29}$$

Furthermore, taking conjugates and using Theorem 3.2.2 (as $f^\infty \in \Gamma_0(X)$), we obtain the desired property:

$$f^\infty = (f^\infty)^{**} = \left(\mathrm{I}_{\mathrm{cl}(\mathrm{dom}\, f^*)}\right)^* = \sigma_{\mathrm{cl}(\mathrm{dom}\, f^*)} = \sigma_{\mathrm{dom}\, f^*}.$$

In the general case, when possibly $f(\theta) \neq 0$, we choose $x_0 \in \mathrm{dom}\, f$ and consider the function

$$g := f(\cdot + x_0) - f(x_0) \in \Gamma_0(X).$$

Then $g(\theta) = 0$, $g^* = f^*(\cdot) - \langle \cdot, x_0 \rangle + f(x_0)$, and the first part of the proof entails $f^\infty = g^\infty = \sigma_{\mathrm{dom}\, g^*} = \sigma_{\mathrm{dom}\, f^*}$; that is, (3.25) holds. Finally, since $f^* \in \Gamma_0(X^*)$ due to Proposition 3.1.4, relation (3.26) results from combining (3.25) and Theorem 3.2.2, $(f^*)^\infty = \sigma_{\mathrm{dom}\, f^{**}} = \sigma_{\mathrm{dom}\, f}$. ∎

Further consequences of Theorem 3.2.2 come in the following sections of this chapter.

3.3 Dual representations of support functions

In this section, we use Theorem 3.2.2 to establish some dual representations of the support function of sublevel sets. The results obtained here will be used in the sequel, specifically in section 4.2, where we develop general schemes of duality in convex optimization. The first result applies Theorem 3.2.2 to write $\sigma_{[f \leq 0]}$ in terms of the conjugate of f.

Theorem 3.3.1 *Given a function $f \in \Gamma_0(X)$ such that $[f \leq 0] \neq \emptyset$, we have*

$$\sigma_{[f \leq 0]} = \mathrm{cl}\left(\inf_{\alpha > 0}(\alpha f)^*\right), \qquad (3.30)$$

and, as a consequence of that,

$$\mathrm{epi}\,\sigma_{[f \leq 0]} = \mathrm{cl}(\mathbb{R}_+ \,\mathrm{epi}\, f^*). \qquad (3.31)$$

Proof. First, using (3.7) and (3.10), Proposition 3.2.6 and Theorem 3.2.2 yield

$$\left(\mathrm{cl}\left(\inf_{\alpha>0}(\alpha f)^*\right)\right)^* = \sup_{\alpha>0}(\alpha f)^{**} = \sup_{\alpha>0}(\alpha f) = \mathrm{I}_{[f \leq 0]}. \qquad (3.32)$$

Consider the function $\varphi : X^* \to \overline{\mathbb{R}}$ defined by

$$\varphi := \inf_{\alpha>0}(\alpha f)^*, \qquad (3.33)$$

so that φ is a marginal of the function

$$(x^*, \alpha) \mapsto (\alpha f)^*(x^*) = \sup_{x \in \text{dom } f} \{\langle x^*, x \rangle - \alpha f(x)\}.$$

This last function is convex because it is the pointwise supremum of the (linear) convex functions

$$(x^*, \alpha) \mapsto \langle x^*, x \rangle - \alpha f(x), \ x \in \text{dom } f,$$

and so φ is also convex. Moreover, since $[f \leq 0] \neq \emptyset$ by assumption, (3.32) also shows that both φ and $\text{cl}\,\varphi$ are proper; that is, in particular, $\text{cl}\,\varphi \in \Gamma_0(X^*)$. Therefore, taking the conjugate in (3.32), Theorem 3.2.2 implies that

$$\sigma_{[f \leq 0]} = \left(\text{cl}\left(\inf_{\alpha>0}(\alpha f)^*\right)\right)^{**} = \text{cl}\,\varphi, \qquad (3.34)$$

showing that (3.30) holds.

To prove (3.31), we easily observe that $\text{epi}_s\,\varphi = \mathbb{R}_+^*\,(\text{epi}_s\,f^*)$, and (3.34) leads us to

$$\text{epi}\,\sigma_{[f \leq 0]} = \text{epi}\,(\text{cl}\,\varphi) = \text{cl}\,(\text{epi}\,\varphi) = \text{cl}\,(\text{epi}_s\,\varphi)$$
$$= \text{cl}\left(\mathbb{R}_+^*\,(\text{epi}_s\,f^*)\right) = \text{cl}\,(\mathbb{R}_+\,(\text{epi}\,f^*)).$$

∎

Theorem 3.3.1 can be easily extended to convex functions which are not necessarily lsc.

Corollary 3.3.2 *The conclusion of Theorem 3.3.1 holds if, instead of $f \in \Gamma_0(X)$, we suppose that $\text{cl}\,f$ is proper and*

$$\text{cl}\,([f \leq 0]) = [\text{cl}\,f \leq 0] \neq \emptyset. \qquad (3.35)$$

Proof. Of course, we have $[\text{cl}\,f \leq 0] \neq \emptyset$ and $\sigma_{[f \leq 0]} = \sigma_{\text{cl}([f \leq 0])} = \sigma_{[\text{cl}\,f \leq 0]}$. Thus, since $(\alpha f)^* = (\alpha(\text{cl}\,f))^*$ for all $\alpha > 0$, by applying Theorem 3.3.1 to $\text{cl}\,f \in \Gamma_0(X)$, we obtain

3.3. DUAL REPRESENTATIONS OF SUPPORT ...

$$\sigma_{[f\leq 0]} = \sigma_{[\text{cl } f\leq 0]} = \text{cl}\left(\inf_{\alpha>0}(\alpha(\text{cl } f))^*\right) = \text{cl}\left(\inf_{\alpha>0}(\alpha f)^*\right)$$

and

$$\text{epi } \sigma_{[f\leq 0]} = \text{epi } \sigma_{[\text{cl } f\leq 0]} = \text{cl}(\mathbb{R}_+ \text{epi}(\text{cl } f)^*) = \text{cl}(\mathbb{R}_+ \text{epi } f^*).$$

∎

We describe below a situation in which hypothesis (3.35) is satisfied.

Lemma 3.3.3 *Let $f : X \to \mathbb{R}_\infty$ be a convex function such that $[f < 0]$ is non-empty. Then we have*

$$[\text{cl } f \leq 0] = \text{cl}\,([f \leq 0]) = \text{cl}\,([f < 0]). \tag{3.36}$$

Proof. It is clear that $[f < 0] \subset [f \leq 0] \subset [\text{cl } f \leq 0]$, and so

$$\text{cl}\,([f < 0]) \subset \text{cl}\,([f \leq 0]) \subset [\text{cl } f \leq 0]. \tag{3.37}$$

For the opposite inclusion, we choose $x_0 \in [f < 0]$. Then, given $x \in [\text{cl } f \leq 0]$ and a net $(\delta_i)_i \subset\ ]0,1[$ such that $\delta_i \downarrow 0$, there exists a net $(x_i)_i \subset X$ such that $x_i \to x$,

$$(\text{cl } f)(x) = \lim_i f(x_i) \quad \text{and} \quad x_i \in [f \leq -\delta_i f(x_0)].$$

Furthermore, the net $(y_i)_i \subset X$ defined by $y_i := \delta_i x_0 + (1-\delta_i)x_i$ also converges to x and satisfies, thanks to the convexity of f,

$$f(y_i) \leq \delta_i f(x_0) + (1-\delta_i)f(x_i) \leq \delta_i f(x_0) - \delta_i(1-\delta_i)f(x_0) = \delta_i^2 f(x_0) < 0,$$

showing that $(y_i)_i \subset [f < 0]$; that is, $x \in \text{cl}\,([f < 0])$, as we wanted to prove. ∎

We proceed by giving a refinement of Theorem 3.3.1 under the non-emptiness of the strict sublevel set $[f < 0]$; this condition will be exploited in later sections such as 4.2 and 8.2, where it is called Slater condition.

Theorem 3.3.4 *For every convex function $f : X \to \mathbb{R}_\infty$ such that $\inf_X f < 0$, we have*

$$\sigma_{[f\leq 0]} = \min_{\alpha\geq 0}(\alpha f)^*. \tag{3.38}$$

If, in addition, $\inf_X f > -\infty$, then we also have

$$\sigma_{[f\le 0]} = \inf_{\alpha>0}(\alpha f)^*. \tag{3.39}$$

Proof. We can assume that $\operatorname{cl} f \in \Gamma_0(X)$; otherwise, the function $\operatorname{cl} f$ would be non-proper and we would have $\operatorname{epi} f^* = \emptyset$ (as $\operatorname{dom}(\operatorname{cl} f) \supset \operatorname{dom} f \ne \emptyset$, $\operatorname{cl} f$ must take the value $-\infty$, so $f^* = (\operatorname{cl} f)^* \equiv +\infty$). The conclusion in this case follows as demonstrated in Exercise 27.

Furthermore, since $[\operatorname{cl} f \le 0] = \operatorname{cl}([f \le 0])$ due to Lemma 3.3.3, Corollary 3.3.2 implies that

$$\sigma_{[f\le 0]} = \operatorname{cl}\left(\inf_{\alpha>0}(\alpha f)^*\right). \tag{3.40}$$

Let us denote $\varphi = \inf_{\alpha>0}(\alpha f)^*$. Therefore, the functions $\operatorname{cl}\varphi$ and φ are proper as well as their associated conjugates, by Proposition 3.1.4. Now we fix $u^* \in \operatorname{dom}(\operatorname{cl}\varphi)$. By (3.40), there exist nets $(u_i^*)_i \subset \varphi^{-1}(\mathbb{R})$ and $(\alpha_i)_i \subset \mathbb{R}_+^*$ such that $(\alpha_i f)^*(u_i^*) \in \mathbb{R}$, for all i, $u_i^* \to u^*$ and

$$\sigma_{[f\le 0]}(u^*) = (\operatorname{cl}\varphi)(u^*) = \lim_i (\alpha_i f)^*(u_i^*). \tag{3.41}$$

The net $(\alpha_i)_i$ must be bounded, otherwise θ would be a w^*-cluster point of the net $(\alpha_i^{-1} u_i^*)_i$, and (3.41) together with the w^*-lower semi-continuity of f^* would lead us to a contradiction:

$$0 < -\inf f = f^*(\theta) \le \liminf_i f^*(\alpha_i^{-1} u_i^*) = \liminf_i \alpha_i^{-1}(\alpha_i f)^*(u_i^*) = 0.$$

Consequently, we may assume without loss of generality that $(\alpha_i)_i$ converges to some $\alpha_0 \ge 0$. In addition, for all $x \in \operatorname{dom} f$, due to (3.41) the scalar α_0 satisfies

$$\langle u^*, x\rangle - \alpha_0 f(x) = \lim_i (\langle u_i^*, x\rangle - \alpha_i f(x))$$
$$\le \limsup_i (\alpha_i f)^*(u_i^*) = \lim_i (\alpha_i f)^*(u_i^*) = (\operatorname{cl}\varphi)(u^*); \tag{3.42}$$

that is, taking the supremum over $x \in \operatorname{dom} f$,

$$(\alpha_0 f)^*(u^*) \le (\operatorname{cl}\varphi)(u^*). \tag{3.43}$$

At this step, we distinguish two cases: First, if $\alpha_0 > 0$, then (3.43) entails

$$\varphi(u^*) = \inf_{\alpha>0}(\alpha f)^*(u^*) \le (\alpha_0 f)^*(u^*) \le (\operatorname{cl}\varphi)(u^*) \le \varphi(u^*).$$

Hence, (3.40) gives us

3.3. DUAL REPRESENTATIONS OF SUPPORT ...

$$\sigma_{[f\leq 0]}(u^*) = (\operatorname{cl}\varphi)(u^*) = \varphi(u^*) = \inf_{\alpha>0}(\alpha f)^*(u^*) = (\alpha_0 f)^*(u^*); \quad (3.44)$$

that is, $\sigma_{[f\leq 0]}(u^*) = \min_{\alpha>0}(\alpha f)^*(u^*)$. Therefore, since

$$\sigma_{[f\leq 0]}(u^*) \leq \sigma_{\operatorname{dom}f}(u^*) = (\operatorname{I}_{\operatorname{dom}f})^*(u^*) = (0f)^*(u^*), \quad (3.45)$$

we conclude that

$$\sigma_{[f\leq 0]}(u^*) = \inf_{\alpha>0}(\alpha f)^*(u^*) = \min_{\alpha>0}(\alpha f)^*(u^*) = \min_{\alpha\geq 0}(\alpha f)^*(u^*), \quad (3.46)$$

and both (3.38) and (3.39) are valid.

Second, if $\alpha_0 = 0$, then (3.43) and (3.40) produce

$$\sigma_{\operatorname{dom}f}(u^*) = (0f)^*(u^*) \leq (\operatorname{cl}\varphi)(u^*) = \sigma_{[f\leq 0]}(u^*) \leq \sigma_{\operatorname{dom}f}(u^*), \quad (3.47)$$

so that

$$\sigma_{[f\leq 0]}(u^*) = (0f)^*(u^*) = (\operatorname{cl}\varphi)(u^*) = \operatorname{cl}\left(\inf_{\alpha>0}(\alpha f)^*\right)(u^*) \leq \inf_{\alpha>0}(\alpha f)^*(u^*), \quad (3.48)$$

and (3.38) follows.

Under the supplementary condition $\inf_X f > -\infty$ (so, $-\infty < \inf_X f < 0$) we also have that

$$\inf_{\alpha>0}(\alpha f)^*(u^*) = \inf_{\alpha>0} \sup_{x\in\operatorname{dom}f}\{\langle u^*, x\rangle - \alpha f(x)\}$$
$$\leq \inf_{\alpha>0}\left(\sigma_{\operatorname{dom}f}(u^*) - \alpha\inf_X f\right) = \sigma_{\operatorname{dom}f}(u^*),$$

and (3.39) follows by combining (3.47) and (3.48).

To finish the proof, we observe, thanks to (3.40), that for all $u^* \notin \operatorname{dom}(\operatorname{cl}\varphi)$

$$+\infty = (\operatorname{cl}\varphi)(u^*) = \sigma_{[f\leq 0]}(u^*) = \operatorname{cl}\left(\inf_{\alpha>0}(\alpha f)^*\right)(u^*) \leq \inf_{\alpha>0}(\alpha f)^*(u^*).$$

In addition, (3.45) implies that $(0f)^*(u^*) = +\infty$ and both (3.38) and (3.39) trivially hold. ∎

The following corollary gives the geometric counterpart to Theorem 3.3.4.

Corollary 3.3.5 *For every convex function $f : X \to \mathbb{R}_\infty$ such that $\inf_X f < 0$, we have*

$$\operatorname{epi}\sigma_{[f\leq 0]} = \left(\mathbb{R}_+^* \operatorname{epi} f^*\right) \cup \operatorname{epi}\sigma_{\operatorname{dom} f}. \tag{3.49}$$

Proof. We may assume, as in the proof of Theorem 3.3.4, that $\operatorname{cl} f \in \Gamma_0(X)$. We fix $(u^*,\lambda) \in \operatorname{epi}\sigma_{[f\leq 0]}$. Then, by Theorem 3.3.4, there exists $\alpha_0 \geq 0$ such that $\sigma_{[f\leq 0]}(u^*) = (\alpha_0 f)^*(u^*) \leq \lambda$. If $\alpha_0 > 0$, then $\alpha_0 f^*(\alpha_0^{-1} u^*) \leq \lambda$ and we get $(u^*,\lambda) \in \alpha_0 \operatorname{epi} f^* \subset \mathbb{R}_+^* \operatorname{epi} f^*$. Otherwise, if $\alpha_0 = 0$, then $\sigma_{[f\leq 0]}(u^*) = (0f)^*(u^*) = \sigma_{\operatorname{dom} f}(u^*) \leq \lambda$, and thus $(u^*,\lambda) \in \operatorname{epi}\sigma_{\operatorname{dom} f}$. Consequently, $(u^*,\lambda) \in \left(\mathbb{R}_+^* \operatorname{epi} f^*\right) \cup \operatorname{epi}\sigma_{\operatorname{dom} f}$, and the inclusion "$\subset$" in (3.49) follows.

To show the opposite inclusion we first observe, by the Fenchel inequality, that for all $u^* \in X^*$

$$\sigma_{[f\leq 0]}(u^*) = \sup_{f(u)\leq 0} \langle u^*, u\rangle \leq \sup_{f(u)\leq 0} (f^*(u^*) + f(u)) \leq f^*(u^*),$$

entailing the relation $\operatorname{epi} f^* \subset \operatorname{epi}\sigma_{[f\leq 0]}$. Hence, because $\operatorname{epi}\sigma_{[f\leq 0]}$ is a cone, $\mathbb{R}_+^* \operatorname{epi} f^* \subset \mathbb{R}_+^* \operatorname{epi}\sigma_{[f\leq 0]} = \operatorname{epi}\sigma_{[f\leq 0]}$. Moreover, since $\operatorname{epi}\sigma_{[f\leq 0]}$ is closed and both functions $\operatorname{cl} f$ and f^* are proper (see Proposition 3.1.4), the last relation together with (3.26) entails

$$\operatorname{epi}\sigma_{\operatorname{dom} f} = \operatorname{epi}\sigma_{\operatorname{dom}(\operatorname{cl} f)} = \operatorname{epi}(f^*)^\infty$$
$$= [\operatorname{epi} f^*]_\infty \subset [\operatorname{epi}\sigma_{[f\leq 0]}]_\infty = \operatorname{epi}\sigma_{[f\leq 0]},$$

showing that the desired inclusion holds. ∎

The following corollary is a consequence of Theorem 3.3.4. In fact, (3.50) and (3.51) are the well-known *bipolar* and *Farkas theorems*, respectively.

Corollary 3.3.6 *For every non-empty set $A \subset X$, we have*

$$A^{\circ\circ} := (A^\circ)^\circ = \overline{\operatorname{co}}(A \cup \{\theta\}), \tag{3.50}$$

and

$$A^{--} := (A^-)^- = \overline{\operatorname{co}}(\operatorname{cone}(A)). \tag{3.51}$$

Consequently,

$$\operatorname{cl}(\operatorname{dom}\sigma_A) = ([\overline{\operatorname{co}} A]_\infty)^-. \tag{3.52}$$

Proof. Since $A^\circ = (A \cup \{\theta\})^\circ$, we can assume, without loss of generality, that $\theta \in A$. To prove (3.50), we consider the function

$$f := \sigma_{A^\circ} - 1 \in \Gamma_0(X).$$

3.3. DUAL REPRESENTATIONS OF SUPPORT ...

So, $f(\theta) = -1 < 0$ and $A^{\circ\circ} = [\sigma_{A^\circ} \leq 1] = [f \leq 0]$. Moreover, using Theorem 3.2.2, we have $\inf_X f = -f^*(\theta) = -1 > -\infty$. Then again by Theorem 3.2.2, (3.39) entails

$$\sigma_{A^{\circ\circ}} = \sigma_{[f \leq 0]} = \inf_{\alpha > 0}(\alpha f)^* = \inf_{\alpha > 0}((\sigma_{\alpha A^\circ})^* + \alpha) = \inf_{\alpha > 0}(I_{\alpha A^\circ} + \alpha).$$

Thus, $\sigma_{A^{\circ\circ}} = \sigma_A$ by Exercise 6 (using the assumption $\theta \in A$), and the conclusion follows from Corollary 3.2.9.

Relation (3.51) comes from (3.50) noting that

$$A^{--} = (\overline{\text{co}}(\text{cone } A))^{\circ\circ}.$$

Finally, thanks to (2.49) and (3.51), (3.52) is equivalent to $[\overline{\text{co}}A]_\infty = (\text{dom }\sigma_A)^-$. So, we are done because (2.62) and (3.25) yield $I_{[\overline{\text{co}}A]_\infty} = (I_{\overline{\text{co}}A})^\infty = \sigma_{\text{dom}\sigma_{\overline{\text{co}}A}} = \sigma_{\text{dom}\sigma_A} = I_{(\text{dom }\sigma_A)^-}$. ∎

The following result completes Proposition 3.1.3 by discussing the effect of imposing the w^*-continuity of the conjugate function.

Proposition 3.3.7 (i) *Let $f: X \to \mathbb{R}_\infty$ be convex. If f^* is finite and w^*-continuous somewhere in X^*, then $\text{dom } f$ is a finite-dimensional set.*

(ii) *The space X is of finite dimension if and only if there exists a convex function $f: X \to \mathbb{R}_\infty$, which is finite and continuous somewhere, whose conjugate is finite and w^*-continuous somewhere.*

Proof. (i) Since f^* is proper and $f^* = (\text{cl } f)^*$, the function $\text{cl } f$ is also (convex, lsc, and) proper by Proposition 3.1.4. Let $x_0^* \in X^*$ be a w^*-continuity point of f^*; by the current assumption, and let vectors $x_1, \ldots, x_k \in X$ such that $V := (\{x_i, i = 1, \ldots, k\})^\circ$ is a w^*-neighborhood of $\theta \in X^*$ and $f^*(x_0^* + x^*) \leq f^*(x_0^*) + 1$ for all $x^* \in V$. In particular, for all $x^* \in (\{x_i, i = 1, \ldots, k\})^\perp \subset V$ we have that $\mathbb{R}_+ x^* \subset V$ and so, taking into account (3.26),

$$\sigma_{\text{dom}f}(x^*) = \sigma_{\text{cl(dom }f)}(x^*) = \sigma_{\text{dom(cl }f)}(x^*) = ((\text{cl } f)^*)^\infty(x^*)$$
$$= (f^*)^\infty(x^*) = \sup_{\alpha > 0} \alpha^{-1}(f^*(x_0^* + \alpha x^*) - f^*(x_0^*))$$
$$= \lim_{\alpha \uparrow +\infty} \alpha^{-1}(f^*(x_0^* + \alpha x^*) - f^*(x_0^*)) \leq \lim_{\alpha \uparrow +\infty} \alpha^{-1} = 0.$$

Consequently, due to Corollary 3.3.6, $\text{dom } f \subset \text{span}\{x_i, i = 1, \ldots, k\}$, and $\text{dom } f$ is a finite-dimensional subset of X.

(ii) Suppose that there exists a convex function $f: X \to \mathbb{R}_\infty$ which is finite and continuous somewhere and such that f^* is finite and w^*-

continuous somewhere too. Then, by (i), the (non-empty) effective domain of f is finite-dimensional. Thus, due to the continuity assumption of f, $\operatorname{dom} f$ contains a closed ball which is finite-dimensional, implying that X is of finite dimension. The converse statement is clear: When X is of finite dimension, the convex function $f := I_{B_X}$ is continuous at $\theta \in X$ and its conjugate $f^* = \|\cdot\|$ is $(norm-)$ w^*-continuous on X^*. ∎

We close this section by giving another illustration of Theorem 3.2.2 and Corollary 3.3.6, providing a slight extension of the Cauchy–Schwarz inequality. Relation (3.53) exhibits a duality between the support and the gauge functions.

Corollary 3.3.8 *For every non-empty closed convex set C containing θ, we have*

$$p_C = \sigma_{C^\circ}, \qquad (3.53)$$

and consequently, for all $x \in X$ and $x^* \in X^*$,

$$\langle x^*, x \rangle \leq \sigma_C(x^*) p_C(x) = \sigma_C(x^*) \sigma_{C^\circ}(x). \qquad (3.54)$$

Proof. We have, for every $x \in X$,

$$p_C(x) = \inf\{\lambda \geq 0 : x \in \lambda C\} = \inf\{I_{\lambda C}(x) + \lambda : \lambda \geq 0\},$$

and (3.10) entails

$$(p_C)^* = \sup_{\lambda \geq 0} (\sigma_{\lambda C} - \lambda)$$

$$= \max\left\{\sup_{\lambda > 0} \lambda(\sigma_C - 1), 0\right\} = \max\{I_{[\sigma_C \leq 1]}, 0\} = I_{[\sigma_C \leq 1]} = I_{C^\circ}.$$

Thus, since $p_C \in \Gamma_0(X)$ (see (2.53) and Proposition 2.3.1), (3.53) follows by applying Theorem 3.2.2.

To show (3.54), fix $x \in X$, $x^* \in X^*$ and $\varepsilon > 0$. Then, since C and C° are non-empty closed convex sets containing θ, we have

$$x \in p_C(x) C \subset (p_C(x) + \varepsilon) C \text{ and } x^* \in p_{C^\circ}(x^*) C^\circ \subset (p_{C^\circ}(x^*) + \varepsilon) C^\circ.$$

Thus, $\langle (p_{C^\circ}(x^*) + \varepsilon)^{-1} x^*, (p_C(x) + \varepsilon)^{-1} x \rangle \leq 1$ and we get

$$\langle x^*, x \rangle \leq (p_{C^\circ}(x^*) + \varepsilon)(p_C(x) + \varepsilon).$$

Finally, taking into account (3.53) and (3.50), the inequality in (3.54) follows by letting $\varepsilon \downarrow 0$. ∎

3.4 Minimax theory

In this section, we use Theorem 3.2.2 to obtain different variants of the *minimax theorem*. We consider two lcs X, Y with respective topological duals X^* and Y^*, both endowed with locally convex topologies making the dual pairs (X, X^*) and (Y, Y^*) compatible. The associated duality pairings in X and Y are denoted by $\langle \cdot, \cdot \rangle$. For a function $f : X \times Y \to \overline{\mathbb{R}}$ and non-empty convex sets $A \subset X$ and $B \subset Y$, our goal is to see that, given appropriate convexity/concavity conditions on f, we can ensure the equality

$$\sup_{x \in A} \inf_{y \in B} f(x,y) = \inf_{y \in B} \sup_{x \in A} f(x,y) \qquad (3.55)$$

or equivalently, as the opposite of the following inequality always holds,

$$\sup_{x \in A} \inf_{y \in B} f(x,y) \geq \inf_{y \in B} \sup_{x \in A} f(x,y).$$

Additional compactness conditions on the sets A, B, and the upper/lower semicontinuity of the functions $f(\cdot, y)$ and $f(x, \cdot)$ will allow handling $\max_A$ and $\min_B$ instead of supremum and infimum, respectively. Theorems that allow such an interchange between the minimum and the supremum are called minimax theorems. They were at the origin of the duality theory and continue to play a fundamental role in convex analysis and optimization.

We start by giving a topological minimax result, which encloses the essential elements behind the proof of minimax theorems. Then convexity will come into play to replace the monotonicity assumption in this result.

Proposition 3.4.1 *Let $(I, \preccurlyeq)$ be a directed set, let $\{\varphi_i : X \to \overline{\mathbb{R}}\}$ be a family of usc functions, which is non-increasing (that is, $i_1 \preccurlyeq i_2 \Rightarrow \varphi_{i_2} \leq \varphi_{i_1}$ for all i_1, i_2), and let $A \subset X$ be a non-empty compact set. Then we have*

$$\inf_i \max_{x \in A} \varphi_i(x) = \max_{x \in A} \inf_i \varphi_i(x). \qquad (3.56)$$

Proof. Since the function $\inf_i \varphi_i$ is usc, both it and the usc function φ_i reach their maxima on the compact set A, justifying the use of the maxima instead of suprema in (3.56). Consequently, there exists a net $(x_i)_i \subset A$ such that $\varphi_i(x_i) = \max_{x \in A} \varphi_i(x)$. In fact, again because of the compactness of A, we assume that $(x_i)_i$ converges to some $\bar{x} \in A$. Therefore, given any index j, since the functions $i \mapsto \varphi_i(x)$, $x \in A$, are non-increasing by assumption, we get

$$\varphi_j(\bar{x}) \geq \limsup_{i,\, j \nprec i} \varphi_j(x_i) \geq \limsup_{i} \varphi_i(x_i)$$
$$= \limsup_{i} \max_{x \in A} \varphi_i(x) \geq \inf_{i} \max_{x \in A} \varphi_i(x).$$

Thus, by the arbitrariness of j,

$$\max_{x \in A} \inf_{i} \varphi_i(x) \geq \inf_{j} \varphi_j(\bar{x}) \geq \inf_{i} \max_{x \in A} \varphi_i(x),$$

and the inequality "$\leq$" in (3.56) is proved. The proof is over since the opposite inequality is obvious. ∎

We will need the following technical lemma in the proof of the minimax theorems below.

Lemma 3.4.2 *Given a function $f : X \times Y \to \overline{\mathbb{R}}$ and non-empty convex sets $A \subset X$ and $B \subset Y$, we consider the function $g : Y^* \to \overline{\mathbb{R}}$ defined by*

$$g(y^*) := \inf_{x \in A} \left(f(x, \cdot) + I_B(\cdot)\right)^* (y^*). \tag{3.57}$$

If A is compact and the functions $f(\cdot, y)$, $y \in B$, are concave and usc, then g is convex and lsc.

Proof. Under the current assumptions, for each $y \in B$, the function $(x, y^*) \in X \times Y^* \mapsto \langle y, y^* \rangle - f(x, y)$ is convex and lsc, and so is the supremum function

$$(x, y^*) \in X \times Y^* \mapsto \sup_{y \in B} \{\langle y, y^* \rangle - f(x, y)\} = \left(f(x, \cdot) + I_B(\cdot)\right)^* (y^*). \tag{3.58}$$

Therefore, g is convex because it is the marginal of the last convex function. To show that g is also lsc, we fix $y^* \in X^*$ and take a net $(y_i^*)_i \subset \operatorname{dom} g$ such that $y_i^* \to y^*$. Then, for each net $\alpha_i \downarrow 0$, we can find another net $(x_i)_i \subset A$ such that $(f(x_i, \cdot) + I_B)^* (y_i^*) \leq g(y_i^*) + \alpha_i$; that is, for each $y \in B$, we have $\langle y, y_i^* \rangle - f(x_i, y) \leq g(y_i^*) + \alpha_i$ for all i. But A is compact, we may assume, without loss of generality, that $(x_i)_i$ converges to some $\bar{x} \in A$. Thus, taking limits in the inequality above and using the upper semicontinuity of the functions $f(\cdot, y)$, $y \in B$, we get

$$\langle y, y^* \rangle - f(\bar{x}, y) \leq \liminf_{i}(\langle y, y_i^* \rangle - f(x_i, y))$$
$$\leq \liminf_{i} g(y_i^*) \text{ for all } y \in B.$$

3.4. MINIMAX THEORY

Hence, passing to the supremum over $y \in B$, we get $(f(\bar{x}, \cdot) + I_B)^*(y^*)$ $\leq \liminf_i g(y_i^*)$ and, finally, we deduce that

$$g(y^*) = \inf_{x \in A} (f(x, \cdot) + I_B)^*(y^*) \leq (f(\bar{x}, \cdot) + I_B)^*(y^*) \leq \liminf_i g(y_i^*);$$

that is, g is lsc at y^*. ∎

We give the first minimax theorem. Note that we do not use here the condition $A \times B \subset f^{-1}(\mathbb{R})$, which is usually required in minimax theorems.

Theorem 3.4.3 *Given a function $f : X \times Y \to \overline{\mathbb{R}}$ and non-empty convex sets $A \subset X$ and $B \subset Y$, we assume the following conditions:*
(i) The set A compact.
(ii) The functions $f(\cdot, y)$, $y \in B$, are concave and usc.
Then we have

$$\max_{x \in A} \inf_{y \in B} f(x, y) \geq \inf_{y \in B} \sup_{x \in A_0} f(x, y), \qquad (3.59)$$

where

$$A_0 := \{x \in A : f(x, \cdot) + I_B(\cdot) \in \Gamma_0(Y)\}. \qquad (3.60)$$

Proof. First of all, we may assume that $A_0 \neq \emptyset$ because, otherwise, (3.59) trivially holds (as $\sup_{x \in A_0} f(x, y) = -\infty$). Note that (3.59) also holds if, in addition, we have $\sup_{x \in A_0} f(x, y) = +\infty$ for all $y \in B$ (see Exercise 35). Thus, in what follows, we assume the non-emptiness of A_0 together with the existence of some $y_0 \in B$ such that

$$\sup_{x \in A_0} f(x, y_0) < +\infty. \qquad (3.61)$$

In addition, since the functions $f(\cdot, y)$, $y \in B$, are usc and concave by (ii), the pointwise infimum $\inf_{y \in B} f(\cdot, y)$ is also usc and concave, so that all these functions achieve their suprema over the compact convex set A. Furthermore, using the definition of the conjugate, we write

$$\max_{x \in A} \inf_{y \in B} f(x, y) = \max_{x \in A} \inf_{y \in Y} (f(x, y) + I_B(y))$$

$$= \max_{x \in A} (-(f(x, \cdot) + I_B)^*(\theta))$$

$$= -\inf_{x \in A} (f(x, \cdot) + I_B)^*(\theta), \qquad (3.62)$$

and Lemma 3.4.2 together with the fact that $A_0 \subset A$ yields

$$\max_{x \in A} \inf_{y \in B} f(x,y) = -\overline{\mathrm{co}} \left(\inf_{x \in A} (f(x,\cdot) + I_B)^* \right)(\theta)$$

$$\geq -\overline{\mathrm{co}} \left(\inf_{x \in A_0} (f(x,\cdot) + I_B)^* \right)(\theta).$$

But we have $f(x,\cdot) + I_B \in \Gamma_0(Y)$, for all $x \in A_0$, and $\sup_{x \in A_0}(f(x,\cdot) + I_B) \in \Gamma_0(Y)$ by (3.61), so Proposition 3.2.6 yields

$$\max_{x \in A} \inf_{y \in B} f(x,y) \geq - \left(\sup_{x \in A_0} (f(x,\cdot) + I_B) \right)^* (\theta)$$

$$= \inf_{y \in Y} \sup_{x \in A_0} (f(x,y) + I_B(y)) = \inf_{y \in B} \sup_{x \in A_0} f(x,y),$$

which is the desired inequality. ∎

Consequently, to obtain an equality like (3.55), we need more properties on the set A_0. The proof of the following corollary is direct from Theorem 3.4.3.

Corollary 3.4.4 *In addition to the assumptions of Theorem 3.4.3, we suppose that the set A_0 is not empty and satisfies*

$$\inf_{y \in B} \sup_{x \in A_0} f(x,y) \geq \inf_{y \in B} \max_{x \in A} f(x,y). \qquad (3.63)$$

Then we have

$$\max_{x \in A} \inf_{y \in B} f(x,y) = \inf_{y \in B} \max_{x \in A} f(x,y).$$

The following example presents a typical situation where Corollary 3.4.4 applies, requiring the convexity of the function $f(x,\cdot)$ only for points $x \in \mathrm{int}\, A$. The finite-dimensional version of this result follows similarly by using the relative interior of A instead of the interior.

Example 3.4.5 *Given a function $f : X \times Y \to \overline{\mathbb{R}}$ and non-empty convex sets $A \subset X$ and $B \subset Y$, we assume the following conditions:*
(i) The set A compact.
(ii) The functions $f(\cdot,y)$, $y \in B$, are concave and usc.
(iii) $(\mathrm{int}\, A) \cap \{x \in X : f(x,y) > -\infty\} \neq \emptyset$ for all $y \in B$.
(iv) $f(x,\cdot) + I_B \in \Gamma_0(Y)$ for all $x \in \mathrm{int}\, A$.
Observe that condition (iii) ensures that (see Exercise 14)

$$\sup_{x \in \mathrm{int}\, A} f(x,y) = \sup_{x \in A} f(x,y) \text{ for all } y \in B.$$

3.4. MINIMAX THEORY

Thus, since (iv) implies that

$$\text{int } A \subset A_0 := \{x \in A : f(x, \cdot) + I_B \in \Gamma_0(Y)\} \subset A,$$

we deduce that

$$\sup_{x \in A_0} f(x, y) = \sup_{x \in A} f(x, y) \text{ for all } y \in B.$$

Consequently, condition (3.63) holds, and Corollary 3.4.4 gives rise to

$$\max_{x \in A} \inf_{y \in B} f(x, y) = \inf_{y \in B} \max_{x \in A} f(x, y).$$

A simpler form of Theorem 3.4.3 is presented below.

Corollary 3.4.6 *Given a function $f : X \times Y \to \overline{\mathbb{R}}$ and non-empty convex sets $A \subset X$ and $B \subset Y$, we assume the following conditions:*
(i) The set $A \subset X$ is compact and the set B is closed.
(ii) For every $y \in B$, $f(\cdot, y)$ is an usc concave function.
(ii) For every $x \in A$, $f(x, \cdot) \in \Gamma_0(Y)$ and $B \cap \text{dom } f(x, \cdot) \neq \emptyset$.
Then we have

$$\max_{x \in A} \inf_{y \in B} f(x, y) = \inf_{y \in B} \max_{x \in A} f(x, y).$$

We give another variant of the minimax theorem, dropping out the lower semicontinuity condition of the functions $f(x, \cdot) + I_B$, $x \in A$, used in Theorem 3.4.3. Instead, we use here the condition that the function f is finite-valued on the set $A \times B$.

Theorem 3.4.7 *Given a function $f : X \times Y \to \overline{\mathbb{R}}$ and non-empty convex sets $A \subset X$ and $B \subset Y$, we assume the following conditions:*
(i) The set A is compact and $A \times B \subset f^{-1}(\mathbb{R})$.
(ii) The functions $f(\cdot, y)$, $y \in B$, are concave and usc.
Then we have

$$\max_{x \in A} \inf_{y \in B} f(x, y) \geq \inf_{y \in B} \sup_{x \in A_1} f(x, y),$$

where $A_1 := \{x \in A : f(x, \cdot) \text{ is convex}\}$.

Proof. Let us first note that the relation $A \times B \subset f^{-1}(\mathbb{R})$ together with condition (ii) entails

$$\sup_{x \in A} f(x, y) = \max_{x \in A} f(x, y) < +\infty \text{ for every } y \in B,$$

implying that
$$B \subset \operatorname{dom}\left(\sup_{x \in A} f(x, \cdot)\right). \tag{3.64}$$

We denote

$$\mathcal{F}^B := \{\text{finite-dimensional linear subspaces of } Y \text{ that intersect } B\},$$

and fix $L \in \mathcal{F}^B$. Then, arguing as in (3.62) and using (3.7), we write

$$\begin{aligned}
\max_{x \in A} \inf_{y \in L \cap B} f(x, y) &= \max_{x \in A} \inf_{y \in Y} (f(x, y) + \mathrm{I}_{L \cap B}(y)) \\
&= \max_{x \in A} [-(f(x, \cdot) + \mathrm{I}_{L \cap B})^*(\theta)] \\
&= \max_{x \in A} [-(\mathrm{cl}_y(f(x, \cdot) + \mathrm{I}_{L \cap B}(\cdot)))^*(\theta)] \\
&= -\inf_{x \in A} [(\mathrm{cl}_y(f(x, \cdot) + \mathrm{I}_{L \cap B}(\cdot)))^*(\theta)], \tag{3.65}
\end{aligned}$$

where cl_y denotes the closure with respect to the variable y. Furthermore, since the function $g_L : Y^* \to \overline{\mathbb{R}}$ defined by

$$g_L(y^*) := \inf_{x \in A} (\mathrm{cl}_y(f(x, \cdot) + \mathrm{I}_{L \cap B}(\cdot)))^*(y^*) = \inf_{x \in A} (f(x, \cdot) + \mathrm{I}_{L \cap B}(\cdot))^*(y^*),$$

is convex and lsc by Lemma 3.4.2, because $A_1 \subset A$ the inequality in (3.65) is also written

$$\begin{aligned}
\max_{x \in A} \inf_{y \in L \cap B} f(x, y) &= -\overline{\mathrm{co}} \left(\inf_{x \in A} (\mathrm{cl}_y(f(x, \cdot) + \mathrm{I}_{L \cap B}(\cdot)))^* \right)(\theta) \\
&\geq -\overline{\mathrm{co}} \left(\inf_{x \in A_1} (\mathrm{cl}_y(f(x, \cdot) + \mathrm{I}_{L \cap B}(\cdot)))^* \right)(\theta). \tag{3.66}
\end{aligned}$$

To remove the closure cl_y from the last inequality, we first note that the functions $\mathrm{cl}_y(f(x, \cdot) + \mathrm{I}_{L \cap B}(\cdot))$, $x \in A_1$, are (lsc) proper and convex (since each of the functions $f(x, \cdot) + \mathrm{I}_{L \cap B}(\cdot)$ is proper, convex, and has a finite-dimensional domain; see Exercise 39). In particular, the function $\varphi := \sup_{x \in A_1} (\mathrm{cl}_y(f(x, \cdot) + \mathrm{I}_{L \cap B}(\cdot)))$ does not take the value $-\infty$. Moreover, by (3.64), we have

$$\emptyset \neq L \cap B \subset \operatorname{dom}\left(\max_{x \in A} f(x, \cdot) + \mathrm{I}_{L \cap B}(\cdot)\right) \subset \operatorname{dom}\left(\sup_{x \in A_1} (f(x, \cdot) + \mathrm{I}_{L \cap B}(\cdot))\right), \tag{3.67}$$

which shows that for all $x \in A_1$

3.4. MINIMAX THEORY

$$\operatorname{dom}\left(\sup_{x \in A_1} (f(x, \cdot) + \mathrm{I}_{L \cap B}(\cdot))\right) = \operatorname{dom}(f(x, \cdot) + \mathrm{I}_{L \cap B}(\cdot)) = L \cap B; \tag{3.68}$$

that is, $L \cap B \subset \operatorname{dom} \varphi$ and the function φ is proper. Consequently, Proposition 3.2.6 entails

$$\varphi^*(\theta) = \overline{\operatorname{co}}\left(\inf_{x \in A_1} (\operatorname{cl}_y(f(x, \cdot) + \mathrm{I}_{L \cap B}(\cdot)))^*\right)(\theta),$$

and (3.66) gives rise to

$$\max_{x \in A} \inf_{y \in L \cap B} f(x, y) \geq -\left(\sup_{x \in A_1} (\operatorname{cl}_y(f(x, \cdot) + \mathrm{I}_{L \cap B}(\cdot)))\right)^*(\theta) = \inf_{y \in Y} \varphi(y). \tag{3.69}$$

Moreover, due to (3.68), for each $x \in A_1$, the set

$$\operatorname{dom}(f(x, \cdot) + \mathrm{I}_{L \cap B}(\cdot)) = L \cap B$$

is finite-dimensional, and therefore, Proposition 5.2.4 (iv) entails

$$\varphi = \sup_{x \in A_1} (\operatorname{cl}_y(f(x, \cdot) + \mathrm{I}_{L \cap B}(\cdot))) = \operatorname{cl}_y \left(\sup_{x \in A_1} (f(x, \cdot) + \mathrm{I}_{L \cap B}(\cdot))\right).$$

Thus, (3.69) and (2.37) yield

$$\begin{aligned}
\max_{x \in A} \inf_{y \in L \cap B} f(x, y) &\geq \inf_{y \in Y} \left(\operatorname{cl}_y \left(\sup_{x \in A_1} (f(x, \cdot) + \mathrm{I}_{L \cap B}(\cdot))\right)\right) \\
&= \inf_{y \in Y} \sup_{x \in A_1} (f(x, \cdot) + \mathrm{I}_{L \cap B}(\cdot)) \\
&\geq \inf_{y \in Y} \sup_{x \in A_1} (f(x, \cdot) + \mathrm{I}_B(\cdot)) \\
&= \inf_{y \in B} \sup_{x \in A_1} f(x, \cdot). \tag{3.70}
\end{aligned}$$

Consequently, endowing the family $\mathcal{F}^B$ with the partial order given by ascending inclusions, and applying Proposition 3.4.1 with $(I, \preccurlyeq) \equiv (\mathcal{F}^B, \subset)$ to the usc concave functions $\varphi_L := \inf_{y \in L \cap B} f(\cdot, y)$, $L \in \mathcal{F}^B$, (3.70) yields

$$\begin{aligned}
\max_{x \in A} \inf_{y \in B} f(x, y) &= \max_{x \in A} \inf_{L \in \mathcal{F}^B} \varphi_L(x) \\
&= \inf_{L \in \mathcal{F}^B} \max_{x \in A} \varphi_L(x) \geq \inf_{y \in B} \sup_{x \in A_1} f(x, \cdot),
\end{aligned}$$

showing that the desired inequality holds. ∎

The following theorem, commonly called the minimax theorem, is a particular case of Theorem 3.4.7. It corresponds to taking $A_1 = A$ in Theorem 3.4.7; that is, assuming that all functions $f(x, \cdot)$, $x \in A$, are convex.

Theorem 3.4.8 *Given a function $f : X \times Y \to \overline{\mathbb{R}}$ and non-empty convex sets $A \subset X$ and $B \subset Y$, we assume the following conditions:*
(i) The set A is compact and $A \times B \subset f^{-1}(\mathbb{R})$.
(ii) The functions $f(\cdot, y)$, $y \in B$, are concave and usc.
(iii) The functions $f(x, \cdot)$, $x \in A$, are convex.
Then we have

$$\max_{x \in A} \inf_{y \in B} f(x, y) = \inf_{y \in B} \max_{x \in A} f(x, y).$$

We give a useful application of Theorem 3.4.8.

Corollary 3.4.9 *Given a collection of proper convex functions $f_k : X \to \mathbb{R}_\infty$, $1 \le k \le n$, we assume that $f := \max_{1 \le k \le n} f_k$ is proper. Then we have*

$$\inf_{x \in X} f(x) = \max_{\lambda \in \Delta_n} \inf_{x \in X} \sum_{1 \le k \le n} \lambda_k f_k(x).$$

Proof. We consider the function $\varphi : \mathbb{R}^n \times X \to \overline{\mathbb{R}}$ defined as

$$\varphi(\lambda, x) := \sum_{1 \le k \le n} \lambda_k f_k(x) - \mathrm{I}_{\mathbb{R}^n_+}(\lambda), \quad \lambda \in \mathbb{R}^n, \ x \in X, \tag{3.71}$$

and denote

$$A := \Delta_n \subset \mathbb{R}^n, \ B := \mathrm{dom}\, f \subset X.$$

It is clear that A is compact and $A \times B \subset \varphi^{-1}(\mathbb{R})$. In addition, the functions $\varphi(\cdot, x)$, $x \in B$, are usc and concave, whereas the functions $\varphi(\lambda, \cdot)$, $\lambda \in A$, are convex. In other words, the conditions of Theorem 3.4.8 are fulfilled, and we deduce that

$$\max_{\lambda \in A} \inf_{x \in B} \varphi(\lambda, x) = \inf_{x \in B} \max_{\lambda \in A} \varphi(\lambda, x);$$

that is,

$$\max_{\lambda \in \Delta_n} \inf_{x \in X} \sum_{1 \le k \le n} \lambda_k f_k(x) = \max_{\lambda \in \Delta_n} \inf_{x \in \mathrm{dom}\, f} \sum_{1 \le k \le n} \lambda_k f_k(x)$$
$$= \max_{\lambda \in \Delta_n} \inf_{x \in \mathrm{dom}\, f} \varphi(\lambda, x)$$
$$= \inf_{x \in \mathrm{dom}\, f} \max_{\lambda \in \Delta_n} \varphi(\lambda, x) = \inf_{x \in X} f(x),$$

where the last equality comes from (2.45). ∎

3.5 Exercises

Exercise 21 *Let $f : X \to \mathbb{R}_\infty$ be such that f^* is proper. To establish the equality $\sigma_{\mathrm{dom}\, f} = \sigma_{\mathrm{dom}\, f^{**}}$, show that it is sufficient to assume $f^*(\theta) = 0$ and $f \in \Gamma_0(X)$. This exercise serves to complete the proof of Proposition 3.1.8, which is part of the proof of the Fenchel–Moreau–Rockafellar theorem (Theorem 3.2.2).*

Exercise 22 *If $A_1, \ldots, A_m \subset X$ are non-empty sets ($m \ge 2$) and $1 \le k < m$, prove that*

$$\sigma_{A_1} + \ldots + \sigma_{A_k} + \max\{\sigma_{A_{k+1}}, \ldots, \sigma_{A_m}\} = \sigma_{A_1 + \ldots + A_k + (A_{k+1} \cup \ldots \cup A_m)},$$

together with

$$[\overline{\mathrm{co}}(A_1 + \ldots + A_m)]_\infty = [\overline{\mathrm{co}}(A_1 \cup \ldots \cup A_m)]_\infty$$
$$= [\overline{\mathrm{co}}(A_1 + \ldots + A_k + (A_{k+1} \cup \ldots \cup A_m))]_\infty. \tag{3.72}$$

Exercise 23 *Consider a family of non-empty sets $\{A_t,\ t \in T_1 \cup T_2\} \subset X$, where T_1 and T_2 are disjoint non-empty sets. Prove that, for every $\rho > 0$,*

$$\left[\overline{\mathrm{co}}\left(\bigcup_{t \in T_1 \cup T_2} A_t\right)\right]_\infty = \left[\overline{\mathrm{co}}\left(\left(\bigcup_{t \in T_1} A_t\right) \cup \left(\bigcup_{t \in T_2} \rho A_t\right)\right)\right]_\infty$$
$$= \left[\overline{\mathrm{co}}\left(\bigcup_{t_1 \in T_1,\, t_2 \in T_2} (A_{t_1} + \rho A_{t_2})\right)\right]_\infty. \tag{3.73}$$

Exercise 24 *Given a non-empty set $A \subset X$, for every $x \in \mathrm{dom}\, \sigma_A$ and $\varepsilon \ge 0$ prove that*

$$\mathrm{N}^\varepsilon_{\mathrm{dom}\, \sigma_A}(x) = \mathrm{N}^\varepsilon_{([\overline{\mathrm{co}}A]_\infty)^-}(x) = \{x^* \in [\overline{\mathrm{co}}A]_\infty : -\varepsilon \le \langle x^*, x \rangle \le 0\}. \tag{3.74}$$

Exercise 25 *Assume that X is a normed space. Given a non-empty set $A \subset X$, we consider the function $f_A : X \to \mathbb{R}_\infty$ given by (see (8.89))*

$$f_A(x) := \mathrm{I}_A(x) + \frac{1}{2}\|x\|^2. \tag{3.75}$$

Prove the following statements:
(i) f_A is convex if and only if A is convex.
(ii) f_A is lsc if and only if A is closed.
(iii) $\mathrm{cl}\, f_A$ is convex if and only if $\mathrm{cl}\, A$ is convex.
(iv) If the weak closure of f, $\mathrm{cl}^w(f_A)$, is convex, then the set $\mathrm{cl}^w A$ is convex.
(v) $(\partial f_A)^{-1}(x) = \pi_A(x)$ for all $x \in X$, where ∂f_A is the convex subdifferential of f_A (see Definition 4.1.2).

Exercise 26 *Let $f : X \to \mathbb{R}_\infty$ be a proper function bounded below by a continuous affine mapping. Prove that $f \square \sigma_{U^\circ}$ is continuous on X, for every $U \in \mathcal{N}_X$, and $\mathrm{cl}\, f = \sup_{U \in \mathcal{N}_X}(f \square \sigma_{U^\circ})$.*

Exercise 27 *Given a convex function $f : X \to \mathbb{R}_\infty$ such that $\mathrm{dom}\, f \neq \emptyset$, we assume that $\mathrm{cl}\, f$ is not proper. Prove that $\sigma_{[f \leq 0]} = \min_{\alpha \geq 0}(\alpha f)^*$, and as a consequence of that,*

$$\mathrm{epi}\, \sigma_{[f \leq 0]} = \left(\mathbb{R}_+^* \, \mathrm{epi}\, f^*\right) \cup \mathrm{epi}\, \sigma_{\mathrm{dom}\, f}.$$

Exercise 28 *Assume that X is a normed space, and let $A, B, C \subset X$ be non-empty sets such that A is bounded and $A + B \subset A + C$. Prove that $B \subset \overline{\mathrm{co}}(C)$.*

Exercise 29 *Let $X = \ell_1$ (see the Glossary of notations for the definition) and*

$$A_1 := \{ie_1 + 2ie_i : i \geq 1\}, \quad A_2 := \{-ie_i : i \geq 1\},$$

where $(e_i)_{i \geq 1}$ is the sequence with all the terms being equal to zero except the ith component which is one. Prove that $[\overline{\mathrm{co}}(A_1 + A_2)]_\infty = \mathbb{R}_+\{e_1\}$.

Exercise 30 *Given a non-empty set $A \subset X$, prove that for every $z \in \mathrm{dom}\, \sigma_A$ we have*

$$\mathrm{N}_{\mathrm{dom}\,\sigma_A}(z) = [\overline{\mathrm{co}} A]_\infty \cap \{z\}^\perp, \tag{3.76}$$

and consequently,

$$(\mathrm{dom}\, \sigma_A)^- = [\overline{\mathrm{co}} A]_\infty. \tag{3.77}$$

3.5. EXERCISES

Exercise 31 *Given an arbitrary family of functions $\{f_t,\ t \in T\} \subset \Gamma_0(X)$, we denote $f := \sup_{t \in T} f_t$. Prove that, for every $x \in \operatorname{dom} f$,*

$$N_{\operatorname{dom} f}(x) = \left\{ x^* \in X^* : (x^*, \langle x^*, x \rangle) \in \left[\overline{\operatorname{co}} \left(\bigcup_{t \in T} \operatorname{gph} f_t^* \right) \right]_\infty \right\} \tag{3.78}$$

$$= \left\{ x^* \in X^* : (x^*, \langle x^*, x \rangle) \in \left[\overline{\operatorname{co}} \left(\bigcup_{t \in T} \operatorname{epi} f_t^* \right) \right]_\infty \right\}. \tag{3.79}$$

Exercise 32 *Let $f \in \Gamma_0(X)$ and $F = \{x \in X : f(x) \leq 0\}$. Prove that the following statements hold:*
  (a) $F \neq \emptyset \iff (\theta, -1) \notin \operatorname{cl}(\operatorname{cone} \operatorname{epi} f^*)$.
  (b) *If $F \neq \emptyset$, then $\operatorname{epi} \sigma_F = \operatorname{cl}(\operatorname{cone} \operatorname{epi} f^*)$.*

Exercise 33 *Given a function $f \in \Gamma_0(X)$, prove that*

$$(f^+)^* = \operatorname{cl} \left(\inf_{\lambda \in [0,1]} (\lambda f)^* \right),$$

where $f^+ := \max\{f, 0\}$ is the positive part of f.

Exercise 34 *Given $a_i \in X^*$, $b_i \in \mathbb{R}$, $1 \leq i \leq k$, with $k \geq 1$, we consider the function $f := \max\{\langle a_i, \cdot \rangle - b_i : 1 \leq i \leq k\}$. Prove that, for each $x^* \in X^*$,*

$$f^*(x^*) = \min \left\{ \sum_{1 \leq i \leq k} \gamma_i b_i : \sum_{1 \leq i \leq k} \gamma_i a_i = x^*,\ \gamma \in \Delta_k \right\}. \tag{3.80}$$

Exercise 35 *Consider a function $f : X \times Y \to \overline{\mathbb{R}}$ defined on the Cartesian product of two lcs X and Y, a non-empty compact convex set $A \subset X$, and a non-empty convex set $B \subset Y$. Assume that the functions $f(\cdot, y)$, $y \in B$, are concave and usc. Denote*

$$A_0 := \{x \in A : f(x, \cdot) + I_B \in \Gamma_0(Y)\},$$

and assume that $A_0 \neq \emptyset$ and $\sup_{x \in A_0} f(x, y) = +\infty$ for all $y \in B$. Prove that

$$\max_{x \in A} \inf_{y \in B} f(x, y) = \inf_{y \in B} \sup_{x \in A_0} f(x, y) = +\infty.$$

3.6 Bibliographical notes

The conducting wire of this chapter is the celebrated Fenchel–Moreau–Rockafellar theorem 3.2.2 ([161]), which is considered the cornerstone of convex analysis. The proof given here is new and, like the original proof by Moreau, is based on the Hahn–Banach separation theorem. In the current chapter, as well as the upcoming chapter 4, we show that many fundamental results in the theory of convex functions admit new proofs which are based on Theorem 3.2.2. This is the case of Proposition 3.2.5 which goes back to [174, Corollary 13.5.1]; also see [149, 150] for a general concept of perspective functions. Theorems 3.3.1 and 3.3.4, giving dual representations of the support function of sublevel sets, provide an instance of a simple convex duality scheme. Corollary 3.3.6, which is the classical bipolar theorem, is proved here via Theorem 3.2.2. Theorem 3.4.3 gives a slightly weaker variant of the minimax theorem. Other concave-like and convex-like variants can be considered as in the Fan and Sion minimax theorems ([25, 83] and [183], respectively). Other minimax results involving dense sets can be found in [128].

Exercises 22 and 31 are in [103]. The case $\varepsilon = 0$ in Exercise 24 can be found in [100, (8), page 835]. Exercise 28 is the so-called *Radström cancellation law*. Exercise 29 was partially suggested to us in a referee report for [102]. Exercise 32 extends Lemma 3.1 in [118] to infinite dimensions.

Chapter 4

Fundamental topics in convex analysis

This chapter accounts for the most relevant developments of convex analysis in relation to the contents of this book. Specifically, we emphasize the role played by the concept of ε-subgradients of a convex function. Here, X is a locally convex space (lcs) and X^* is its topological dual space. Unless otherwise stated, we assume that X^* (as well as any other involved dual lcs) is endowed with a compatible topology, in particular, the topologies $\sigma(X^*, X)$ and $\tau(X^*, X)$, or the dual norm topology when X is a reflexive Banach space. The associated bilinear form is represented by $\langle \cdot, \cdot \rangle$.

4.1 Subdifferential theory

Given a convex function $f : X \to \overline{\mathbb{R}}$, a point $x \in X$ where f is finite, and a direction $u \in X$, we have shown in section 2.2 that the function $s \mapsto s^{-1}(f(x+su) - f(x))$ is non-decreasing on $\mathbb{R}_+^*$, so the directional derivative of f at x in the direction u turns out to be (see (2.30) and (2.39))

$$f'(x;u) = \lim_{s\downarrow 0} \frac{f(x+su) - f(x)}{s} = \inf_{s>0} \frac{f(x+su) - f(x)}{s} \in \overline{\mathbb{R}}.$$

It also makes sense to define, for each $\varepsilon \geq 0$, the *ε-directional derivative* of f at x in the direction $u \in X$ by

$$f'_\varepsilon(x;u) := \inf_{s>0} \frac{f(x+su) - f(x) + \varepsilon}{s},$$

with $f'_0(x;u) \equiv f'(x;u)$. Note that, for all $\varepsilon \geq 0$,

$$\operatorname{dom} f'_\varepsilon(x,\cdot) = \mathbb{R}_+ \left(\operatorname{dom} f - x\right). \tag{4.1}$$

Furthermore, if we consider the function $h_\varepsilon : X \to \overline{\mathbb{R}}$ defined by

$$h_\varepsilon := f'_\varepsilon(x;\cdot), \tag{4.2}$$

then its Fenchel conjugate is expressed as (see Exercise 37)

$$(h_\varepsilon)^*(x^*) = \sup_{s>0} \frac{f(x) + f^*(x^*) - \langle x, x^* \rangle - \varepsilon}{s}. \tag{4.3}$$

In other words, $(h_\varepsilon)^*$ is nothing else but the indicator function of the (possibly empty) $(w^*\text{-})$ closed convex subset of X^* given by

$$\partial_\varepsilon f(x) := \{x^* \in X^* : f(x) + f^*(x^*) \leq \langle x, x^* \rangle + \varepsilon\}. \tag{4.4}$$

Using the definition of f^*, the last set is equivalently written as

$$\partial_\varepsilon f(x) := \{x^* \in X^* : f(y) \geq f(x) + \langle x^*, y - x \rangle - \varepsilon \text{ for all } y \in X\}, \tag{4.5}$$

showing that an element $x^* \in X^*$ belongs to $\partial_\varepsilon f(x)$ if and only if the continuous affine function

$$y \mapsto f(x) + \langle x^*, y - x \rangle - \varepsilon$$

is below f and differs only by ε from the value of f at x. In particular, when $\varepsilon = 0$, such a continuous affine function coincides with f at x. Therefore, (4.3) is written as

$$(h_\varepsilon)^* = I_{\partial_\varepsilon f(x)}, \tag{4.6}$$

and we can also define $\partial_\varepsilon f(x)$ as

4.1. SUBDIFFERENTIAL THEORY

$$\partial_\varepsilon f(x) := \{x^* \in X^* : \langle x^*, u \rangle \leq h_\varepsilon(u) \text{ for all } u \in X\}. \quad (4.7)$$

It is clear that the operator $\partial_\varepsilon f : X \to 2^{X^*}$ satisfies the ε-*Fermat rule*, stating that

$$\theta \in \partial_\varepsilon f(x) \Leftrightarrow x \in \varepsilon\text{- argmin } f, \quad (4.8)$$

where

$$\varepsilon\text{- argmin } f := \left\{x \in X : f(x) \leq \inf_X f + \varepsilon\right\}$$

is the set of ε-*minima* of the function f. Furthermore, as long as $\partial_\varepsilon f(x) \neq \emptyset$, we can verify that (Exercise 38)

$$[\partial_\varepsilon f(x)]_\infty = \mathrm{N}_{\mathrm{dom}\,f}(x). \quad (4.9)$$

In addition, under the assumption $\partial_\varepsilon f(x) \neq \emptyset$, the function f admits a continuous affine minorant and thus, thanks to the convexity of h_ε, (4.6) together with Theorem 3.2.2 gives rise to

$$\mathrm{cl}\, h_\varepsilon = (\mathrm{I}_{\partial_\varepsilon f(x)})^* = \sigma_{\partial_\varepsilon f(x)} \quad (4.10)$$

(a more precise relationship will be given in (4.28), showing that the closure in $\mathrm{cl}\, h_\varepsilon$ can be removed when $f \in \Gamma_0(X)$ and $\varepsilon > 0$). Therefore, in particular, $\mathrm{cl}\, h_\varepsilon$ is a sublinear function. More precisely, as we quote in the following proposition for later references, the function h_ε itself is sublinear. The positive homogeneity of h_ε easily follows from its definition, whereas the subadditivity comes from the convexity of f. In particular, h_ε is a convex function. This useful behavior of h_ε parallels the role of the classical differential in providing linear approximations for differentiable functions, while in our convex framework the desired approximations are furnished by sublinear functions.

Proposition 4.1.1 *Given a convex function* $f : X \to \mathbb{R}_\infty$ *and* $x \in \mathrm{dom}\, f$, *for every* $\varepsilon \geq 0$ *the mapping* $u \in X \mapsto f'_\varepsilon(x; u)$ *is sublinear.*

We extend the notation of $\partial_\varepsilon f(x)$ to the whole space X and to the negative values of the parameters ε, stating

$$\partial_\varepsilon f(x) := \emptyset \text{ when } f(x) \notin \mathbb{R} \text{ or } \varepsilon < 0. \quad (4.11)$$

Definition 4.1.2 *Given a function* $f : X \to \overline{\mathbb{R}}$ *and* $\varepsilon \in \mathbb{R}$, *the set* $\partial_\varepsilon f(x)$ *is called* ε-*subdifferential of* f *at* $x \in X$, *while the elements of* $\partial_\varepsilon f(x)$ *are called* ε-*subgradients of* f *at* x.

In particular, the set $\partial f(x) \equiv \partial_0 f(x)$ is called (exact) *subdifferential* of f at x, and its elements are the *subgradients* of f at x. We also call $\partial_\varepsilon f(x)$ *approximate subdifferential* when it is not necessary to specify the value of ε. The function f is said to be ε-*subdifferentiable* at x when $\partial_\varepsilon f(x)$ is not empty. Subdifferentiability is defined in a similar way.

It is clear that, for every set $C \subset X$, $\varepsilon \geq 0$ and $x \in C$, we have

$$\partial_\varepsilon I_C(x) = \mathrm{N}_C^\varepsilon(x),$$

where

$$\mathrm{N}_C^\varepsilon(x) = \{x^* \in X^* : \langle x^*, y - x \rangle \leq \varepsilon \text{ for all } y \in C\}$$

is the ε-normal set of C at $x \in C$ (see (2.24)). Furthermore, from the definition of $\partial_\varepsilon f$ we get

$$x^* \in \partial_\varepsilon f(x) \Leftrightarrow (x^*, -1) \in \mathrm{N}_{\mathrm{epi}\, f}^\varepsilon(x, f(x)).$$

We will use the following comparison result for approximate subdifferentials. It shows that a function and its closure have the same approximate subdifferential up to an appropriate adjustment of the parameter ε.

Remark 1 *For every pair of functions $f, g : X \to \overline{\mathbb{R}}$ such that $g \leq f$, $x \in \mathrm{dom}\, f \cap \mathrm{dom}\, g$, and $\varepsilon \in \mathbb{R}$, we have $\partial_{\varepsilon + g(x)} g(x) \subset \partial_{\varepsilon + f(x)} f(x)$. Applying this when $g = \bar{f}$, we obtain the equality*

$$\partial_{\varepsilon + \bar{f}(x)} \bar{f}(x) = \partial_{\varepsilon + f(x)} f(x).$$

The following example is related to Proposition 2.2.5 (and (2.40)), and provides estimates for the subdifferential of convex functions defined on $\mathbb{R}$ via the right and left derivatives.

Example 4.1.3 *Given a proper convex function $f : \mathbb{R} \to \mathbb{R}_\infty$, the right and left derivatives of f at $t \in \mathrm{dom}\, f$, $f'_+(t)$, $f'_-(t)$, exist in $\overline{\mathbb{R}}$ (and belong to $\mathbb{R}$ when $t \in \mathrm{int}(\mathrm{dom}\, f)$). Furthermore, for all $t \in \mathrm{dom}\, f$, we have $f'_-(t) \leq f'_+(t)$ and*

$$\partial f(t) = [f'_-(t), f'_+(t)] \cap \mathbb{R}.$$

Indeed, the existence of the derivatives in question stems from the fact that

4.1. SUBDIFFERENTIAL THEORY

$$f'_+(t) = f'(t;1) = \inf_{s>t} \frac{f(s)-f(t)}{s-t} = \lim_{s\downarrow t} \frac{f(s)-f(t)}{s-t} \in \overline{\mathbb{R}}$$

and

$$f'_-(t) = -f'(t;-1) = \sup_{s<t} \frac{f(s)-f(t)}{s-t} = \lim_{s\uparrow t} \frac{f(s)-f(t)}{s-t} \in \overline{\mathbb{R}}.$$

Combining Proposition 2.2.6 and Corollary 2.2.9, we deduce that $f'_-(t), f'_+(t) \in \mathbb{R}$ when $t \in \operatorname{int}(\operatorname{dom} f)$. Now, taking into account Proposition 4.1.1, for every $q \in [f'_-(t), f'_+(t)] \cap \mathbb{R}$ one has

$$q(s-t) \leq (s-t)f'_+(t) = (s-t)f'(t;1)$$
$$= f'(t;s-t) \leq f(s) - f(t) \quad \text{for all } s \geq t,$$

and

$$q(s-t) \leq (s-t)f'_-(t) = -f'(t;-1)(s-t)$$
$$= f'(t;s-t) \leq f(s) - f(t) \quad \text{for all } t > s,$$

showing that $q \in \partial f(t)$. Conversely, if $q \in \mathbb{R}$ is such that $q(s-t) \leq f(s) - f(t)$ for all $s, t \in \mathbb{R}$, then we get

$$f'_-(t) = \lim_{s\uparrow t} \frac{f(s)-f(t)}{s-t} \leq q \leq \lim_{s\downarrow t} \frac{f(s)-f(t)}{s-t} = f'_+(t).$$

Another feature of the ε-subdifferential that can be easily verified is that it is ε-*cyclically monotone*; that is,

$$\sum_{i=0}^{n} \langle x_i^*, x_{i+1} - x_i \rangle + \langle x_{n+1}^*, x_0 - x_{n+1} \rangle \leq \varepsilon, \tag{4.12}$$

for all $x_i \in X$, $x_i^* \in \partial_{\varepsilon_i} f(x_i)$, $\varepsilon_i \geq 0$, $i = 0, 1, \ldots, n+1$, $n \geq 1$, such that $\sum_{i=0}^{n+1} \varepsilon_i \leq \varepsilon$. In particular, the subdifferential is monotone and cyclically monotone (see (2.28) and (2.29), respectively, for the definition of these concepts).

The first result of this section establishes a useful fact relating the ε-subdifferentiability of a general convex function and the lower semicontinuity of its ε-directional derivative, which is a sublinear function by Proposition 4.1.1.

Proposition 4.1.4 *Given a convex function $f : X \to \overline{\mathbb{R}}$, $x \in f^{-1}(\mathbb{R})$ and $\varepsilon \geq 0$, the following assertions are equivalent:*
  (i) f is ε-subdifferentiable at x.
  (ii) $f'_\varepsilon(x; \cdot)$ is ε-subdifferentiable at θ.
  (iii) $f'_\varepsilon(x; \cdot)$ is lsc at θ.

Proof. Remember that $h = f'_\varepsilon(x; \cdot)$, so $h(\theta) = 0$. If $f : X \to \overline{\mathbb{R}}$ is ε-subdifferentiable at x, then for each $x_0^* \in \partial_\varepsilon f(x)$ relation (4.7) yields

$$\langle x_0^*, u \rangle \leq f'_\varepsilon(x; u) = h(u) - h(\theta) \text{ for all } u \in X.$$

This shows that $x_0^* \in \partial h(\theta) = \partial_\varepsilon h(\theta)$ (since h is sublinear, by Proposition 4.1.1), and $(i) \Rightarrow (ii)$.

If $x_0^* \in \partial_\varepsilon h(\theta) = \partial h(\theta)$, then the convexity of h entails $(\operatorname{cl} h)(\theta) = h(\theta)$; that is, h is lsc at θ (Exercise 62). Therefore, $(ii) \Rightarrow (iii)$.

To show that $(iii) \Rightarrow (i)$, we assume that h is lsc at θ. Since $\operatorname{cl} h$ is convex and $(\operatorname{cl} h)(\theta) = h(\theta) = 0$, the function $\operatorname{cl} h$ is proper. Furthermore, by Proposition 3.1.4, there are $x_0^* \in X^*$ and $\alpha_0 \in \mathbb{R}$ such that

$$h(\lambda u) \geq (\operatorname{cl} h)(\lambda u) \geq \langle x_0^*, \lambda u \rangle + \alpha_0 \text{ for all } u \in X \text{ and } \lambda > 0.$$

But h is sublinear, then $h(u) \geq \langle x_0^*, u \rangle$ for all $u \in X$; that is, $x_0^* \in \partial_\varepsilon f(\theta)$ by (4.7). ∎

A notable difference between the exact and approximate subdifferentials of a convex function is the fact that the first one is a local notion; that is, if two convex functions f and g coincide in a neighborhood of x, then $\partial f(x) = \partial g(x)$, but it may happen that $\partial_\varepsilon f(x) \neq \partial_\varepsilon g(x)$ for all $\varepsilon > 0$. This can be seen in the following example.

Example 4.1.5 *Consider the convex functions $f, g_k : \mathbb{R} \to \mathbb{R}_\infty$, $k \geq 1$, defined by*

$$f(x) := x^2 \text{ and } g_k(x) := x^2 + I_{[-1/k, 1/k]}(x).$$

So, f and g_k coincide locally at 0 for all $k \geq 1$. At the same time, we verify that $\partial_\varepsilon f(0) \subsetneq \partial_\varepsilon g_k(0)$ for every $\varepsilon > 0$ and all $k > 2(2-\sqrt{2})/\sqrt{\varepsilon}$ (see Exercise 42).

It is also important to study how the ε-subdifferential behaves with respect to operations with functions. The following proposition establishes several elementary rules that follow from the definition of ε-subdifferential.

4.1. SUBDIFFERENTIAL THEORY

Proposition 4.1.6 *The following rules are valid, for any function $f : X \to \overline{\mathbb{R}}$ and every $x \in X$:*

(i) $\partial_\varepsilon(f + x_0^* + \alpha)(x) = \partial_\varepsilon f(x) + x_0^*$ for all $x_0^* \in X^*$, $\alpha \in \mathbb{R}$ and $\varepsilon \geq 0$.

(ii) $\lambda \partial_\varepsilon f(x) = \partial_{\lambda\varepsilon}(\lambda f)(x)$ for all $\lambda > 0$ and $\varepsilon \geq 0$.

(iii) If $g : X \to \overline{\mathbb{R}}$ is another convex function, then $\partial_{\varepsilon_1} f(x) + \partial_{\varepsilon_2} g(x) \subset \partial_{\varepsilon_1+\varepsilon_2}(f+g)(x)$ for all $\varepsilon_1, \varepsilon_2 \geq 0$.

(iv) $\partial_{\varepsilon_1} f(x) \subset \partial_{\varepsilon_2} f(x)$ for all $0 \leq \varepsilon_1 \leq \varepsilon_2$.

(v) $x^* \in \partial_\varepsilon f(x)$ if and only if $f(x) + f^*(x^*) \leq \langle x^*, x \rangle + \varepsilon$, for all $\varepsilon \geq 0$. When $\varepsilon = 0$ the last inequality becomes an equality.

(vi) $x^* \in \partial f(x)$ whenever there exist nets $x_i^* \to^{w^*} x^*$ and $\varepsilon_i \downarrow 0$ such that $x_i^* \in \partial_{\varepsilon_i} f(x)$ for all i.

(vii) $\partial_\varepsilon f(x) \times \{-1\} = \mathrm{N}^\varepsilon_{\mathrm{epi}\, f}(x, f(x)) \cap (X^* \times \{-1\})$ for every $\varepsilon \geq 0$.

(viii) If $A : Y \to X$ is a linear mapping defined on another lcs Y with adjoint A^*, then $A^* \partial_\varepsilon f(Ay) \subset \partial_\varepsilon (f \circ A)(y)$ for all $y \in Y$ and $\varepsilon \geq 0$.

(ix) If $\varepsilon \geq 0$, f is lsc at x, and the nets $(x_i)_i \subset X$, $(x_i^*)_i \subset X^*$ are such that $x_i \to x$, $x_i^* \to^{w^*} x^* \in X^*$, $(x_i^*)_i \subset U^\circ$ for some $U \in \mathcal{N}_X$, and $x_i^* \in \partial_\varepsilon f(x_i)$ for all i, then we have $x^* \in \partial_\varepsilon f(x)$.

To illustrate statement (v) in Proposition 4.1.6, we consider a nonempty closed convex set $C \subset X^*$. Since $\sigma_C^* = \mathrm{I}_C$, by (3.17), we obtain

$$\partial \sigma_C(\theta) = \{x^* \in X^* : \sigma_C(\theta) + \mathrm{I}_C(x^*) = 0\} = C. \quad (4.13)$$

In particular, if $(X, \|\cdot\|)$ is a normed space and B_{X^*} is the closed unit dual ball, then the norm coincides with $\sigma_{B_{X^*}}$ by (3.22). Thus, (4.13) entails

$$\partial \|\cdot\|(\theta) = \partial \sigma_{B_{X^*}}(\theta) = B_{X^*}. \quad (4.14)$$

We can also add to the above list the following properties that are easily verified. In fact, for every $x \in f^{-1}(\mathbb{R})$ and $\varepsilon \geq 0$, we have that

$$f'_\varepsilon(x; \cdot) = \inf_{\delta > \varepsilon} f'_\delta(x; \cdot)$$

and

$$\partial_\varepsilon f(x) = \bigcap_{\delta > \varepsilon} \partial_\delta f(x). \quad (4.15)$$

In addition, we can verify that

$$\partial f(x) = \bigcap_{L \in \mathcal{F}(x)} \partial(f + I_L)(x). \tag{4.16}$$

Regarding the relationship between the ε-subdifferential of a (possibly non-convex) function and that of its conjugate, we have

$$(\partial_\varepsilon f)^{-1} \subset \partial_\varepsilon f^*, \tag{4.17}$$

for every function $f : X \to \overline{\mathbb{R}}$, $x \in X$ and $\varepsilon \geq 0$. More precisely, when $f \in \Gamma_0(X)$, Theorem 3.2.2 gives rise to

$$(\partial_\varepsilon f)^{-1} = \partial_\varepsilon f^*. \tag{4.18}$$

The following result gives some geometrical insight to the ε-subdifferential concept.

Proposition 4.1.7 *Let $f : X \to \mathbb{R}_\infty$ be a convex function, which is finite and continuous at $x \in X$. Then, for every $\varepsilon \geq 0$, there exists some $U \in \mathcal{N}_X$ such that, associated with every $V \in \mathcal{N}_{X^*}$, there is $\lambda_V > 0$ satisfying*

$$\partial_\varepsilon f(x + y) \subset \lambda_V V \quad \text{for all } y \in U. \tag{4.19}$$

Consequently, the same neighborhood U satisfies
(i)
$$\partial_\varepsilon f(x + y) \subset U^\circ \quad \text{for all } y \in U. \tag{4.20}$$

(ii) $\partial_\varepsilon f(x + y)$ is w^*-compact for all $y \in U$.
(iii) For every w^*-open set $W \subset X^*$ such that $\partial_\varepsilon f(x) \subset W$, there exists $U_0 \in \mathcal{N}_X$ such that

$$\partial_\varepsilon f(x + y) \subset W \quad \text{for all } y \in U_0.$$

Proof. Fix $\varepsilon \geq 0$ and let $U \in \mathcal{N}_X$ such that, using Proposition 2.2.6,

$$|f(y) - f(z)| \leq 1 \text{ for all } y, z \in x + 2(1 + \varepsilon)U.$$

Then, for every $y \in x + (1 + \varepsilon)U$, $y^* \in \partial_\varepsilon f(y)$ and $u \in (1 + \varepsilon)U$, we have $u + y \in x + 2(1 + \varepsilon)U$ and, so,

$$\langle y^*, u \rangle = \langle y^*, (u + y) - y \rangle \leq f(u + y) - f(y) + \varepsilon \leq 1 + \varepsilon;$$

that is, $y^* \in U^\circ$. Take $V \in \mathcal{N}_{X^*}$ and let $V_0 \subset V$ be an open neighborhood of θ, so that $X^* = \cup_{\lambda > 0}(\lambda V_0)$ (since V_0 is absorbing). Given that

4.1. SUBDIFFERENTIAL THEORY

U° is w^*-compact by Theorem 2.1.9, from the inclusion $U^\circ \subset X^* = \cup_{\lambda>0}(\lambda V_0)$ we find $\lambda_1, \ldots, \lambda_k > 0$, $k \geq 0$, such that

$$U^\circ \subset \bigcup_{1 \leq i \leq k} (\lambda_i V_0) \subset \bigcup_{1 \leq i \leq k} (\lambda_i V) \subset \lambda_V V,$$

where $\lambda_V := \max_{1 \leq i \leq k} \lambda_i > 0$. Therefore, $\partial_\varepsilon f(x+y) \subset U^\circ \subset \lambda_V V$ for all $y \in U$ ($\subset (1+\varepsilon)\bar{U}$), showing that (4.19) and (4.20) hold. In particular, since $\partial_\varepsilon f(y)$ is w^*-closed and U° is w^*-compact, (4.20) implies that $\partial_\varepsilon f(x+y)$ is w^*-compact for all $y \in U$; that is, (ii) follows.

Assertion (iii) remains to be verified. Fix a w^*-open set $W \subset X^*$ such that $\partial_\varepsilon f(x) \subset W$, and let $U \in \mathcal{N}_X$ be as in (4.20). Then, proceeding by contradiction, we assume the existence of nets $(x_i)_i \subset x+U$ and $(x_i^*)_i \subset X^*$ such that $x_i \to x$ and $x_i^* \in \partial_\varepsilon f(x_i) \setminus W$ for all i. Then $(x_i^*)_i \subset U^\circ$ and Theorem 2.1.9 entails, without loss of generality that, $x_i^* \to^{w^*} x^*$ for some $x^* \in X^*$. Thus, by Proposition 4.1.6(ix) we deduce that $x^* \in \partial_\varepsilon f(x) \setminus W_0$, and this yields a contradiction. ∎

The following known result deals with the Fréchet and Gâteaux-differentiability of a convex function f, defined on a Banach space X. The characterizations here are given in terms of continuous *selections* of the subdifferential mapping ∂f; that is, mappings s $: X \to X^*$ with the property s$(x) \in \partial f(x)$ for all $x \in X$. We say that a selection s is $(norm, w^*)$-continuous at $x \in X$, if s$(x_i) \to$ s(x) in the w^*-topology provided that $(x_i)_i \subset X$ is a net such that $x_i \to x$ in the norm topology. The $(norm, norm)$-continuity is defined in a similar way by replacing the w^*-topology with the *norm*-topology in X^*.

Proposition 4.1.8 *Assume that X is a Banach space. For every convex function $f : X \to \mathbb{R}$ that is continuous at $x \in X$, the following assertions hold:*

(i) The function f is Gâteaux-differentiable at x if and only if there exists a selection of ∂f which is $(norm, w^*)$-continuous at x.

(ii) The function f is Fréchet-differentiable at x if and only if every selection of ∂f is $(norm, norm)$-continuous at x.

Relation (4.15) produces outer estimates of $\partial_\varepsilon f$ by means of approximate subdifferentials with larger parameters. Alternatively, the following result, which is used later on, gives inner estimates for $\partial_\varepsilon f(x)$ by using approximate subdifferentials with smaller parameters.

Proposition 4.1.9 *Given a function $f : X \to \mathbb{R}_\infty$, $x \in \text{dom } f$, and $\varepsilon > 0$, we have*

$$\partial_\varepsilon f(x) = \mathrm{cl}\left(\bigcup_{0<\delta<\varepsilon} \partial_\delta f(x)\right), \qquad (4.21)$$

as long as the last set is not empty.

Proof. By the current assumption, there exists $\gamma \in {]}0,\varepsilon[$ such that $\partial_\gamma f(x) \neq \emptyset$. If we take $x_\gamma^* \in \partial_\gamma f(x)$, then the function $g: X \to \mathbb{R}_\infty$, defined by

$$g(\cdot) := f(\cdot) - f(x) - \langle x_\gamma^*, \cdot - x \rangle + \gamma,$$

is nonnegative and, due to Proposition 4.1.6(*i*), satisfies

$$\partial_\mu g(x) = \partial_\mu f(x) - x_\gamma^* \text{ for all } \mu \geq 0. \qquad (4.22)$$

We choose $x^* \in \partial_\varepsilon f(x)$ so that $x^* - x_\gamma^* \in \partial_\varepsilon g(x)$. Thus, by Proposition 4.1.6(*ii*), for every fixed $\alpha \in {]}0,1[$ we obtain

$$\alpha(x^* - x_\gamma^*) \in \alpha \partial_\varepsilon g(x) = \partial_{\alpha\varepsilon}(\alpha g)(x). \qquad (4.23)$$

Therefore, since $\alpha g \leq g$ and $g(x) = \gamma$, for every $z^* \in \partial_{\alpha\varepsilon}(\alpha g)(x)$ and $z \in X$, we have

$$\langle z^*, z - x \rangle \leq (\alpha g)(z) - (\alpha g)(x) + \alpha\varepsilon$$
$$\leq g(z) - \alpha\gamma + \alpha\varepsilon = g(z) - g(x) + (1-\alpha)\gamma + \alpha\varepsilon,$$

showing that $z^* \in \partial_{(1-\alpha)\gamma+\alpha\varepsilon} g(x)$; that is, $\partial_{\alpha\varepsilon}(\alpha g)(x) \subset \partial_{(1-\alpha)\gamma+\alpha\varepsilon} g(x)$. But $(1-\alpha)\gamma + \alpha\varepsilon < \varepsilon$, so (4.22) and (4.23) yield

$$\alpha(x^* - x_\gamma^*) \in \partial_{\alpha\varepsilon}(\alpha g)(x) \subset \partial_{(1-\alpha)\gamma+\alpha\varepsilon} g(x)$$
$$= \partial_{(1-\alpha)\gamma+\alpha\varepsilon} f(x) - x_\gamma^* \subset \left(\bigcup_{0<\delta<\varepsilon} \partial_\delta f(x)\right) - x_\gamma^*,$$

and we infer that $\alpha x^* + (1-\alpha) x_\gamma^* \in \bigcup_{0<\delta<\varepsilon} \partial_\delta f(x)$. Consequently, $x^* \in \mathrm{cl}\left(\bigcup_{0<\delta<\varepsilon} \partial_\delta f(x)\right)$ when $\alpha \uparrow 1$, and the inclusion "$\subset$" in (4.21) follows. The proof is finished because the opposite inclusion in (4.21) is obvious. ∎

The following result, which is an immediate consequence of Theorem 3.2.2, is one of the main important properties of the ε-subdifferential. It asserts that every function from $\Gamma_0(X)$ is ε-subdifferentiable in its effective domain whenever $\varepsilon > 0$. Obviously, this is not the case for the exact subdifferential as shown by the lsc convex function $f: \mathbb{R} \to \mathbb{R}_\infty$ defined by $f := -\sqrt{x} + I_{\mathbb{R}_+}(x)$. In this case, we have

4.1. SUBDIFFERENTIAL THEORY

$$\partial_\varepsilon f(x) =]-\infty, -1/(4\varepsilon)] \text{ for all } \varepsilon > 0,$$

while

$$\partial f(0) = \bigcap_{\varepsilon>0} \partial_\varepsilon f(x) = \bigcap_{\varepsilon>0}]-\infty, -1/(4\varepsilon)] = \emptyset. \qquad (4.24)$$

Proposition 4.1.10 *Every function $f \in \Gamma_0(X)$ is ε-subdifferentiable in dom f for $\varepsilon > 0$.*

Proof. Given a function $f \in \Gamma_0(X)$, we fix $x \in \text{dom } f$ and $\varepsilon > 0$. Then, by Theorem 3.2.2, we have

$$-\infty < f(x) = f^{**}(x) = \sup\{\langle x^*, x\rangle - f^*(x^*) : x^* \in \text{dom } f^*\} < +\infty,$$

and there exists some $x^* \in X^*$ such that

$$f(x) < \langle x^*, x\rangle - f^*(x^*) + \varepsilon; \qquad (4.25)$$

that is, $x^* \in \partial_\varepsilon f(x)$. ∎

Thanks to Proposition 4.1.10, Proposition 4.1.9 takes a simpler form when applied to functions from $\Gamma_0(X)$.

Corollary 4.1.11 *Given $f \in \Gamma_0(X)$, $x \in \text{dom } f$, and $\varepsilon > 0$, we have that*

$$\partial_\varepsilon f(x) = \text{cl}\left(\bigcup_{0<\delta<\varepsilon} \partial_\delta f(x)\right) \qquad (4.26)$$

and, consequently,

$$f'_\varepsilon(x; u) = \sup_{0<\delta<\varepsilon} f'_\delta(x; u) \text{ for all } u \in X. \qquad (4.27)$$

Proof. Since $\partial_\delta f(x) \neq \emptyset$ for all $\delta \in]0, \varepsilon[$, due to Proposition 4.1.10, (4.26) follows by Proposition 4.1.9. The second statement of the corollary is an immediate consequence of (4.26) and the definition of $f'_\varepsilon(x; u)$. ∎

Relation (4.26) is not necessarily true when the convex function f is not lsc at $x \in \text{dom } f$. In fact, in that case we have

$$\partial_\varepsilon f(x) = \emptyset \text{ for } \varepsilon \in [0, f(x) - (\text{cl } f)(x)[.$$

To see this, suppose that $x^*_\varepsilon \in \partial_\varepsilon f(x)$ for some $\varepsilon \geq 0$. Then we get

$$\langle x^*_\varepsilon, y - x\rangle \leq f(y) - f(x) + \varepsilon \text{ for all } y \in X,$$

implying that

$$\langle x_\varepsilon^*, y-x\rangle \leq (\mathrm{cl}\, f)(y) - f(x) + \varepsilon \text{ for all } y \in X.$$

Therefore, taking $y = x$ in this last inequality, we obtain $\varepsilon \geq f(x) - (\mathrm{cl}\, f)(x)$ and, consequently,

$$\mathrm{cl}\left(\bigcup_{0<\delta<f(x)-(\mathrm{cl}\, f)(x)} \partial_\delta f(x)\right) = \emptyset.$$

At the same time, we can have $\partial_{f(x)-(\mathrm{cl}\, f)(x)} f(x) \neq \emptyset$, leading us in such a case to the strict inclusion

$$\mathrm{cl}\left(\bigcup_{0<\delta<f(x)-(\mathrm{cl}\, f)(x)} \partial_\delta f(x)\right) \subset \partial_{f(x)-(\mathrm{cl}\, f)(x)} f(x).$$

The following proposition provides additional information about the relationship between $\sigma_{\partial_\varepsilon f(x)}$ and the ε-directional derivative of f, which further extends and improves (4.10). Next we denote by $q_{x,u} : \mathbb{R}_+ \to \mathbb{R}_\infty$, $x \in \mathrm{dom}\, f$, and $u \in X$, the function defined as

$$q_{x,u}(s) := \begin{cases} s(f(x+s^{-1}u) - f(x)), & \text{if } s > 0, \\ \sigma_{\mathrm{dom}\, f^*}(u), & \text{if } s = 0. \end{cases}$$

Proposition 4.1.12 *Given function $f \in \Gamma_0(X)$, for all $x \in \mathrm{dom}\, f$, $u \in X$, and $\varepsilon > 0$, we have*

$$\sigma_{\partial_\varepsilon f(x)} = f_\varepsilon'(x;\cdot). \tag{4.28}$$

More precisely, one has

$$\sigma_{\partial_\varepsilon f(x)}(u) = \min_{s \geq 0} (q_{x,u}(s) + s\varepsilon) \tag{4.29}$$

and, consequently,

$$\mathrm{epi}\, f_\varepsilon'(x;\cdot) = \{\mathbb{R}_+^* (\mathrm{epi}\, f - (x, f(x) - \varepsilon))\} \cup \mathrm{epi}\, f^\infty. \tag{4.30}$$

Proof. Suppose first that $x = \theta$ and $f(\theta) = 0$; the general case will be examined in the second part of the proof. We introduce the function $g := f^* - \varepsilon$, which belongs to $\Gamma_0(X^*)$, thanks to Proposition 3.1.4. Moreover, using (4.4) and Proposition 4.1.10, we get

4.1. SUBDIFFERENTIAL THEORY

$$[g \leq 0] = [f^* \leq \varepsilon] = \partial_\varepsilon f(\theta) \neq \emptyset. \tag{4.31}$$

In addition, again by Proposition 4.1.10, we have

$$[g < 0] = [f^* < \varepsilon] \supset [f^* \leq \delta] = \partial_\delta f(\theta) \neq \emptyset \text{ for all } \delta < \varepsilon, \tag{4.32}$$

while the Fenchel inequality entails

$$\inf_{X^*} g = \inf_{X^*} (f^* + f(\theta) - \varepsilon) \geq -\varepsilon. \tag{4.33}$$

Let us now verify that

$$(sg)^* = (s * f) + s\varepsilon \text{ for all } s \geq 0. \tag{4.34}$$

Indeed, when $s > 0$, Theorem 3.2.2 yields

$$\begin{aligned}(sg)^*(\cdot) &= (sf^*)^*(\cdot) + s\varepsilon = sf^{**}(s^{-1}\cdot) + s\varepsilon \\ &= sf(s^{-1}\cdot) + s\varepsilon = (s * f)(\cdot) + s\varepsilon.\end{aligned} \tag{4.35}$$

Similarly, when $s = 0$ we get

$$(0g)^* = (0f^*)^* = (I_{\text{dom } f^*})^* = \sigma_{\text{dom } f^*} = 0 * f = 0 * f + 0\varepsilon.$$

Thus, applying Theorem 3.3.4 (since $-\infty < \inf_X g < 0$, by (4.32) and (4.33)), relation (4.34) reads

$$\begin{aligned}\sigma_{[g \leq 0]}(u) &= \inf_{s > 0}(sg)^*(u) = \min_{s \geq 0}(sg)^*(u) \\ &= \min_{s \geq 0}((s * f)(u) + s\varepsilon) = \min_{s \geq 0}(q_{\theta,u}(s) + s\varepsilon),\end{aligned} \tag{4.36}$$

and (4.28) together with (4.29) follows because $[g < 0] = \partial_\varepsilon f(\theta)$ and

$$\inf_{s > 0}(sg)^* = \inf_{s > 0} s\left(f(s^{-1}\cdot) + \varepsilon\right) = f'_\varepsilon(\theta; \cdot).$$

Therefore, using (4.28), (4.31), Corollary 3.3.5, and Theorem 3.2.2, (4.30) follows as

$$\begin{aligned}\text{epi } f'_\varepsilon(\theta; \cdot) &= \text{epi } \sigma_{\partial_\varepsilon f(x)} = \text{epi } \sigma_{[g \leq 0]} = (\mathbb{R}^*_+ \text{ epi } g^*) \cup [\text{epi } g^*]_\infty \\ &= \mathbb{R}^*_+(\text{epi } f - (\theta, -\varepsilon)) \cup [\text{epi } f - (\theta, -\varepsilon)]_\infty \\ &= \mathbb{R}^*_+(\text{epi } f - (\theta, -\varepsilon)) \cup \text{epi } f^\infty,\end{aligned}$$

and the proof is done when $x = \theta$ and $f(\theta) = 0$.

Finally, dealing with any $x \in \operatorname{dom} f$, we consider the function $\varphi := f(\cdot + x) - f(x) \in \Gamma_0(X)$, which satisfies $\varphi(\theta) = 0$. Notice that $\partial_\varepsilon \varphi(\theta) = \partial_\varepsilon f(x)$, $\varphi'_\varepsilon(\theta; \cdot) = f'_\varepsilon(x; \cdot)$ and for all $u \in X$ and $s > 0$

$$(s * \varphi)(u) = s\varphi(s^{-1}u) = s(f(x + s^{-1}u) - f(x)).$$

Furthermore, since $\varphi^* = f^*(\cdot) - \langle \cdot, x \rangle + f(x)$, we also have $(0 * \varphi)(u) = \sigma_{\operatorname{dom} \varphi^*} = \sigma_{\operatorname{dom} f^*}$; that is, $(s * \varphi)(u) = q_{x,u}(s)$ for all $u \in X$. Thus, applying the first part of the proof to the function φ we obtain $\sigma_{\partial_\varepsilon f(x)} = \sigma_{\partial_\varepsilon \varphi(\theta)} = \varphi'_\varepsilon(\theta; \cdot) = f'_\varepsilon(x; \cdot)$ (see (4.36)) and

$$\sigma_{\partial_\varepsilon f(x)}(u) = \sigma_{\partial_\varepsilon \varphi(\theta)} = \min_{s \geq 0}((s * \varphi)(u) + s\varepsilon) = \min_{s \geq 0}(q_{x,u}(s) + s\varepsilon).$$

Moreover, since $\operatorname{epi} \varphi = \operatorname{epi} f - (x, f(x))$ and $(\operatorname{epi} \varphi)_\infty = (\operatorname{epi} f)_\infty$, we also get

$$\operatorname{epi} f'_\varepsilon(x; \cdot) = \operatorname{epi} \varphi'_\varepsilon(\theta; \cdot) = \{\mathbb{R}^*_+(\operatorname{epi} \varphi - (\theta, -\varepsilon))\} \cup \operatorname{epi} \varphi^\infty$$
$$= \{\mathbb{R}^*_+(\operatorname{epi} f - (x, f(x) - \varepsilon))\} \cup \operatorname{epi} f^\infty.$$

∎

Relation (4.28) could also be proved in a more direct way: since $f \in \Gamma_0(X)$, f is ε-subdifferentiable at x by Proposition 4.1.10, and Proposition 4.1.4 implies that $f'_\varepsilon(x; \cdot)$ is lsc at θ. Thus, (4.28) follows from (4.10).

The ε-subdifferential allows us to establish calculus rules without additional continuity assumptions. This is an important feature of the ε-subdifferential, which facilitates the formulation of optimality conditions for the convex optimization problems studied in section 8.2.

Proposition 4.1.13 below provides alternative formulas for the ε-subdifferential of the inf-convolution. Such an operation is frequently used in optimization and variational analysis, especially in regularization processes (e.g., Corollary 5.1.14). The characterizations given here are in terms of ε-subdifferentials of the involved functions, evaluated at almost attaining points; that is, points $x_1, x_2 \in X$ satisfying

$$f_1(x_1) + f_2(x_2) < (f_1 \square f_2)(x) + \alpha \quad \text{for } \alpha > 0 \text{ small.}$$

Formula (4.38) also involves such attaining points, but implicitly, since the non-emptiness of the intersection $\partial_{\varepsilon_1} f_1(x_1) \cap \partial_{\varepsilon_2} f_2(x_2)$ guarantees that x_1 and x_2 are also almost attaining points. It is precisely the existence of such points that supports the success of the

4.1. SUBDIFFERENTIAL THEORY

ε-subdifferential in producing explicit calculus rules. This is in opposition to the exact subdifferential which would require the existence of exact attaining points (equivalently, the exactness of the inf-convolution at x). In fact, this exactness requirement motivates the need for additional qualifying conditions (e.g., Proposition 4.1.20 and subsequent results).

Formula (4.39) is a sharp version of formulas (4.37) and (4.38) when $\varepsilon > 0$, using approximate subdifferentials of the data with parameters whose sum does not exceed ε.

Proposition 4.1.13 *Given the functions $f_1, f_2 : X \to \mathbb{R}_\infty$, for every $x \in (f_1 \square f_2)^{-1}(\mathbb{R})$ and $\varepsilon \geq 0$ we have*

$$\partial_\varepsilon (f_1 \square f_2)(x) = \bigcap_{\alpha > 0} \bigcup_{\substack{x_1 + x_2 = x \\ \varepsilon_1, \varepsilon_2 > 0, \ \varepsilon_1 + \varepsilon_2 = \varepsilon + \alpha \\ f_1(x_1) + f_2(x_2) < (f_1 \square f_2)(x) + \alpha}} \partial_{\varepsilon_1} f_1(x_1) \cap \partial_{\varepsilon_2} f_2(x_2) \tag{4.37}$$

$$= \bigcap_{\alpha > 0} \bigcup_{\substack{x_1 + x_2 = x \\ \varepsilon_1, \varepsilon_2 > 0, \ \varepsilon_1 + \varepsilon_2 = \varepsilon + \alpha}} \partial_{\varepsilon_1} f_1(x_1) \cap \partial_{\varepsilon_2} f_2(x_2). \tag{4.38}$$

Moreover, whenever $\varepsilon > 0$ and

$$\bigcup_{0 < \delta < \varepsilon} \partial_\delta (f_1 \square f_2)(x) \neq \emptyset,$$

we have

$$\partial_\varepsilon (f_1 \square f_2)(x) = \operatorname{cl} \left(\bigcup_{\substack{x_1 + x_2 = x \\ \varepsilon_1, \varepsilon_2 > 0, \ \varepsilon_1 + \varepsilon_2 = \varepsilon}} \partial_{\varepsilon_1} f_1(x_1) \cap \partial_{\varepsilon_2} f_2(x_2) \right). \tag{4.39}$$

Proof. We only need to prove the inclusions "$\subset$" in (4.37) and (4.39), since all the others are straightforward. Given $x^* \in \partial_\varepsilon (f_1 \square f_2)(x)$, we choose $\alpha > 0$ and $y \in X$ such that $f_1(y) + f_2(x - y) < (f_1 \square f_2)(x) + \alpha$. Let us also denote

$$\delta_1 := f_1(y) + f_1^*(x^*) - \langle x^*, y \rangle, \quad \delta_2 := f_2(x - y) + f_2^*(x^*) - \langle x^*, x - y \rangle,$$

so that $\delta_1, \delta_2 \geq 0$ by the Fenchel inequality. Moreover, since $(f_1 \square f_2)^* = f_1^* + f_2^*$ by Proposition 3.1.13, we also have

$$\delta_1 + \delta_2 = f_1(y) + f_2(x - y) + (f_1 \square f_2)^*(x^*) - \langle x^*, x \rangle$$
$$< (f_1 \square f_2)(x) + (f_1 \square f_2)^*(x^*) - \langle x^*, x \rangle + \alpha$$

and there exists $\rho > 0$ such that (as $x^* \in \partial_\varepsilon(f_1 \square f_2)(x)$)

$$\delta_1 + \delta_2 + \rho < (f_1 \square f_2)(x) + (f_1 \square f_2)^*(x^*) - \langle x^*, x \rangle + \alpha \le \varepsilon + \alpha.$$

Take $\varepsilon_1 := \delta_1 + \rho/2$ and $\varepsilon_2' := \delta_2 + \rho/2$, so that $\varepsilon_1, \varepsilon_2' \ge \rho/2 > 0$, $\varepsilon_1 + \varepsilon_2' = \delta_1 + \delta_2 + \rho < \varepsilon + \alpha$ and

$$x^* \in \partial_{\delta_1} f_1(y) \cap \partial_{\delta_2} f_2(x - y) \subset \partial_{\varepsilon_1} f_1(y) \cap \partial_{\varepsilon_2'} f_2(x - y)$$
$$\subset \partial_{\varepsilon_1} f_1(y) \cap \partial_{\varepsilon + \alpha - \varepsilon_1} f_2(x - y),$$

as $\varepsilon_2' < \varepsilon + \alpha - \varepsilon_1$. The desired inclusion in (4.37) follows by taking $\varepsilon_2 := \varepsilon + \alpha - \varepsilon_1$, and then intersecting over $\alpha > 0$.

Suppose now that $\varepsilon > 0$ and $\cup_{0 < \delta < \varepsilon} \partial_\delta(f_1 \square f_2)(x) \ne \emptyset$. Then, by Proposition 4.1.9 and (4.38),

$$\partial_\varepsilon(f_1 \square f_2)(x) = \mathrm{cl}\left(\bigcup_{0 < \delta < \varepsilon} \partial_\delta(f_1 \square f_2)(x) \right)$$
$$= \mathrm{cl}\left(\bigcup_{0 < \delta < \varepsilon} \bigcap_{\alpha > 0} \bigcup_{\substack{x_1 + x_2 = x \\ \varepsilon_1, \varepsilon_2 > 0,\ \varepsilon_1 + \varepsilon_2 = \delta + \alpha}} \partial_{\varepsilon_1} f_1(x_1) \cap \partial_{\varepsilon_2} f_2(x_2) \right)$$

and, taking $\alpha = \varepsilon - \delta$,

$$\partial_\varepsilon(f_1 \square f_2)(x) \subset \mathrm{cl}\left(\bigcup_{\substack{x_1 + x_2 = x \\ \varepsilon_1, \varepsilon_2 > 0,\ \varepsilon_1 + \varepsilon_2 = \varepsilon}} \partial_{\varepsilon_1} f_1(x_1) \cap \partial_{\varepsilon_2} f_2(x_2) \right),$$

which is the non-trivial inclusion in (4.39). ∎

The following result expresses the ε-subdifferential of $\mathrm{cl}\, f$ by means of the ε-subdifferential of f at nearby points.

Proposition 4.1.14 *Let $f : X \to \mathbb{R}_\infty$ be a proper function. Then we have*

$$\partial_\varepsilon(\mathrm{cl}\, f)(x) = \bigcap_{\delta > 0,\ U \in \mathcal{N}_x} \bigcup_{y \in U} \partial_{\varepsilon + \delta} f(x + y) \text{ for all } x \in X \text{ and } \varepsilon \ge 0$$
(4.40)

and, equivalently,

$$(\partial_\varepsilon(\mathrm{cl}\, f))^{-1}(x^*) = \bigcap_{\delta > 0} \mathrm{cl}((\partial_{\varepsilon + \delta} f)^{-1}(x^*)) \text{ for all } x^* \in X^*.$$

Proof. Fix $x \in X$, $\varepsilon \ge 0$ and take $x^* \in \partial_\varepsilon(\mathrm{cl}\, f)(x)$. So, for every $\eta > 0$,

$$f^*(x^*) + (\mathrm{cl}\, f)(x) = (\mathrm{cl}\, f)^*(x^*) + (\mathrm{cl}\, f)(x) \le \langle x^*, x \rangle + \varepsilon + \eta/2.$$
(4.41)

4.1. SUBDIFFERENTIAL THEORY

Then, given $\delta > 0$ and $U \in \mathcal{N}_X$, we choose $U_0 \in \mathcal{N}_X$ such that $U_0 \subset U$, $\langle x^*, x \rangle \le \langle x^*, x+y \rangle + \frac{\delta}{4}$, for all $y \in U_0$, and (by (2.34))

$$\inf_{y \in U_0} f(x+y) < (\operatorname{cl} f)(x) + \frac{\delta}{4}.$$

Therefore, taking $\eta = \delta$ in (4.41), there exists some $y \in U_0 \subset U$ such that

$$f^*(x^*) + f(x+y) < f^*(x^*) + (\operatorname{cl} f)(x) + \frac{\delta}{4}$$
$$\le \langle x^*, x \rangle + \varepsilon + \frac{\delta}{2} + \frac{\delta}{4}$$
$$\le \langle x^*, x+y \rangle + \frac{\delta}{4} + \varepsilon + \frac{\delta}{2} + \frac{\delta}{4} = \langle x^*, x+y \rangle + \varepsilon + \delta;$$

that is, $x^* \in \partial_{\varepsilon+\delta} f(x+y)$ and the proof of the inclusion "$\subset$" in (4.40) is done. The opposite inclusion there is straightforward.

To prove the second statement, we use (4.40) to see that $x \in (\partial_\varepsilon(\operatorname{cl} f))^{-1}(x^*)$ if and only if

$$x^* \in \partial_\varepsilon(\operatorname{cl} f)(x) = \bigcap_{\delta > 0, \ U \in \mathcal{N}_X} \bigcup_{y \in U} \partial_{\varepsilon+\delta} f(x+y);$$

in other words, if and only if for all $\delta > 0$ and $U \in \mathcal{N}_X$ there exists some $y \in U$ such that $x^* \in \partial_{\varepsilon+\delta} f(x-y)$ (recall that U is balanced). Equivalently, $x \in (\partial_{\varepsilon+\delta} f)^{-1}(x^*) + y$, and we deduce that $x \in (\partial_{\varepsilon+\delta} f)^{-1}(x^*) + U$ for all $\delta > 0$ and $U \in \mathcal{N}_X$. Setting $\delta > 0$, we deduce that $x \in \operatorname{cl}((\partial_{\varepsilon+\delta} f)^{-1}(x^*))$ and end up intersecting over $\delta > 0$. ∎

Continuing with the developments of subdifferential calculus, we characterize the ε-subdifferential of the convex and closed convex hulls. Given $z \in X$, $\varepsilon, \delta \ge 0$, $k \ge 1$, and $\lambda \in \Delta_k$, we consider the set in $X \times \mathbb{R}_+$ defined as

$$E(z, \varepsilon, \delta, k, \lambda) := \left\{ (x_i, \varepsilon_i)_{1 \le i \le k} : \begin{array}{l} \sum_{1 \le i \le k} \lambda_i x_i = z, \ \sum_{1 \le i \le k} \lambda_i \varepsilon_i \le \varepsilon + \delta \\ \text{and } \sum_{1 \le i \le k} \lambda_i f(x_i) \le (\operatorname{co} f)(z) + \delta \end{array} \right\}.$$

Proposition 4.1.15 *Given a proper function $f : X \to \mathbb{R}_\infty$, for every $x \in \operatorname{dom}(\operatorname{co} f)$ and $\varepsilon \ge 0$ we have*

$$\partial_\varepsilon(\operatorname{co} f)(x) = \bigcap_{\delta > 0} \bigcup_{\substack{(x_i, \varepsilon_i)_{1 \le i \le k} \in E(x, \varepsilon, \delta, k, \lambda) \\ \lambda \in \Delta_k, \ k \ge 1}} \bigcap_{1 \le i \le k} \partial_{\varepsilon_i} f(x_i), \qquad (4.42)$$

and, when co f is proper,

$$\partial_\varepsilon(\overline{\operatorname{co}}f)(x) = \bigcap_{\substack{\delta>0 \\ U\in\mathcal{N}_x}} \bigcup_{\substack{(x_i,\varepsilon_i)_{1\leq i\leq k}\in E(x+y,\varepsilon,\delta,k,\lambda) \\ y\in U,\ \lambda\in\Delta_k,\ k\geq 1}} \bigcap_{1\leq i\leq k} \partial_{\varepsilon_i}f(x_i). \qquad (4.43)$$

Proof. Take $x^* \in \partial_\varepsilon(\operatorname{co}f)(x)$ and $\delta > 0$, so that

$$(\operatorname{co}f)(x) + (\operatorname{co}f)^*(x^*) = (\operatorname{co}f)(x) + f^*(x^*) \leq \langle x^*, x\rangle + \varepsilon < \langle x^*, x\rangle + \varepsilon + \delta/2.$$

Then there are $k \in \mathbb{N}$, $x_i \in X$, $1 \leq i \leq k$, and $\lambda \in \Delta_k$ such that

$$\sum_{1\leq i\leq k} \lambda_i x_i = x, \quad \sum_{1\leq i\leq k} \lambda_i f(x_i) \leq (\operatorname{co}f)(x) + \delta/2 < (\operatorname{co}f)(x) + \delta,$$

and

$$\sum_{1\leq i\leq k} \lambda_i f(x_i) + f^*(x^*) \leq (\operatorname{co}f)(x) + f^*(x^*) + \delta/2 < \langle x^*, x\rangle + \varepsilon + \delta.$$

Therefore, denoting $\varepsilon_i := f(x_i) + f^*(x^*) - \langle x^*, x_i\rangle$, we have $\varepsilon_i \geq 0$, $x^* \in \partial_{\varepsilon_i}f(x_i)$ and

$$\sum_{1\leq i\leq k} \lambda_i \varepsilon_i = \sum_{1\leq i\leq k} \lambda_i(f(x_i) + f^*(x^*) - \langle x^*, x_i\rangle)$$
$$= \sum_{1\leq i\leq k} \lambda_i f(x_i) + f^*(x^*) - \langle x^*, x\rangle < \varepsilon + \delta,$$

which show that

$$x^* \in \bigcap_{1\leq i\leq k} \partial_{\varepsilon_i}f(x_i) \subset \bigcup_{\substack{(x_i,\varepsilon_i)_{1\leq i\leq k}\in E(x,\varepsilon,\delta,k,\lambda) \\ \lambda\in\Delta_k,\ k\geq 1}} \bigcap_{1\leq i\leq k} \partial_{\varepsilon_i}f(x_i).$$

The inclusion "⊂" in (4.42) follows by intersecting over $\delta > 0$, and the opposite inclusion there is easily checked. Finally, (4.43) is satisfied by combining (4.42) and Proposition 4.1.14. ∎

We provide below a calculus rule for the approximate subdifferential and the conjugate of the sum and the composition of convex functions with continuous linear mappings. Everything is done without additional continuity assumptions on the involved functions. In particular, formula (4.44) transforms the operation of composition into a post-composition of the conjugate function and the adjoint of the given linear mapping.

4.1. SUBDIFFERENTIAL THEORY

Proposition 4.1.16 *Let Y be another lcs, $f \in \Gamma_0(X)$, $g \in \Gamma_0(Y)$, and let $A : X \to Y$ be a continuous linear mapping with continuous adjoint A^*. Then, provided that $\operatorname{dom} f \cap A^{-1}(\operatorname{dom} g) \neq \emptyset$, we have*

$$(f + g \circ A)^* = \operatorname{cl}(f^* \square (A^* g^*)), \tag{4.44}$$

and, for every $x \in X$ and $\varepsilon \geq 0$,

$$\partial_\varepsilon (f + g \circ A)(x) = \bigcap_{\delta > 0} \operatorname{cl} \left(\bigcup_{\substack{\varepsilon_1 + \varepsilon_2 = \varepsilon + \delta \\ \varepsilon_1, \varepsilon_2 \geq 0}} (\partial_{\varepsilon_1} f(x) + A^* \partial_{\varepsilon_2} g(Ax)) \right). \tag{4.45}$$

In particular, when $\varepsilon > 0$ we also have

$$\partial_\varepsilon (f + g \circ A)(x) = \operatorname{cl} \left(\bigcup_{\substack{\varepsilon_1 + \varepsilon_2 = \varepsilon \\ \varepsilon_1, \varepsilon_2 \geq 0}} (\partial_{\varepsilon_1} f(x) + A^* \partial_{\varepsilon_2} g(Ax)) \right). \tag{4.46}$$

Proof. We fix $x_0 \in \operatorname{dom} f \cap A^{-1}(\operatorname{dom} g)$. Given Proposition 3.1.13, the Fenchel inequality and Theorem 3.2.2 imply that, for all $x^* \in X^*$,

$$(f^* \square (A^* g^*))(x^*) + (f + g \circ A)(x_0) = (f^* \square (A^* g^*))(x^*) + (f^{**} + g^{**} \circ A)(x_0)$$
$$= (f^* \square (A^* g^*))(x^*) + (f^* \square (A^* g^*))^*(x_0) \geq \langle x^*, x_0 \rangle;$$

in other words,

$$(f^* \square (A^* g^*))(x^*) \geq \langle x^*, x_0 \rangle - f(x_0) - g(Ax_0) \quad \text{for all } x^* \in X^*,$$

and taking the closed hull we get

$$f^* \square (A^* g^*) \geq \operatorname{cl}(f^* \square (A^* g^*)) \geq \langle \cdot, x_0 \rangle - f(x_0) - g(Ax_0).$$

Furthermore, using Proposition 3.1.4, by (2.56) and (2.61) we get

$$\emptyset \neq \operatorname{dom} f^* + A^*(\operatorname{dom} g^*) = \operatorname{dom}(f^* \square (A^* g^*)) \subset \operatorname{dom}(\operatorname{cl}(f^* \square (A^* g^*))),$$

and $\operatorname{cl}(f^* \square (A^* g^*)) \in \Gamma_0(X)$; that is, due to (3.7) and Theorem 3.2.2,

$$(f^* \square (A^* g^*))^{**} = (\operatorname{cl}(f^* \square (A^* g^*)))^{**} = \operatorname{cl}(f^* \square (A^* g^*)). \tag{4.47}$$

Finally, formula (4.44) follows because Proposition 3.1.13 and Theorem 3.2.2 imply that

$$(f^*\square(A^*g^*))^{**} = ((f^*\square(A^*g^*))^*)^* = (f^{**} + (A^*g^*)^*)^*$$
$$= (f^{**} + g^{**} \circ A^{**})^* = (f + g \circ A)^*,$$

where we use the fact that $A^{**} = A$ (see (3.16)).

To show the non-trivial inclusion "⊂" in (4.45), we fix $x \in X$, $\varepsilon \geq 0$ and take $x^* \in \partial_\varepsilon (f + g \circ A)(x)$; hence, $x \in \mathrm{dom}\, f \cap A^{-1}(\mathrm{dom}\, g)$. Moreover, by (4.44), for every given $\delta > 0$ we have

$$(f + g \circ A)(x) + (\mathrm{cl}(f^*\square(A^*g^*)))(x^*) = (f + g \circ A)(x) + (f + g \circ A)^*(x^*)$$
$$\leq \langle x^*, x \rangle + \varepsilon < \langle x^*, x \rangle + \varepsilon + \delta,$$

and then there exist nets $(x_i^*)_i$, $(y_i^*)_i \subset X^*$, and $(z_i^*)_i \subset Y^*$ such that $x_i^* \to x^*$, $A^* z_i^* = x_i^* - y_i^*$ and, for all i,

$$f^*(y_i^*) + g^*(z_i^*) \leq \langle x_i^*, x \rangle - (f + g \circ A)(x) + \varepsilon + \delta. \tag{4.48}$$

Let us denote

$$\varepsilon_{1,i} := f^*(y_i^*) + f(x) - \langle y_i^*, x \rangle \geq 0, \quad \varepsilon_{2,i} := g^*(z_i^*) + g(Ax) - \langle z_i^*, Ax \rangle \geq 0,$$

so that $y_i^* \in \partial_{\varepsilon_{1,i}} f(x)$, $z_i^* \in \partial_{\varepsilon_{2,i}} g(Ax)$ and, due to (4.48),

$$\varepsilon_{1,i} + \varepsilon_{2,i} = f^*(y_i^*) + g^*(z_i^*) + f(x) + g(Ax) - \langle y_i^* + A^* z_i^*, x \rangle$$
$$= f^*(y_i^*) + g^*(z_i^*) + f(x) + g(Ax) - \langle x_i^*, x \rangle \leq \varepsilon + \delta.$$

Therefore, writing

$$x_i^* = y_i^* + (x_i^* - y_i^*) = y_i^* + A^* z_i^*$$
$$\in \partial_{\varepsilon_{1,i}} f(x) + A^* \partial_{\varepsilon_{2,i}} g(Ax) \subset \partial_{\varepsilon_{1,i}} f(x) + A^* \partial_{\varepsilon + \delta - \varepsilon_{1,i}} g(Ax)$$
$$\subset \bigcup_{\substack{\varepsilon_1 + \varepsilon_2 = \varepsilon + \delta \\ \varepsilon_1, \varepsilon_2 \geq 0}} (\partial_{\varepsilon_1} f(x) + A^* \partial_{\varepsilon_2} g(Ax)),$$

the inclusion "⊂" in (4.45) follows by, successively, passing to the limit on i and intersecting over $\delta > 0$.

To verify the non-trivial inclusion "⊂" in (4.46), we fix $\varepsilon > 0$ and $x \in \mathrm{dom}\, f \cap A^{-1}(\mathrm{dom}\, g)$. So, taking into account Corollary 4.1.11, formula (4.45) results in

4.1. SUBDIFFERENTIAL THEORY

$$\partial_\varepsilon(f+g\circ A)(x) = \mathrm{cl}\left(\bigcup_{0<\alpha<\varepsilon}\partial_\alpha(f+g\circ A)(x)\right)$$

$$= \mathrm{cl}\left(\bigcup_{0<\alpha<\varepsilon}\bigcap_{\delta>0}\mathrm{cl}\left(\bigcup_{\substack{\varepsilon_1+\varepsilon_2=\alpha+\delta\\ \varepsilon_1,\varepsilon_2\geq 0}}(\partial_{\varepsilon_1}f(x)+A^*\partial_{\varepsilon_2}g(Ax))\right)\right).$$

So, taking $\delta = \varepsilon - \alpha$, we get

$$\partial_\varepsilon(f+g\circ A)(x) \subset \mathrm{cl}\left(\bigcup_{\substack{\varepsilon_1+\varepsilon_2=\varepsilon\\ \varepsilon_1,\varepsilon_2\geq 0}}\partial_{\varepsilon_1}f(x)+A^*\partial_{\varepsilon_2}g(Ax)\right),$$

and we are done since the opposite inclusion is straightforward. ∎

As an illustration of Proposition 4.1.16, we give the following corollary which extends Proposition 3.2.8. Now, the given family is not necessarily non-decreasing, but it satisfies a more general property that we will later call closedness for convex combinations in Definition 5.1.3.

Corollary 4.1.17 *Given a family of functions* $\{f_t,\ t\in T\}\subset \Gamma_0(X)$, *we assume that* $f := \sup_{t\in T} f_t$ *is proper and, for all* $\lambda \in \Delta(T)$, *there exists some* $s \in T$ *such that*

$$\sum_{t\in\mathrm{supp}\,\lambda}\lambda_t f_t \leq f_s. \tag{4.49}$$

Then we have $f^* = \mathrm{cl}\left(\inf_{t\in T} f_t^*\right).$

Proof. By Proposition 3.2.6, we have

$$f^* = \overline{\mathrm{co}}\left(\inf_{t\in T} f_t^*\right) \leq \mathrm{cl}\left(\inf_{t\in T} f_t^*\right),$$

so we just need to check that, for each fixed $x^* \in \mathrm{dom}\, f^*$,

$$\mathrm{cl}\left(\inf_{t\in T} f_t^*\right)(x^*) \leq f^*(x^*).$$

Let $\alpha \in \mathbb{R}$ such that $f^*(x^*) = \overline{\mathrm{co}}\left(\inf_{t\in T} f_t^*\right)(x^*) < \alpha$, and take a net $x_i^* \to x^*$ satisfying

$$f^*(x^*) = \lim_i\left(\mathrm{co}\left(\inf_{t\in T} f_t^*\right)\right)(x_i^*).$$

Since f^* is also proper, due to Proposition 3.1.4, we may assume that $(\text{co}\,(\inf_{t \in T} f_t^*))(x_i^*) \in \mathbb{R}$ for all i. Then there are associated $(\lambda_i)_i \subset \Delta(T)$ and $x_{i,t}^* \in X^*$, $t \in \text{supp}\,\lambda_i$, such that $\sum_{t \in T} \lambda_{i,t} x_{i,t}^* = x_i^*$ and

$$f^*(x^*) = \lim_i \sum_{t \in \text{supp}\,\lambda_i} \lambda_{i,t} \left(\inf_{s \in T} f_s^* \right)(x_{i,t}^*) < \alpha.$$

More precisely, we can find associated $f_{s_{i,t}}$, $s_{i,t} \in T$, such that (without loss of generality) $\sum_{t \in \text{supp}\,\lambda_i} \lambda_{i,t} f_{s_{i,t}}^*(x_{i,t}^*) < \alpha$ for all i. Then, for each i, by assumption we choose $s_i \in T$ such that $\sum_{t \in \text{supp}\,\lambda_i} \lambda_{i,t} f_{s_{i,t}} \leq f_{s_i}$. But $\left(\sum_{t \in \text{supp}\,\lambda_i} \lambda_{i,t} f_{s_{i,t}} \right)^*$ is the closed hull of the inf-convolution of the functions $(\lambda_{i,t} f_{s_{i,t}})^*$ (see Proposition 4.1.16), so we get

$$f_{s_i}^*(x_i^*) \leq \left(\sum_{t \in \text{supp}\,\lambda_i} \lambda_{i,t} f_{s_{i,t}} \right)^*(x_i^*)$$
$$\leq \sum_{t \in \text{supp}\,\lambda_i} (\lambda_{i,t} f_{s_{i,t}})^*(\lambda_{i,t} x_{i,t}^*) = \sum_{t \in \text{supp}\,\lambda_i} \lambda_{i,t} f_{s_{i,t}}^*(x_{i,t}^*) < \alpha.$$

In other words,

$$\text{cl}\left(\inf_{t \in T} f_t^* \right)(x^*) \leq \liminf_i \left(\inf_{t \in T} f_t^* \right)(x_i^*) \leq \liminf_i f_{s_i}^*(x_i^*) \leq \alpha,$$

and the desired inequality, $\text{cl}\,(\inf_{t \in T} f_t^*)(x^*) \leq f^*(x^*)$, follows when $\alpha \downarrow f^*(x^*)$. ∎

We give an example of families of functions satisfying the property used in Corollary 4.1.17; in fact, as we show here, any family can be adjusted to satisfy such a property.

Example 4.1.18 *For any family $\{f_t,\, t \in T\} \subset \Gamma_0(X)$ such that $f := \sup_{t \in T} f_t$ is proper, we have*

$$f^* = \text{cl}\left(\inf_{\lambda \in \Delta(T)} f_\lambda^* \right),$$

where $f_\lambda := \sum \lambda_t f_t$. Consequently, the function $\inf_{\lambda \in \Delta(T)} f_\lambda^*$ and its closure are convex and proper.

Indeed, by (2.45), we have

$$f = \sup_{t \in T} f_t = \sup_{\lambda \in \Delta(T)} f_\lambda. \tag{4.50}$$

In addition, for any $\lambda_1, \ldots, \lambda_k \in \Delta(T)$, $k \geq 1$ and $\alpha := (\alpha_1, \ldots, \alpha_k) \in \Delta_k^*$, we have

4.1. SUBDIFFERENTIAL THEORY

$$g := \sum_{1 \le i \le k} \alpha_i f_{\lambda_i} = \sum_{t \in \text{supp } \lambda_i, 1 \le i \le k} (\alpha_i \lambda_{i,t}) f_t = \sum_{t \in T} \left(\sum_{1 \le i \le k} \alpha_i \lambda_{i,t} \right) f_t.$$

Furthermore, it is clear that the element $\tilde{\lambda} \in \mathbb{R}_+^{(T)}$ defined by $\tilde{\lambda}_t := \sum_{1 \le i \le k} \alpha_i \lambda_{i,t}$ satisfies $\sum_{t \in T} \tilde{\lambda}_t = \sum_{1 \le i \le k} \alpha_i = 1$ and $g = f_{\tilde{\lambda}}$; that is, the family $\{f_\lambda, \lambda \in \Delta(T)\}$ satisfies condition (4.49). Thus, we conclude by applying Corollary 4.1.17.

The operations of addition and composition in Proposition 4.1.16 can be put in the general form $\varphi(x) := F(x, \theta)$ that we present in the following corollary. For instance, the operation $f + g \circ A$ corresponds to the function φ for $F : X \times Y \to \mathbb{R}_\infty$ being defined by

$$F(x, y) := f(x) + g(Ax + y). \tag{4.51}$$

Corollary 4.1.19 *Let Y be another lcs, $F \in \Gamma_0(X, Y)$ and let $\varphi : X \to \mathbb{R}_\infty$ be defined as $\varphi(x) := F(x, \theta)$. Then, provided that $\{x \in X : (x, \theta)\} \in \text{dom } F\} \ne \emptyset$, we have*

$$\varphi^* = \text{cl} \left(\inf_{y^* \in Y^*} F^*(\cdot, y^*) \right) \tag{4.52}$$

and, for every $x \in X$ and $\varepsilon > 0$,

$$\partial_\varepsilon \varphi(x) = \text{cl} \{x^* \in X^* : \exists \, y^* \in Y^* \text{ such that } (x^*, y^*) \in \partial_\varepsilon F(x, \theta)\}. \tag{4.53}$$

Proof. We introduce the continuous linear mapping $A : X \to X \times Y$ defined as $Ax = (x, \theta)$, so that $\varphi = F \circ A$. Since $A^{-1}(\text{dom } F) = \{x \in X : (x, \theta)\} \in \text{dom } F\} \ne \emptyset$ and the adjoint mapping of A, $A^* : X^* \times Y^* \to X^*$, is defined by $A^*(x^*, y^*) = x^*$, (4.44) entails $\varphi^* = (F \circ A)^* = \text{cl}(A^* F^*)$. Thus, (4.52) follows because

$$(A^* F^*)(x^*) = \inf\{F^*(u^*, y^*) : A^*(u^*, y^*) = x^*, \ (u^*, y^*) \in X^* \times Y^*\}$$
$$= \inf\{F^*(x^*, y^*) : y^* \in Y^*\}.$$

To verify (4.53), we fix $x \in X$ and $\varepsilon > 0$. Then (4.46) yields

$$\partial_\varepsilon \varphi(x) = \text{cl}(A^* \partial_\varepsilon F(x, \theta))$$
$$= \text{cl}\{x^* \in X^* : \exists \, y^* \in Y^* \text{ such that } (x^*, y^*) \in \partial_\varepsilon F(x, \theta)\},$$

and we are done. ■

The formulas of Proposition 4.1.16 use limiting processes (since they involve closures), so they are called approximate or fuzzy rules. To get more precise characterizations, additional conditions are usually needed as happens in Proposition 4.1.20 below. Given a continuous linear mapping $A : X \to Y$, where Y is another lcs, we consider again the mapping $\hat{A} : X \times \mathbb{R} \to Y \times \mathbb{R}$ defined by (see (2.59))

$$\hat{A}(x, \alpha) := (Ax, \alpha). \tag{4.54}$$

Then we easily verify that the adjoint of $\hat{A}$, $\hat{A}^* : Y^* \times \mathbb{R} \to X^* \times \mathbb{R}$, is given by

$$\hat{A}^*(y^*, \lambda) = (A^*y^*, \lambda). \tag{4.55}$$

Proposition 4.1.20 *Given convex functions $f : X \to \overline{\mathbb{R}}$, $g : Y \to \overline{\mathbb{R}}$, and a linear continuous mapping $A : X \to Y$ with continuous adjoint A^*, we suppose that g is finite and continuous at some point in $A(\operatorname{dom} f)$. Then the following assertions hold true:*
(i)

$$(f + g \circ A)^* = f^* \square (A^* g^*), \tag{4.56}$$

where both the inf-convolution and the post-composition are exact.
(ii)

$$\operatorname{epi}(f + g \circ A)^* = \operatorname{epi} f^* + \hat{A}^*(\operatorname{epi} g^*), \tag{4.57}$$

where $\hat{A}^*$ comes from (4.55).
(iii) *For every $x \in X$ and $\varepsilon \geq 0$,*

$$\partial_\varepsilon (f + g \circ A)(x) = \bigcup_{0 \leq \varepsilon_1, \varepsilon_2,\ \varepsilon_1 + \varepsilon_2 \leq \varepsilon} \left(\partial_{\varepsilon_1} f(x) + A^* \partial_{\varepsilon_2} g(Ax) \right). \tag{4.58}$$

Proof. (i) We may assume that $\theta \in \operatorname{dom} f \cap (A^{-1}(\operatorname{dom} g)) = \operatorname{dom}(f + g \circ A)$ and g is continuous at $\theta \in A(\operatorname{dom} f)$.

Let us proceed with the proof of (4.56), assuming first that $f \in \Gamma_0(X)$ and $g \in \Gamma_0(Y)$. We fix $x^* \in X^*$. If $(f + g \circ A)^*(x^*) = +\infty$, then Proposition 4.1.16 yields

$$+\infty = (f + g \circ A)^*(x^*) = \operatorname{cl}(f^* \square (A^* g^*))(x^*) \leq (f^* \square (A^* g^*))(x^*),$$

and (4.56) obviously holds. Therefore, we can assume that $(f + g \circ A)^*(x^*) < +\infty$. Consequently, since $(f + g \circ A)^* \in \Gamma_0(X^*)$ by Proposition 3.1.4, we again use Proposition 4.1.16 to find a net $(x_i^*) \subset X^*$ that w^*-converges to x^* and satisfies

4.1. SUBDIFFERENTIAL THEORY

$$(f+g\circ A)^*(x^*) = \text{cl}(f^*\square(A^*g^*))(x^*) = \lim_i (f^*\square(A^*g^*))(x_i^*) \in \mathbb{R}. \tag{4.59}$$

Furthermore, we choose nets $(z_i^*) \subset X^*$ and $(y_i^*) \subset Y^*$ such that $A^*y_i^* = x_i^* - z_i^*$ and

$$(f+g\circ A)^*(x^*) = \lim_i (f^*(z_i^*) + g^*(y_i^*)). \tag{4.60}$$

Note that $f^*(z_i^*) \geq -f(\theta) > -\infty$ and $g^*(y_i^*) \geq -g(\theta) > -\infty$. Then (4.60) implies the existence of some $m \geq 0$ such that $g^*(y_i^*) \leq m$ eventually for i. Therefore, by Proposition 3.1.3, we may assume that $(y_i^*)_i$ is w^*-convergent to some $\bar{y}^* \in Y^*$, so $(z_i^*)_i$ is also w^*-convergent to $x^* - A^*\bar{y}^*$. Consequently, taking the limits in (4.60), the w^*-lower semicontinuity of the conjugate function and (4.59) imply that

$$\begin{aligned}
f^*(x^* - A^*\bar{y}^*) + g^*(\bar{y}^*) &\leq \liminf_i f^*(z_i^*) + \liminf_i g^*(y_i^*) \\
&\leq \liminf_i (f^*(z_i^*) + g^*(y_i^*)) \\
&= (f+g\circ A)^*(x^*) = \text{cl}(f^*\square(A^*g^*))(x^*) \\
&\leq (f^*\square(A^*g^*))(x^*) \leq f^*(x^* - A^*\bar{y}^*) + g^*(\bar{y}^*).
\end{aligned}$$

In other words, (4.56) follows under the current assumption that $f \in \Gamma_0(X)$ and $g \in \Gamma_0(Y)$.

Now, we show (4.56) in the general case where f and g are convex functions, possibly not lsc. We observe that $\text{cl}\, g \in \Gamma_0(Y)$ as g is finite and continuous at θ. Furthermore, by Proposition 2.2.11, the current continuity assumption yields

$$\text{cl}(f + g \circ A) = (\text{cl}\, f) + (\text{cl}\, g) \circ A. \tag{4.61}$$

If $\text{cl}\, f \notin \Gamma_0(X)$, then $\text{cl}\, f$ is not proper and (4.61) together with the relation $(\text{cl}\, f)(\theta) = -\infty$ (see (40)) implies that

$$\text{cl}(f + g \circ A)(\theta) = (\text{cl}\, f)(\theta) + (\text{cl}\, g)(A\theta) = (\text{cl}\, f)(\theta) + g(\theta) = -\infty.$$

Consequently,

$$(f+g\circ A)^* = (\text{cl}(f+g\circ A))^* \equiv +\infty \equiv (\text{cl}\, f)^* = f^*$$

and, at the same time, for all $x^* \in X^*$ we have

$$(f^*\square(A^*g^*))(x^*) = \inf\{f^*(x_1^*) + (A^*g^*)(x_2^*) : x_1^* + x_2^* = x^*\} = +\infty.$$

Therefore, $(f+g\circ A)^* = f^*\square(A^*g^*) \equiv +\infty$ and (4.56) also holds with exact inf-convolution and exact post-composition.

We now assume that $\mathrm{cl}\, f \in \Gamma_0(X)$. So, taking into account (4.61) and the fact that $\mathrm{cl}\, g$ is also proper, by the reasoning before we get

$$(f+g\circ A)^* = (\mathrm{cl}(f+g\circ A))^* = ((\mathrm{cl}\, f) + (\mathrm{cl}\, g)\circ A)^*$$
$$= (\mathrm{cl}\, f)^*\square(A^*(\mathrm{cl}\, g)^*) = f^*\square(A^*g^*),$$

and, for each $x^* \in X^*$, there exist $\bar{z}^* \in X^*$ and $\bar{y}^* \in Y^*$ such that $\bar{z}^* + A^*\bar{y}^* = x^*$ and

$$(f+g\circ A)^*(x^*) = ((\mathrm{cl}\, f) + (\mathrm{cl}\, g)\circ A)^*(x^*)$$
$$= (\mathrm{cl}\, f)^*(\bar{z}^*) + (\mathrm{cl}\, g)^*(\bar{y}^*) = f^*(\bar{z}^*) + g^*(\bar{y}^*).$$

Consequently, both the inf-convolution and the post-composition in $f^*\square(A^*g^*)$ are exact, and we conclude (4.56) in all its generality.

(ii) The inclusion "$\supset$" in (4.57) follows as we show next. First, the continuity of $\hat{A}^*$ yields

$$\mathrm{epi}\, f^* + \hat{A}^*(\mathrm{epi}\, g^*) \subset \mathrm{epi}\, f^* + \hat{A}^*(\mathrm{cl}(\mathrm{epi}_s\, g^*)) \subset \mathrm{cl}(\mathrm{epi}\, f^*) + \mathrm{cl}(\hat{A}^*(\mathrm{epi}_s\, g^*)).$$

Second, because of (2.9) and (2.60), we have

$$\mathrm{cl}(\mathrm{epi}\, f^*) + \mathrm{cl}(\hat{A}^*(\mathrm{epi}_s\, g^*)) \subset \mathrm{cl}(\mathrm{epi}\, f^* + \hat{A}^*(\mathrm{epi}_s\, g^*))$$
$$= \mathrm{cl}(\mathrm{epi}\, f^* + \mathrm{epi}_s(A^*g^*)) = \mathrm{cl}(\mathrm{epi}\, f^* + \mathrm{epi}(A^*g^*)).$$

Third, by (2.58) we have $\mathrm{cl}(\mathrm{epi}\, f^* + \mathrm{epi}(A^*g^*)) = \mathrm{epi}(\mathrm{cl}(f^*\square(A^*g^*))$, and so (4.56) produces

$$\mathrm{epi}\, f^* + \hat{A}^*(\mathrm{epi}\, g^*) \subset \mathrm{epi}(\mathrm{cl}(f^*\square(A^*g^*)) = \mathrm{epi}(f+g\circ A)^*. \quad (4.62)$$

To prove the converse inclusion, observe that if $\mathrm{cl}\, f$ is not proper, then $(f+g\circ A)^* \equiv +\infty$ as shown above. So, $\mathrm{epi}(f+g\circ A)^* = \emptyset$ and (4.62) becomes an equality as required. Suppose now that $\mathrm{cl}\, f$ is proper and take $(x^*,\lambda) \in \mathrm{epi}(f+g\circ A)^*$. Then, by (4.56), we find some $y^* \in \mathrm{dom}\, g^*$ such that $f^*(x^* - A^*y^*) + g^*(y^*) = (f+g\circ A)^*(x^*) \leq \lambda$; that is, $g^*(y^*), f^*(x^* - A^*y^*) \in \mathbb{R}$ (due to Proposition 3.1.4). Therefore,

$$(x^*,\lambda) = (x^* - A^*y^*, f^*(x^* - A^*y^*) + (\lambda - f^*(x^* - A^*y^*) - g^*(y^*)))$$
$$+ (A^*y^*, g^*(y^*)) \subset \mathrm{epi}\, f^* + \hat{A}^*(\mathrm{epi}\, g^*),$$

and the desired inclusion follows.

4.1. SUBDIFFERENTIAL THEORY   121

(*iii*) We verify the non-trivial inclusion "$\subset$" in (4.58). Fix $x \in X$, $\varepsilon \geq 0$ and take $x^* \in \partial_\varepsilon (f + g \circ A)(x)$, so that

$$(f + g \circ A)(x) + (f + g \circ A)^*(x^*) \leq \langle x^*, x \rangle + \varepsilon.$$

Thus, (4.56) gives rise to the existence of some $y^* \in \operatorname{dom} g^*$ such that $f(x) + g(Ax) + f^*(x^* - A^*y^*) + g^*(y^*) \leq \langle x^*, x \rangle + \varepsilon$. Let us denote

$$\varepsilon_1 := f(x) + f^*(x^* - A^*y^*) - \langle x^* - A^*y^*, x \rangle, \ \varepsilon_2 := g(Ax) + g^*(y^*) - \langle y^*, Ax \rangle,$$

so that $\varepsilon_1, \varepsilon_2 \geq 0$, $x^* - A^*y^* \in \partial_{\varepsilon_1} f(x)$, $y^* \in \partial_{\varepsilon_2} g(Ax)$ and

$$\varepsilon_1 + \varepsilon_2 = f(x) + g(Ax) + f^*(x^* - A^*y^*) + g^*(y^*) - \langle x^*, x \rangle \leq \varepsilon.$$

Therefore, $x^* = (x^* - A^*y^*) + A^*y^* \in \partial_{\varepsilon_1} f(x) + A^* \partial_{\varepsilon_2} g(Ax)$, and the desired inclusion follows. ∎

We illustrate the previous result by means of a simple example.

Example 4.1.21 *Let X, Y be two normed spaces, and let $A : X \to Y$ be a continuous linear mapping. Consider the function $\varphi(x) := \|Ax\|$, $x \in X$, where $\|\cdot\|$ denotes the norm in Y. Then, for all $x \in X$, we have*

$$\partial \varphi(x) = A^* \partial \|Ax\|. \tag{4.63}$$

In particular, if $Y = \mathbb{R}$ and $A(\cdot) := \langle x_0^*, \cdot \rangle$ for some $x_0^* \in X^*$, then

$$\partial \varphi(x) = \begin{cases} [-1, 1] x_0^*, & \text{if } \langle x_0^*, x \rangle = 0, \\ \operatorname{sign}(\langle x_0^*, x \rangle) x_0^*, & \text{otherwise.} \end{cases} \tag{4.64}$$

In fact, (4.63) follows by applying Proposition 4.1.20 to $f \equiv 0$ and $g := \|\cdot\|$, since g is obviously continuous. In the second case, the adjoint of A is given by $A^* \alpha := \alpha x_0^*$, $\alpha \in \mathbb{R}$, and (4.63) reduces to $\partial \varphi(x) = A^* \partial |Ax| = (\partial |\langle x_0^*, x \rangle|) x_0^*$. Then (4.64) follows because

$$\partial |\cdot|(\beta) = \begin{cases} [-1, 1], & \text{if } \beta = 0, \\ \operatorname{sign}(\beta), & \text{otherwise.} \end{cases}$$

The following proposition is an important consequence of Proposition 4.1.20 on the subdifferentiability of convex functions, which extends (4.28) from positive to any nonnegative ε. It shows, in particular, that convex functions inherit a lot from the linear ones, proving that every continuous convex function is subdifferentiable.

Proposition 4.1.22 *Let $f : X \to \mathbb{R}_\infty$ be a convex function, which is finite and continuous at $x \in \operatorname{dom} f$. Then, for all $\varepsilon \geq 0$, $\partial_\varepsilon f(x)$ is a non-empty w^*-compact set satisfying*

$$f'_\varepsilon(x; \cdot) = \sigma_{\partial_\varepsilon f(x)}.$$

Proof. Fix $\varepsilon \geq 0$. We apply Proposition 4.1.20(i) to the convex functions f and $g := \mathrm{I}_{\{x\}}$ to conclude the existence of some $x^* \in X^*$ such that

$$-f(x) = (f + \mathrm{I}_{\{x\}})^*(\theta) = (f^* \square \sigma_{\{x\}})(\theta) = f^*(x^*) + \langle x, -x^* \rangle \, ;$$

that is, $x^* \in \partial f(x) \subset \partial_\varepsilon f(x)$. Moreover, since the w^*-lsc function $h := f^*(\cdot) - \langle \cdot, x \rangle$ is w^*-inf-compact by Proposition 3.1.3, the set $\partial_\varepsilon f(x) = [h \leq \varepsilon - f(x)]$ is w^*-compact. Moreover, if $U \subset X$ is an open neighborhood of θ such that, for all $u \in U$,

$$|f(x+u) - f(x)| \leq 1, \tag{4.65}$$

then $f'_\varepsilon(x; u) \leq f(x+u) - f(x) + \varepsilon \leq 1 + \varepsilon$. Now $f'_\varepsilon(x; \cdot)$ is sublinear according to Proposition 4.1.1 and, therefore, Proposition 2.2.6 entails its continuity on X. Thus, (4.10) leads us to $\sigma_{\partial_\varepsilon f(x)} = \operatorname{cl}(f'_\varepsilon(x; \cdot)) = f'_\varepsilon(x; \cdot)$. ∎

Using Proposition 4.1.22, we get the following result which is in line with (2.43). In what follows Y is another lcs.

Proposition 4.1.23 *Let $F : X \times Y \to \overline{\mathbb{R}}$ be a convex function such that $F(x_0, \cdot)$ is finite and continuous at θ for some $x_0 \in X$. Then we have*

$$\inf_{x \in X} F(x, \theta) = \inf_{x \in X} (\operatorname{cl} F)(x, \theta) \tag{4.66}$$

and, consequently,

$$(F(\cdot, \theta))^* = ((\operatorname{cl} F)(\cdot, \theta))^*. \tag{4.67}$$

Proof. We have

$$\varphi(x) := F(x, \theta) \geq (\operatorname{cl} F)(x, \theta), \text{ for all } x \in X, \tag{4.68}$$

so the inequality "$\geq$" in (4.66) follows. To prove the opposite inequality, consider the (convex) marginal function $h : Y \to \overline{\mathbb{R}}$ defined by

$$h(y) := \inf_{x \in X} F(x, y).$$

4.1. SUBDIFFERENTIAL THEORY

So, thanks to (2.37), we have

$$h(\theta) \geq \inf_{x \in X}(\operatorname{cl}\varphi)(x) = \inf_{x \in X}\varphi(x). \tag{4.69}$$

In particular, if $h(\theta) = -\infty$, then the inequality in (4.69) holds as an equality, and (4.68) gives

$$-\infty = h(\theta) = \inf_{x \in X}\varphi(x) \geq \inf_{x \in X}(\operatorname{cl} F)(x,\theta);$$

that is, (4.66) follows in this case. Thus, in the rest of the proof we assume that $h(\theta) > -\infty$. Moreover, if $U \in \mathcal{N}_Y$ is such that

$$F(x_0, y) \leq F(x_0, \theta) + 1 \text{ for all } y \in U,$$

then

$$h(y) \leq F(x_0, y) \leq F(x_0, \theta) + 1 \text{ for all } y \in U, \tag{4.70}$$

and, in particular, h is uniformly bounded from above on U. Consequently, h is proper; otherwise, there would exist some $y_0 \in U$ such that $h(y_0) = -\infty$. Then $-y_0 \in U$ and the convexity of h leads us to the contradiction

$$-\infty < h(\theta) = h((1/2)y_0 - (1/2)y_0)$$
$$\leq (1/2)h(y_0) + (1/2)h(-y_0) = -\infty.$$

Now, since h is confirmed to be proper, (4.70) and Proposition 2.2.6 imply that h is continuous at θ. Therefore, according to Proposition 4.1.22, it is subdifferentiable at θ and there exists some subgradient $y_0^* \in \partial h(\theta)$ such that, for all $u, y \in X$,

$$\inf_{x \in X}\varphi(x) = h(\theta) \leq h(y) - \langle y_0^*, y \rangle \leq F(u, y) - \langle y_0^*, y \rangle.$$

Next, by taking the closure in both sides, we get $\inf_{x \in X}\varphi(x) \leq (\operatorname{cl} F)(u, y) - \langle y_0^*, y \rangle$ for all $u, y \in X$, and we deduce that

$$\inf_{x \in X}\varphi(x) \leq (\operatorname{cl} F)(u, \theta), \text{ for all } u \in X.$$

Hence, by taking the infimum over $u \in X$, we obtain the desired inequality that yields (4.66).

To establish (4.67), by (2.37) we obtain that for each $x^* \in X^*$

$$(F(\cdot, \theta))^*(x^*) = - \inf_{x \in X} \{F(x, \theta) - \langle x^*, x \rangle\} = - \inf_{x \in X} \operatorname{cl}(\tilde{F}(\cdot, \theta)),$$

where $\tilde{F} : X \times Y \to \overline{\mathbb{R}}$ is the convex function defined by $\tilde{F}(x, y) := F(x, y) - \langle x^*, x \rangle$. Since $\tilde{F}$ satisfies the same continuity assumptions as F, (4.66) gives rise to

$$(F(\cdot, \theta))^*(x^*) = - \inf_{x \in X} (\operatorname{cl} \tilde{F})(\cdot, \theta)$$
$$= - \inf_{x \in X} \{(\operatorname{cl} F)(\cdot, \theta) - \langle x^*, x \rangle\} = ((\operatorname{cl} F)(\cdot, \theta))^*(x^*),$$

showing that (4.67) also holds. ∎

We now give the counterpart to Corollary 4.1.19 under additional continuity conditions on the function F. It is worth noting that Proposition 4.1.20 can also be derived from the following result (see Exercise 51).

Proposition 4.1.24 *Let $F : X \times Y \to \mathbb{R}_\infty$ be a convex function, and denote $\varphi := F(\cdot, \theta)$. Assume that $F(x_0, \cdot)$ is finite and continuous at θ for some $x_0 \in X$. Then the following assertions hold true:*
(i)

$$\varphi^* = \min_{y^* \in Y^*} F^*(\cdot, y^*). \qquad (4.71)$$

(ii)

$$\operatorname{epi} \varphi^* = \{(x^*, \lambda) \in X^* \times \mathbb{R} : \exists \, y^* \in Y^* \text{ such that } (x^*, y^*, \lambda) \in \operatorname{epi} F^*\}. \qquad (4.72)$$

(iii) *For every $x \in X$ and $\varepsilon \geq 0$,*

$$\partial_\varepsilon \varphi(x) = \{x^* \in X^* : \exists \, y^* \in Y^*, \, (x^*, y^*) \in \partial_\varepsilon F(x, \theta)\}. \qquad (4.73)$$

Proof. (i) We start by assuming that $F \in \Gamma_0(X \times Y)$, so that $\varphi \in \Gamma_0(X)$ and

$$\varphi^* = \operatorname{cl}\left(\inf_{y^* \in Y^*} F^*(\cdot, y^*)\right), \qquad (4.74)$$

by Corollary 4.1.19. So, to show (4.71), we just need to prove the inequality "$\geq$" there for a given $x^* \in \operatorname{dom} \varphi^*$. Then, by (4.74) and the fact that $\varphi^* \in \Gamma_0(X^*)$ (thanks to Proposition 3.1.4), we take nets $x_i^* \to x^*$ and $(y_i^*)_i \in Y^*$ such that $F^*(x_i^*, y_i^*) \in \mathbb{R}$ (for all i) and

4.1. SUBDIFFERENTIAL THEORY

$$\varphi^*(x^*) = \lim_i F^*(x_i^*, y_i^*) \in \mathbb{R}. \tag{4.75}$$

Thus, without loss of generality, for all $y \in Y$ we have

$$\langle x_i^*, x_0 \rangle + \langle y_i^*, y \rangle - F(x_0, y) \leq F^*(x_i^*, y_i^*) \leq \varphi^*(x^*) + 1 \text{ for all } i,$$

and so, as $x_i^* \to x^*$, there exists some constant $m \in \mathbb{R}$ such that $(F(x_0, \cdot))^*(y_i^*) \leq m$ for all i (without loss of generality). Consequently, using Proposition 3.1.3 and the current continuity assumption, we can assume that $y_i^* \to y^*$ (in the w^*-topology) for some $y^* \in Y^*$. Thus, (4.75) and (4.74) imply that

$$\inf_{v^* \in Y^*} F^*(x^*, v^*) \geq \varphi^*(x^*) \geq F^*(x^*, y^*),$$

and (4.71) follows when $F \in \Gamma_0(X \times Y)$.

To show (4.71) in the general case, we fix $x^* \in \operatorname{dom} \varphi^*$. Then we distinguish two situations: First, we assume that $\operatorname{cl} F$ is proper. Then, applying the previous paragraph to $\operatorname{cl} F \in \Gamma_0(X \times Y)$, by (3.7) and the fact that $(F(\cdot, \theta))^* = ((\operatorname{cl} F)(\cdot, \theta))^*$, coming from (4.67), we deduce that

$$\varphi^*(x^*) = ((\operatorname{cl} F)(\cdot, \theta))^*(x^*) = \min_{y^* \in Y^*} (\operatorname{cl} F)^*(x^*, y^*) = \min_{y^* \in Y^*} F^*(x^*, y^*),$$

and (4.71) follows in the first case. Second, if $\operatorname{cl} F$ is not proper (so, it takes the value $-\infty$ somewhere), then $F^* \equiv +\infty$ and we obtain

$$\min_{y^* \in Y^*} F^*(x^*, y^*) = F^*(x^*, y^*) = +\infty \text{ for all } y^* \in X^*.$$

At the same time, the condition $F(x_0, \theta) \in \mathbb{R}$ implies that $(\operatorname{cl} F)(x_0, \theta) = -\infty$ and, using again (4.67), we infer that $\varphi^*(x^*) = (F(\cdot, \theta))^*(x^*) = ((\operatorname{cl} F)(\cdot, \theta))^* = +\infty$. Thus, (4.71) also holds, and the proof of (i) is finished.

(ii) If $(x^*, \lambda) \in \operatorname{epi} \varphi^*$, then (4.71) yields some $y^* \in Y^*$ such that $\varphi^*(x^*) = F^*(x^*, y^*) \leq \lambda$, and we deduce that $(x^*, y^*, \lambda) \in \operatorname{epi} F^*$. Conversely, if $(x^*, \lambda) \in X^* \times \mathbb{R}$ is such that $(x^*, y^*, \lambda) \in \operatorname{epi} F^*$ for some $y^* \in Y^*$, then (4.74) gives rise to

$$\varphi^*(x^*) \leq \inf_{v^* \in Y^*} F^*(x^*, u^*) \leq F^*(x^*, y^*) \leq \lambda;$$

that is, $(x^*, \lambda) \in \operatorname{epi} \varphi^*$ and (4.72) follows.

(iii) We take $x^* \in \partial_\varepsilon \varphi(x)$. So, by (4.71), there exists some $y^* \in Y^*$ such that

$$F(x, \theta) + F^*(x^*, y^*) = \varphi(x) + \varphi^*(x^*)$$
$$\leq \langle x^*, x \rangle + \varepsilon = \langle (x^*, y^*), (x, \theta) \rangle + \varepsilon;$$

that is, $(x^*, y^*) \in \partial_\varepsilon F(x, \theta)$ and the non-trivial inclusion "$\subset$" in (4.73) follows. ■

The continuity assumptions in Proposition 4.1.20 are often used in convex analysis, although they are not always necessary. The following example demonstrates this fact and, at the same time, illustrates what can be gained from Proposition 4.1.24 rather than Proposition 4.1.20.

Example 4.1.25 *Let $F : X \times Y \to \overline{\mathbb{R}}$ be a convex function such that $F(x_0, \cdot)$ is finite and continuous at θ for some $x_0 \in X$. Then the functions F and $g := I_{X \times \{\theta\}}$ satisfy the conclusion of Proposition 4.1.20; that is, we have that*

$$(F + g)^* = F^* \square g^* \; (= F^* \square \sigma_{X \times \{\theta\}}) \qquad (4.76)$$

(with an exact inf-convolution),

$$\mathrm{epi}(F + g)^* = \mathrm{epi}\, F^* + \mathrm{epi}\, g^* \; (= \mathrm{epi}\, F^* + \{\theta\} \times Y^* \times \mathbb{R}_+), \qquad (4.77)$$

and, for each $x \in X$ and $\varepsilon \geq 0$,

$$\partial_\varepsilon (F + g)(x) = \bigcup_{\substack{\varepsilon_1 + \varepsilon_2 = \varepsilon \\ \varepsilon_1, \varepsilon_2 \geq 0}} (\partial_{\varepsilon_1} F(x) + \partial_{\varepsilon_2} g(x)) \; (= \partial_\varepsilon F(x) + \{\theta\} \times Y^*).$$
$$(4.78)$$

In fact, by Proposition 4.1.24(i), for each $(x^*, y^*) \in X^* \times Y^*$ we have that

$$(F + g)^*(x^*, y^*) = \sup_{x \in X} \{\langle x^*, x \rangle - F(x, \theta)\}$$
$$= (F(\cdot, \theta))^*(x^*) = \min_{v^* \in Y^*} F^*(x^*, v^*)$$
$$= \min_{u^*, v^* \in Y^*} \left(F^*(u^*, v^*) + \sigma_{X \times \{\theta\}}(x^* - u^*, y^* - v^*) \right)$$
$$= (F^* \square g^*)(x^*, y^*),$$

and (4.76) holds. Finally, relations (4.77) and (4.78) are derived from (4.76) as in the proof of statements (ii)–(iii) in Proposition 4.1.24.

4.1. SUBDIFFERENTIAL THEORY

The following result gives a finite-dimensional counterpart to Proposition 4.1.20. The unilateral continuity condition used there is now replaced with a symmetric relation that requires the sets $A(\operatorname{dom} f)$ and $\operatorname{dom} g$ to overlap sufficiently. Notice that $A(\operatorname{ri}(\operatorname{dom} f)) = \operatorname{ri}(A(\operatorname{dom} f))$, due to (2.13).

Proposition 4.1.26 *Consider two convex functions $f : \mathbb{R}^m \to \mathbb{R}_\infty$, $g : \mathbb{R}^n \to \mathbb{R}_\infty$, and let $A : \mathbb{R}^m \to \mathbb{R}^n$ be a linear mapping. Assume that $A(\operatorname{ri}(\operatorname{dom} f)) \cap \operatorname{ri}(\operatorname{dom} g) \neq \emptyset$. Then (4.56), (4.57), and (4.58) hold.*

Proof. For the sake of simplicity, we only prove (4.58) when $\varepsilon = 0$, $m = n$, and A is the identity mapping; the general case is made by arguing in a similar way. We may suppose that $\theta \in \operatorname{dom} f \cap \operatorname{dom} g$, without loss of generality, and denote $E := \operatorname{span}(\operatorname{dom} g)$. We consider the respective restrictions $\tilde{f}$ and $\tilde{g}$ of the functions $f + I_E$ and g to E. Based on the current assumption, we choose $x_0 \in \operatorname{ri}(\operatorname{dom} f) \cap \operatorname{ri}(\operatorname{dom} g)$, so that $x_0 \in E$, $\tilde{f}(x_0) = f(x_0) \in \mathbb{R}$, and Corollary 2.2.9 entails that $\tilde{g}$ is (finite and) continuous at $x_0 \in \tilde{f}^{-1}(\mathbb{R})$. Therefore, by Proposition 4.1.20, we have $\partial(\tilde{f} + \tilde{g})(x) = \partial \tilde{f}(x) + \partial \tilde{g}(x)$ which in turn yields (see Exercise 55(ii))

$$\partial(f+g)(x) = \partial(f + g + I_E)(x) = \partial(f + I_E)(x) + \partial g(x) + E^\perp. \tag{4.79}$$

Let us also set $F := \operatorname{span}(\operatorname{dom} f)$ and consider the respective restrictions $\hat{f}$ and $\hat{I}_{E \cap F}$ of the functions f and $I_{F \cap E}$ to F. Then, by Corollary 2.2.9, $\hat{f}$ is continuous at $x_0 \in E \cap F = \operatorname{dom} \hat{I}_{E \cap F}$ and, again, Proposition 4.1.20 gives us $\partial(\hat{f} + \hat{I}_{E \cap F})(x) = \partial \hat{f}(x) + \partial \hat{I}_{E \cap F}(x)$; that is to say (Exercise 55(ii)),

$$\begin{aligned}\partial(f + I_E)(x) &= \partial(f + I_{E \cap F})(x) \\ &= \partial f(x) + \partial I_{E \cap F}(x) + F^\perp = \partial f(x) + (E \cap F)^\perp + F^\perp.\end{aligned} \tag{4.80}$$

Therefore, combining the last relation together with (4.79) and observing that $(E \cap F)^\perp = \operatorname{cl}(E^\perp + F^\perp) = E^\perp + F^\perp$ (Exercise 53), we get

$$\partial(f+g)(x) = \partial f(x) + \partial g(x) + E^\perp + F^\perp = \partial f(x) + \partial g(x),$$

and (4.58) follows.

The proof of (4.56) follows the same reasoning as the proof of (4.58) above and is left as exercise (see Exercise 56). Relation (4.57) comes from (4.56) as in the proof of Proposition 4.1.20. ∎

Remark 2 *The subdifferential rule in Proposition 4.1.26 is also valid in more general situations. For example, for a function f defined on (the lcs) X instead of $\mathbb{R}^m$, and a linear mapping $A : X \to \mathbb{R}^n$, the formula $\partial(f + g \circ A)(x) = \partial f(x) + A^*\partial g(Ax)$ holds whenever*

$$A(\text{qri}(\text{dom } f)) \cap \text{ri}(\text{dom } g) \neq \emptyset.$$

Here, the set $\text{qri}(\text{dom } f)$ represents the quasi-relative interior of $\text{dom } f$; that is, the set of points $u \in \text{dom } f$ such that $\text{cl}(\mathbb{R}_+(\text{dom } f - u))$ is a linear subspace of X.

We proceed by giving a couple of applications based on Proposition 4.1.20, which are also needed in what follows.

Corollary 4.1.27 (i) *Let $f, g, h : X \to \overline{\mathbb{R}}$ be convex functions such that $\theta \in \text{dom } f$ and $(\text{dom } h) \cap \text{ri}(\text{dom } f) \neq \emptyset$. If $g \leq f$ and the restriction of f to $E := \text{aff}(\text{dom } f)$ is finite and continuous on $\text{ri}(\text{dom } f)$, then for all $x \in \text{dom } f \cap \text{dom } g \cap \text{dom } h$ we have that*

$$\partial(g + I_{\text{dom } f} + h)(x) = \partial(g + I_{\text{dom } f})(x) + \partial(h + I_E)(x).$$

(ii) *If $C \subset X$ is a convex set, then, for every $x \in C$ and $L \in \mathcal{F}(x)$ such that $L \cap \text{ri}(C)$ is non-empty, we have*

$$N_{L \cap C}(x) = \text{cl}(L^\perp + N_C(x)).$$

Proof. (i) Fix $x \in \text{dom } f \cap \text{dom } g \cap \text{dom } h$ and denote by $\tilde{f}, \tilde{g}$, and $\tilde{h}$ the respective restrictions to E of the functions f, $g + I_{\text{dom } f}$, and $h + I_E$. Since $g + I_{\text{dom } f} \leq f$, we have $\tilde{g} \leq \tilde{f}$ and the continuity of $\tilde{f}$ on $\text{ri}(\text{dom } f)$ implies that $\tilde{g}$ is locally uniformly bounded from above on $\text{ri}(\text{dom } f)$. We distinguish two cases: First, if $\tilde{g}$ is finite somewhere in $(\text{dom } h) \cap \text{ri}(\text{dom } f)$ $(\subset (\text{dom } \tilde{h}) \cap \text{ri}(\text{dom } f))$, then it is proper and Proposition 2.2.6 implies its continuity on $\text{ri}(\text{dom } f)$. Thus, applying Proposition 4.1.20 in the lcs E, we get

$$\partial(\tilde{g} + \tilde{h})(x) = \partial\tilde{g}(x) + \partial\tilde{h}(x). \tag{4.81}$$

Second, if $\tilde{g}$ were not finite at every point in $(\text{dom } h) \cap \text{ri}(\text{dom } f)$, then the condition $g \leq f$ and the continuity assumption on f imply that $\tilde{g}$ and $\tilde{g} + \tilde{h}$ are identically $-\infty$ in $(\text{dom } h) \cap \text{ri}(\text{dom } f)$. This entails that $\partial(\tilde{g} + \tilde{h}) \equiv \partial\tilde{g} \equiv \emptyset$, and (4.81) also holds.

Now, taking into account Exercise 55(ii), (4.81) gives rise to

4.1. SUBDIFFERENTIAL THEORY

$$\partial(g + \mathrm{I}_{\mathrm{dom}\,f} + h)(x) = \partial(g + \mathrm{I}_{\mathrm{dom}\,f})(x) + \partial(h + \mathrm{I}_E)(x) + E^\perp$$
$$= \partial(g + \mathrm{I}_{\mathrm{dom}\,f})(x) + \partial(h + \mathrm{I}_E)(x),$$

and (i) follows.

(ii) The convex functions $f \equiv \mathrm{I}_C$, $g \equiv 0$, and $h \equiv \mathrm{I}_L$ satisfy the conditions of statement (i) for $E := \mathrm{aff}(C)$, so we obtain

$$\partial(\mathrm{I}_C + \mathrm{I}_L)(x) = \mathrm{N}_C(x) + \partial(\mathrm{I}_L + \mathrm{I}_E)(x)$$
$$= \mathrm{N}_C(x) + (E \cap L)^\perp = \mathrm{N}_C(x) + \mathrm{cl}(L^\perp + E^\perp),$$

where the last equality comes from Exercise 53. Therefore,

$$\mathrm{N}_{L \cap C}(x) \subset \mathrm{N}_C(x) + \mathrm{cl}(L^\perp + \mathrm{N}_C(x)) = \mathrm{cl}(L^\perp + \mathrm{N}_C(x)),$$

and we conclude the proof of (ii) since the opposite of the last inclusion is straightforward. ∎

A second application of Proposition 4.1.20 comes next. Relation (4.82) is a technical result that will be used in the proof of certain formulas of the subdifferential of the supremum function in section 5.3.

Proposition 4.1.28 *Let $B \subset X$ convex, $x \in B$, and let $(A_i)_i \subset X^*$ be a non-increasing net of non-empty convex sets; that is, $A_{i_2} \subset A_{i_1}$ whenever i_2 is posterior to i_1, such that*

$$\left(\bigcup_i \mathrm{dom}\,\sigma_{A_i}\right) \cap \mathrm{int}(\mathrm{cone}(B - x)) \neq \emptyset.$$

Then we have

$$\bigcap_i \mathrm{cl}\,(A_i + \mathrm{N}_B(x)) = \mathrm{N}_B(x) + \bigcap_i \mathrm{cl}\,A_i. \tag{4.82}$$

Proof. Only the inclusion "⊂" needs to be proved. Let C denote the closed convex set on the left-hand side of (4.82), and consider the nontrivial case $C \neq \emptyset$. Then, denoting $g := \inf_i \sigma_{A_i}$ and $h := \sigma_{\mathrm{N}_B(x)}$, we have

$$\sigma_C \leq \inf_i \sigma_{A_i + \mathrm{N}_B(x)} = \left(\inf_i \sigma_{A_i}\right) + \sigma_{\mathrm{N}_B(x)} = g + h, \tag{4.83}$$

and both functions g and h are sublinear (Exercise 7). Then we choose any $z_0 \in (\cup_i \mathrm{dom}\,\sigma_{A_i}) \cap \mathrm{int}(\mathrm{cone}(B - x))$ ($\subset \mathrm{dom}\,g$). So, from the inequality

$$h(y-x) = \sigma_{N_B(x)}(y-x) \leq 0 \text{ for all } y \in B,$$

Proposition 2.2.6 implies that h is continuous at z_0. Therefore, using Proposition 2.2.11, (4.83) implies that

$$\sigma_C \leq \operatorname{cl}(g+h) = (\operatorname{cl} g) + (\operatorname{cl} h) = (\operatorname{cl}(\inf_i \sigma_{A_i})) + \sigma_{N_B(x)}$$

and both functions σ_C and $(\operatorname{cl}(\inf_i \sigma_{A_i})) + \sigma_{N_B(x)}$ coincide at θ. Consequently, taking into account (2.50) and remembering that h is continuous at $z_0 \in \operatorname{dom} g \subset \operatorname{dom}(\operatorname{cl} g)$, Proposition 4.1.20 and (4.13) produce

$$\emptyset \neq C = \partial \sigma_C(\theta) \subset \partial((\operatorname{cl} g) + h)(\theta)$$
$$= \partial(\operatorname{cl} g)(\theta) + \partial h(\theta) = \partial(\operatorname{cl}(\inf_i \sigma_{A_i}))(\theta) + N_B(x). \quad (4.84)$$

In particular, the function $\operatorname{cl}(\inf_i \sigma_{A_i})$ is proper for having a non-empty subdifferential at θ. Then, thanks to Proposition 3.2.8(ii) applied to the net $(\operatorname{cl} A_i)_i$, which is also non-increasing, we have that $A := \cap_i \operatorname{cl} A_i \neq \emptyset$ and $\operatorname{cl}(\inf_i \sigma_{A_i}) = \operatorname{cl}(\inf_i \sigma_{\operatorname{cl} A_i}) = \sigma_A$. Consequently, again by (4.13), (4.84) leads us to $C \subset \partial \sigma_A(\theta) + N_B(x) = A + N_B(x)$, which is the desired inclusion. ■

In the following proposition we avoid the continuity assumptions in Proposition 4.1.20, replacing f and g with the respective augmented functions $f + I_{\operatorname{dom} g}$ and $g + I_{\operatorname{dom} f}$. More general formulas will be established in section 7.2. Remember that $\mathcal{F}(x)$ is defined in (1.4).

Proposition 4.1.29 *Let $f, g : X \to \mathbb{R}_\infty$ be proper convex functions. Then, for every $x \in X$,*

$$\partial(f+g)(x) = \bigcap_{L \in \mathcal{F}(x)} \left(\partial(f + I_{L \cap \operatorname{dom} g})(x) + \partial(g + I_{L \cap \operatorname{dom} f})(x) \right). \quad (4.85)$$

Additionally, if $\operatorname{ri}(\operatorname{dom} f \cap \operatorname{dom} g) \neq \emptyset$ and the restriction of g to $\operatorname{aff}(\operatorname{dom} f \cap \operatorname{dom} g)$ is continuous on $\operatorname{ri}(\operatorname{dom} f \cap \operatorname{dom} g)$, then

$$\partial(f+g)(x) = \partial(f + I_{\operatorname{dom} g})(x) + \partial(g + I_{\operatorname{dom} f})(x). \quad (4.86)$$

Proof. If $x \notin \operatorname{dom} f \cap \operatorname{dom} g$, (4.85) and (4.86) trivially hold. We proceed by showing (4.86) when $x \in \operatorname{dom} f \cap \operatorname{dom} g$ and assuming, without loss of generality, that $\theta \in \operatorname{dom} f \cap \operatorname{dom} g$; so, $E := \operatorname{aff}(\operatorname{dom} f \cap \operatorname{dom} g)$ is a closed subspace of X. Let us introduce the proper convex functions $\tilde{f}, \tilde{g} : E \to \mathbb{R}_\infty$ defined by

$$\tilde{f} = (f + \mathrm{I}_{\mathrm{dom}\,g})_{|E}, \ \tilde{g} = (g + \mathrm{I}_{\mathrm{dom}\,f})_{|E}.$$

Notice that $\mathrm{dom}\,\tilde{f} = \mathrm{dom}\,\tilde{g} = \mathrm{dom}\,f \cap \mathrm{dom}\,g$. Then, by Proposition 4.1.20, the additional continuity assumption yields $\partial(\tilde{f} + \tilde{g})(x) = \partial \tilde{f}(x) + \partial \tilde{g}(x)$. So, we get (see Exercise 55(ii))

$$\partial(\tilde{f} + \tilde{g})(x) = \partial(f + \mathrm{I}_{\mathrm{dom}\,g})(x) + \partial(g + \mathrm{I}_{\mathrm{dom}\,f})(x) + E^{\perp},$$

which easily leads us to (4.86).

To show (4.85) we observe that, using (4.16),

$$\begin{aligned}\partial(f+g)(x) &= \bigcap_{L \in \mathcal{F}(x)} \partial(f+g+\mathrm{I}_L)(x) \\ &= \bigcap_{L \in \mathcal{F}(x)} \partial((f + \mathrm{I}_{L \cap \mathrm{dom}\,g}) + (g + \mathrm{I}_{L \cap \mathrm{dom}\,f}))(x), \quad (4.87)\end{aligned}$$

where each pair of functions $f + \mathrm{I}_{L \cap \mathrm{dom}\,g}$ and $g + \mathrm{I}_{L \cap \mathrm{dom}\,f}$ satisfies the assumptions of the paragraph above. Thus,

$$\partial((f + \mathrm{I}_{L \cap \mathrm{dom}\,g}) + (g + \mathrm{I}_{L \cap \mathrm{dom}\,f}))(x) = \partial(f + \mathrm{I}_{L \cap \mathrm{dom}\,g})(x) + \partial(g + \mathrm{I}_{L \cap \mathrm{dom}\,f})(x),$$

and (4.85) follows from (4.87). ∎

4.2 Convex duality

In the present section, we use the previous dual representations of the support function of sublevel sets to provide a new and unified approach to *convex duality*, covering *Lagrange* and *Fenchel* duality. The results here will be applied later, in section 8.2, to develop a duality theory for convex infinite optimization. We begin by extending Theorem 3.3.1 by providing dual representations of the function $p_0 : X^* \to \overline{\mathbb{R}}$ defined by

$$p_0(x^*) := -\inf_{x \in [f \leq 0]} \left(f_0(x) - \langle x^*, x\rangle\right), \quad (4.88)$$

by means of the conjugates of f_0 and f. Remember that the statement of Theorem 3.3.1 corresponds to $f_0 \equiv 0$, where (4.88) takes the form

$$p_0(x^*) = -\inf_{x \in [f \leq 0]} \left(-\langle x^*, x\rangle\right) = \sigma_{[f \leq 0]}(x^*).$$

The function $p_0 : X^* \to \overline{\mathbb{R}}$ gives a particular example of what we will later call (linear) perturbation functions of the optimization problem $\inf_{[f \leq 0]} f_0$.

Theorem 4.2.1 *Let functions $f_0, f \in \Gamma_0(X)$ such that*

$$[f \leq 0] \cap \operatorname{dom} f_0 \neq \emptyset.$$

Then we have

$$p_0 = \operatorname{cl}\left(\inf_{\alpha > 0} (f_0^* \square (\alpha f)^*)\right) \qquad (4.89)$$

and

$$\operatorname{epi} p_0 = \operatorname{cl}(\operatorname{epi} f_0^* + \mathbb{R}_+ \operatorname{epi} f^*). \qquad (4.90)$$

Proof. Fix $x^* \in X^*$. Since $\inf_{[f \leq 0]} f_0 < +\infty$ and $\operatorname{dom} f_0 \cap \operatorname{dom} \mathrm{I}_{[f \leq 0]} \neq \emptyset$ by the current assumption, Proposition 4.1.16 yields

$$\inf_{[f \leq 0]} (f_0 - x^*) = \inf_X (f_0 + \mathrm{I}_{[f \leq 0]} - x^*)$$
$$= -(f_0 + \mathrm{I}_{[f \leq 0]})^*(x^*) = -\operatorname{cl}(f_0^* \square \sigma_{[f \leq 0]})(x^*). \qquad (4.91)$$

Thus, since $[f \leq 0] \neq \emptyset$, by applying (3.30) to $\sigma_{[f \leq 0]}$ we obtain

$$p_0(x^*) = \operatorname{cl}(f_0^* \square \operatorname{cl}(\inf_{\alpha>0}(\alpha f)^*))(x^*). \qquad (4.92)$$

Now we proceed by showing that the inner closure in (4.92) can be dropped out. Indeed, the convex functions $f_0^* \square \operatorname{cl}(\inf_{\alpha>0}(\alpha f)^*)$ and $f_0^* \square (\inf_{\alpha>0}(\alpha f)^*)$ have the same (proper) conjugate which is, by Theorem 3.2.2, given by

$$f_0^{**} + \sup_{\alpha>0}(\alpha f)^{**} = f_0 + \sup_{\alpha>0}(\alpha f) = f_0 + \mathrm{I}_{[f \leq 0]}.$$

So, again by Theorem 3.2.2, we have $\operatorname{cl}(f_0^* \square \operatorname{cl}(\inf_{\alpha>0}(\alpha f)^*)) = \operatorname{cl}(f_0^* \square (\inf_{\alpha>0}(\alpha f)^*))$, and (4.89) is derived from (4.92).

Now, due to (4.91), we have $p_0 = \operatorname{cl}(f_0^* \square \sigma_{[f \leq 0]})$ and so, applying (3.31), we obtain

$$\operatorname{epi} p_0 = \operatorname{cl}(\operatorname{epi}(f_0^* \square \sigma_{[f \leq 0]})) = \operatorname{cl}(\operatorname{epi} f_0^* + \operatorname{epi} \sigma_{[f \leq 0]})$$
$$= \operatorname{cl}(\operatorname{epi} f_0^* + \operatorname{cl}(\mathbb{R}_+ \operatorname{epi} f^*)) = \operatorname{cl}(\operatorname{epi} f_0^* + \mathbb{R}_+ \operatorname{epi} f^*);$$

that is, (4.90) also holds. ∎

4.2. CONVEX DUALITY

The following result simplifies the representation of the value function p_0 given in (4.89) by removing the closure there.

Theorem 4.2.2 *Given convex functions $f_0, f : X \to \mathbb{R}_\infty$, we assume that*
$$[f < 0] \cap \operatorname{dom} f_0 \neq \emptyset.$$
Then we have
$$\inf_{[f \leq 0]} f_0 = \max_{\alpha \geq 0} \inf_{x \in X} (f_0(x) + \alpha f(x)). \tag{4.93}$$

Moreover, if one of the functions f_0 and f is finite and continuous at some point in the domain of the other (or, more generally, under any condition ensuring that $(f_0 + \alpha f)^* = f_0^* \square (\alpha f)^*$ with exact inf-convolution), then
$$\inf_{[f \leq 0]} f_0 = - \min_{x^* \in X^*,\, \alpha \geq 0} (f_0^*(x^*) + (\alpha f)^*(-x^*)). \tag{4.94}$$

Proof. First, if $\inf_{[f \leq 0]} f_0 = -\infty$, then we get
$$\inf_{x \in X} (f_0(x) + \alpha f(x)) \leq \inf_{[f \leq 0]} f_0(x) = -\infty \text{ for all } \alpha \geq 0,$$

and (4.93) holds trivially. Thus, taking into account the current assumption, we suppose in the rest of the proof that $-\infty < \inf_{[f \leq 0]} f_0 < +\infty$. Let us express $\inf_{[f \leq 0]} f_0$ as the following optimization problem, posed in $X \times \mathbb{R}$:
$$\inf_{[f \leq 0]} f_0 = \inf_{\substack{(x, \gamma) \in X \times \mathbb{R},\, f(x) \leq 0,\\ f_0(x) - \gamma \leq 0,}} \gamma = -\sigma_{[h \leq 0]}(\theta, -1), \tag{4.95}$$

where $h : X \times \mathbb{R} \to \mathbb{R}_\infty$ is the convex function defined by
$$h(x, \gamma) := \max\{f(x), f_0(x) - \gamma\}. \tag{4.96}$$

Next, taking any point $x_0 \in [f < 0] \cap \operatorname{dom} f_0$, we observe that $h(x_0, f_0(x_0) + 1) = \max\{f(x_0), -1\} < 0$; that is, $[h < 0] \neq \emptyset$. Therefore, according to (3.38) in Theorem 3.3.4, relation (4.95) gives rise to
$$\inf_{[f \leq 0]} f_0 = -\min_{\beta \geq 0} (\beta h)^*(\theta, -1) = \max_{\beta \geq 0} (-(\beta h)^*(\theta, -1)). \tag{4.97}$$

Furthermore, taking into account Corollary 3.4.9 and (2.45), for each given $\beta \geq 0$ we have

$$\begin{aligned}
-(\beta h)^*(\theta, -1) &= \inf_{x \in X, \gamma \in \mathbb{R}} \{\gamma + \beta \max\{f(x), f_0(x) - \gamma\}\} \\
&= \inf_{(x,\gamma) \in X \times \mathbb{R}} \max_{\eta \in [0,1]} \{\gamma + \beta(1-\eta)f(x) + \beta\eta f_0(x) - \beta\eta\gamma\} \\
&= \max_{\eta \in [0,1]} \inf_{x \in \operatorname{dom} f \cap \operatorname{dom} f_0} \inf_{\gamma \in \mathbb{R}} \psi(\gamma), \quad (4.98)
\end{aligned}$$

where $\psi : \mathbb{R} \to \mathbb{R}$ is the function (depending on $x \in \operatorname{dom} f \cap \operatorname{dom} f_0$, $\beta \geq 0$ and $\eta \in [0,1]$) defined as

$$\psi(\gamma) := (1 - \eta\beta)\gamma + \eta\beta f_0(x) + \beta(1-\eta)f(x);$$

that is, (4.97) reads

$$\inf_{[f \leq 0]} f_0 = \max_{\beta \geq 0, \eta \in [0,1]} \inf_{x \in \operatorname{dom} f \cap \operatorname{dom} f_0} \inf_{\gamma \in \mathbb{R}} \psi(\gamma).$$

Since $\inf_{\gamma \in \mathbb{R}} \psi(\gamma)$ is equal to 0 or $-\infty$ (because ψ is affine) and $\inf_{[f \leq 0]} f_0 \in \mathbb{R}$, the maximum in the last expression is reached when $\eta\beta = 1$; that is, $\beta \geq 1$ and

$$\begin{aligned}
\inf_{[f \leq 0]} f_0 &= \max_{\beta \geq 1} \inf_{x \in \operatorname{dom} f \cap \operatorname{dom} f_0} (f_0(x) + (\beta - 1)f(x)) \\
&= \max_{\alpha \geq 0} \inf_{x \in X} (f_0(x) + \alpha f(x)),
\end{aligned}$$

showing that (4.93) holds.

Finally, taking into account (4.93) together with Proposition 4.1.20 and the fact that the functions f_0 and αf satisfy the same continuity assumptions as those imposed on the functions f_0 and f, (4.94) follows as

$$\begin{aligned}
\inf_{[f \leq 0]} f_0 &= \max_{\alpha \geq 0} \inf_{x \in X} (f_0(x) + \alpha f(x)) = -\min_{\alpha \geq 0} (f_0 + \alpha f)^*(\theta) \\
&= -\min_{\alpha \geq 0,\, x^* \in X^*} (f_0^*(x^*) + (\alpha f)^*(-x^*)).
\end{aligned}$$

∎

Theorems 4.2.1 and 4.2.2 can be written in a more general form. In fact, defining the function $F : X \times \mathbb{R} \to \mathbb{R}_\infty$ as

$$F(x, y) := f_0(x) + \mathrm{I}_{[f \leq y]}(x),$$

4.2. CONVEX DUALITY

we get
$$\inf_{[f \le 0]} f_0 = \inf_{x \in X} F(x, \theta),$$

and the conclusion in Theorems 4.2.1 and 4.2.2 can be read as a relation involving F and its conjugate. With the choice of F above, the condition $[f \le 0] \cap \operatorname{dom} f_0 \ne \emptyset$ is equivalent to $\inf_x F(x, \theta) < +\infty$, while the existence of a point $x_0 \in [f < 0] \cap \operatorname{dom} f_0$ is nothing other than the continuity of $F(x_0, \cdot)$ at θ (in fact, we have that $F(x_0, \cdot) = f_0(x_0)$ in a neighborhood of θ).

More generally, we can be interested in finding a dual representation of the quantity $\inf_{x \in X} F(x, \theta)$ in terms of the conjugate of F, for any convex function $F: X \times Y \to \mathbb{R}_\infty$. We have the following result whose proof is based on Theorems 4.2.1 and 4.2.2.

Theorem 4.2.3 *Given another lcs Y, we consider a convex function $F: X \times Y \to \mathbb{R}_\infty$ such that $\inf_{x \in X} F(x, \theta) < +\infty$. Then we have*

$$\inf_{x \in X} F(x, \theta) = -\operatorname{cl}\left(\inf_{y^* \in Y^*} F^*(\cdot, y^*)\right)(\theta). \tag{4.99}$$

In addition, if there exists some $x_0 \in X$ such that $F(x_0, \cdot)$ is finite and continuous at θ, then

$$\inf_{x \in X} F(x, \theta) = -\min_{y^* \in Y^*} F^*(\theta, y^*). \tag{4.100}$$

Proof. We consider the convex functions $f_0, f: X \times Y \to \mathbb{R}_\infty$ defined by
$$f_0 := F, \quad f := I_{X \times \{\theta\}} - 1,$$

so that
$$[f \le 0] \cap \operatorname{dom} f_0 = \{(x, \theta) \in X \times Y : F(x, \theta) < +\infty\} \ne \emptyset,$$

and $(\alpha f)^* = \sigma_{X \times \{\theta\}} + \alpha = I_{\{\theta\} \times Y^*} + \alpha$ for all $\alpha > 0$. Therefore, applying Theorem 4.2.1 in $X \times Y$, we obtain

$$\inf_{x\in X} F(x,\theta) = \inf_{f(x,y)\le 0} f_0(x,y) = -\operatorname{cl}\left(\inf_{\alpha>0}\,(f_0^*\square(\alpha f)^*)\right)(\theta)$$

$$= -\operatorname{cl}\left(\inf_{\alpha>0}\,(f_0^*\square(\mathrm{I}_{\{\theta\}\times Y^*}+\alpha))\right)(\theta)$$

$$= -\operatorname{cl}\left(f_0^*\square\mathrm{I}_{\{\theta\}\times Y^*}\right)(\theta) = -\operatorname{cl}\left(\inf_{y^*\in Y^*} F^*(\cdot,y^*)\right)(\theta),$$

and (4.99) follows.

To establish the second assertion, we observe that

$$[f<0]\cap\operatorname{dom} f_0 = [f\le 0]\cap\operatorname{dom} f_0 = \{(x,\theta)\in X\times Y : F(x,\theta)<+\infty\}\ne\emptyset.$$

So, by (4.93),

$$\inf_{x\in X} F(x,\theta) = \inf_{[f\le 0]} f_0 = \max_{\alpha\ge 0}\inf_{x\in X,\,y\in Y}(f_0(x,y)+\alpha f(x,y))$$

$$= \max_{\alpha\ge 0}\inf_{x\in X,\,y\in Y}(F(x,y)+\mathrm{I}_{X\times\{\theta\}}(x,y)-\alpha)$$

$$= \inf_{x\in X,\,y\in Y}(F(x,y)+\mathrm{I}_{X\times\{\theta\}}(x,y)) = -(F+\mathrm{I}_{X\times\{\theta\}})^*(\theta).$$

Moreover, using Example 4.1.25, we have that

$$(F+\mathrm{I}_{X\times\{\theta\}})^*(\theta) = \min_{y^*\in Y^*}\left(F^*(\theta,y^*)+\mathrm{I}_{\{\theta\}\times Y^*}(\theta,-y^*)\right) = \min_{y^*\in Y^*} F^*(\theta,y^*),$$

and the desired relation follows. ■

Anticipating the analysis of duality that we will develop in section 8.2 for infinite optimization, here we rewrite the previous results using the language of the classical duality theory. To do this, given convex functions $f_i : X\to\mathbb{R}_\infty$, $i=0,\ldots,m$, $m\ge 1$, we consider the convex optimization problem

$$(\mathrm{P})\quad \inf_{f_i(x)\le 0,\ i=1,\ldots,m} f_0(x),$$

together with its *Lagrangian dual problem*

$$(\mathrm{D})\quad \sup_{\lambda\in\mathbb{R}^m_+}\inf_{x\in X} L(x,\lambda)$$

given through the so-called *Lagrangian function* $L : X\times\mathbb{R}^m\to\mathbb{R}_\infty$:

$$L(x,\lambda) := f_0(x)+\lambda_1 f_1(x)+\ldots+\lambda_m f_m(x),\ \lambda:=(\lambda_1,\ldots,\lambda_m)\in\mathbb{R}^m.$$

4.2. CONVEX DUALITY

The following result establishes the (Lagrangian) *strong duality* between (P) and (D), asserting that problem (D) has optimal solutions and that the optimal values of (P) and (D) coincide. This is possible thanks to the *Slater condition*, which requires the existence of some $x_0 \in \text{dom } f_0$ such that

$$\max_{1 \leq i \leq m} f_i(x_0) < 0. \tag{4.101}$$

Corollary 4.2.4 *Given convex functions $f_i : X \to \mathbb{R}_\infty$, $i = 1, \ldots, m$, $m \geq 1$, we assume the Slater condition. Then the strong duality between (P) and (D) holds; that is,*

$$\inf_{f_i(x) \leq 0,\, i=1,\ldots,m} f_0(x) = \max_{\lambda \in \mathbb{R}_+^m} \inf_{x \in X} L(x, \lambda).$$

Proof. Applying Theorem 4.2.2 with $f := \max_{1 \leq i \leq m} f_i$, we obtain

$$\inf_{[f \leq 0]} f_0 = \max_{\alpha \geq 0} \inf_{x \in X} (f_0(x) + \alpha f(x)).$$

Thus, by (2.45),

$$\inf_{[f \leq 0]} f_0 = \max_{\alpha \geq 0} \inf_{x \in X} \sup_{\lambda \in \Delta_m} (f_0(x) + \alpha \lambda_1 f_1(x) + \ldots + \alpha \lambda_m f_m(x)),$$

and Corollary 3.4.9 gives rise to

$$\inf_{[f \leq 0]} f_0 = \max_{\alpha \geq 0} \max_{\lambda \in \Delta_m} \inf_{x \in X} (f_0(x) + \alpha \lambda_1 f_1(x) + \ldots + \alpha \lambda_m f_m(x))$$
$$= \max_{\lambda \in \mathbb{R}_+^m} \inf_{x \in X} L(x, \lambda).$$

∎

The above duality result can also be included within the more general framework of *Fenchel duality*. Given another lcs Y and a convex function $F : X \times Y \to \mathbb{R}_\infty$, we introduce the family of problems

$$(\text{P}_y) \quad \inf_{x \in X} F(x, y), \quad y \in Y,$$

and

$$(\text{D}_{x^*}) \quad \inf_{y^* \in Y^*} F^*(x^*, y^*), \quad x^* \in X^*,$$

respectively called *perturbed primal problem* and *perturbed dual problem*. Problem (P$_\theta$) is referred to as *primal problem*, and (D$_\theta$) as *dual problem* (of (P$_\theta$)). Let us represent by $v_\text{P} : Y \to \overline{\mathbb{R}}$ and $v_\text{D} : X^* \to \overline{\mathbb{R}}$ the perturbed optimal value functions of (P$_y$) and (D$_{x^*}$), respectively; that is,

$$v_\text{P}(y) := \inf_{x \in X} F(x, y) \text{ and } v_\text{D}(x^*) := \inf_{y^* \in Y^*} F^*(x^*, y^*).$$

Then v_P and v_D are convex since they are marginal functions of the convex functions F and F^*, respectively, and $v_\text{P}(\theta)$ and $v_\text{D}(\theta)$ coincide with the optimal values of (P$_\theta$) and (D$_\theta$):

$$v_\text{P}(\theta) \equiv v(\text{P}_\theta) := \inf_{x \in X} F(x, \theta) \text{ and } v_\text{D}(\theta) \equiv v(\text{D}_\theta) := \inf_{y^* \in Y^*} F^*(\theta, y^*).$$

It is worth noting that when $F \in \Gamma_0(X \times Y)$, problem (P$_\theta$) can also be seen as a dual problem of (D$_\theta$), considering that we are dealing with the compatible dual pairs (X, X^*) and (Y, Y^*), so that $X^{**} \equiv X$ and $Y^{**} \equiv Y$. In fact, representing (D$_\theta$) as

$$\inf_{y^* \in Y^*} F^*(\theta, y^*) = \inf_{y^* \in Y^*} G(y^*, \theta),$$

where $G \in \Gamma_0(Y^* \times X^*)$ is defined by $G(y^*, x^*) := F^*(x^*, y^*)$, the dual of (D$_\theta$) is given by $\inf_{x \in X} G^*(\theta, x)$. Therefore, applying Theorem 3.2.2, we have

$$G^*(\theta, x) := \sup_{x^* \in X^*,\, y^* \in Y^*} \{\langle x^*, x\rangle - G(y^*, x^*)\}$$
$$= \sup_{x^* \in X^*,\, y^* \in Y^*} \{\langle x^*, x\rangle - F^*(x^*, y^*)\} = F^{**}(x, \theta) = F(x, \theta),$$

and $\inf_{x \in X} G^*(\theta, x) = \inf_{x \in X} F(x, \theta)$. In other words, the dual of (D$_\theta$) is nothing more than the primal problem (P$_\theta$).

The main conclusion of Theorem 4.2.3 implies that

$$v_\text{P}(\theta) = \inf_{x \in X} F(x, \theta) = -\text{cl}\left(\inf_{y^* \in Y^*} F^*(\cdot, y^*)\right)(\theta)$$
$$= -(\text{cl } v_\text{D})(\theta) \geq -v_\text{D}(\theta),$$

and, consequently, we have the *weak duality* between (P$_\theta$) and (D$_\theta$) :

$$v(\text{P}_\theta) + v(\text{D}_\theta) \geq 0.$$

4.2. CONVEX DUALITY

Moreover, when $F(x_0, \cdot)$ is finite and continuous at θ for some $x_0 \in X$, the second conclusion of Theorem 4.2.3 implies that

$$v_{\mathrm{P}}(\theta) = \inf_{x \in X} F(x, \theta) = - \min_{y^* \in Y^*} F^*(\cdot, y^*)(\theta) = -v_{\mathrm{D}}(\theta),$$

giving rise to the *strong duality* between (P_θ) and (D_θ) : the dual problem (D_θ) has optimal solutions and there is a *zero duality gap*; that is, $v(\mathrm{P}_\theta) + v(\mathrm{D}_\theta) = 0$. The following corollary summarizes these facts.

Corollary 4.2.5 *Let* $F : X \times Y \to \mathbb{R}_\infty$ *be a convex function such that* $\inf_{x \in X} F(x, \theta) < +\infty$. *Then the weak duality holds between* (P_θ) *and* (D_θ). *Furthermore, if* $F(x_0, \cdot)$ *is finite and continuous at* θ *for some* $x_0 \in X$, *then the strong duality is also fulfilled.*

It is worth observing that Corollary 4.2.5 can also be used to establish the (Lagrangian) strong duality given in Corollary 4.2.4 (see Exercise 48). We also give the following example, which is a useful illustration of Corollary 4.2.5.

Example 4.2.6 *Given two convex functions* $f, g : X \to \mathbb{R}_\infty$ *such that* $\mathrm{dom}\, f \cap \mathrm{dom}\, g \neq \emptyset$, *we consider the convex function* $F : X \times X \to \mathbb{R}_\infty$ *defined as*

$$F(x, z) := f(x) + g(x + z).$$

So, direct calculations give rise to

$$F^*(x^*, z^*) = f^*(x^* - z^*) + g^*(z^*),$$

and problems (P_z) and (D_{x^*}) are written in the form

$$(\mathrm{P}_z) \quad \inf_{x \in X} \left(f(x) + g(x + z) \right), \ z \in X,$$

$$(\mathrm{D}_{x^*}) \quad \inf_{z^* \in X^*} \left(f^*(x^* - z^*) + g^*(z^*) \right), \ x^* \in X^*.$$

If g is finite and continuous at some point $x_0 \in \mathrm{dom}\, f$, then $F(x_0, \cdot)$ is finite and continuous at θ. Therefore, Corollary 4.2.5 implies the strong duality between (P_θ) and (D_θ); that is, we have

$$\inf_{x \in X} \left(f(x) + g(x) \right) = - \min_{z^* \in X^*} \left(f^*(z^*) + g^*(-z^*) \right),$$

which is nothing other than the conclusion of Proposition 4.1.20(i).

Since the supremum function will be our main concern in the next two chapters, we take advantage here of the duality theory presented above to provide a dual representation for optimization problems written in the form $\inf_X(\max\{f,g\})$.

Example 4.2.7 *Given two convex functions $f, g : X \to \mathbb{R}_\infty$ such that $\operatorname{dom} f \cap \operatorname{dom} g \neq \emptyset$, we assume that g is finite and continuous at some point $x_0 \in \operatorname{dom} f$. We introduce the convex function $F : X \times X \to \mathbb{R}_\infty$ defined as*
$$F(x, z) := \max\{f(x), g(x + z)\},$$
which is also written, using (2.45), as
$$F(x, z) = \max_{\lambda \in [0,1]} F_\lambda(x, z),$$
where $F_\lambda : X \times X \to \mathbb{R}_\infty$ is the convex function defined as $F_\lambda(x, z) := \lambda f(x) + (1 - \lambda)g(x + z)$. Observe that each function $F_\lambda(x_0, \cdot)$ is finite and continuous at θ. Then, using Corollary 3.4.9, we have
$$\inf_{x \in X} F(x, \theta) = \inf_{x \in X} \max_{\lambda \in [0,1]} F_\lambda(x, \theta) = \max_{\lambda \in [0,1]} \inf_{x \in X} F_\lambda(x, \theta),$$
where, thanks to Corollary 4.2.5,
$$\inf_{x \in X} F_\lambda(x, \theta) = - \min_{z^* \in X^*} F_\lambda^*(\theta, z^*) \text{ for all } \lambda \in [0, 1].$$

Consequently,
$$\inf_{x \in X} (\max\{f(x), g(x)\}) = \inf_{x \in X} F(x, \theta) = \max_{\lambda \in [0,1]} \left(- \min_{z^* \in X^*} F_\lambda^*(\theta, z^*) \right)$$
and, since $F_\lambda^*(\theta, z^*) = (\lambda f)^*(-z^*) + ((1 - \lambda)g)^*(z^*)$, we conclude that
$$\inf_{x \in X} (\max\{f(x), g(x)\}) = - \min_{z^* \in X^*,\, \lambda \in [0,1]} ((\lambda f)^*(z^*) + ((1 - \lambda)g)^*(-z^*)).$$

We close this section by introducing another type of duality, called *Singer–Toland duality*, which is also a consequence of Theorem 3.2.2. Note that when $g \equiv 0$, (4.102) reduces to the simple relation $\inf_X f = -f^*(\theta)$.

4.2. CONVEX DUALITY

Proposition 4.2.8 *Consider two functions $f, g : X \to \mathbb{R}_\infty$ such that $f^* \in \Gamma_0(X^*)$ and $g \in \Gamma_0(X)$. Then we have*

$$\inf_{x \in X} (f(x) - g(x)) = \inf_{x^* \in X^*} (g^*(x^*) - f^*(x^*)). \tag{4.102}$$

Proof. We have, for all $x^* \in X^*$,

$$\inf_{x \in X} (f(x) - g(x)) = \inf_{x \in X} (f(x) - \langle x^*, x \rangle + \langle x^*, x \rangle - g(x))$$
$$\leq \inf_{x \in X} (f(x) - \langle x^*, x \rangle) + g^*(x^*) = -f^*(x^*) + g^*(x^*),$$

and, taking into account our conventions, we conclude that

$$\inf_{x \in X} (f(x) - g(x)) \leq \inf_{x^* \in X^*} (g^*(x^*) - f^*(x^*)).$$

Moreover, since $g^* \in \Gamma_0(X)$ by Proposition 3.1.4, arguments similar to the above ones show that

$$\inf_{x^* \in X^*} (g^*(x^*) - f^*(x^*)) \leq \inf_{x \in X} (f^{**}(x) - g^{**}(x)) \leq \inf_{x \in X} (f(x) - g^{**}(x)),$$

and the desired equality follows as $g^{**} = g$, due to Theorem 3.2.2. ∎

Proposition 4.2.8 allows us to express the conjugate of the difference $f - g$ in terms of the conjugates of f and g.

Corollary 4.2.9 *Consider two functions $f, g : X \to \mathbb{R}_\infty$ such that $f^* \in \Gamma_0(X^*)$ and $g \in \Gamma_0(X)$. Then we have that*

$$(f - g)^* = f^* \boxminus g^* := \sup_{z^* \in X^*} (f^*(z^* + \cdot) - g^*(z^*)).$$

Proof. Given $x^* \in X^*$, we have

$$(f - g)^*(x^*) = -\inf_{x \in X} (f(x) - \langle x^*, x \rangle - g(x)),$$

and Proposition 4.2.8 applied to the pair $(f - x^*, g)$ yields

$$(f - g)^*(x^*) = -\inf_{z^* \in X^*} (g^*(z^*) - f^*(z^* + x^*)) = (f^* \boxminus g^*)(x^*).$$

∎

The following result establishes (global) necessary optimality conditions for unconstrained *dc (difference of convex) optimization*. This extends the classical optimality conditions for convex functions (see (4.8)).

Proposition 4.2.10 *Given $f, g \in \Gamma_0(X)$, we assume that x is an ε-minimum of the dc optimization problem $\inf_X(f - g)$. Then*

$$\partial_\delta g(x) \subset \partial_{\delta+\varepsilon} f(x) \text{ for all } \delta \geq 0.$$

Proof. Take $\delta \geq 0$ and $x^* \in \partial_\delta g(x)$. Observe that $f^* \in \Gamma_0(X^*)$, due to Proposition 3.1.4. Then, according to Proposition 4.2.8, for all $x^* \in X^*$ we have

$$f(x) - g(x) \leq \inf_X(f - g) + \varepsilon$$
$$= \inf_{X^*}(g^* - f^*) + \varepsilon \leq g^*(x^*) - f^*(x^*) + \varepsilon.$$

Thus,

$$f(x) + f^*(x^*) \leq g(x) + g^*(x^*) + \varepsilon \leq \langle x, x^* \rangle + \delta + \varepsilon,$$

and we deduce that $x^* \in \partial_{\delta+\varepsilon} f(x)$. ∎

4.3 Convexity in Banach spaces

In this section we adapt some results from previous sections for the setting of Banach spaces. We consider a Banach space X and denote by X^* and X^{**} the associated dual and bidual spaces, respectively. All the given norms are denoted by $\|\cdot\|$ (when no confusion is possible), while $B_X(x, r)$ is the closed ball centered at $x \in X$ with radius $r > 0$. The unit ball is $B_X := B_X(\theta, 1)$. The balls $B_{X^*}(x^*, r)$ and $B_{X^{**}}(z, r)$, when $x^* \in X^*$ and $z \in X^{**}$, as well as the closed unit balls B_{X^*} and $B_{X^{**}}$ are defined similarly. We identify X as a linear subspace of X^{**} by means of the injection $x \equiv \hat{x} := \langle \cdot, x \rangle : X^* \to \mathbb{R}$ (see (2.26)). Along with the norm topology on X^{**}, we will also consider the w^*-topology $w^{**} \equiv \sigma(X^{**}, X^*)$ together with the associated convergence $\to^{w^{**}}$. Remember that $\to^w$ and $\to^{w^*}$ represent the convergence with respect to the weak topologies $w \equiv \sigma(X, X^*)$ and $w^* \equiv \sigma(X^*, X)$, respectively, and that $\to$ denotes the *norm*-convergence in X, X^*, and X^{**}.

The main issue in the current Banach setting is that sometimes one is led to use the *norm*-topology on X^*, even though it is not compatible with the dual pair (X, X^*) outside reflexive spaces. For instance, given a function $f : X \to \overline{\mathbb{R}}$, nothing prevents defining the conjugate $(f^*)^*$ of f^* as a function defined on X^{**} by

4.3. CONVEXITY IN BANACH SPACES

$$(f^*)^*(z) = \sup\{\langle z, x^*\rangle - f^*(x^*) : \ x^* \in X^*\}, \ z \in X^{**}.$$

It follows easily from this definition that the restriction of $(f^*)^*$ to X coincides with the biconjugate of f, f^{**}, introduced above and, therefore, Theorem 3.2.2 implies that the restriction to X of $(f^*)^*$ coincides with the closed convex hull $\overline{\text{co}} f$ under simple conditions satisfied by f. However, a relation of the form $(f^*)^* = \overline{\text{co}} f$ on the whole space X^{**} is obviously out of the scope of this book. Another difficulty that arises in the Banach setting comes from the fact that the subdifferential of f^* is a subset of X^{**} and, thus, $\partial f^*(x^*)$ might contain elements in $X^{**} \setminus X$. The following example illustrates the last observation.

Example 4.3.1 *Take*

$$X = c_0 := \{(x_n)_{n \geq 1} : x_n \in \mathbb{R}, \ x_n \to 0\},$$

endowed with the ℓ_∞-norm, so that $X^* = \ell_1$ and $X^{**} = \ell_\infty$. We consider the function $f := I_{B_X}$, so that $f^* \equiv \|\cdot\|_{\ell_1}$ and f^* is continuous. Moreover, for every $x^* \in \ell_1$ such that $x_n^* > 0$ for all $n \geq 1$, it is known that f^* is Gâteaux-differentiable at x^* (see Exercise 58), so that

$$\partial f^*(x^*) = \{(f^*)'_G(x^*)\} = \{(1, 1, \ldots)\} \subset \ell_\infty \setminus c_0.$$

One of the possibilities to overcome the aforementioned difficulty is to use the compatible dual pair $((X^*, \|\cdot\|_*), (X^{**}, w^{**}))$. To do this, we begin by introducing extensions to the bidual space X^{**} of functions defined on X. Associated with a function $f : X \to \mathbb{R}_\infty$, we consider its extension $\hat{f} : X^{**} \to \mathbb{R}_\infty$ to X^{**} given by

$$\hat{f}(z) := \begin{cases} f(x), & \text{if } z = \hat{x}, \ x \in X, \\ +\infty, & \text{if not}, \end{cases} \qquad (4.103)$$

where $x \in X \mapsto \hat{x} := \langle \cdot, x\rangle \in X^{**}$ is the usual injection mapping. Note that the closure of $\hat{f}$ with respect to the w^{**}-topology, denoted by $\mathrm{cl}^{w^{**}} \hat{f} : X^{**} \to \overline{\mathbb{R}}$, is given by

$$(\mathrm{cl}^{w^{**}} \hat{f})(z) := \liminf_{y \to^{w^{**}} z, \ y \in X^{**}} \hat{f}(y) = \liminf_{x \to^{w^{**}} z, \ x \in X} f(x),$$

so that, for all $x \in X$,

$$(\mathrm{cl}^{w^{**}} \hat{f})_{|X}(x) = (\mathrm{cl}^w f)(x) := \liminf_{y \to^w x, \ y \in X} f(y), \qquad (4.104)$$

where $\operatorname{cl}^w f$ denotes the closure of f with respect to the weak topology $w = \sigma(X, X^*)$. Thus, in the particular case where X is reflexive, we have that $\hat{f} \equiv f$ and, thus, $\operatorname{cl}^{w^{**}} \hat{f} = \operatorname{cl}^w f$. Also, taking into account (3.7), we easily check that

$$f^* = (\hat{f})^* = (\operatorname{cl}^{w^{**}} \hat{f})^* \text{ and } \partial_\varepsilon \hat{f}(x) = \partial_\varepsilon f(x) \text{ for every } x \in X \text{ and } \varepsilon \geq 0.$$

The following lemma, which is used in the proof of Theorem 4.3.3 below, establishes another relation between $\partial(\operatorname{cl}^{w^{**}} \hat{f})$ and $\partial(\operatorname{cl}^w f)$.

Lemma 4.3.2 *For every function $f : X \to \mathbb{R}_\infty$, we have*

$$X \cap (\partial(\operatorname{cl}^{w^{**}} \hat{f}))^{-1} = (\partial(\operatorname{cl}^w f))^{-1}.$$

Proof. Consider the compatible dual pair $((X^{**}, w^{**}), (X^*, \|\cdot\|))$. We have that $x \in X \cap (\partial(\operatorname{cl}^{w^{**}} \hat{f}))^{-1}(x^*)$ if and only if $x \in X$ and, by Proposition 4.1.6(v),

$$(\operatorname{cl}^{w^{**}} \hat{f})(x) + (\operatorname{cl}^{w^{**}} \hat{f})^*(x^*) = \langle x^*, x \rangle.$$

Then, given that $(\operatorname{cl}^{w^{**}} \hat{f})(x) = (\operatorname{cl}^w f)(x)$ because $x \in X$, and $(\operatorname{cl}^{w^{**}} \hat{f})^* = (\hat{f})^* = f^* = (\operatorname{cl}^w f)^*$, due to (3.7), the above relation reads $(\operatorname{cl}^w f)(x) + (\operatorname{cl}^w f)^*(x^*) = \langle x^*, x \rangle$; equivalently, $x \in \partial(\operatorname{cl}^w f))^{-1}(x^*)$. Therefore, $x \in X \cap (\partial(\operatorname{cl}^{w^{**}} \hat{f}))^{-1}(x^*)$ if and only if $x \in (\partial(\operatorname{cl}^w f))^{-1}(x^*)$. ∎

The following result specifies Theorem 3.2.2 to the current Banach spaces framework, where the dual space X^* is endowed with its dual norm topology. When applied to the indicator function of the unit closed ball in X, B_X, this result gives rise to the Goldstein theorem (Exercise 65).

Theorem 4.3.3 *Let $f : X \to \mathbb{R}_\infty$ be a function having a continuous affine minorant. Then*

$$(f^*)^* = \overline{\operatorname{co}}^{w^{**}} (\hat{f}) \qquad (4.105)$$

and, in particular,

$$((f^*)^*)_{|X} \equiv \overline{\operatorname{co}} f. \qquad (4.106)$$

Proof. The function $\hat{f}$ satisfies $(\hat{f})^* = f^*$. Then, considering that $((X^{**}, w^{**}), (X^*, \|\cdot\|))$ is a compatible dual pair, it follows that the conjugate of $(\hat{f})^*$ is nothing other than the biconjugate of $\hat{f}$; that is, $(f^*)^* = ((\hat{f})^*)^* = (\hat{f})^{**}$. Also, using the current assumption, if $x_0^* \in X^*$

4.3. CONVEXITY IN BANACH SPACES

and $\alpha \in \mathbb{R}$ are such that $f(\cdot) \geq l(\cdot) := \langle x_0^*, \cdot \rangle + \alpha$, then any affine continuous extension $\hat{l}$ of the mapping l to X^{**} satisfies $\hat{f}(\cdot) \geq \hat{l}(\cdot)$. Therefore, applying Theorem 3.2.2 in the pair $((X^{**}, w^{**}), (X^*, \|\cdot\|))$, we conclude that

$$(f^*)^* = (\hat{f})^{**} = \overline{\text{co}}^{w^{**}}(\hat{f}),$$

which entails the first assertion of the proposition. Finally, for each $x \in X$, (4.105) gives rise to

$$((f^*)^*)_{|X}(x) = \left(\overline{\text{co}}^{w^{**}}(\hat{f})\right)(x) = \left(\text{cl}^{w^{**}}(\widehat{\text{co}\, f})\right)(x),$$

where $\widehat{\text{co}\, f}$ is the extension of the function $\text{co}\, f$ to X^{**}, as defined in (4.103). Thus, as we can easily verify that $\widehat{\text{co}\, f} = \text{co}\, \hat{f}$, the last equality above yields $((f^*)^*)_{|X}(x) = \left(\text{cl}(\text{co}\, \hat{f})\right)(x) = (\overline{\text{co}} f)(x)$. ∎

Although $\partial f^*(x^*)$ may not have points in X (as in Exercise 58), the following corollary shows that the approximate subdifferential $\partial_\varepsilon f^*(x^*)$ with $\varepsilon > 0$ can be built on its elements in X. Remember that the Mackey topology on X^* is denoted by τ.

Corollary 4.3.4 *Given a function $f \in \Gamma_0(X)$, for every $x^* \in X^*$ and $\varepsilon > 0$, we have*

$$\partial_\varepsilon f^*(x^*) = \text{cl}^{w^{**}}(\partial_\varepsilon f^*(x^*) \cap X) = \text{cl}^{w^{**}}((\partial_\varepsilon f)^{-1}(x^*)), \quad (4.107)$$

and, consequently,

$$\partial f^*(x^*) = \bigcap_{\varepsilon > 0} \text{cl}^{w^{**}}(\partial_\varepsilon f^*(x^*) \cap X) = \bigcap_{\varepsilon > 0} \text{cl}^{w^{**}}((\partial_\varepsilon f)^{-1}(x^*)).$$

In particular, if f^* is τ-continuous at $x^* \in X^*$, then for every $\varepsilon \geq 0$

$$\partial_\varepsilon f^*(x^*) = (\partial_\varepsilon f)^{-1}(x^*). \quad (4.108)$$

Proof. Fix $x^* \in X^*$. Taking into account that $(f^*)^*_{|X} \equiv f$ by Theorem 4.3.3, for all $\varepsilon \geq 0$ we have that

$$\partial_\varepsilon f^*(x^*) \cap X = \{z \in X : f^*(x^*) + (f^*)^*(z) \leq \langle z, x^* \rangle + \varepsilon\}$$
$$= \{x \in X : f^*(x^*) + f(x) \leq \langle x^*, x \rangle + \varepsilon\} = (\partial_\varepsilon f)^{-1}(x^*). \quad (4.109)$$

Furthermore, by applying Proposition 4.1.12 in the pair $((X^*, \|\cdot\|), (X^{**}, w^{**}))$, when $\varepsilon > 0$ we get

$$(f^*)'_\varepsilon(x^*; z^*) = \sigma_{\partial_\varepsilon f^*(x^*)}(z^*) \text{ for all } z^* \in X^*.$$

The same proposition applied in the pair $((X^*, w^*), (X, \|\cdot\|))$ gives rise to

$$(f^*)'_\varepsilon(x^*; z^*) = \sigma_{\partial_\varepsilon f^*(x^*) \cap X}(z^*) \text{ for all } z^* \in X^*.$$

Consequently, combining with (4.109),

$$\sigma_{\partial_\varepsilon f^*(x^*)}(z^*) = \sigma_{\partial_\varepsilon f^*(x^*) \cap X}(z^*) = \sigma_{(\partial_\varepsilon f)^{-1}(x^*)}(z^*) \text{ for all } z^* \in X^*,$$

and Corollary 3.2.9 implies the first statement of the corollary. The second statement is an immediate consequence of the first.

To show the last statement, we assume that f^* is τ-continuous at x^*. Then f^* is also *norm*-continuous at x^* and by applying Proposition 4.1.22, first in the pair $((X^*, \|\cdot\|), (X^{**}, w^{**}))$ and next in the pair $((X^*, \tau), (X, \|\cdot\|))$, we deduce that

$$\sigma_{\partial_\varepsilon f^*(x^*)}(z^*) = (f^*)'_\varepsilon(x^*; z^*)$$
$$= \sigma_{\partial_\varepsilon f^*(x^*) \cap X}(z^*) \text{ for all } z^* \in X^* \text{ and } \varepsilon \geq 0.$$

This shows that $\partial_\varepsilon f^*(x^*) = \text{cl}^{w^{**}}(\partial_\varepsilon f^*(x^*) \cap X)$, due to Corollary 3.2.9. Moreover, since f^* is τ-continuous at x^*, by applying Proposition 4.1.22 to the pair $((X^*, \tau), (X, w))$ it follows that the set $\partial_\varepsilon f^*(x^*) \cap X$ is w-compact and, thus, w^{**}-compact. So, $\partial_\varepsilon f^*(x^*) = \partial_\varepsilon f^*(x^*) \cap X = (\partial_\varepsilon f)^{-1}(x^*)$, due to (4.109). ∎

Remember that (4.108) above holds for every function $f \in \Gamma_0(X)$ when X^* is endowed with the w^*-topology (see (4.18)). As an illustration, we apply it in the following example to the indicator function of the closed unit ball in X, B_X.

Example 4.3.5 *Consider the convex function $f = I_{B_X}$, so that*

$$f^* = \sigma_{B_X} = \sigma_{B_{X^{**}}} = \|\cdot\|_* \text{ (the norm in } X^*),$$

due to the Goldstein theorem. Therefore, f^* is obviously (norm-) continuous in X^* and we have, applying (4.14) in the compatible dual pair $((X^*, \|\cdot\|_*), (X^{**}, w^{**}))$,

$$\partial f^*(\theta) = B_{X^{**}}.$$

At the same time, applying (4.13) in the compatible dual pair $((X^*, w^*), (X, \|\cdot\|))$, we get $\partial f^*(\theta) \cap X = B_X$ and, again, the Goldstein theorem yields

4.3. CONVEXITY IN BANACH SPACES

$$\partial f^*(\theta) = B_{X^{**}} = \mathrm{cl}^{w^{**}}(B_X) = \mathrm{cl}^{w^{**}}(\partial f^*(\theta) \cap X).$$

This formula can also be obtained from (4.107), since $\partial_\varepsilon f^*(\theta) = \partial f^*(\theta)$ by the positive homogeneity of the norm.

Next, we discuss the continuity assumption imposed on f^* in Corollary 4.3.4: If $f^* = \|\cdot\|_*$ is τ-continuous somewhere, then Corollary 2.2.7 would imply that f^* is also τ-continuous at θ. Consequently, (4.108) would lead us to

$$B_{X^{**}} = \partial f^*(\theta) = \partial f^*(\theta) \cap X = B_X,$$

and X would be a reflexive Banach space. But if, instead of the τ-continuity, we assume the (more restrictive) w^*-continuity of f^* at θ (equivalently, somewhere in X^*), then, due to the continuity of $f = I_{B_X}$ at θ, Proposition 3.3.7 would imply that X is a finite-dimensional space.

The following result is the well-known *Ekeland variational principle*, which is stated in the broader framework of complete metric spaces.

Theorem 4.3.6 *Let (X, d) be a complete metric space, and let $f : X \to \mathbb{R}_\infty$ be an lsc function, which is bounded below. Let $x_0 \in X$ and $\varepsilon > 0$ such that*

$$f(x_0) \leq \inf\nolimits_X f + \varepsilon.$$

Then, for every $\lambda > 0$, there exists $x_\lambda \in X$ such that

$$d(x_\lambda, x_0) \leq \lambda, \quad |f(x_\lambda) - f(x_0)| \leq \varepsilon,$$

and

$$f(x_\lambda) < f(x) + \varepsilon \lambda^{-1} d(x_\lambda, x) \text{ for all } x \neq x_\lambda.$$

Proposition 4.3.7 shows that the ε-subdifferential of convex functions defined on Banach spaces is approximated by exact subdifferentials at nearby points.

Proposition 4.3.7 *Given $f \in \Gamma_0(X)$ and $x \in \mathrm{dom}\, f$, for every $\varepsilon > 0$ and $x^* \in \partial_\varepsilon f(x)$ there exist $x_\varepsilon \in B_X(x, \sqrt{\varepsilon})$, $y_\varepsilon^* \in B_{X^*}$, and $\lambda_\varepsilon \in [-1, 1]$ such that*

$$x_\varepsilon^* := x^* + \sqrt{\varepsilon}(y_\varepsilon^* + \lambda_\varepsilon x^*) \in \partial f(x_\varepsilon),$$

and

$$|\langle x_\varepsilon^*, x_\varepsilon - x\rangle| \le \varepsilon + \sqrt{\varepsilon}, \quad |f(x_\varepsilon) - f(x)| \le \varepsilon + \sqrt{\varepsilon}.$$

Proof. Fix $x \in \operatorname{dom} f$, $\varepsilon > 0$, and $x^* \in \partial_\varepsilon f(x)$. We consider the function $g := f(\cdot) - \langle x^*, \cdot \rangle \in \Gamma_0(X)$, so that

$$g(x) \le \inf_X g + \varepsilon;$$

that is, g is bounded below by $g(x) - \varepsilon \in \mathbb{R}$. Then, applying Theorem 4.3.6 with $\lambda = \sqrt{\varepsilon}$ in the Banach space $(X, \|\cdot\|_0)$, where $\|\cdot\|_0 := \|\cdot\| + |\langle x^*, \cdot \rangle|$, we find $x_\varepsilon \in X$ such that $|g(x_\varepsilon) - g(x)| \le \varepsilon$,

$$\|x_\varepsilon - x\|_0 = \|x_\varepsilon - x\| + |\langle x^*, x_\varepsilon - x\rangle| \le \sqrt{\varepsilon} \qquad (4.110)$$

and

$$g(x_\varepsilon) < g(z) + \sqrt{\varepsilon}\,\|z - x_\varepsilon\| + \sqrt{\varepsilon}\,|\langle x^*, z - x_\varepsilon\rangle| \quad \text{for all } z \ne x_\varepsilon.$$

Hence, since $\partial(\|\cdot - x_\varepsilon\|)(x_\varepsilon) = \partial \|\cdot\|(\theta) = B_{X^*}$ by (4.14), and (see Example 4.1.21)

$$\partial |\langle x^*, \cdot - x_\varepsilon\rangle|(x_\varepsilon) = \partial |\langle x^*, \cdot\rangle|(\theta) = [-1,1]x^*,$$

by Proposition 4.1.20 we find some $y_\varepsilon^* \in B_{X^*}$ and $\lambda_\varepsilon \in [-1,1]$ such that

$$\theta \in \partial(g + \sqrt{\varepsilon}\,\|\cdot - x_\varepsilon\| + \sqrt{\varepsilon}\,|\langle x^*, \cdot - x_\varepsilon\rangle|)(x_\varepsilon)$$
$$= \partial g(x_\varepsilon) - \sqrt{\varepsilon}\,y_\varepsilon^* - \sqrt{\varepsilon}\,\lambda_\varepsilon x^* = \partial f(x_\varepsilon) - x^* - \sqrt{\varepsilon}\,y_\varepsilon^* - \sqrt{\varepsilon}\,\lambda_\varepsilon x^*;$$

that is, $x_\varepsilon^* := x^* + \sqrt{\varepsilon}(y_\varepsilon^* + \lambda_\varepsilon x^*) \in \partial f(x_\varepsilon)$. Moreover, using (4.110),

$$\begin{aligned}|\langle x_\varepsilon^* - x^*, x_\varepsilon - x\rangle| &= \sqrt{\varepsilon}\,|\langle y_\varepsilon^* + \lambda_\varepsilon x^*, x_\varepsilon - x\rangle| \\ &\le \sqrt{\varepsilon}(\|y_\varepsilon^*\|\,\|x_\varepsilon - x\| + |\lambda_\varepsilon|\,|\langle x^*, x_\varepsilon - x\rangle|) \\ &\le \sqrt{\varepsilon}\,\|x_\varepsilon - x\|_0 \le \varepsilon.\end{aligned}$$

Thus, again by (4.110), we have $|\langle x_\varepsilon^*, x_\varepsilon - x\rangle| \le |\langle x_\varepsilon^* - x^*, x_\varepsilon - x\rangle| + \sqrt{\varepsilon} \le \varepsilon + \sqrt{\varepsilon}$, and the inequality $|g(x_\varepsilon) - g(x)| \le \varepsilon$ entails

$$|f(x_\varepsilon) - f(x)| \le \varepsilon + |\langle x^*, x - x_\varepsilon\rangle| \le \varepsilon + \sqrt{\varepsilon}.$$

∎

The following result gives another variant of Proposition 4.3.7.

4.3. CONVEXITY IN BANACH SPACES

Proposition 4.3.8 *Given $f \in \Gamma_0(X)$ and $x \in \text{dom}\, f$, for every $\varepsilon > 0$ and $x^* \in \partial_\varepsilon f(x)$ there exist $x_\varepsilon \in B_X(x, \sqrt{\varepsilon})$ and $x_\varepsilon^* \in \partial f(x_\varepsilon)$ such that $\|x_\varepsilon^* - x^*\| \leq \sqrt{\varepsilon}$ and $|f(x_\varepsilon) - f(x) + \langle x^*, x - x_\varepsilon \rangle| \leq 2\varepsilon$.*

Proof. The proof is similar to that of Proposition 4.3.7, but applying Theorem 4.3.6 in the Banach space X (endowed with the original norm instead of $\|\cdot\|_0$). ∎

The following result is a notable consequence of Proposition 4.3.7 which, in particular, shows that a function from $\Gamma_0(X)$ is subdifferentiable in a dense set of its effective domain. A non-convex version of this result is provided in Corollary 8.3.10.

Corollary 4.3.9 *The following assertions hold for every function $f \in \Gamma_0(X)$:*

(i) For every $x \in \text{dom}\, f$, there exists a sequence $(x_n)_n \subset X$ norm-convergent to x such that $f(x_n) \to f(x)$ and $\partial f(x_n) \neq \emptyset$ for all $n \geq 1$; that is, in particular, $\text{cl}(\text{dom}\, f) = \text{cl}(\text{dom}\, \partial f)$.

(ii) For every $x^* \in \text{dom}\, f^*$, there exists a sequence $(x_n^*)_n \subset X^*$ norm-convergent to x^* such that $f^*(x_n^*) \to f^*(x^*)$ and $x_n^* \in \text{Im}\, \partial f$ for all $n \geq 1$; that is, in particular, $\text{cl}^{\|\cdot\|_*}(\text{dom}\, f^*) = \text{cl}^{\|\cdot\|_*}(\text{Im}\, \partial f)$.

Proof. (i) Take $x \in \text{dom}\, f$ and $\varepsilon > 0$, so that $\partial_\varepsilon f(x) \neq \emptyset$ by Proposition 4.1.10. Next, given $x^* \in \partial_\varepsilon f(x)$, Proposition 4.3.7 yields some $x_\varepsilon \in B_X(x, \sqrt{\varepsilon})$ and $x_\varepsilon^* \in \partial f(x_\varepsilon)$ such that $|f(x_\varepsilon) - f(x)| \leq \varepsilon + \sqrt{\varepsilon}$. This proves assertion (i) as well as the relation $\text{cl}(\text{dom}\, f) = \text{cl}(\text{dom}\, \partial f)$.

(ii) The proof of this part also uses Proposition 4.3.7 (see the bibliographical notes of this chapter). ∎

We close this section by giving, without proof, a well-known primal characterization of the differentiability of conjugate functions in the setting of Banach spaces. A non-convex counterpart to the following result is given in section 8.3.

Proposition 4.3.10 *The following assertions are equivalent for every function $f \in \Gamma_0(X)$:*

(i) The function f^* is Fréchet-differentiable at x^*, with Fréchet-derivative $(f^*)'(x^*)$.

(ii) For every sequence $(x_n)_n \subset X$ such that $\langle x_n, x^* \rangle - f(x_n) \to f^*(x^*)$, we have that $(x_n)_n$ norm-converges to $(f^*)'(x^*)$, and so $(f^*)'(x^*) \in X$.

Proposition 4.3.10 applies to the support function and gives rise to the following result.

Corollary 4.3.11 *The following assertions are equivalent for every non-empty closed convex set $C \subset X$:*

(i) The support function $\sigma_C : X^* \to \mathbb{R}_\infty$ is Fréchet-differentiable at $x^* \in X^*$ and $(\sigma_C)'(x^*) \in X$.

(ii) For every sequence $(x_n)_n \subset X$ such that $\langle x_n, x^* \rangle - I_C(x_n) \to \sigma_C(x^*) \in \mathbb{R}$ (equivalently, $\langle x_n, x^* \rangle \to \sigma_C(x^*)$ with $(x_n)_n \subset C$), we have that $(x_n)_n$ norm-converges to $(\sigma_C)'(x^*)$.

We recall below the *Stegall variational principle*, which is valid in Banach spaces with the RNP. A slight extension of this result will be proved in Corollary 8.3.8. We need the following definition.

Definition 4.3.12 *Given a function $f : X \to \mathbb{R}_\infty$ and a set $C \subset \operatorname{dom} f$, a point $x \in C$ is said to be a* strong minimum *of f on C if every sequence $(x_n)_n \subset C$ such that $f(x_n) \to \inf_C f$ converges to x.*

Theorem 4.3.13 *Let X be a Banach space with the RNP, let the function $f : X \to \mathbb{R}_\infty$ be lsc bounded below, and let $C \subset \operatorname{dom} f$ be a closed bounded convex set. Then there exists some $\mathcal{G}_\delta$-dense set $D \subset X^*$ such that every function $f - x^*$, with $x^* \in D$, attains a strong minimum on C.*

4.4 Subdifferential integration

We provide in this section some *integration criteria* for convex functions defined on the lcs X. The first result, given in Theorem 4.4.3, uses an integration criterion that assumes continuity somewhere of the involved functions or their conjugates. The need for such continuity assumptions is necessary for the current setting of locally convex spaces, as there are functions in $\Gamma_0(X)$ with an empty subdifferential everywhere.

We first prove a couple of lemmas related to the integration of functions from $\Gamma_0(\mathbb{R})$. The integral in the first lemma is given in the sense of *Riemann integration*.

Lemma 4.4.1 *Let $\varphi \in \Gamma_0(\mathbb{R})$ such that $J := \operatorname{int}(\operatorname{dom} \varphi) \neq \emptyset$ and take $\alpha \in \mathbb{R}$. Then $\varphi'(t; \alpha) \in \mathbb{R}$ for all $t \in J$, $\varphi'(\cdot; \alpha)$ is non-decreasing on J if $\alpha > 0$, $\varphi'(\cdot; \alpha)$ is non-increasing on J if $\alpha < 0$, and we have*

$$\int_{s_1}^{s_2} \varphi'(t; \alpha) dt = \alpha(\varphi(s_2) - \varphi(s_1)) \text{ for all } s_1 \leq s_2, s_1, s_2 \in J.$$

4.4. SUBDIFFERENTIAL INTEGRATION

Proof. Taking into account the positive homogeneity of the directional derivative, we prove the lemma when $\alpha = 1$ (the case $\alpha = -1$ is similar). Due to the convexity of φ, by Corollary 2.2.8 the function φ is locally Lipschitz on J, and Proposition 2.2.5 implies that $\varphi'(s; 1)$ exists and belongs to $\mathbb{R}$ for every $s \in J$. Moreover, using (2.40), for every $s_1, s_2 \in J$ and $h > 0$ such that $s_1 < s_1 + h < s_2 < s_2 + h$ we have

$$\frac{\varphi(s_1 + h) - \varphi(s_1)}{h} \leq \frac{\varphi(s_2 + h) - \varphi(s_1)}{s_2 + h - s_1} \leq \frac{\varphi(s_2 + h) - \varphi(s_2)}{h}.$$

So, letting $h \downarrow 0$, the non-decreasingness of $\varphi'(\cdot; 1)$ on J follows by Proposition 2.2.5; hence, it is integrable on every set $[s_1, s_2] \subset J$. Next, given $[s_1, s_2] \subset J$ and $\eta > 0$, the integrability of $\varphi'(\cdot; 1)$ yields a subdivision $t_0 = s_1 < \ldots < t_k = s_2$, $k \geq 1$, of $[s_1, s_2]$ such that

$$\int_{s_1}^{s_2} \varphi'(t; 1) dt \leq \sum_{0 \leq i \leq k-1} \varphi'(t_i; 1)(t_{i+1} - t_i) + \eta.$$

Thus, since

$$\varphi'(t_i; 1) = \inf_{h>0} h^{-1}(\varphi(t_i + h) - \varphi(t_i)) \leq (t_{i+1} - t_i)^{-1}(\varphi(t_{i+1}) - \varphi(t_i)),$$

we deduce that

$$\int_{s_1}^{s_2} \varphi'(t; 1) dt \leq \sum_{0 \leq i \leq k-1} (\varphi(t_{i+1}) - \varphi(t_i)) + \eta = \varphi(s_2) - \varphi(s_1) + \eta. \tag{4.111}$$

Similarly, we find another subdivision (denoted in the same way for simplicity) $t_0 = s < \ldots < t_l = s_2$, $l \geq 1$, such that

$$\int_{s_1}^{s_2} \varphi'(t; 1) dt \geq \sum_{0 \leq i \leq l-1} \varphi'(t_{i+1}; 1)(t_{i+1} - t_i) - \eta. \tag{4.112}$$

Moreover, Proposition 2.2.5 entails

$$\varphi'(t_{i+1}; 1) = \lim_{h \downarrow 0} h^{-1}(\varphi(t_{i+1} + h) - \varphi(t_{i+1}))$$
$$\geq (t_i - t_{i+1})^{-1}(\varphi(t_{i+1} + (t_i - t_{i+1})) - \varphi(t_{i+1}))$$
$$= (t_i - t_{i+1})^{-1}(\varphi(t_i) - \varphi(t_{i+1})),$$

and (4.112) gives rise to

$$\int_{s_1}^{s_2} \varphi'(t;1)dt \geq \sum_{0 \leq i \leq l-1} (\varphi(t_{i+1}) - \varphi(t_i)) - \eta = \varphi(s_2) - \varphi(s_1) - \eta.$$
(4.113)

Therefore, (4.111) and (4.113) lead us to

$$\varphi(s_2) - \varphi(s_1) - \eta \leq \int_{s_1}^{s_2} \varphi'(t;1)dt \leq \varphi(s_2) - \varphi(s_1) + \eta,$$

and we conclude the proof when $\eta \downarrow 0$. ∎

Lemma 4.4.2 *Let functions $\varphi, \psi \in \Gamma_0(\mathbb{R})$ such that $J := \text{int}(\text{dom}\,\varphi) \neq \emptyset$ and*

$$\partial\varphi(s) \subset \partial\psi(s) \text{ for all } s \in \mathbb{R}. \tag{4.114}$$

Then φ and ψ are equal up to some additive constant.

Proof. According to Proposition 4.1.22, (4.114) entails the following relationship between the directional derivatives of φ and ψ, for every $s \in J$ ($\subset \text{int}(\text{dom}\,\psi)$),

$$\varphi'(s;\alpha) = \max_{s^* \in \partial\varphi(s)} \alpha s^* \leq \max_{s^* \in \partial\psi(s)} \alpha s^* = \psi'(s;\alpha), \text{ for all } \alpha \in \mathbb{R}.$$

Then Lemma 4.4.1 gives as a result, for all $s_1 \leq s_2$, $s_1, s_2 \in J, \alpha \in \mathbb{R}$,

$$\alpha(\varphi(s_2) - \varphi(s_1)) = \int_{s_1}^{s_2} \varphi'(t;\alpha)dt \leq \int_{s_1}^{s_2} \psi'(t;\alpha)dt = \alpha(\psi(s_2) - \psi(s_1)),$$

and the lower semicontinuity of φ and ψ implies, due to Corollary 2.2.10, that

$$\varphi(s_2) - \varphi(s_1) = \psi(s_2) - \psi(s_1) \text{ for all } s_1, s_2 \in \text{dom}\,\varphi. \tag{4.115}$$

At the same time, due to (4.18), we also have

$$\partial\varphi^*(s) = (\partial\varphi)^{-1}(s) \subset (\partial\psi)^{-1}(s) = \partial\psi^*(s) \text{ for all } s \in \mathbb{R}. \tag{4.116}$$

If $\text{int}(\text{dom}\,\varphi^*) \neq \emptyset$, then the reasoning above also yields

$$\varphi^*(s_2) - \varphi^*(s_1) = \psi^*(s_2) - \psi^*(s_1) \text{ for all } s_1, s_2 \in \text{dom}\,\varphi^*, \tag{4.117}$$

so that $\psi^* \leq \varphi^* + (\psi^*(s_1) - \varphi^*(s_1))$ for any fixed $s_1 \in \text{dom}\,\varphi^*$ (the last set is non-empty and we have $\psi^*(s_1), \varphi^*(s_1) \in \mathbb{R}$, by Proposition 3.1.4 and the fact $\text{dom}\,\varphi^* \subset \text{dom}\,\psi^*$ coming from (4.117)). Thus, by Theorem 3.2.2, we infer that $\varphi \leq \psi + (\psi^*(s_1) - \varphi^*(s_1))$ and, so,

4.4. SUBDIFFERENTIAL INTEGRATION

dom $\psi \subset$ dom φ. Consequently, the conclusion of the lemma follows from (4.115). Otherwise, if int(dom φ^*) = $\emptyset$, then the properness of φ^* (also coming from Proposition 3.1.4) ensures that dom $\varphi^* = \{s^*\}$ for some $s^* \in \mathbb{R}$. So, $\varphi^* = I_{\{s^*\}} + \varphi^*(s^*)$ and, again, Theorem 3.2.2 implies that $\varphi(s) = s^*s - \varphi^*(s^*)$ for all $s \in \mathbb{R}$. In other words, dom $\varphi = \mathbb{R}$ and (4.115) also leads us to the desired conclusion. ∎

Theorem 4.4.3 *Given a function* $f \in \Gamma_0(X)$ *and an lsc function* $g : X \to \mathbb{R}_\infty$, *we assume that* (i) *f is continuous somewhere or* (ii) f^* *is τ-continuous somewhere and* $g \in \Gamma_0(X)$. *If*

$$\partial f(x) \subset \partial g(x) \text{ for all } x \in X, \tag{4.118}$$

then f and g are equal up to some additive constant.

Proof. The continuity assumption on f in (i) implies, due to Proposition 4.1.22, that $\partial f(x) \neq \emptyset$ for all $x \in \text{int}(\text{dom } f)$. Thus, assuming without loss of generality that $\theta \in \text{int}(\text{dom } f)$ and $f(\theta) = 0$, we get the continuity of f at θ together with $\emptyset \neq \partial f(\theta) \subset \partial g(\theta) = \partial(\overline{co}g)(\theta)$. Therefore, $\theta \in \text{int}(\text{dom } g)$ and we obtain that $(\overline{co}g)(\theta) = g(\theta)$; hence, we may also suppose that $g(\theta) = 0$.

Let us start by supposing that g is convex, so that $g \in \Gamma_0(X)$ as $\partial g(\theta) \neq \emptyset$. We also fix $u \in X$ and consider the functions $\varphi, \psi \in \Gamma_0(\mathbb{R})$ defined as

$$\varphi(s) := f(su) \text{ and } \psi(s) := g(su), \ s \in \mathbb{R}. \tag{4.119}$$

Notice that, for all sufficiently small $s>0$, we have that $su \in \text{int}(\text{dom } f)$ by (2.15) and, so, the continuity of f at θ ensures that φ is finite and continuous at such small s. Then, the conditions of Proposition 4.1.20 are fulfilled and we obtain, for all $s \in \mathbb{R}$,

$$\partial \varphi(s) = \{\langle x^*, u \rangle : \ x^* \in \partial f(su)\} \subset \{\langle x^*, u \rangle : \ x^* \in \partial g(su)\} \subset \partial \psi(s) \tag{4.120}$$

Consequently, since int(dom $\varphi) \neq \emptyset$ and $f(\theta) = g(\theta) = 0$, Lemma 4.4.2 entails that

$$f(su) = g(su) + f(\theta) + g(\theta) = g(su) \text{ for all } s \in \mathbb{R},$$

showing that $f(u) = g(u)$ for all $u \in X$.

To finish the proof of case (i), we consider that g is any lsc function; that is, g is not necessarily convex but satisfies $\overline{co}g \in \Gamma_0(X)$, as a consequence of (4.118). Since $\partial f \subset \partial g \subset \partial(\overline{co}g)$, by (4.118), and $(\overline{co}g)(\theta) = f(\theta) = 0$, the reasoning above implies that $f = \overline{co}g$. More precisely, given any $x \in X$ such that $\partial f(x) \neq \emptyset$, (4.118) implies that $\partial g(x) \neq \emptyset$ and, so, $f(x) = (\overline{co}g)(x) = g(x)$. Moreover, if $x \in \text{dom}\, f \subset \text{cl}(\text{int}(\text{dom}\, f))$, then by (2.15) there exists a sequence $(x_n)_n \subset \text{int}(\text{dom}\, f)$ such that $x_n \to x$ and $f(x_n) \to f(x)$. Hence, Proposition 4.1.22 ensures that $f(x_n) = g(x_n)$ for all $n \geq 1$ and, taking limits as $n \to \infty$, the equality $f = \overline{co}g$ together with the lower semicontinuity of g entails

$$f(x) = \lim_n f(x_n) = \liminf_n g(x_n) \geq g(x) \geq (\overline{co}g)(x) = f(x);$$

that is, $f(x) = g(x)$. The last equality also holds when $x \notin \text{dom}\, f$, as $+\infty = f(x) = (\overline{co}g)(x) \leq g(x)$, and the proof is over under condition (i).

Finally, under condition (ii), (4.18) and (4.118) entail

$$\partial f^*(x^*) = (\partial f)^{-1}(x^*) \subset (\partial g)^{-1}(x^*) = \partial g^*(x^*) \text{ for all } x^* \in X^*,$$

and we are in case (i). Then, applying the paragraph above in the pair $((X^*, \tau), (X, \mathfrak{T}_X))$ yields $f^* = g^* - c$ for some c. Therefore, according to Theorem 3.2.2, we have $f = f^{**} = g^{**} + c = g + c$ and the conclusion follows. ∎

The following corollary uses an integration criterion slightly weaker than (4.118), which requires comparison of $\partial f(x)$ and $\partial g(x)$ only on the set $\text{int}(\text{dom}\, f)$.

Corollary 4.4.4 *Given a function $f \in \Gamma_0(X)$, which is continuous somewhere, and an lsc function $g := X \to \mathbb{R}_\infty$, we assume that*

$$\partial f(x) \subset \partial g(x) \text{ for all } x \in \text{int}(\text{dom}\, f).$$

Then there exists some $c \in \mathbb{R}$ such that $f = g + c$ on $\text{cl}(\text{dom}\, f)$.

Proof. First, we assume that g is convex, so that $g \in \Gamma_0(X)$ because $\partial g(x) \supset \partial f(x) \neq \emptyset$ for all $x \in \text{int}(\text{dom}\, f)$ ($\neq \emptyset$) due to Proposition 4.1.22. Also, we assume without loss of generality that $\theta \in \text{int}(\text{dom}\, f)$. We fix $\lambda \in \,]0,1[$ and denote $A_\lambda := \lambda\, \text{id}_X$ (a multiple of the identity mapping on X), which is an auto-adjoint mapping (i.e., $A_\lambda^* = A_\lambda$). Then, using (2.15), we have that $\lambda\, \text{cl}(\text{dom}\, f) \subset \text{int}(\text{dom}\, f)$, and the current assumption yields

$$A_\lambda \partial f(A_\lambda x) \subset A_\lambda \partial g(A_\lambda x) \text{ for all } x \in \text{cl}(\text{dom}\, f).$$

4.4. SUBDIFFERENTIAL INTEGRATION

Thus, thanks to (4.58) in Proposition 4.1.20, we have for all $x \in \operatorname{cl}(\operatorname{dom} f)$

$$\partial \left(f \circ A_\lambda + I_{\operatorname{cl}(\operatorname{dom} f)}\right)(x) = A_\lambda^* \partial f(A_\lambda x) + N_{\operatorname{cl}(\operatorname{dom} f)}(x)$$
$$\subset A_\lambda^* \partial g(A_\lambda x) + N_{\operatorname{cl}(\operatorname{dom} f)}(x)$$
$$\subset \partial \left(g \circ A_\lambda + I_{\operatorname{cl}(\operatorname{dom} f)}\right)(x).$$

This inclusion also trivially holds when $x \notin \operatorname{cl}(\operatorname{dom} f)$. Therefore, because the involved functions are in $\Gamma_0(X)$ and the convex function $f \circ A_\lambda + I_{\operatorname{cl}(\operatorname{dom} f)}$ is continuous somewhere in X, Theorem 4.4.3 yields some $c_\lambda \in \mathbb{R}$ such that, for all $x \in X$,

$$f(\lambda x) + I_{\operatorname{cl}(\operatorname{dom} f)}(x) = g(\lambda x) + I_{\operatorname{cl}(\operatorname{dom} f)}(x) + c_\lambda; \qquad (4.121)$$

in fact, we have that $c_\lambda = f(\theta) - g(\theta) =: c$ as $\theta \in \operatorname{dom} f$. Thus, taking the limits as $\lambda \uparrow 1$ in (4.121), the lower semicontinuity of the given functions and Proposition 2.2.10 imply that

$$f(x) = g(x) + c \text{ for all } x \in \operatorname{cl}(\operatorname{dom} f),$$

and we are done when g is convex.

Finally, we suppose that g is any lsc function (possibly not convex). Since $\partial f \subset \partial g \subset \partial(\overline{\operatorname{co}} g)$, by the first part we conclude that

$$f(x) = (\overline{\operatorname{co}} g)(x) + c \text{ for all } x \in \operatorname{cl}(\operatorname{dom} f). \qquad (4.122)$$

In particular, for all $x \in \operatorname{int}(\operatorname{dom} f)$ we have $\emptyset \neq \partial f(x) \subset \partial g(x)$ and, so, $f(x) = (\overline{\operatorname{co}} g)(x) + c = g(x) + c$. If $x \in \operatorname{dom} f$, then by (2.15) we find a sequence $(x_n)_n \subset \operatorname{int}(\operatorname{dom} f)$ such that $x_n \to x$ and $f(x_n) \to f(x)$. Therefore, using (4.122), we get

$$f(x) = \lim_n f(x_n) = \liminf_n g(x_n) + c$$
$$\geq g(x) + c \geq (\overline{\operatorname{co}} g)(x) + c = f(x);$$

that is, $f(x) = g(x) + c$. Finally, if $x \in \operatorname{cl}(\operatorname{dom} f) \setminus \operatorname{dom} f$, then (4.122) yields $+\infty = f(x) = (\overline{\operatorname{co}} g)(x) + c \leq g(x) + c$, and the relation $f \equiv g + c$ holds on $\operatorname{cl}(\operatorname{dom} f)$. ∎

In the following proposition, we modify the integration criterion in Corollary 4.4.4 requiring only that the subdifferentials of the functions involved intersect over the set $\operatorname{int}(\operatorname{dom} f) \cup \operatorname{int}(\operatorname{dom} g)$.

Proposition 4.4.5 *Let function $f, g \in \Gamma_0(X)$ be such that*

$$\partial f(x) \cap \partial g(x) \neq \emptyset \quad \text{for all } x \in \operatorname{int}(\operatorname{dom} f) \cup \operatorname{int}(\operatorname{dom} g).$$

Then f and g are equal up to some additive constant, provided that one of them is continuous somewhere.

Proof. The current assumption implies that $\operatorname{int}(\operatorname{dom} f) \subset \operatorname{dom} g$ and $\operatorname{int}(\operatorname{dom} g) \subset \operatorname{dom} f$; that is, $\operatorname{int}(\operatorname{dom} f) = \operatorname{int}(\operatorname{dom} g)$. Thus, denoting $h := f \square g$, the current assumption implies that $\operatorname{int}(\operatorname{dom} h) = 2\operatorname{int}(\operatorname{dom} f) = 2\operatorname{int}(\operatorname{dom} g)$, $\operatorname{cl}(\operatorname{dom} h) = 2\operatorname{cl}(\operatorname{dom} f) = 2\operatorname{cl}(\operatorname{dom} g)$ and for all $x \in \operatorname{int}(\operatorname{dom} f)$ (see Exercise 49)

$$h(2x) = f(x) + g(x) \text{ and } \partial h(2x) = \partial f(x) \cap \partial g(x) \neq \emptyset. \quad (4.123)$$

In particular, the convex function h is proper and continuous on the interior of its effective domain. Moreover, given $x \in \operatorname{cl}(\operatorname{dom} f)$, for every fixed $x_0 \in \operatorname{int}(\operatorname{dom} f) \ (\subset \operatorname{dom} g)$ and $\lambda \in \,]0,1[$ we have that $x_\lambda := \lambda x_0 + (1-\lambda)x \in \operatorname{int}(\operatorname{dom} f)$, by (2.15), and (4.123) entails

$$\begin{aligned}
f(x) + g(x) &\leq \liminf\nolimits_{\lambda \downarrow 0}(f(x_\lambda) + g(x_\lambda)) = \liminf\nolimits_{\lambda \downarrow 0} h(2x_\lambda) \\
&\leq \liminf\nolimits_{\lambda \downarrow 0}(\lambda h(2x_0) + (1-\lambda)h(2x)) \\
&= \liminf\nolimits_{\lambda \downarrow 0}(\lambda f(x_0) + \lambda g(x_0) + (1-\lambda)h(2x)) \\
&= h(2x) \leq f(x) + g(x),
\end{aligned}$$

showing that $h(2x) = f(x) + g(x)$ for all $x \in \operatorname{cl}(\operatorname{dom} f)$. In addition, if $x \notin \operatorname{cl}(\operatorname{dom} f)$, then $2x \notin \operatorname{cl}(\operatorname{dom} h)$ and we obtain that $+\infty = h(2x) \leq f(x) + g(x) = +\infty$. Therefore, $h(2\cdot) = f + g$ and Proposition 4.1.20 together with Exercise 49 yields, for all $x \in X$,

$$\partial(h(2\cdot))(x) = 2\partial h(2x) \subset 2(\partial f(x) \cap \partial g(x)) = \partial(2f)(x) \cap \partial(2g)(x).$$

Then, according to Theorem 4.4.3, we find some constants $c_1, c_2 \in \mathbb{R}$ such that

$$2f(x) = h(2x) + c_1 = 2g(x) + c_2 \quad \text{for all } x \in X,$$

and the proof is finished. ∎

To avoid the continuity assumption used in previous integration criteria, we proceed by giving other conditions that use the ε-subdifferential.

Proposition 4.4.6 *Given $f \in \Gamma_0(X)$ and a function $g : X \to \mathbb{R}_\infty$, we consider a function $\zeta : \mathbb{R}_+^* \to \mathbb{R}_+^*$ such that $\limsup_{\varepsilon \downarrow 0} \zeta(\varepsilon) = 0$ and*

4.4. SUBDIFFERENTIAL INTEGRATION

assume the existence of some $\varepsilon_0 > 0$ such that, for all $x \in X$ and $\varepsilon \in \,]0, \varepsilon_0]$,

$$\partial_\varepsilon f(x) \subset \partial_{\zeta(\varepsilon)} g(x).$$

Then f and g are equal up to some additive constant.

Proof. Let us first suppose that $g \in \Gamma_0(X)$. In addition, since $\emptyset \neq \partial_\varepsilon f(x) \subset \partial_{\zeta(\varepsilon)} g(x)$ due to Proposition 4.1.10, we deduce that $\emptyset \neq \mathrm{dom}\, f \subset \mathrm{dom}\, g$. So, we may assume that $\theta \in \mathrm{dom}\, f \cap \mathrm{dom}\, g$ and $f(\theta) = g(\theta) = 0$. Next, we fix $u \in \mathrm{dom}\, f$ ($\subset \mathrm{dom}\, g$) and consider, as in (4.119), the functions $\varphi, \psi \in \Gamma_0(\mathbb{R})$ defined by

$$\varphi(s) := f(su) \text{ and } \psi(s) := g(su).$$

Then, due to Proposition 4.1.16, the current assumption entails for every $s \in \mathbb{R}$

$$\begin{aligned}\partial \varphi(s) &= \bigcap_{\varepsilon > 0} \mathrm{cl}\left(\{\langle x^*, u \rangle : x^* \in \partial_\varepsilon f(su)\}\right) \\ &\subset \bigcap_{\varepsilon > 0} \mathrm{cl}\left(\{\langle x^*, u \rangle : x^* \in \partial_{\zeta(\varepsilon)} g(su)\}\right) \\ &\subset \bigcap_{\varepsilon > 0} \mathrm{cl}\left(\{\langle x^*, u \rangle : x^* \in \partial_\varepsilon g(su)\}\right) = \partial \psi(s). \end{aligned} \quad (4.124)$$

Note that if $\emptyset \neq \mathrm{int}(\mathrm{dom}\, \varphi)$ ($\subset \mathbb{R}$), then the function φ is continuous on $\mathrm{int}(\mathrm{dom}\, \varphi)$ (Corollary 2.2.9) and Theorem 4.4.3 yields some constant $c \in \mathbb{R}$ such that

$$f(su) = \varphi(s) = \psi(s) + c = g(su) + c \text{ for all } s \in \mathbb{R};$$

indeed, $c = f(\theta) - f(\theta) = 0$ and the last equality reduces to

$$f(u) = g(u) \text{ for all } u \in \mathrm{dom}\, f. \quad (4.125)$$

Otherwise, if $\mathrm{dom}\, \varphi = \{s_0\}$ for some $s_0 \in \mathbb{R}$, then $\varphi = \mathrm{I}_{\{s_0\}} + \varphi(s_0)$, and the conjugate of φ, given by $\varphi^*(s) = s_0 s - \varphi(s_0)$, $s \in \mathbb{R}$, is obviously continuous. Therefore, again due to Theorem 4.4.3, (4.124) also leads to (4.125), which in turn implies that

$$g(u) \leq f(u) \text{ for all } u \in X. \quad (4.126)$$

At the same time, taking into account (4.18), the current assumption also implies that, for each $x^* \in X^*$ and $\varepsilon \in \,]0, \varepsilon_0]$,

$$\partial_\varepsilon f^*(x^*) = (\partial_\varepsilon f)^{-1}(x^*) \subset (\partial_{\zeta(\varepsilon)} g)^{-1}(x^*) = \partial_{\zeta(\varepsilon)} g^*(x^*).$$

Note that $f^*, g^* \in \Gamma_0(X^*)$ by Proposition 3.1.4; hence, we may also assume that $f^*(\theta) = g^*(\theta) = 0$. Therefore, the paragraph above (see (4.126)) ensures that $g^*(u^*) \leq f^*(u)$ for all $u^* \in X^*$, and Theorem 3.2.2 gives rise to the opposite inequality of (4.126). In other words, the conclusion of the theorem follows when $g \in \Gamma_0(X)$.

Assume now that g is any (proper) function so that, by the current assumption, for each $x \in X$ and $\varepsilon \in \,]0, \varepsilon_0]$,

$$\partial_\varepsilon f(x) \subset \partial_{\zeta(\varepsilon)} g(x) \subset \partial_{\zeta(\varepsilon)} (\overline{\mathrm{co}} g)(x).$$

Thus, since $\overline{\mathrm{co}}(g) \in \Gamma_0(X)$ as a consequence of Proposition 4.1.10, by the paragraph above there exists some $c \in \mathbb{R}$ such that

$$f(x) = (\overline{\mathrm{co}} g)(x) + c \text{ for all } x \in X. \tag{4.127}$$

More precisely, if $x \in \mathrm{dom}\, f$, then for all $\varepsilon \in \,]0, \varepsilon_0]$ we have $\emptyset \neq \partial_\varepsilon f(x) \subset \partial_{\zeta(\varepsilon)} g(x)$ by Proposition 4.1.10, and the condition $\limsup_{\delta \downarrow 0} \zeta(\delta) = 0$ implies that $g(x) = (\overline{\mathrm{co}} g)(x)$; that is, by (4.127),

$$f(x) = g(x) + c \text{ for all } x \in \mathrm{dom}\, f. \tag{4.128}$$

Finally, if $x \notin \mathrm{dom}\, f$, then (4.127) yields $+\infty = f(x) = (\overline{\mathrm{co}} g)(x) + c \leq g(x) + c$ and (4.128) also holds outside $\mathrm{dom}\, f$. ∎

The following corollary is a simple version of Proposition 4.4.6.

Corollary 4.4.7 *Given functions $f, g \in \Gamma_0(X)$, we assume the existence of some $\varepsilon_0 > 0$ such that, for all $x \in \mathrm{dom}\, f \cup \mathrm{dom}\, g$ and $\varepsilon \in \,]0, \varepsilon_0]$,*

$$\partial_\varepsilon f(x) \cap \partial_\varepsilon g(x) \neq \emptyset.$$

Then f and g are equal up to some additive constant.

Proof. We denote $h := f \square g$. Then, since the current assumption implies that $\mathrm{dom}\, f = \mathrm{dom}\, g$ (by Proposition 4.1.10), we have that

$$\mathrm{dom}\, h = \mathrm{dom}\, f + \mathrm{dom}\, g = \mathrm{dom}\, f + \mathrm{dom}\, f = 2\,\mathrm{dom}\, f,$$

where the last equality comes from the convexity of $\mathrm{dom}\, f$. Next we fix $x \in \frac{1}{2}\mathrm{dom}\, h$ so that $x \in \mathrm{dom}\, f = \mathrm{dom}\, g$. So, by the current assumption, for each $\varepsilon \in \,]0, \varepsilon_0]$ we have

4.4. SUBDIFFERENTIAL INTEGRATION

$$\emptyset \neq \partial_\varepsilon f(x) \cap \partial_\varepsilon g(x) \subset \partial_{2\varepsilon} h(2x)$$

(Exercise 49), showing that $h(2x) \geq f(x) + g(x) - 2\varepsilon$; that is, $h(2x) = f(x) + g(x)$. Moreover, for $x \notin \frac{1}{2} \operatorname{dom} h$ we have that $+\infty = h(2x) \leq f(x) + g(x)$, and we conclude that $h = f(\cdot/2) + g(\cdot/2)$; that is, $h \in \Gamma_0(X)$ and

$$\partial_\varepsilon h(x) \subset \partial_\varepsilon f(x/2) \cap \partial_\varepsilon g(x/2) \text{ for all } x \in X \text{ and } \varepsilon \in \,]0, \varepsilon_0]. \quad (4.129)$$

Consequently, according to Proposition 4.4.6, we find constants $c_1, c_2 \in \mathbb{R}$ such that $f(x) = h(2x) + c_1 = g(x) + c_2$ for all $x \in X$, and we are done. ∎

In the setting of Banach spaces, the subdifferential on its own allows us to recover convex functions, without any further continuity assumptions on the involved functions.

Proposition 4.4.8 *Suppose X is Banach and let functions f, g : $X \to \mathbb{R}_\infty$ be such that*

$$\partial f(x) \subset \partial g(x) \quad \text{for all } x \in X.$$

If $f \in \Gamma_0(X)$ and g is lsc, then f and g are equal up to some additive constant.

Proof. Assuming first that $g \in \Gamma_0(X)$, we choose $k_0 \in \mathbb{N}$ such that

$$\operatorname{dom} f \cap (k_0 B_X) \neq \emptyset, \; \operatorname{dom} g \cap (k_0 B_X) \neq \emptyset,$$

where B_X is the closed unit ball in X, so that for each fixed $k \geq k_0$

$$f_k := f + I_{kB_X} \in \Gamma_0(X) \text{ and } g_k := g + I_{kB_X} \in \Gamma_0(X).$$

Furthermore, taking into account Proposition 4.1.20, we get the relations $f_k^* = f^* \square \sigma_{kB_X}$ and

$$\partial f_k(x) = \partial f(x) + N_{kB_X}(x) \subset \partial g(x) + N_{kB_X}(x) \subset \partial g_k(x) \text{ for all } x \in X;$$

in particular, the function f_k^* is *norm*-continuous on X^*. Moreover, for any fixed $\varepsilon > 0$ and each $x \in X$, thanks to Proposition 4.3.8 we have that

$$\partial_\varepsilon f_k(x) \subset \bigcup_{y \in B_X(x,\sqrt{\varepsilon})} \partial f_k(y) + \sqrt{\varepsilon} B_{X^*} \subset \bigcup_{y \in B_X(x,\sqrt{\varepsilon})} \partial g_k(y) + \sqrt{\varepsilon} B_{X^*},$$

which leads us, using (4.18), for all $x^* \in X^*$ to

$$\partial_\varepsilon f_k^*(x^*) \cap X = (\partial_\varepsilon f_k)^{-1}(x^*) \subset \bigcup_{y^* \in B_{X^*}(x^*, \sqrt{\varepsilon})} (\partial f_k(y^*))^{-1} + \sqrt{\varepsilon} B_X$$

$$\subset \bigcup_{y^* \in B_{X^*}(x^*, \sqrt{\varepsilon})} (\partial g_k(y^*))^{-1} + \sqrt{\varepsilon} B_X$$

$$\subset \bigcup_{y^* \in B_{X^*}(x^*, \sqrt{\varepsilon})} \partial g_k^*(y^*) + \sqrt{\varepsilon} B_X.$$

Furthermore, since $\partial_\varepsilon f_k^*(x^*)$ is bounded in $(X^{**}, \sigma(X^{**}, X^*))$ by Proposition 4.1.7, it is also *norm*-bounded (as X is Banach), so there exists some bounded set $D \subset X^{**}$ (possibly depending on x^*) such that the above inclusions read

$$\partial_\varepsilon f_k^*(x^*) \cap X \subset \bigcup_{y^* \in B_{X^*}(x^*, \sqrt{\varepsilon})} (D \cap \partial g_k^*(y^*)) + \sqrt{\varepsilon} B_X.$$

Consequently, taking into account the Goldstein theorem and the fact that $B_{X^{**}}$ is $\sigma(X^{**}, X^*)$-compact by Theorem 2.1.9, Corollary 4.3.4 entails

$$\partial f_k^*(x^*) = \bigcap_{\varepsilon > 0} \mathrm{cl}^{w^{**}} (\partial_\varepsilon f_k^*(x^*) \cap X)$$

$$\subset \bigcap_{\varepsilon > 0} \mathrm{cl}^{w^{**}} \left(\bigcup_{y^* \in B_{X^*}(x^*, \sqrt{\varepsilon})} (D \cap \partial g_k^*(y^*)) + \sqrt{\varepsilon} B_X \right)$$

$$= \bigcap_{\varepsilon > 0} \mathrm{cl}^{w^{**}} \left(\bigcup_{y^* \in B_{X^*}(x^*, \sqrt{\varepsilon})} (D \cap \partial g_k^*(y^*)) + \sqrt{\varepsilon} B_{X^{**}} \right)$$

$$= \bigcap_{\varepsilon > 0} \mathrm{cl}^{w^{**}} \left(\bigcup_{y^* \in B_{X^*}(x^*, \sqrt{\varepsilon})} (D \cap \partial g_k^*(y^*)) \right) \subset \partial g_k^*(x^*),$$

where the last equality comes from Proposition 4.1.6(ix). In other words, we have shown that

$$\partial f_k^*(x^*) \subset \partial g_k^*(x^*) \text{ for all } x^* \in X^*,$$

and, therefore, Theorem 4.4.3 implies that $f_k^* = g_k^* - c_k$, for some $c_k \in \mathbb{R}$; that is, thanks to Theorem 3.2.2, we have that

$$f + I_{kB_X} = f_k = g_k + c_k = g + I_{kB_X} + c_k.$$

More precisely, given any $x_0 \in \mathrm{dom}\, f \cap \mathrm{dom}\, g$, we can take $c_k = f(x_0) - g(x_0) =: c$ for all k large enough, and we deduce that $f = g + c$.

4.4. SUBDIFFERENTIAL INTEGRATION

Let us now consider the general case where g is any function. Using the relation $\partial g \subset \partial(\overline{co}g)$, and the fact that $\overline{co}g \in \Gamma_0(X)$ that comes from the combination of the current assumption and Corollary 4.3.9(i), the current assumption entails $\partial f(x) \subset \partial(\overline{co}g)(x)$ for all $x \in X$. Therefore, by the previous paragraph, there exists some $c \in \mathbb{R}$ such that

$$f = \overline{co}g + c \leq g + c; \qquad (4.130)$$

in particular, $f(x) = g(x) + c = +\infty$ for all $x \notin \mathrm{dom}\, f$. Furthermore, if $x \in X$ is such that $\partial f(x) \neq \emptyset$, then $\partial g(x) \neq \emptyset$ and we deduce that $(\overline{co}g)(x) = g(x)$. Thus, the equality in (4.130) implies that $f(x) = (\overline{co}g)(x) + c = g(x) + c$. More generally, if $x \in \mathrm{dom}\, f$, then Corollary 4.3.9(i) gives rise to some sequence $(x_n)_n \subset X$ such that $x_n \to x$, $f(x_n) \to f(x)$, and $\partial f(x_n) \neq \emptyset$ for all $n \geq 1$. Thus, from the last paragraph, we infer that $f(x_n) = g(x_n) + c$ for all $n \geq 1$, and by taking limits for $n \to +\infty$ and using the lower semicontinuity of g we infer that

$$f(x) = \lim_n f(x_n) = \lim_n g(x_n) + c = \liminf_n g(x_n) + c \geq g(x) + c.$$

Consequently, the desired relationship between f and g follows taking into account (4.130). ∎

The following corollary provides a process for constructing convex functions from their subgradients. Below we use the convention $\sum_{i=0}^{0} := 0$.

Corollary 4.4.9 *Given a function $f \in \Gamma_0(X)$, we suppose that either X is Banach or at least one of the functions f and f^* is continuous somewhere. Then $(\partial f)^{-1}(X^*) \neq \emptyset$ and, for each $x_0 \in (\partial f)^{-1}(X^*)$, we have for all $x \in X$,*

$$f(x) = f(x_0) + \sup \left\{ \sum_{i=0}^{n-1} \langle x_i^*, x_{i+1} - x_i \rangle + \langle x_n^*, x - x_n \rangle \right\}, \qquad (4.131)$$

where the supremum is taken over $n \in \mathbb{N}$, $x_i \in X$, and $x_i^* \in \partial f(x_i)$, $i = 1, \ldots, n$.

Proof. We denote by g the function on the right-hand side of (4.131), so that $g \leq f$ by the own definition of the subdifferential. Furthermore, there must exist some $z_0 \in X$ such that $\partial f(z_0) \neq \emptyset$, and so $g \in \Gamma_0(X)$. In fact, if X is Banach, then the existence of such a z_0 follows from Corollary 4.3.9. Also, if f is continuous at some $z_0 \in X$, then $\partial f(z_0) \neq \emptyset$, thanks to Proposition 4.1.22. Finally, if f^* is continuous at some $z_0^* \in X^*$, then it is τ-continuous there and (4.108) together with Proposition

4.1.22 implies that $\emptyset \neq \partial f^*(z_0^*) = (\partial f)^{-1}(z_0^*)$; that is, there exists some $z_0 \in X$ such that $z_0^* \in \partial f(z_0)$ and $\partial f(z_0)$ is non-empty.

Let us fix $z \in X$ and pick $z^* \in \partial f(z)$. Then $g(z) \in \mathbb{R}$ and so, for every $\delta > 0$, there exist some $n \in \mathbb{N}$, $x_i \in X$, and $x_i^* \in \partial f(x_i)$, $i = 0, 1, \ldots, n$, such that

$$g(z) - \delta < f(x_0) + \sum_{i=0}^{n-1} \langle x_i^*, x_{i+1} - x_i \rangle + \langle x_n^*, z - x_n \rangle.$$

Thus, for $x_{n+1} := z$ we have that $z^* \in \partial f(x_{n+1})$ and, for every $x \in X$,

$$\langle z^*, x - z \rangle + g(z) - \delta < f(x_0) + \sum_{i=0}^{n-1} \langle x_i^*, x_{i+1} - x_i \rangle + \langle x_n^*, z - x_n \rangle + \langle z^*, x - z \rangle$$
$$= f(x_0) + \sum_{i=0}^{n} \langle x_i^*, x_{i+1} - x_i \rangle + \langle z^*, x - z \rangle \leq g(x).$$

In other words, $z^* \in \cap_{\delta > 0} \partial_\delta g(z) = \partial g(z)$, and we infer that $\partial f \subset \partial g$. Consequently, using Proposition 4.4.8 and Theorem 4.4.3, there exists some constant $c \in \mathbb{R}$ such that $f = g + c$. More precisely, taking $n = 1$ and $(x_1, x_1^*) = (x_0, x_0^*)$ in (4.131), we get

$$f(x_0) = f(x_0) + \langle x_0^*, x_1 - x_0 \rangle + \langle x_1^*, x_0 - x_1 \rangle \leq g(x_0) \leq f(x_0),$$

which shows that $c = f(x_0) - g(x_0) = 0$ and, finally, $f = g$. ∎

The main difference between the following result and Corollary 4.4.9 is that now the supremum in (4.132) is only evaluated for elements $x_i \in \text{int}(\text{dom } f)$ when this latter set is not empty. In this case, convex functions are constructed from their subgradients when evaluated within the interior of the effective domain.

Corollary 4.4.10 *Assume that $f \in \Gamma_0(X)$ is continuous at $x_0 \in \text{int}(\text{dom } f)$. Then, for every $x \in \text{cl}(\text{dom } f)$, we have that*

$$f(x) = f(x_0) + \sup \left\{ \sum_{i=0}^{n-1} \langle x_i^*, x_{i+1} - x_i \rangle + \langle x_n^*, x - x_n \rangle \right\}, \quad (4.132)$$

where the supremum is taken over $n \in \mathbb{N}$, $x_i \in \text{int}(\text{dom } f)$ and $x_i^* \in \partial f(x_i)$, $i = 0, 1, \ldots, n$.

Proof. We proceed as in the proof of Corollary 4.4.9 and denote by g the function on the right-hand side of (4.132), so that $g \leq f$, $g \in \Gamma_0(X)$ and we verify that $\partial f(z) \subset \partial g(z)$ for all $z \in \text{int}(\text{dom } f)$. Consequently, by Corollary 4.4.4, there exists some constant $c \in \mathbb{R}$ such that $f = g + c$ on $\text{cl}(\text{dom } f)$; indeed, like in the proof of Corollary 4.4.9 we can verify that $c = 0$. ∎

4.4. SUBDIFFERENTIAL INTEGRATION

Since the functions in $\Gamma_0(X)$ may have no subgradients, as X is a general lcs, the point x_0 in the following corollary is taken in the effective domain of f instead of $(\partial f)^{-1}(X^*)$ as in Corollary 4.4.10. This choice ensures that $\partial_\varepsilon f(x_0) \neq \emptyset$ for all $\varepsilon > 0$.

Corollary 4.4.11 *Given function $f \in \Gamma_0(X)$, we fix $x_0 \in \operatorname{dom} f$ and $\bar{\varepsilon} > 0$. Then, for every $x \in X$, we have that*

$$f(x) = f(x_0) + \sup\left\{\sum_{i=0}^{n-1}\langle x_i^*, x_{i+1} - x_i\rangle + \langle x_n^*, x - x_n\rangle - \sum_{i=0}^{n}\varepsilon_i\right\}, \tag{4.133}$$

where the supremum is taken over $n \in \mathbb{N}$, $x_i \in X$, $0 < \varepsilon_i < \bar{\varepsilon}$, and $x_i^* \in \partial_{\varepsilon_i} f(x_i)$, $i = 0, 1, \ldots, n$.

Proof. We denote by g the function on the right-hand side of (4.133) so that, for all $x \in X$,

$$g(x) \leq f(x_0) + \sup\left\{\sum_{i=0}^{n-1}(f(x_{i+1}) - f(x_i) + \varepsilon_i) + (f(x) - f(x_n) + \varepsilon_n) - \sum_{i=0}^{n}\varepsilon_i\right\} = f(x),$$

and hence $g \in \Gamma_0(X)$, since $\partial_\varepsilon f(x) \neq \emptyset$ for all $x \in \operatorname{dom} f$ and $\varepsilon > 0$ due to Proposition 4.1.10. Now take $\varepsilon \in\,]0, \bar{\varepsilon}[$, $z \in X$ and pick $z^* \in \partial_\varepsilon f(z)$. Then $g(z) \in \mathbb{R}$ and so, for every $\delta > 0$, there exist some $n \in \mathbb{N}$, $x_i \in X$, $0 < \varepsilon_i < \bar{\varepsilon}$, and $x_i^* \in \partial_{\varepsilon_i} f(x_i)$, $i = 0, 1, \ldots, n$, such that

$$g(z) - \delta < f(x_0) + \sum_{i=0}^{n-1}\langle x_i^*, x_{i+1} - x_i\rangle + \langle x_n^*, z - x_n\rangle - \sum_{i=0}^{n}\varepsilon_i.$$

Thus, as in the proof of Corollary 4.4.9, we denote $x_{n+1} := z$ and $\varepsilon_{n+1} = \varepsilon$, so that $z^* \in \partial_\varepsilon f(x_{n+1})$ and, for every $x \in X$,

$$\langle z^*, x - z\rangle + g(z) - \delta < f(x_0) + \sum_{i=0}^{n-1}\langle x_i^*, x_{i+1} - x_i\rangle + \langle x_n^*, z - x_n\rangle$$
$$+ \langle z^*, x - z\rangle - \sum_{i=0}^{n+1}\varepsilon_i + \varepsilon$$
$$= f(x_0) + \sum_{i=0}^{n}\langle x_i^*, x_{i+1} - x_i\rangle + \langle z^*, x - z\rangle$$
$$- \sum_{i=0}^{n+1}\varepsilon_i + \varepsilon \leq g(x) + \varepsilon;$$

in other words, $z^* \in \cap_{\delta > 0}\partial_{\delta+\varepsilon}g(z) = \partial_\varepsilon g(z)$, and we infer that $\partial_\varepsilon f(z) \subset \partial_\varepsilon g(z)$ for all $z \in X$ and $\varepsilon \in\,]0, \bar{\varepsilon}[$. Therefore, by Proposition 4.4.6, we conclude that $f = g + c$ for some $c \in \mathbb{R}$. More precisely, taking $n = 1$, $x_1 = x_0 = x$ and $x_1^* = x_0^* \in \partial_\delta f(x_0)$ for $\delta > 0$ (this last set is not empty by Proposition 4.1.10) in the right-hand side of (4.133), we obtain that

$$f(x_0) + \sup\{\langle x_0^*, x_0 - x_0 \rangle + \langle x_0^*, x_0 - x_0 \rangle - 2\delta\} \le g(x_0);$$

that is, $f(x_0) - 2\delta \le g(x_0) \le f(x_0)$ for all $\delta > 0$, and we derive that $c = f(x_0) - g(x_0) = 0$. ∎

We have seen in (4.12) that the subdifferential operator is monotone and cyclically monotone (see (2.28) and (2.29), respectively). The following proposition shows that it is also maximally cyclically monotone.

Proposition 4.4.12 *Given function $f \in \Gamma_0(X)$, we assume that either X is Banach or at least one of the functions f and f^* is continuous somewhere. Then the mapping ∂f is maximally cyclically monotone.*

Proof. Assuming the existence of some cyclically monotone operator A such that $\mathrm{gph}\, A \supset \mathrm{gph}(\partial f)$, we fix $x_0 \in X$, $x_0^* \in \partial f(x_0)$ (such points exist by Corollary 4.4.9) and define the function $g : X \to \mathbb{R}_\infty$ as

$$g(x) = \sup\left\{\sum_{i=0}^{n-1} \langle x_i^*, x_{i+1} - x_i \rangle + \langle x_n^*, x - x_n \rangle\right\},$$

where the supremum is taken over $n \in \mathbb{N}$, $x_i \in A^{-1}(X^*)$, $x_i^* \in A(x_i)$, $i = 1, \ldots, n$. Then g is convex and lsc. In fact, we have that $g \in \Gamma_0(X)$, since $g(x_0) \le 0$ by the cyclic monotonicity of A. Furthermore, given $z \in X$, $z^* \in Az$, $\delta > 0$ and $m \ge 1$, we choose $n \in \mathbb{N}$ and $x_i^* \in A(x_i)$, $i = 1, \ldots, n$, such that

$$\min\{g(z), m\} - \delta < \sum_{i=0}^{n-1} \langle x_i^*, x_{i+1} - x_i \rangle + \langle x_n^*, z - x_n \rangle.$$

Thus, taking $x_{n+1} := z$, we obtain that $z^* \in A(x_{n+1})$ and so, for every $x \in X$,

$$\langle z^*, x - z \rangle + \min\{g(z), m\} - \delta < \sum_{i=0}^{n-1} \langle x_i^*, x_{i+1} - x_i \rangle + \langle x_n^*, z - x_n \rangle + \langle z^*, x - z \rangle$$
$$= \sum_{i=0}^{n} \langle x_i^*, x_{i+1} - x_i \rangle + \langle z^*, x - x_{n+1} \rangle \le g(x);$$

in other words, as $m \uparrow +\infty$, we get

$$\langle z^*, x - z \rangle + g(z) - \delta \leq g(x) \text{ for all } x \in X.$$

Therefore, $z \in \operatorname{dom} g$ and $z^* \in \cap_{\delta > 0} \partial_\delta g(z) = \partial g(z)$, and we deduce that $\operatorname{gph}(\partial f) \subset \operatorname{gph} A \subset \operatorname{gph}(\partial g)$. Consequently, using Proposition 4.4.8 and Theorem 4.4.3, we find some constant $c \in \mathbb{R}$ such that $f = g + c$, showing that $\partial f \subset A \subset \partial g = \partial f$. ∎

The maximal monotonicity of the subdifferential of functions in $\Gamma 1_0(X)$, where X is Banach, is also a well-known property.

Proposition 4.4.13 *Assume that X is Banach and take $f \in \Gamma_0(X)$. Then ∂f is maximally monotone.*

4.5 Exercises

Exercise 36 *Given function $f : X \to \overline{\mathbb{R}}$ and $\varepsilon \geq 0$, prove that $\operatorname{cone}(\operatorname{dom} f^*) \subset \operatorname{dom}(\sigma_{\varepsilon\text{-}\operatorname{argmin} f})$, where ε-argmin $f = \emptyset$ if $\inf_X f \notin \mathbb{R}$.*

Exercise 37 *Given a function $f : X \to \overline{\mathbb{R}}$, $x \in f^{-1}(\mathbb{R})$, and $\varepsilon \geq 0$, prove that*

$$(f'_\varepsilon(x; \cdot))^* = \sup_{s > 0} \frac{f(x) + f^*(x^*) - \langle x, x^* \rangle - \varepsilon}{s}.$$

Exercise 38 *Let $x \in X$ and $\varepsilon \geq 0$ such that $\partial_\varepsilon f(x) \neq \emptyset$. Prove that $[\partial_\varepsilon f(x)]_\infty = \mathrm{N}_{\operatorname{dom} f}(x)$.*

Exercise 39 *Let $f : X \to \mathbb{R}_\infty$ be a convex function such that $\operatorname{dom} f \neq \emptyset$. Prove that if $X = \mathbb{R}^n$, then f is proper if $\operatorname{cl} f$ is proper. Give a counterexample when X is infinite-dimensional; in other words, find a proper convex function f such that $\operatorname{cl} f$ is not proper.*

Exercise 40 *Let $f, g : X \to \overline{\mathbb{R}}$ be two convex functions. Prove that f and g have the same subdifferential at $x \in f^{-1}(\mathbb{R}) \cap g^{-1}(\mathbb{R})$ in each one of the following cases:*
  (i) f and g coincide in a neighborhood of x.
  (ii) f is lsc at x and $g := \max\{f, f(x) - 1\}$.

Exercise 41 *Given another lcs Y, functions $f_1, \ldots, f_m \in \Gamma_0(X)$, $g \in \Gamma_0(Y)$, with $m \geq 1$, and a continuous linear mapping $A : X \to Y$, prove the following statements:*

(i) For every $x \in \text{dom}\,(f_1 + \ldots + f_m + (g \circ A))$ and all $\varepsilon_1, \ldots, \varepsilon_m$, $\varepsilon > 0$ we have

$$\text{N}_{\text{dom}(f_1 + \ldots + f_m + g \circ A)}(x) = [\text{cl}\,(\partial_{\varepsilon_1} f_1(x) + \ldots + \partial_{\varepsilon_m} f_m(x) + A^*(\partial_\varepsilon g(Ax)))]_\infty.$$

(ii) If $k < m$, for all $x \in \text{dom}\,(f_1 + \ldots + f_m + g \circ A)$ and all $\varepsilon_1, \ldots, \varepsilon_m, \varepsilon > 0$ one has

$$\text{N}_{\text{dom}(f_1 + \ldots + f_m + g \circ A)}(x) = \begin{bmatrix} \text{cl}\,(\partial_{\varepsilon_1} f_1(x) + \ldots + \partial_{\varepsilon_k} f_k(x) \\ + \text{co}\,(\partial_{\varepsilon_{k+1}} f_{k+1}(x) \cup \ldots \cup A^*\,(\partial_\varepsilon g(Ax)))) \end{bmatrix}_\infty.$$

Exercise 42 *Given the functions $f, g_k : \mathbb{R} \to \mathbb{R}_\infty$, $k \geq 1$, with $f(x) = x^2$ and $g_k := f + \text{I}_{[-1/k, 1/k]}$, despite the fact that f and g_k coincide locally at 0, prove that for every $\varepsilon > 0$ we have $\partial_\varepsilon f(0) \subsetneq \partial_\varepsilon g_k(0)$, provided that $k > \frac{2(2-\sqrt{2})}{\sqrt{\varepsilon}}$.*

Exercise 43 *Consider a convex function $f : X \to \mathbb{R}_\infty$ and let $x \in \text{dom}\,f$. Prove that $\text{N}_{\text{dom}\,f}(x) = \left[\text{N}^\varepsilon_{\text{dom}\,f}(x)\right]_\infty$ for every $\varepsilon \geq 0$.*

Exercise 44 *Let f be a proper convex function defined on X. If $x \in X$ is such that $f(x) \in \mathbb{R}$ and $\mathbb{R}_+(\text{epi}\,f - (x, f(x)))$ is closed, prove that $\text{cl}\,f$ is proper and*

$$\text{epi}\,f'(x; \cdot) = \mathbb{R}_+(\text{epi}\,f - (x, f(x))). \tag{4.134}$$

Consequently, show that $\partial f(x) \neq \emptyset$, $f'(x; \cdot) = (\text{cl}\,f)'(x; \cdot) = \sigma_{\partial f(x)}$, and that f is lsc at x.

Exercise 45 *Let $f, g : X \to \mathbb{R}_\infty$ be two convex functions, and let $x \in X$ such that $f(x)$ and $g(x)$ are finite. If the sets $\mathbb{R}_+(\text{epi}\,f - (x, f(x)))$ and $\mathbb{R}_+(\text{epi}\,g - (x, g(x)))$ are closed, prove that $\mathbb{R}_+(\text{epi}(f + g) - (x, f(x) + g(x)))$ is also closed.*

Exercise 46 *Consider function $f \in \Gamma_0(Y)$ and a continuous linear mapping $A : X \to Y$, where Y is another lcs. Let the functions $g, h : X \times Y \to \mathbb{R}_\infty$ be defined as*

$$g(x, y) := f(y) \text{ and } h(x, y) := \text{I}_{\text{gph}\,A}(x, y).$$

Given $x \in A^{-1}(\text{dom}\,f)$, prove that $x^* \in \partial(f \circ A)(x)$ if and only if $(x^*, \theta) \in \partial(g + h)(x, Ax)$.

4.5. EXERCISES

Exercise 47 Let $f: X \to \mathbb{R}_\infty$ be a proper convex function, and let $x \in \operatorname{dom} f$. Prove that

$$\partial_\varepsilon f(x) = \partial(f'_\varepsilon(x; \cdot))(\theta) \text{ and } \operatorname{dom} f'_\varepsilon(x; \cdot) = \mathbb{R}_+(\operatorname{dom} f - x).$$

Exercise 48 Given functions $f_i \in \Gamma_0(X)$, $i = 1, \ldots, m$, $m \geq 1$, we assume that the Slater condition is fulfilled in $\operatorname{dom} f_0$; that is, $\max_{1 \leq i \leq m} f_i(x_0) < 0$ for some $x_0 \in \operatorname{dom} f_0$. Use Corollary 4.2.5 to prove that

$$\inf_{[\max_{1 \leq i \leq m} f_i \leq 0]} f_0 = \max_{\lambda_1, \ldots \lambda_m \geq 0} \inf_{x \in X} \left(f_0(x) + \lambda_1 f_1(x) + \ldots + \lambda_m f_m(x) \right).$$

Exercise 49 Given two functions $f, g: X \to \overline{\mathbb{R}}$ and $\varepsilon_1, \varepsilon_2 \geq 0$, prove that

$$\partial_{\varepsilon_1} f(x) \cap \partial_{\varepsilon_2} g(y) \subseteq \partial_{\varepsilon_1 + \varepsilon_2}(f \square g)(x + y) \text{ for all } x, y \in X, \quad (4.135)$$

$$\partial_{\varepsilon_1} f(x) \cap \partial_{\varepsilon_2} g(y) \neq \emptyset \implies (f \square g)(x + y) \geq f(x) + g(y) - \varepsilon_1 - \varepsilon_2,$$

and, for all $x \in \operatorname{dom} f$, $y \in \operatorname{dom} g$ and $\varepsilon, \delta \geq 0$,

$$(f \square g)(x + y) \geq f(x) + g(y) - \varepsilon$$
$$\implies \partial_\delta (f \square g)(x + y) \subset \partial_{\delta + \varepsilon} f(x) \cap \partial_{\delta + \varepsilon} g(x).$$

In addition, provided that $\partial f(x) \cap \partial g(y) \neq \emptyset$, prove that

$$\partial(f \square g)(x + y) = \partial f(x) \cap \partial g(y).$$

Exercise 50 Given a Banach space X, we endow X^{**} with the w^{**}-topology. Given a proper function $f: X \to \overline{\mathbb{R}}$, prove that for every $z \in X^{**}$

$$\partial(\operatorname{cl}^{w^{**}} \hat{f})(z) = \bigcap_{\varepsilon > 0, \, U \in \mathcal{N}_{X^{**}}} \bigcup_{y \in z + U} \partial_\varepsilon f(y),$$

where $\mathcal{N}_{X^{**}}$ denotes the collection of the θ-neighborhoods in (X^{**}, w^{**}) and, consequently, for every $x^* \in X^*$

$$\left(\partial(\operatorname{cl}^{w^{**}} \hat{f}) \right)^{-1}(x^*) = \bigcap_{\varepsilon > 0} \operatorname{cl}^{w^{**}}((\partial_\varepsilon f)^{-1}(x^*)).$$

Exercise 51 Give a proof of (4.56) based on Proposition 4.1.24.

Exercise 52 Let Y be another lcs, $f \in \Gamma_0(Y)$, and let $A_0 : X \to Y$ be a continuous linear mapping such that $A_0^{-1}(\mathrm{dom}\, f) \neq \emptyset$. If $Ax := A_0 x + b$, with $b \in Y$, prove that

$$\partial (f \circ A)(x) = \bigcap_{\delta > 0} \mathrm{cl}\, (A_0^* \partial_\delta f(Ax)), \quad \text{for every } x \in X, \quad (4.136)$$

and

$$\partial_\varepsilon (f \circ A)(x) = \mathrm{cl}\, (A_0^* \partial_\varepsilon f(Ax)), \quad \text{for every } x \in X \text{ and } \varepsilon > 0, \quad (4.137)$$

where A_0^* is the adjoint of A_0.

Exercise 53 Let Y be another lcs, $L \subset X$ and $M \subset Y$ two closed convex cones, and $A : X \to Y$ be a continuous linear mapping with adjoint A^*. Use Proposition 4.1.16 to prove that

$$(L \cap A^{-1}(M))^- = \mathrm{cl}(L^- + A^*(M^-)).$$

Exercise 54 Consider the support function σ_A of a non-empty set $A \subset \mathbb{R}^n$. Prove that $\mathrm{N}_{\mathrm{dom}\, \sigma_A}(x) = (\overline{\mathrm{co}} A)_\infty \cap \{x\}^\perp$, for every $x \in \mathrm{dom}\, \sigma_A$.

Exercise 55 (i) Given function $f : X \to \mathbb{R}_\infty$ and linear subspace $L \subset X$, for every $x \in L$ and $\varepsilon \geq 0$ prove that

$$\{x^*_{|L} \in L^* : x^* \in \partial_\varepsilon f(x)\} \subset \partial_\varepsilon f_{|L}(x), \quad (4.138)$$

with equality when $\mathrm{dom}\, f \subset L$. In particular, prove that

$$\partial_\varepsilon (f + \mathrm{I}_L)_{|L}(x) = \{x^*_{|L} \in L^* : x^* \in \partial_\varepsilon (f + \mathrm{I}_L)(x)\}, \quad (4.139)$$

and so

$$\partial_\varepsilon (f + \mathrm{I}_L)(x) = \{x^* + L^\perp : x^*_{|L} \in \partial_\varepsilon (f + \mathrm{I}_L)_{|L}(x)\}. \quad (4.140)$$

(ii) Let $g : X \to \mathbb{R}_\infty$ be another function and $x \in X$. If

$$\partial (f_{|L} + g_{|L})(x) = \partial (f_{|L})(x) + \partial (g_{|L})(x),$$

prove that

$$\partial (f + g + \mathrm{I}_L)(x) = \partial (f + \mathrm{I}_L)(x) + \partial (g + \mathrm{I}_L)(x) + L^\perp.$$

4.5. EXERCISES

(iii) Let $M \subset X$ be another linear subspace. If

$$\partial((f + I_L + g)_{|M})(x) = \partial(f + I_L)_{|M}(x) + \partial g_{|M}(x) \qquad (4.141)$$

and

$$\partial\left((f + I_M)_{|L}\right)(x) = \partial(f + I_L)_{|L}(x) + \partial\left(I_M\right)_{|L}(x), \qquad (4.142)$$

prove that

$$\partial(f + g + I_{L \cap M})(x) = \partial\left(f + I_L\right)(x) + \partial\left(g + I_M\right)(x) + \mathrm{cl}(M^\perp + L^\perp).$$

(iv) Given a non-empty set $A \subset X$ and $x \in A$, apply (i) to show that, for every linear subspace $L \subset X$ such that $x \in L$, we have

$$\{x^*_{|L} \in L^* : x^* \in \mathrm{N}^\varepsilon_A(x)\} \subset \tilde{\mathrm{N}}^\varepsilon_{A \cap L}(x),$$

where $\tilde{\mathrm{N}}^\varepsilon_{A \cap L}(x)$ is the ε-normal set to A at x relative to the subspace L; that is,

$$\tilde{\mathrm{N}}^\varepsilon_{A \cap L}(x) := \{x^* \in L^* : \langle x^*, y - x \rangle \leq \varepsilon \text{ for all } y \in A \cap L\}.$$

Moreover, if $A \subset L$, then prove that the following equality holds:

$$\{x^*_{|L} \in L^* : x^* \in \mathrm{N}^\varepsilon_A(x)\} = \tilde{\mathrm{N}}^\varepsilon_{A \cap L}(x).$$

Exercise 56 *Starting with Proposition 4.1.20, complete the proof of Proposition 4.1.26 by verifying (4.56).*

Exercise 57 *Given non-empty convex sets $A, C \subset X$ and $x \in C$, we assume that A is closed and $\mathrm{int}(C - x) \cap \mathrm{dom}\, \sigma_A \neq \emptyset$. Prove that the set $A + \mathrm{N}_C(x)$ is closed.*

Exercise 58 *Let $g := \|\cdot\|_{\ell_1}$. Prove that $\partial g(x) = \{(1, 1, \ldots)\}$ for all $x \in \ell_1$ such that $x_n > 0$ for all $n \geq 1$.*

Exercise 59 *Let $f \in \Gamma_0(X)$ be such that f^* is $\tau(X^*, X)$-continuous at some point in $\mathrm{dom}\, f$. Assume that $x^*_0 \in \partial_\varepsilon f(x_0) \cap \mathrm{int}(\mathrm{dom}\, f^*)$, for some $\varepsilon \geq 0$ and $x_0 \in X$. Prove that, for every $\beta \geq 0$, every continuous seminorm p in X, and every $\lambda > 0$, there are $x_\varepsilon \in X$, $y^*_\varepsilon \in [p \leq 1]^\circ$, and $\lambda_\varepsilon \in [-1, 1]$ such that:*

$$p(x_0 - x_\varepsilon) + \beta \left| \langle x^*_0, x_0 - x_\varepsilon \rangle \right| \leq \lambda, \quad \left| \langle x^*_\varepsilon, x_0 - x_\varepsilon \rangle \right| \leq \varepsilon + \lambda/\beta,$$

$$|f(x_0) - f(x_\varepsilon)| \leq \varepsilon + \lambda/\beta,$$

and

$$x_\varepsilon^* := x_0^* + \frac{\varepsilon}{\lambda}(y_\varepsilon^* + \beta\lambda_\varepsilon x_0^*) \in \partial f(x_\varepsilon) \cap \partial_{2\varepsilon} f(x_0).$$

Exercise 60 *Let $f, g \in \Gamma_0(X)$ be given.*
(i) We assume that f is finite and continuous at some point, and $\partial f(x) \cap \partial g(x) \neq \emptyset$ for all $x \in \text{int}(\text{dom } f)$. Prove that f and $g + \text{I}_{\overline{\text{dom } f}}$ are equal up to some additive constant.
(ii) We assume that there exists some $\varepsilon_0 > 0$ such that $\partial_\varepsilon f(x) \cap \partial_\varepsilon g(x) \neq \emptyset$ for all $x \in \overline{\text{dom } f}$ and $\varepsilon \in \,]0, \varepsilon_0]$. Prove that f and $g + \text{I}_{\overline{\text{dom } f}}$ are equal up to some additive constant.

Exercise 61 *Let $f, g \in \Gamma_0(X)$ be such that $\partial f(x) \subset \partial g(x)$ for all $x \in X$. Provided that $\text{dom } f$ is finite-dimensional, prove that the functions f and $g + \text{I}_{\text{aff}(\text{dom } f)}$ are equal up to some additive constant.*

Exercise 62 *Let f be a function defined on X and ε-subdifferentiable at $x \in X$ with $\varepsilon \geq 0$. Prove the following assertions:*

$$f(x) \geq (\text{cl } f)(x) \geq (\overline{\text{co}} f)(x) \geq f(x) - \varepsilon,$$

and

$$\partial_\delta f(x) \subset \partial_\delta (\text{cl } f)(x) \subset \partial_{\delta+\varepsilon} f(x),$$

$$\partial_\delta f(x) \subset \partial_\delta (\overline{\text{co}} f)(x) \subset \partial_{\delta+\varepsilon} f(x) \text{ for all } \delta \geq 0.$$

In particular, if $\varepsilon = 0$, then $f(x) = (\text{cl } f)(x) = (\overline{\text{co}} f)(x)$ and

$$\partial_\delta f(x) = \partial_\delta (\text{cl } f)(x) = \partial_\delta (\overline{\text{co}} f)(x) \text{ for all } \delta \geq 0.$$

Exercise 63 *Let $f : X \to \overline{\mathbb{R}}$ be a function such that $\overline{\text{co}} f$ is proper, and let $x \in \text{dom } f$. Prove that $\partial_\varepsilon f(x) = \partial_{(\overline{\text{co}} f)(x) - f(x) + \varepsilon}(\overline{\text{co}} f)(x)$. Consequently, prove that f is ε-subdifferentiable for each $\varepsilon > f(x) - (\overline{\text{co}} f)(x)$, while $f(x) = (\overline{\text{co}} f)(x)$ if and only if $\partial_\varepsilon f(x) \neq \emptyset$ for all $\varepsilon > 0$.*

Exercise 64 *Consider a family of functions $\{f_t, t \in T\} \subset \Gamma_0(X)$, $x \in \text{dom } f$, where $f = \sup_{t \in T} f_t$, and $\varepsilon > 0$. We define the function $h := \inf_{t \in T}(f_t^*(\cdot) - \langle \cdot, x \rangle + f(x))$ and the set $T_\varepsilon(x) := \{t \in T : f_t(x) \geq f(x) - \varepsilon\}$. Prove that $h \geq 0$, $h^* = \sup_{t \in T}(f_t(x + \cdot) - f(x))$, and*

$$[h < \varepsilon] \subset \bigcup_{t \in T_\varepsilon(x)} \partial_\varepsilon f_t(x). \tag{4.143}$$

Exercise 65 Let X be a Banach space. Apply Theorem 4.3.3 to prove that $B_{X^{**}} = \mathrm{cl}^{w^*}(B_X)$ (Goldstein's theorem).

Exercise 66 Given two functions $f, g : X \to \mathbb{R}_\infty$ with f being convex and having a proper conjugate, we assume the existence of some $\delta > 0$ such that

$$\partial_\varepsilon f(x) \subset \partial_\varepsilon g(x) \text{ for all } x \in X \text{ and } \varepsilon \in \,]0, \delta[.$$

Then the functions $\mathrm{cl}\, f$ and $\mathrm{cl}\, g$ are equal up to an additive constant.

4.6 Bibliographical notes

The proofs of Propositions 4.1.1 and 4.1.8 can be found in [201, Theorems 2.1.14 and 3.3.2, respectively]. Other similar expressions to the one in Proposition 4.1.13 are given in [110, Theorem 6.1] for the case of convex functions. A related result to formula (4.40) can be found in [198, Theorem 2.2 and Remark 2.4].

Theorems 3.3.1 and 3.3.4, together with Corollary 3.3.5, lead to the well-established Fenchel and Lagrange dualities given in Theorem 4.2.3 and subsequent corollaries. Proposition 4.1.10, which is well-known, is proved here by means of the Fenchel–Moreau–Rockafellar theorem. Proposition 4.1.12 is a new result which reinforces the fact that the support function of the ε-subdifferential coincides with the ε-directional derivative. Propositions 4.1.16, 4.1.20, and 4.1.22 are well-known results (see, e.g., [161]). Other conditions ensuring the sum and the composition rules can be found, e.g., in [201, Theorem 2.8.3] (see, also, [31, Proposition 4.1]). The subdifferential rules in Proposition 4.1.26 are given in [174, Theorems 23.8 and 23.9], where the author applied the rule for the conjugate of the sum in [174, Theorem 16.4]. Our proof here is new and is based on the infinite-dimensional results established in Proposition 4.1.20. The subdifferential rule given in Remark 2 can be found in [19, Corollary 4.4]. More details on the notion of the quasi-relative interior can be found in [19] (see, also, [26] and [201]). Proposition 4.1.15 was also given in Corollaries 10.1 and 10.2 of [110]. Proposition 4.1.16 gives the Hiriart-Urruty–Phelps formulas ([111]) for the ε-subdifferential of the sum and composition.

Theorem 4.3.6 is the well-known Ekeland variational principle [79]. Proposition 4.3.7 gathers many facts related to the Borwein ([17]), the Brøndsted–Rockafellar ([30]), and the Bishop–Phelps ([168]) theorems. Proposition 4.3.8 can be found in [188]. Corollary 4.3.9 is [201, Theorem 3.1.4]. Proposition 4.3.10 is [5, Corollary 5]. Theorem 4.3.13 is [23, Corollary 6.6.17] and constitutes a version of the Stegall variational principle. The integration criterion given in Theorem 4.4.3 is not new (see, e.g., [153]) and the proof here is based on a reduction to the one-dimensional setting. Other integration results in locally convex spaces can be found in [41]. Similar results to Proposition 4.4.6 can be found in [148]. Proposition 4.4.5 is new and extends the result of [126], obtained when the involved functions are continuous everywhere. Other integration criteria in the line of Proposition 4.4.5, and using the ε-subdifferential, are given in [104]. The maximality of the subdifferential was first investigated in [153] for certain classes of convex functions in Hilbert spaces, and then in [160]. Propositions 4.4.12 and 4.4.13 are the most general results about the maximal (cyclic-) monotonicity of the subdifferential mapping (given for Banach spaces in [173]; see, also, [175]). This discovery allows laying a bridge between convex analysis and operator theory. Standard modern references on the theory of monotone operators are [12], [168], and [182], among others. The first part in Exercise 39 is [23, Exercise 2.4.8], and the second part is [23, Exercise 4.1.2(c)]. Exercise 59 is [60, Theorem 4.2].

Chapter 5

Supremum of convex functions

In this chapter, we characterize the subdifferential of pointwise suprema of arbitrarily indexed families of convex functions, defined on a separated locally convex space X. First, some subdifferential formulas are derived in section 5.1 via classical tools of convex analysis, mainly based on conjugation processes through the Fenchel–Moreau–Rockafellar theorem (Theorem 3.2.2). This direct approach leads to formulas which are written in terms of sums and/or maxima of finite subfamilies. The formulas in section 5.2 provide more geometric insight as they involve the ε-subdifferential of the data functions that are almost active at the reference point, together with the normal cone to finite-dimensional sections of the domain of the supremum function. Our analysis makes use of a mere closedness condition, but without assuming any restriction on the set of indices or on the behavior of the data functions in relation to their dependence on the corresponding index. However, additional continuity assumptions are used in section 5.3 to derive simpler formulas that recover, extend, and unify some classical results on the subject. In this case, finite-dimensional sections of the domain are not necessary. In order to have a complete picture of the present formulas in which both (almost) active and non-active data functions are involved, we establish in section 6.1 some equivalent representations of the normal cone to the domain of the supremum. Then we derive, from the main formula of section 5.2, an alternative (symmetric) formula for the subdifferential in which only the data functions are involved, both the (almost) active and non-active functions. In sections 6.2 and 6.4, additional structures relying on arguments of compactness and compactification will lead us to formulas expressed in terms of the exact subdifferential of the data functions.

Remember that X stands for an lcs and X^* for its topological dual space. Unless otherwise stated, we assume that X^* is endowed with a compatible topology, in particular, the topologies $\sigma(X^*, X)$ and $\tau(X^*, X)$, or the dual norm topology when X is a reflexive Banach space. The associated bilinear form is represented by $\langle \cdot, \cdot \rangle$.

5.1 Conjugacy-based approach

In this section, we present some particular instances of formulas of the approximate subdifferential of the supremum function through classic tools of convex analysis. More precisely, we use a conjugate representation of the supremum function, which is a consequence of the Fenchel–Moreau–Rockafellar theorem (see Theorem 3.2.2), to give formulas that involve convex combinations of the data functions. This section is a sample of how much we can directly obtain from Theorem 3.2.2.

A standard and important example of supremum function is the support function σ_C of a given non-empty set $C \subset X^*$. Thanks to Theorem 3.2.2, we know that the conjugate of σ_C is nothing more than the indicator function of the closed convex hull of the set C. Thus, with the help of the Fenchel–Moreau inequality (4.4), we obtain the following characterization of $\partial_\varepsilon \sigma_C$, which constitutes an extension of (4.13).

Proposition 5.1.1 *Let $C \subset X^*$ be a non-empty set. Then, for every $x \in X$ and $\varepsilon \geq 0$,*

$$\partial_\varepsilon \sigma_C(x) = \{x^* \in \overline{\text{co}}(C) : \langle x^*, x \rangle \geq \sigma_C(x) - \varepsilon\} \qquad (5.1)$$

and consequently, provided that C is a non-empty closed convex set,

$$\partial_\varepsilon \sigma_C(x) = \{x^* \in C : \langle x^*, x \rangle \geq \sigma_C(x) - \varepsilon\}. \qquad (5.2)$$

Despite the general appearance of formula (5.1), valid for sets that are not necessarily convex or closed, observe that it is equivalent to (5.2) as a consequence of the relation $\sigma_C = \sigma_{\overline{\text{co}}(C)}$. Starting from the

5.1. CONJUGACY-BASED APPROACH

last expression, we analyze in the following result the ε-subdifferential of the supremum of affine functions. The proof of this result will follow from formula (5.1) with the help of Proposition 4.1.16.

Proposition 5.1.2 *Let*

$$C := \{(a_t, b_t) \in X^* \times \mathbb{R},\ t \in T\}$$

be a non-empty set, and denote

$$f := \sup_{t \in T}(\langle a_t, \cdot \rangle - b_t).$$

Then, for every $x \in \operatorname{dom} f$ and $\varepsilon > 0$, we have

$$\partial_\varepsilon f(x) = \operatorname{cl}\left\{x^* \in X^* : (x^*, \beta) \in \operatorname{co}(C),\ \beta \in \mathbb{R},\ f(x) \geq \langle x^*, x \rangle - \beta \geq f(x) - \varepsilon\right\}, \tag{5.3}$$

and, particularly,

$$\partial f(x) = \left\{x^* \in X^* : (x^*, \beta) \in \overline{\operatorname{co}}(C),\ \beta \in \mathbb{R},\ \langle x^*, x \rangle - \beta = f(x)\right\}. \tag{5.4}$$

Proof. Let us write $f = \sigma_C \circ (A_0 + (\theta, -1))$, where $A_0 : X \to X \times \mathbb{R}$ is the continuous linear mapping defined by $A_0(x) := (x, 0)$, and fix $x \in \operatorname{dom} f$ and $\varepsilon > 0$. Then, since the adjoint mapping $A_0^* : X^* \times \mathbb{R} \to X^*$ of A_0 is given by $A_0^*(x^*, \beta) = x^*$, by making use of relation (5.1) and formula (4.46) in Proposition 4.1.16 we obtain $\partial_\varepsilon f(x) = \operatorname{cl}(A_0^*(\partial_\varepsilon \sigma_C(x, -1))) = \operatorname{cl}(B_\varepsilon)$, where

$$B_\varepsilon := \left\{x^* \in X^* : (x^*, \beta) \in \overline{\operatorname{co}}(C),\ \beta \in \mathbb{R},\ f(x) \geq \langle x^*, x \rangle - \beta \geq f(x) - \varepsilon\right\}.$$

Take nets $(x_i^*)_i \subset B_\varepsilon$ and $(\beta_i) \subset \mathbb{R}$ such that $x_i^* \to x^* \in X^*$ and

$$(x_i^*, \beta_i) \in \overline{\operatorname{co}}(C),\ f(x) \geq \langle x_i^*, x \rangle - \beta_i \geq f(x) - \varepsilon, \tag{5.5}$$

so that $\langle x_i^*, x \rangle - f(x) \leq \beta_i \leq \langle x_i^*, x \rangle - f(x) + \varepsilon$. So, taking into account that $(x_i^*)_i$ converges, we may assume that $(\beta)_i$ converges to some β that satisfies $(x^*, \beta) \in \overline{\operatorname{co}}(C)$ and $\langle x^*, x \rangle - \beta \geq f(x) - \varepsilon$, due to (5.5). In other words, $x^* \in B_\varepsilon$ and we deduce that B_ε is closed. Consequently,

$$\partial_\varepsilon f(x) = B_\varepsilon. \tag{5.6}$$

Next, we claim that the given $x \in \operatorname{dom} f$ and $\varepsilon > 0$ satisfy

$$\partial_\varepsilon f(x) = \bigcap_{\delta > \varepsilon} \mathrm{cl}\,(E_\delta), \tag{5.7}$$

where

$$E_\delta := \{x^* \in X^* : (x^*, \beta) \in \mathrm{co}(C),\ \beta \in \mathbb{R},\ f(x) \geq \langle x^*, x\rangle - \beta \geq f(x) - \delta\}.$$

Indeed, given $x^* \in \partial_\varepsilon f(x)$ and $\delta > \delta_1 > \varepsilon \geq 0$, we have $x^* \in \partial_{\delta_1} f(x)$ and (5.6) gives rise to the existence of some $\beta \in \mathbb{R}$ such that $(x^*, \beta) \in \overline{\mathrm{co}}(C)$ and $\langle x^*, x\rangle - \beta \geq f(x) - \delta_1 > f(x) - \delta$. Thus, we can find a net $(x_i^*, \beta_i)_i \subset \mathrm{co}(C)$ converging to (x^*, β) such that $\langle x_i^*, x\rangle - \beta_i > f(x) - \delta$ for all i; that is, $x_i^* \in E_\delta$ and we deduce that $x^* \in \mathrm{cl}\,(E_\delta)$. In other words, by the arbitrariness of $\delta > \varepsilon$, the inclusion "$\subset$" in (5.7) holds, and the claim is proved because the opposite inclusion also follows easily from (5.6),

$$\bigcap_{\delta > \varepsilon} \mathrm{cl}\,(E_\delta) \subset \bigcap_{\delta > \varepsilon} \mathrm{cl}\,(B_\delta) = \bigcap_{\delta > \varepsilon} \mathrm{cl}\,(\partial_\delta f(x)) = \bigcap_{\delta > \varepsilon} \partial_\delta f(x) = \partial_\varepsilon f(x).$$

Now, taking into account Corollary 4.1.11 and the fact that $f \in \Gamma_0(X)$, (5.7) yields

$$\partial_\varepsilon f(x) = \mathrm{cl}\left(\bigcup_{0 < \gamma < \varepsilon} \partial_\gamma f(x)\right) = \mathrm{cl}\left(\bigcup_{0 < \gamma < \varepsilon\delta > \gamma} \mathrm{cl}\,(E_\delta)\right) \subset \mathrm{cl}\,(E_\varepsilon) \subset \mathrm{cl}\,(B_\varepsilon) = \partial_\varepsilon f(x),$$

and (5.3) follows. Moreover, using again (5.6), we have that $\partial f(x) = \bigcap_{\delta > 0} \partial_\delta f(x) = \bigcap_{\delta > 0} B_\delta$. Thus, given any $x^* \in \partial f(x)$, for every $\delta > 0$, there exists some $\beta_\delta \in \mathbb{R}$ such that $(x^*, \beta_\delta) \in \overline{\mathrm{co}}(C)$ and $\langle x^*, x\rangle - \beta_\delta \geq f(x) - \delta$. In particular, proceeding as in the paragraph above, we see that $\langle x^*, x\rangle - f(x) \leq \beta_\delta \leq \langle x^*, x\rangle - f(x) + \delta$, which implies that $\beta_\delta \to \beta = \langle x^*, x\rangle - f(x)$ as $\delta \downarrow 0$. Thus, $(x^*, \beta) \in \overline{\mathrm{co}}(C)$ and $\langle x^*, x\rangle - \beta = f(x)$, and the non-trivial inclusion "$\subset$" in (5.4) follows. ■

With the aim of gathering the previous results on the supremum of affine and linear functions, we introduce the following convexity-like concept about the dependence of data functions on indices. We shall use the notation

$$\Delta(S) := \left\{\lambda \in \mathbb{R}_+^{(S)} : \sum_{t \in S} \lambda_t = 1\right\}, \text{ for sets } S \subset T,$$

(already introduced in (2.46)) and

$$S_0 := \{t \in T : \bar{f}_t \in \Gamma_0(X)\}. \tag{5.8}$$

5.1. CONJUGACY-BASED APPROACH

Definition 5.1.3 *A family of convex functions* $\{f_t : X \to \overline{\mathbb{R}}, t \in T\}$ *is said to be* closed for convex combinations *if, for each* $\lambda \in \Delta(S_0)$, *there exists some* $s \in S_0$ *such that*

$$f_\lambda := \sum_{t \in \text{supp}\,\lambda} \lambda_t f_t \leq f_s. \tag{5.9}$$

The family $\{f_c := \langle c, \cdot \rangle, c \in C\}$ for a convex subset $C \subset X^*$, associated with the support function σ_C, provides a basic example of families that are closed for convex combinations. Also, families that are nondecreasing with respect to the associated indices (see Example 5.1.6 for the definition) is another useful example of such families.

Theorem 5.1.4 below gives the main result of the current section. It provides characterizations of the ε-subdifferential of the supremum function f of a family $\{f_t : X \to \overline{\mathbb{R}}, t \in T\}$ of convex functions that is closed for convex combinations and satisfies the following *closedness criterion*

$$\text{cl}\,f = \sup_{t \in T}(\text{cl}\,f_t). \tag{5.10}$$

More details on this condition will be given in section 5.2. The resulting formula of $\partial_\varepsilon f$ is given by means of the approximate subdifferential of the data functions augmented by the indicator of convex sets $D \subset X$ that satisfy

$$\text{dom}\,f \subset D \subset \bigcap_{t \in T \setminus S_0} \text{cl}\,(\text{dom}\,f_t) \tag{5.11}$$

(with the convention $\cap_{t \in \emptyset} \text{cl}\,(\text{dom}\,f_t) = X$). Thus, in particular, when all the $\bar{f}_t$'s are in $\Gamma_0(X)$ we take $D = X$, so that the resulting formula writes $\partial_\varepsilon f$ by means exclusively of the data functions.

Theorem 5.1.4 *Let* $\{f_t : X \to \overline{\mathbb{R}}, t \in T\}$ *be a family of convex functions which is closed for convex combinations,* $D \subset X$ *a convex set satisfying (5.11), and denote* $f := \sup_{t \in T} f_t$. *If (5.10) holds, then for all* $x \in \text{dom}\,f$ *and* $\varepsilon > 0$

$$\partial_\varepsilon f(x) = \text{cl}\left\{ \bigcup_{t \in S_0} \partial_{(\varepsilon + f_t(x) - f(x))}(f_t + I_D)(x) \right\}, \tag{5.12}$$

and $\partial f(x)$ *is obtained by intersecting these sets over* $\varepsilon > 0$.

Proof. We give the proof in the case where $\{f_t, t \in T\} \subset \Gamma_0(X)$ and $D = X\,(= \cap_{t \in \emptyset} \text{cl}\,(\text{dom}\,f_t))$; the general case is proved in Exercise 71.

178   CHAPTER 5.   SUPREMUM OF CONVEX FUNCTIONS

First, we denote $A := \{(x^*, f_t^*(x^*)) : x^* \in \text{dom } f_t^*, \, t \in T\}$ so that, by Theorem 3.2.2,

$$f = \sup_{t \in T} f_t = \sup_{t \in T} f_t^{**} = \sup_{t \in T, \, x^* \in \text{dom } f_t^*} \{\langle x^*, \cdot \rangle - f_t^*(x^*)\}$$

$$= \sup_{(a_t^*, b_t) \in A, \, t \in T} \{\langle a_t^*, \cdot \rangle - b_t\}.$$

So, by (5.3), for every $x \in \text{dom } f$ and $\varepsilon > 0$ we obtain

$$\partial_\varepsilon f(x) = \text{cl}\left\{x^* \in X^* : (x^*, \alpha^*) \in \text{co}(A), \, \langle x^*, x \rangle - \alpha^* \geq f(x) - \varepsilon\right\}.$$

Moreover, if $(x^*, \alpha^*) \in \text{co}(A)$ satisfies $\langle x^*, x \rangle - \alpha^* \geq f(x) - \varepsilon$, then there are $t_1, \ldots, t_k \in T$, $x_i^* \in \text{dom } f_{t_i}^*$, $i = 1, \ldots, k$, $k \geq 1$, and $\lambda \in \Delta_k^*$ such that $(x^*, \alpha^*) = \sum_{1 \leq i \leq k} \lambda_i (x_i^*, f_{t_i}^*(x_i^*))$ and

$$f_\lambda(x) := \sum_{1 \leq i \leq k} \lambda_i f_{t_i}(x) \geq \sum_{1 \leq i \leq k} \langle \lambda_i x_i^*, x \rangle - \sum_{1 \leq i \leq k} \lambda_i f_{t_i}^*(x_i^*)$$

$$= \left\langle \sum_{1 \leq i \leq k} \lambda_i (x_i^*, f_{t_i}^*(x_i^*)), (x, -1) \right\rangle = \langle x^*, x \rangle - \alpha^* \geq f(x) - \varepsilon. \tag{5.13}$$

Next, we set $\varepsilon_i := f_{t_i}(x) + f_{t_i}^*(x_i^*) - \langle x_i^*, x \rangle \geq 0$, $i = 1, \ldots, k$, and, by the current closedness assumption, we choose $s \in T$ $(= S_0)$ such that $f_\lambda \leq f_s$. Then, by applying (5.13) we get

$$\sum_{1 \leq i \leq k} \lambda_i \varepsilon_i + f_s(x) - f_\lambda(x) = f_s(x) + \sum_{1 \leq i \leq k} \lambda_i \left(f_{t_i}^*(x_i^*) - \langle x_i^*, x \rangle\right) \leq f_s(x) - f(x) + \varepsilon$$

and, in particular, $\sum_{1 \leq i \leq k} \lambda_i \varepsilon_i \leq f_\lambda(x) - f(x) + \varepsilon \leq \varepsilon$. Consequently, using Proposition 4.1.6,

$$x^* = \sum_{1 \leq i \leq k} \lambda_i x_i^* \in \sum_{1 \leq i \leq k} \lambda_i \partial_{\varepsilon_i} f_{t_i}(x) = \sum_{1 \leq i \leq k} \partial_{\lambda_i \varepsilon_i}(\lambda_i f_{t_i})(x)$$

$$\subset \partial_{(\Sigma_{1 \leq i \leq k} \lambda_i \varepsilon_i)} f_\lambda(x) \subset \partial_{(\Sigma_{1 \leq i \leq k} \lambda_i \varepsilon_i + f_s(x) - f_\lambda(x))} f_s(x) \subset \partial_{(\varepsilon + f_s(x) - f(x))} f_s(x).$$

Therefore, the inclusion "$\subset$" in (5.12) holds, and the proof is finished as the converse inclusion is straightforward. ∎

Being an easy consequence of Theorem 5.1.4, formula (5.14) below is given for the purpose of comparison with the results of the next section (more precisely, Theorem 5.2.2), where we deal with families

5.1. CONJUGACY-BASED APPROACH

that are not necessarily closed for convex combinations. This formula confirms that $\partial_\varepsilon f(x)$ only involve those data functions that are indexed in the set of almost active indices at x; i.e., $T_\varepsilon(x) := \{t \in T : f_t(x) \geq f(x) - \varepsilon\}$, $\varepsilon \geq 0$. Furthermore, only data functions that have proper closures are explicitly used within $\partial_\varepsilon f(x)$. The impact of the other data functions, having non-proper closures, is only decisive when choosing the set D in (5.11).

Corollary 5.1.5 *With the assumptions of Theorem 5.1.4 we have, for all $x \in \mathrm{dom}\, f$ and $\varepsilon > 0$,*

$$\partial_\varepsilon f(x) = \mathrm{cl}\left\{\bigcup_{t \in T_{\varepsilon-\alpha}(x),\, \alpha \geq 0} \partial_\alpha (f_t + \mathrm{I}_D)(x)\right\}. \tag{5.14}$$

Proof. By a simple change of parameters in (5.12), putting $\alpha := \varepsilon + f_t(x) - f(x)$, we obtain $\alpha \leq \varepsilon$ and $f_t(x) - f(x) = \alpha - \varepsilon$. Hence, $t \in T_{\varepsilon-\alpha}(x)$ and the inclusion "$\subset$" in (5.14) holds (because $\partial_\alpha f_t(x) = \emptyset$ when $\alpha < 0$). The opposite inclusion in (5.14) is easily checked. ∎

In the following example, we illustrate the application of Theorem 5.1.4 and Corollary 5.1.5 by analyzing the supremum of non-decreasing nets. See Exercise 74 for an alternative proof of this result.

Example 5.1.6 *Let $f_t : X \to \overline{\mathbb{R}}$, $t \in T$, be a non-decreasing net of convex functions satisfying (5.10), where $(T, \preccurlyeq)$ is a given directed set, and suppose that $f := \sup_{t \in T} f_t$ has a proper closure.* In order to compute $\partial_\varepsilon f(x)$, we first observe that $\{f_t, t \in T\}$ is closed for convex combinations. Indeed, for every $\lambda \in \Delta(S_0)$ with $\mathrm{supp}\,\lambda := (\lambda_{t_1}, \ldots, \lambda_{t_k})$, $\lambda_{t_1} \preccurlyeq \ldots \preccurlyeq \lambda_{t_k}$, and $t_1, \ldots, t_k \in T$, we have that $f_\lambda \leq f_{t_k}$. This implies that $\sum_{1 \leq i \leq k} \lambda_i \, \mathrm{cl}(f_{t_i}) \leq \mathrm{cl}\left(\sum_{1 \leq i \leq k} \lambda_i f_{t_i}\right) \leq \mathrm{cl}(f_{t_k})$ and $\mathrm{cl}(f_{t_k})$ is necessarily proper; that is, $t_k \in S_0$.

Next, we verify the existence of some $t_0 \in T$ such that $\mathrm{cl}(f_t)$ is proper for all $t \subset T$ with $t_0 \preccurlyeq t$. Otherwise, if not, then $\mathrm{cl}(f_t)$ would be frequently non-proper, and this would imply that $\mathrm{cl}\, f$ is non-proper too, a contradiction with our assumption. Consequently, the family $\{f_s : t_0 \preccurlyeq s,\, s \in T\}$ also satisfies (5.10), since the net $(\mathrm{cl}\, f_t)_{t \in T}$ is also non-decreasing and, so, $\mathrm{cl}\, f = \sup_{t \in T}(\mathrm{cl}\, f_t) = \sup_{t_0 \preccurlyeq s,\, s \in T}(\mathrm{cl}\, f_s)$. Therefore, as $f = \sup_{t_0 \preccurlyeq s,\, s \in T} f_s$, by applying (5.12) with $D = X$ and using the fact that $(f_t)_{t \in T}$ is non-decreasing, for all $x \in \mathrm{dom}\, f$ and $\varepsilon > 0$ we obtain

$$\partial_\varepsilon f(x) = \mathrm{cl}\left\{\bigcup_{t_0 \preccurlyeq s,\, s \in T} \partial_{(\varepsilon + f_s(x) - f(x))} f_s(x)\right\} \subset \mathrm{cl}\left\{\bigcup_{s \in T} \bigcap_{s \preccurlyeq t} \partial_\varepsilon f_t(x)\right\},$$

showing that

$$\partial_\varepsilon f(x) = \mathrm{cl}\left\{\bigcup_{t\in T}\partial_{(\varepsilon+f_t(x)-f(x))}f_t(x)\right\} = \mathrm{cl}\left(\bigcup_{s\in T}\bigcap_{s\preccurlyeq t}\partial_\varepsilon f_t(x)\right). \quad (5.15)$$

It is not difficult to verify that the convex combinations closedness condition used in Theorem 5.1.4 does not constitute any loss of generality, because we can replace the family $\{f_t,\,t\in T\}$ by the new one

$$\left\{f_\lambda := \sum_{t\in\mathrm{supp}\,\lambda}\lambda_t f_t,\,\lambda\in\Delta(S_0);\,f_t,\,t\in T\setminus S_0\right\},$$

where $S_0 = \{t\in T:\bar f_t\in\Gamma_0(X)\}$ (see (5.8)). Due to the easy relation $\mathrm{cl}(f_\lambda) \ge \sum_{t\in\mathrm{supp}\,\lambda}\lambda_t\,\mathrm{cl}(f_t)$, we see that $\mathrm{cl}(f_\lambda)$ is proper for all $\lambda\in\Delta(S_0)$ and, therefore, we can easily verify that the new family above is closed for convex combinations. The point here is that the value of the supremum is not altered with this change, both the old and the new families have the same supremum. Consequently, applying formula (5.12) to this new family, we obtain the following result.

Theorem 5.1.7 *Let $\{f_t,t\in T\}$ be a family of convex functions and denote $f := \sup_{t\in T} f_t$. If (5.10) holds, then, for all $x\in \mathrm{dom}\,f$ and $\varepsilon > 0$, we have*

$$\partial_\varepsilon f(x) = \mathrm{cl}\left\{\bigcup_{\lambda\in\Delta(S_0)}\partial_{(\varepsilon+f_\lambda(x)-f(x))}\left(f_\lambda + \mathrm{I}_D\right)(x)\right\}, \quad (5.16)$$

where $D\subset X$ is any convex set satisfying (5.11).

Proof. Following the previous theorem, it is enough to apply (5.12) to the family of convex functions $\{f_\lambda,\lambda\in\Delta(S_0);\,f_t,\,t\in T\setminus S_0\}$ once we have verified that this family satisfies condition (5.10); that is,

$$\bar f = \max\left\{\sup_{\lambda\in\Delta(S_0)}\mathrm{cl}(f_\lambda),\,\sup_{t\in T\setminus S_0}\mathrm{cl}(f_t)\right\}. \quad (5.17)$$

In fact, the inequality "$\ge$" always holds, while the opposite one follows because the given new family contains the original one that is supposed to satisfy (5.10). ∎

Remark 3 *(i) We obtain the expression for $\partial f(x)$ in Theorem 5.1.7 just by intersecting (5.16) on $\varepsilon > 0$. Alternatively, we easily check the*

5.1. CONJUGACY-BASED APPROACH

following formula that gives a characterization of $\partial_\varepsilon f(x)$ by covering together both cases, $\varepsilon > 0$ and $\varepsilon = 0$:

$$\partial_\varepsilon f(x) = \bigcap_{\delta > \varepsilon} \mathrm{cl} \left\{ \bigcup_{\lambda \in \Delta(S_0)} \partial_{(\delta + f_\lambda(x) - f(x))} \left(f_\lambda + \mathrm{I}_D \right)(x) \right\}, \forall x \in \mathrm{dom}\, f, \forall \varepsilon \geq 0.$$

(ii) If, additionally, we assume in theorem 5.1.7 that the family $\{f_t, t \in T\}$ is closed for convex combinations, then (5.16) implies that, for all $x \in \mathrm{dom}\, f$ and $\varepsilon > 0$,

$$\partial_\varepsilon f(x) = \mathrm{cl} \left\{ \bigcup_{t \in T} \partial_{(\varepsilon + f_t(x) - f(x))} \left(f_t + \mathrm{I}_D \right)(x) \right\},$$

and, taking into account the remark in (i), we deduce that, for all $x \in \mathrm{dom}\, f$ and $\varepsilon \geq 0$,

$$\partial_\varepsilon f(x) = \bigcap_{\delta > \varepsilon} \mathrm{cl} \left\{ \bigcup_{t \in T} \partial_{(\delta + f_t(x) - f(x))} \left(f_t + \mathrm{I}_D \right)(x) \right\}.$$

Next, with the purpose of illustrating the scope of Theorems 5.1.4 and 5.1.7, we give an example dealing with the conjugate function. In turn, the result provided in this example (and its non-convex version given in Exercise 75) is so general that it allows us to rediscover formula (5.12) and its consequences given above. To do this, one can consider the function $g := \inf_{t \in T} f_t^*$ whose conjugate is nothing other than our supremum function $f = \sup_{t \in T} f_t = g^*$.

Example 5.1.8 *Given a convex function $f : X \to \mathbb{R}_\infty$, we verify that (5.14) gives rise to the following characterization, completing relation (4.18) and Corollary 4.3.4,*

$$\partial_\varepsilon f^*(x^*) = \mathrm{cl}((\partial_\varepsilon f)^{-1}(x^*)) \text{ for all } x^* \in X^* \text{ and } \varepsilon > 0. \qquad (5.18)$$

In particular, if f is additionally lsc, then the set $(\partial_\varepsilon f)^{-1}(x^*)$ is closed and the last relation reduces to the well-known expression, $\partial_\varepsilon f^* = (\partial_\varepsilon f)^{-1}$ (see (4.18)). Of course, we may additionally assume that f^* is proper (and, hence, f is proper too), otherwise (5.18) always holds, since all the involved sets would be empty.

To establish (5.18), we first observe that $\{f_x := \langle \cdot, x \rangle - f(x),\ x \in \mathrm{dom}\, f\}$ is closed for convex combinations. In fact, for every $\lambda \in \Delta(\mathrm{dom}\, f)$, we have $\bar{x} := \sum_{x \in \mathrm{supp}\, \lambda} \lambda_x x \in \mathrm{dom}\, f$ and

$$\sum_{x\in \mathrm{supp}\,\lambda} \lambda_x f_x \leq \langle \cdot, \bar{x}\rangle - f(\bar{x}) =: f_{\bar{x}}.$$

Next, since $f^* = \sup_{x\in \mathrm{dom}\, f} f_x$ and $\{f_x,\ x\in \mathrm{dom}\,f\} \subset \Gamma_0(X)$, (5.14) yields

$$\partial_\varepsilon f^*(x^*) = \mathrm{cl}\left\{\bigcup_{x\in T_{\varepsilon-\alpha}(x^*),\ \alpha\geq 0} \partial_\alpha f_x(x^*)\right\}$$

$$= \mathrm{cl}\left\{\bigcup_{\alpha\geq 0} T_{\varepsilon-\alpha}(x^*)\right\} = \mathrm{cl}(T_\varepsilon(x^*)) = \mathrm{cl}((\partial_\varepsilon f)^{-1}(x^*)),$$

as we wanted to prove.

Remark 4 *It was also possible to obtain Example 5.1.8 as a consequence of its non-convex counterpart given in Exercise 75, since the convexity of f implies that, for all $\lambda \in \Delta(\mathrm{dom}\, f)$ and $\varepsilon_x \geq 0$ (see (5.84)),*

$$\sum_{x\in \mathrm{dom}\, f} \lambda_x (\partial_{\varepsilon_x} f)^{-1}(x^*) \subset (\partial_\varepsilon f)^{-1}(x^*).$$

It is important to point out that, when X is a reflexive Banach space, the closure operation previously used in (5.12), (5.14), and (5.16) can be taken with respect to the norm topology in X^*. This fact comes from certain convexity properties of the sets between the curly brackets which appear there and from the Mazur theorem (see Exercise 72).

In the special situation where T is finite we do not require any lower semicontinuity-like condition, because the family $\left\{\sum_{1\leq k\leq n}\lambda_k f_k,\ \lambda \in \Delta_n\right\}$ satisfies condition (5.10) for free as we show in the following corollary.

Corollary 5.1.9 *Given a finite family of convex functions $\{f_k,\ 1\leq k\leq n\}$ and $f := \max_{1\leq k\leq n} f_k$, for every $x\in \mathrm{dom}\, f$ and $\varepsilon \geq 0$ we have*

$$\partial_\varepsilon f(x) = \bigcup_{\lambda\in\Delta_n} \partial_{(\varepsilon + \sum_{1\leq k\leq n}\lambda_k f_k(x) - f(x))}\left(\sum_{1\leq k\leq n}\lambda_k f_k\right)(x) \qquad (5.19)$$

and, particularly, for $\varepsilon = 0$,

$$\partial f(x) = \left\{\bigcup \partial\left(\sum_{1\leq k\leq n}\lambda_k f_k\right)(x) : \lambda \in \Delta_n,\ \sum_{1\leq k\leq n}\lambda_k f_k(x) = f(x)\right\}.$$

5.1. CONJUGACY-BASED APPROACH

Proof. We start by proving that the family $\{g_\lambda := \sum_{1 \leq k \leq n} \lambda_k f_k, \lambda \in \Delta_n\}$, whose supremum coincides with f, satisfies condition (5.10); that is, $\operatorname{cl} f = \max_{\lambda \in \Delta_n} \operatorname{cl}(g_\lambda)$. Indeed, for every $z \in X$, we have

$$(\operatorname{cl} f)(z) = \sup_{U \in \mathcal{N}_X} \inf_{y \in U} f(z+y) = \sup_{U \in \mathcal{N}_X} \inf_{y \in U} \max_{\lambda \in \Delta_n} g_\lambda(z+y),$$

and the minimax theorem (Theorem 3.4.3 and Corollary 3.4.9) yields

$$(\operatorname{cl} f)(z) = \sup_{U \in \mathcal{N}_X} \max_{\lambda \in \Delta_n} \inf_{y \in U} \sum_{1 \leq k \leq n} \lambda_k f_k(z+y) = \max_{\lambda \in \Delta_n} \sup_{U \in \mathcal{N}_X} \inf_{y \in U} \sum_{1 \leq k \leq n} \lambda_k f_k(z+y).$$

Therefore,

$$(\operatorname{cl} f)(z) = \max_{\lambda \in \Delta_n} \sup_{U \in \mathcal{N}_X} \inf_{y \in U} g_\lambda(z+y) = \max_{\lambda \in \Delta_n} \operatorname{cl}(g_\lambda)(z),$$

and condition (5.10) is satisfied.

Now, since $\{g_\lambda, \lambda \in \Delta_n\}$ is obviously closed for convex combinations, applying Theorem 5.1.7 (or, more specifically, Remark 3(ii)) with $D = \operatorname{dom} f = \operatorname{dom} g_\lambda$, for all $\lambda \in \Delta_n$, for every $x \in \operatorname{dom} f$ and $\varepsilon \geq 0$ we obtain

$$\partial_\varepsilon f(x) = \bigcap_{\delta > \varepsilon} \operatorname{cl} \left\{ \bigcup_{\lambda \in \Delta_n} \partial_{(\delta + g_\lambda(x) - f(x))} (g_\lambda + I_D)(x) \right\}.$$

Therefore, given $x^* \in \partial_\varepsilon f(x)$ and $\delta_i \downarrow \varepsilon$ ($\delta_i \equiv \varepsilon$ when $\varepsilon > 0$), we find $\lambda_i \in \Delta_n$ and $x_i^* \in \partial_{(\delta_i + g_{\lambda_i}(x) - f(x))}(g_{\lambda_i} + I_D)(x)$ such that $x_i^* \to x^*$. We may assume without loss of generality that $\lambda_i \to \bar{\lambda} \in \Delta_n$ and, thus, we can easily prove that $x^* \in \partial_{(\varepsilon + g_{\bar{\lambda}}(x) - f(x))}(g_{\bar{\lambda}} + I_D)(x)$, yielding the non-trivial inclusion in (5.19). ∎

Next, we apply Corollary 5.1.9 to the positive part function, $f^+ = \max\{0, f\}$.

Example 5.1.10 *Consider the convex function $f : X \to \overline{\mathbb{R}}$. Then, for every $x \in \operatorname{dom} f$ and $\varepsilon \geq 0$, Corollary 5.1.9 yields*

$$\partial_\varepsilon f^+(x) = \bigcup_{0 \leq \lambda \leq 1} \partial_{(\varepsilon + \lambda f(x) - f^+(x))}(\lambda f)(x). \tag{5.20}$$

In particular, for $\varepsilon = 0$ we get

$$\partial f^+(x) = \left\{ \bigcup_{0 \le \lambda \le 1} \partial(\lambda f)(x) : \lambda f(x) = f^+(x) \right\};$$

that is,

$$\partial f^+(x) = \begin{cases} \partial f(x), & \text{if } f(x) > 0, \\ \bigcup_{0 < \lambda \le 1} \lambda \partial f(x) \cup N_{\text{dom } f}(x), & \text{if } f(x) = 0, \\ N_{\text{dom } f}(x), & \text{if } f(x) < 0. \end{cases}$$

Another way to make the family $\{f_t, t \in T\}$ closed for convex combinations is to take maxima of finite subfamilies; i.e.,

$$f_J := \max\{f_t, t \in J\}, \, J \in \mathcal{T} := \{S \subset T : |S| < +\infty\}. \tag{5.21}$$

Therefore, in contrast to the previous approach using convex combinations, the resulting formulas of $\partial_\varepsilon f(x)$ now involve approximate subdifferentials of the functions f_J, reducing the problem to that of computing the subdifferential of the supremum of a finite family. Furthermore, in Exercise 77, the expression of $\partial_\varepsilon f(x)$ is further simplified when the underlying space is a reflexive Banach space.

Corollary 5.1.11 *Let $\{f_t, t \in T\}$ be a family of convex functions satisfying (5.10), and denote $f := \sup_{t \in T} f_t$. Then, for all $x \in \text{dom } f$ and $\varepsilon > 0$,*

$$\partial_\varepsilon f(x) = \text{cl} \left\{ \bigcup_{J \in \mathcal{T}, \, J \subset S_0} \partial_{(\varepsilon + f_J(x) - f(x))} (f_J + I_D)(x) \right\}, \tag{5.22}$$

where $D \subset X$ is any convex set satisfying (5.11).

Proof. We use Theorem 5.1.4 to calculate the subdifferential of the supremum of $\{f_J, \, J \in \mathcal{T}, \, J \subset S_0; \, f_t, \, t \in T \setminus S_0\}$, which has the same supremum as the original family $\{f_t, \, t \in T\}$. Also, it is clear that $\text{cl}(f_J) \in \Gamma_0(X)$ for all $J \in \mathcal{T}, \, J \subset S_0$, because $\text{cl}(f_J) \ge \max\{\text{cl}(f_t), \, t \in J\}$ and $\text{cl}(f_t) \in \Gamma_0(X)$ for all $t \in J \subset S_0$. At the same time, this new family is easily checked to be closed for convex combinations, and to satisfy condition (5.10); that is,

$$\bar{f} = \max \left\{ \sup_{J \in \mathcal{T}, \, J \subset S_0} \text{cl}(f_J); \, \sup_{t \in T \setminus S_0} \bar{f}_t \right\}.$$

■

5.1. CONJUGACY-BASED APPROACH

In the formulas given above the functions whose closures are proper stand out, while the others with non-proper closures are present via the set D. If this distinction needs not to be made, then Example 5.1.6 could directly be applied to the net $(f_J)_{J \in \mathcal{T}}$ by considering $\mathcal{T}$ as a set directed by ascending inclusions. Then, since we can easily verify that $f = \sup_{J \in \mathcal{T}} f_J$ and $\operatorname{cl} f = \sup_{J \in \mathcal{T}} (\operatorname{cl} f_J)$ (under (5.10)), Example 5.1.6 gives rise, for all $x \in \operatorname{dom} f$ and $\varepsilon > 0$, to

$$\partial_\varepsilon f(x) = \operatorname{cl}\left\{\bigcup_{J \in \mathcal{T}} \partial_{(\varepsilon + f_J(x) - f(x))} f_J(x)\right\}. \tag{5.23}$$

It is worth observing that, when X is a reflexive Banach space, the closure in this expression can be removed provided that, instead of $\mathcal{T}$, we consider the collection of countable subsets of T (see Exercise 77).

The need to make explicit characterizations of the normal cone to sublevel sets is of great interest in convex analysis (see, for instance, the proof of Theorem 8.2.2). The following two examples provide precise characterizations of the normal cone which are immediate consequences of Theorem 5.1.4.

Example 5.1.12 *Consider a convex function $f : X \to \overline{\mathbb{R}}$ such that*

$$\operatorname{cl}([f \leq 0]) = [\bar{f} \leq 0]. \tag{5.24}$$

We are going to prove that, for every $x \in [f \leq 0]$ and every $\varepsilon > 0$,

$$\mathrm{N}_{[f \leq 0]}^\varepsilon(x) = \operatorname{cl}\left(\bigcup_{t>0} t\partial_{(\frac{\varepsilon}{t} + f(x))} f(x)\right); \tag{5.25}$$

consequently, $\mathrm{N}_{[f \leq 0]}(x)$ is obtained by intersecting the last sets over $\varepsilon > 0$.

To this aim, we define the functions

$$g_t := tf, \ t > 0, \text{ and } g := \sup_{t>0} g_t.$$

Obviously, $\{g_t, \ t > 0\}$ is closed for convex combinations. Therefore, because $g = \mathrm{I}_{[f \leq 0]}$ and (5.10) holds (as a consequence of (5.24)), by (5.12) we infer that

$$\mathrm{N}_{[f \leq 0]}^\varepsilon(x) = \partial_\varepsilon g(x) = \operatorname{cl}\left(\bigcup_{t>0} \partial_{(\varepsilon + tf(x))}(tf)(x)\right) = \operatorname{cl}\left(\bigcup_{t>0} t\partial_{(\frac{\varepsilon}{t} + f(x))} f(x)\right).$$

Our analysis also covers the case where the function that defines the sublevel set in Example 5.1.12 is a supremum.

Example 5.1.13 *Consider a family of convex functions $\{f_t, t \in T\}$ such that $f := \sup_{t \in T} f_t$ and*

$$\mathrm{cl}\,([f \leq 0]) = \left[\sup_{t \in T} \bar{f}_t \leq 0\right].$$

Then, by combining Example 5.1.12 and Theorem 5.1.7 we obtain, for all $x \in [f \leq 0]$ and $\varepsilon > 0$,

$$\mathrm{N}^\varepsilon_{[f \leq 0]}(x) = \mathrm{cl}\left(\bigcup_{\mu > 0} \mu \partial_{\frac{\varepsilon}{\mu} + f(x)} f(x)\right) = \mathrm{cl}\left(\bigcup_{\mu > 0,\, \lambda \in \Delta(S_0)} \mu \partial_{\left(\frac{\varepsilon}{\mu} + f_\lambda(x)\right)} (f_\lambda + \mathrm{I}_D)(x)\right),$$

where $S_0 = \{t \in T : \bar{f}_t \in \Gamma_0(X)\}$, $f_\lambda = \sum_{t \in \mathrm{supp}\,\lambda} \lambda_t f_t$, and $D \subset X$ is any convex set satisfying (5.11).

We close this section by showing how Theorems 5.1.4 and 5.1.7 can be useful for performing regularization procedures that are used for developing subdifferential calculus rules. More results in this direction will be given in chapter 7. In fact, for every function $f \in \Gamma_0(X)$ defined on a Banach space X, we know that the Moreau envelope of f, which is given by $f^\gamma := f \square \frac{1}{2\gamma} \|\cdot\|^2$, non-decreases to f as $\gamma \downarrow 0$. This can also be extended to finite sums of convex functions $f_1, \ldots, f_k \in \Gamma_0(X)$, so that $f_1^\gamma + \ldots + f_k^\gamma \nearrow f_1 + \ldots + f$ as $\gamma \downarrow 0$. So, we can apply our previous results to write $\partial_\varepsilon(f + g)$ by means of $\partial_\varepsilon(f^\gamma + g^\gamma)$. The advantage here is that f^γ and g^γ are regular enough to guarantee that $\partial_\varepsilon(f^\gamma + g^\gamma)$ can be decomposed. We have the following corollary.

Corollary 5.1.14 *Assume that X is Banach. Then the following assertions hold true:*

(i) Given $f, g \in \Gamma_0(X)$, for all $x \in \mathrm{dom}\, f \cap \mathrm{dom}\, g$ and $\varepsilon > 0$, we have

$$\partial_\varepsilon(f + g)(x) = \mathrm{cl}\left\{\bigcup_{\gamma > 0,\, \alpha \in \mathbb{R}} \partial_{(\varepsilon - \alpha + f^\gamma(x) - f(x))} f^\gamma(x) + \partial_{(\alpha + g^\gamma(x) - g(x))} g^\gamma(x)\right\}.$$

(ii) Let $\{f_t, t \in T\} \subset \Gamma_0(X)$ and denote $f := \sup_{t \in T} f_t$. Then, for all $x \in \mathrm{dom}\, f$ and $\varepsilon > 0$, we have

$$\partial_\varepsilon f(x) = \mathrm{cl}\left\{\bigcup_{\gamma > 0,\, \alpha \in \Delta(T)} \partial_{\left(\varepsilon + \sum_{t \in \mathrm{supp}\,\alpha} \alpha_t f_t^\gamma(x) - f(x)\right)} \left(\sum_{t \in \mathrm{supp}\,\alpha} \alpha_t f_t^\gamma\right)(x)\right\}.$$

Proof. (i) Since $f, g, f^\gamma, g^\gamma \in \Gamma_0(X)$, and $f^\gamma + g^\gamma \nearrow f + g$, Theorem 5.1.4 applies and yields

$$\partial_\varepsilon (f+g)(x) = \mathrm{cl}\left\{\bigcup_{\gamma>0} \partial_{(\varepsilon+f^\gamma(x)+g^\gamma(x)-f(x)-g(x))}(f^\gamma+g^\gamma)(x)\right\},$$

and the desired result follows by Proposition 4.1.20(iii).

(ii) We have $\sup_{t\in T} f_t^\gamma \nearrow f$ and, again, Theorem 5.1.4 yields

$$\partial_\varepsilon f(x) = \mathrm{cl}\left\{\bigcup_{\gamma>0} \partial_{(\varepsilon+\sup_{t\in T} f_t^\gamma(x)-f(x))}\left(\sup_{t\in T} f_t^\gamma\right)(x)\right\}.$$

Thus, applying Theorem 5.1.7 (formula (5.16)) with $D = X$ (as the functions involved are in $\Gamma_0(X)$),

$$\partial_\varepsilon f(x) = \mathrm{cl}\left\{\bigcup_{\gamma>0} \mathrm{cl}\left(\bigcup_{\alpha\in\Delta(T)} \partial_{\left(\varepsilon+\sum_{t\in\mathrm{supp}\,\alpha}\alpha_t f_t^\gamma(x)-f(x)\right)}\left(\sum_{t\in\mathrm{supp}\,\alpha}\alpha_t f_t^\gamma\right)(x)\right)\right\},$$

and we are done. ∎

In the subsequent sections, we derive different characterizations of the subdifferential of pointwise suprema, which provide more geometrical insight, and highlight the role played by (almost) active and non-active functions. For instance, in the case of the support function σ_C, the forthcoming characterizations will rely on the set C and not on its closed convex hull, as in (5.1). Also, in the setting of Theorem 5.1.4, we shall characterize the subdifferential of the supremum function f by appealing to the (almost) active functions and making implicit use of the remaining functions through the normal cone to the effective domain of f (or equivalent representations of it).

5.2 Main subdifferential formulas

We deal with an arbitrary family of convex functions $f_t : X \to \overline{\mathbb{R}}, t \in T$, defined on the real (separated) locally convex space X. The index set T is a fixed arbitrary (possibly, infinite) set. Our aim in this section is to express in the most general framework the subdifferential of the associated supremum function $f := \sup_{t\in T} f_t$, at any reference point $x \in X$. The formulas that we give rely on the ε-active functions at x, i.e., $T_\varepsilon(x) := \{t \in T : f_t(x) \geq f(x) - \varepsilon\}$, $\varepsilon \geq 0$. More precisely, our

188   CHAPTER 5. SUPREMUM OF CONVEX FUNCTIONS

objective is to express the subdifferential of f at x by using exclusively the data functions f_t, $t \in T_\varepsilon(x)$, for small positive ε's. Only a weak closedness condition, which is satisfied by many families of functions, is assumed in this section; but no hypothesis affecting the set of indices or the continuity behavior of the functions involved is required.

We start with the following simple example illustrating the impossibility of writing in general the set ∂f by means of ∂f_t, $t \in T$.

Example 5.2.1 *Let $f_1, f_2 : \mathbb{R} \to \mathbb{R}_\infty$ be defined as*

$$f_1(x) = -\sqrt{|x|} + I_{[0,+\infty[} \text{ and } f_2(x) = f_1(-x),$$

so that $f_1(0) = f_2(0) = 0$ and $f := \max\{f_1, f_2\} = I_{\{0\}}$. Then it turns out that $\partial f(0) = N_{\{0\}}(0) = \mathbb{R}$, while $\partial f_1(0) = \partial f_2(0) = \emptyset$.

Theorem 5.2.2 below provides the main formula of the subdifferential of the supremum function f, under the closedness condition (5.10); that is, $\mathrm{cl}\, f = \sup_{t \in T}(\mathrm{cl}\, f_t)$. This condition is satisfied not only for lsc functions but also for wider families of functions as it is shown, for instance, in Proposition 5.2.4 below. It is worth observing that the inequality "$\geq$" in (5.10) always holds, due to the following obvious relation $f \geq f_t \geq \mathrm{cl}(f_t)$. Hence, the lsc function $\sup_{t \in T}(\mathrm{cl}\, f_t)$ is majorized by f and, thus, also by its closure $\mathrm{cl}\, f$.

Due to the generality of our infinite-dimensional setting, we shall make use of the family $\mathcal{F}(x)$ (see (1.4)).

Theorem 5.2.2 *Assume that the convex functions $f_t : X \to \overline{\mathbb{R}}$, $t \in T$, satisfy condition (5.10). Then, for every $x \in X$,*

$$\partial f(x) = \bigcap_{\varepsilon > 0,\ L \in \mathcal{F}(x)} \overline{\mathrm{co}} \left(\bigcup_{t \in T_\varepsilon(x)} \partial_\varepsilon f_t(x) + N_{L \cap \mathrm{dom}\, f}(x) \right). \tag{5.26}$$

Proof. For the sake of brevity, here we only give the crucial part of the proof corresponding to the case when $f_t \in \Gamma_0(X)$, for all $t \in T$, so that condition (5.10) trivially holds. The proof of the general case is completed in Exercises 80 and 81, applying the arguments below to updated families of functions related to $\{\mathrm{cl}\, f_t, \ t \in T\}$ (while distinguishing among proper and non-proper functions).

Given $x \in X$, we denote by A the set on the right-hand side of (5.26), and start by proving the inclusion $A \subset \partial f(x)$. To this aim we prove that

$$\bigcup_{t \in T_\varepsilon(x)} \partial_\varepsilon f_t(x) + N_{L \cap \mathrm{dom}\, f}(x) \subset \partial_{2\varepsilon}(f + I_{L \cap \mathrm{dom}\, f})(x),$$

5.2. MAIN SUBDIFFERENTIAL FORMULAS

for each $\varepsilon > 0$ and $L \in \mathcal{F}(x)$. Indeed, for every $t \in T_\varepsilon(x)$ we have that

$$\partial_\varepsilon f_t(x) + \mathrm{N}_{L \cap \mathrm{dom}\, f}(x) \subset \partial_\varepsilon (f_t + \mathrm{I}_{L \cap \mathrm{dom}\, f})(x) \subset \partial_{2\varepsilon}(f + \mathrm{I}_{L \cap \mathrm{dom}\, f})(x), \tag{5.27}$$

and so (4.15) and (4.16) give rise to

$$A \subset \bigcap_{L \in \mathcal{F}(x)} \bigcap_{\varepsilon > 0} \partial_{2\varepsilon}(f + \mathrm{I}_{L \cap \mathrm{dom}\, f})(x) = \partial f(x).$$

Next, we shall prove the converse inclusion $\partial f(x) \subset A$, assuming that $x = \theta$, $\partial f(\theta) = \partial(\mathrm{cl}\, f)(\theta) \neq \emptyset$, $f(\theta) = (\mathrm{cl}\, f)(\theta) = 0$ (Exercise 78).

Given $x^* \in X^* \setminus A$, we shall prove that $x^* \in X^* \setminus \partial f(\theta)$. Then there are some $\varepsilon > 0$ and $L \in \mathcal{F}(\theta)$ such that

$$x^* \notin \mathrm{cl}\,(\mathrm{co}\,(A_\varepsilon) + \mathrm{N}_{L \cap \mathrm{dom}\, f}(\theta)),$$

where $A_\varepsilon := \cup_{t \in T_\varepsilon(\theta)} \partial_\varepsilon f_t(\theta)$ is non-empty because the functions f_t are assumed to be lsc. Next, the separation theorem gives rise to some $\bar{x} \in X$ and $\gamma < 0$ such that

$$\langle \bar{x}, x^* \rangle + \gamma > \langle \bar{x}, u^* + \alpha v^* \rangle \text{ for all } u^* \in A_\varepsilon,\ v^* \in \mathrm{N}_{L \cap \mathrm{dom}\, f}(\theta) \text{ and } \alpha > 0. \tag{5.28}$$

Hence, dividing by α and making $\alpha \uparrow \infty$, we get

$$\bar{x} \in (\mathrm{N}_{L \cap \mathrm{dom}\, f}(\theta))^- = \mathrm{cl}(\mathrm{cone}(L \cap \mathrm{dom}\, f)). \tag{5.29}$$

Notice that we have

$$\sigma_{A_\varepsilon}(z) = \sup_{u^* \in A_\varepsilon} \langle z, u^* \rangle \leq f(z) + 2\varepsilon \quad \text{for all } z \in X,$$

which easily comes from the following inequalities:

$$\langle z, u^* \rangle \leq f_t(z) - f_t(\theta) + \varepsilon \leq f(z) + 2\varepsilon \text{ for all } z \in X,\ t \in T_\varepsilon(\theta),\ u^* \in \partial_\varepsilon f_t(\theta).$$

Hence, $\mathrm{dom}\, f \subset \mathrm{dom}\, \sigma_{A_\varepsilon}$ and we have that

$$C := \mathrm{cone}(L \cap \mathrm{dom}\, f) \subset \mathrm{dom}\, \sigma_{A_\varepsilon} = \mathrm{dom}(\sigma_{A_\varepsilon} - x^*). \tag{5.30}$$

In particular, we have that $\bar{x} \in \mathrm{cl}(C)$, by (5.29), and $\mathrm{ri}\, C \neq \emptyset$ because $\mathrm{aff}\, C \subset L$ and L is finite-dimensional. Next, by taking $v^* = \theta$ in (5.28) we get

$$\gamma > \langle \bar{x}, u^* - x^* \rangle \text{ for all } u^* \in A_\varepsilon,$$

which gives rise to (applying Exercise 14 to the convex proper function $\sigma_{A_\varepsilon} - x^*$ satisfying $(\operatorname{ri} C) \cap \operatorname{dom}(\sigma_{A_\varepsilon} - x^*) = \operatorname{ri} C \neq \emptyset$, by (5.30))

$$\gamma/2 > \gamma \geq (\sigma_{A_\varepsilon} - x^*)(\bar{x}) \geq \inf_{z \in \operatorname{cl}(C)} (\sigma_{A_\varepsilon} - x^*)(z) = \inf_{z \in C} (\sigma_{A_\varepsilon} - x^*)(z).$$

Therefore, we find $\bar{z} \in L \cap \operatorname{dom} f$ and $\bar{\lambda} > 0$ such that

$$0 > \gamma/2 > (\sigma_{A_\varepsilon} - x^*)(\bar{\lambda}\bar{z}) \geq \langle \bar{\lambda}\bar{z}, u^* - x^* \rangle \text{ for all } u^* \in A_\varepsilon,$$

or, equivalently, $\gamma/(2\bar{\lambda}) > \langle \bar{z}, u^* - x^* \rangle$ for all $u^* \in A_\varepsilon$. Thus, for all $u^* \in A_\varepsilon$ and $v^* \in N_{L \cap \operatorname{dom} f}(\theta)$, $\bar{z}$ satisfies

$$\langle \bar{z}, x^* \rangle + \gamma/(2\bar{\lambda}) > \langle \bar{z}, u^* \rangle \geq \langle \bar{z}, u^* + v^* \rangle.$$

In other words, up to an adjustment of the parameter γ, $\bar{z}$ plays the same role as the one of $\bar{x}$ in (5.28); therefore, we can suppose that

$$\bar{x} \in L \cap \operatorname{dom} f. \tag{5.31}$$

Now, we consider the function $h := \inf_{t \in T} f_t^*$ so that, by (3.10) and Theorem 3.2.2,

$$h^* = \sup_{t \in T} f_t^{**} = \sup_{t \in T} f_t = f. \tag{5.32}$$

Observe that, due to the Fenchel inequality, for all $z^* \in X^*$

$$h(z^*) = \inf_{t \in T} f_t^*(z^*) = \inf_{t \in T} (f_t^*(z^*) + f(\theta) - \langle \theta, z^* \rangle)$$
$$\geq \inf_{t \in T} (f_t^*(z^*) + f_t(\theta) - \langle \theta, z^* \rangle) \geq 0. \tag{5.33}$$

Let us check that the function h additionally satisfies

$$h(z^*) \geq \varepsilon \text{ for all } z^* \in X^* \setminus A_\varepsilon. \tag{5.34}$$

Indeed, given $z^* \in X^* \setminus A_\varepsilon$, so that $z^* \notin \partial_\varepsilon f_t(\theta)$ for all $t \in T_\varepsilon(\theta)$, by using again the Fenchel inequality we obtain that, for all $t \in T_\varepsilon(\theta)$,

$$f_t^*(z^*) = f_t^*(z^*) + f(\theta) - \langle \theta, z^* \rangle \geq f_t^*(z^*) + f_t(\theta) - \langle \theta, z^* \rangle > \varepsilon.$$

Also, for all $t \in T \setminus T_\varepsilon(\theta)$ we have that

$$f_t^*(z^*) = f_t^*(z^*) + f(\theta) - \langle \theta, z^* \rangle > f_t^*(z^*) + f_t(\theta) + \varepsilon - \langle \theta, z^* \rangle \geq \varepsilon,$$

5.2. MAIN SUBDIFFERENTIAL FORMULAS

and the two inequalities above yield $h(z^*) = \inf_{t \in T} f_t^*(z^*) \geq \varepsilon$, which is (5.34). Next, by (5.32), for every given $\lambda \in \,]0,1[$ we have that

$$f(\lambda \bar{x}) = h^{**}(\lambda \bar{x}) = \sup\{\langle \lambda \bar{x}, z^* \rangle - h(z^*) : z^* \in X^*\}$$

$$= \max\left\{\sup_{z^* \in A_\varepsilon} [\langle \lambda \bar{x}, z^* \rangle - h(z^*)], \sup_{z^* \in X^* \setminus A_\varepsilon} [\langle \lambda \bar{x}, z^* \rangle - h(z^*)]\right\}. \tag{5.35}$$

Then, on the one hand, we have that $\langle \bar{x}, x^* \rangle + \gamma \geq \sigma_{A_\varepsilon}(\bar{x})$, by (5.28), and so, due to the fact that $h^* \geq 0$ (see (5.33)),

$$\sup_{z^* \in A_\varepsilon} [\langle \lambda \bar{x}, z^* \rangle - h(z^*)] \leq \sup_{z^* \in A_\varepsilon} \langle \lambda \bar{x}, z^* \rangle = \lambda \sigma_{A_\varepsilon}(\bar{x}) \leq \lambda(\gamma + \langle \bar{x}, x^* \rangle) < \langle \lambda \bar{x}, x^* \rangle. \tag{5.36}$$

On the other hand, (5.34) yields

$$\sup_{z^* \in X^* \setminus A_\varepsilon} [\langle \lambda \bar{x}, z^* \rangle - h(z^*)] \leq \sup_{z^* \in X^* \setminus A_\varepsilon} \lambda [\langle \bar{x}, z^* \rangle - h(z^*)] + \sup_{z^* \in X^* \setminus A_\varepsilon} (1-\lambda)[-h(z^*)]$$

$$\leq \lambda h^*(\bar{x}) - (1-\lambda)\varepsilon = \lambda f(\bar{x}) - (1-\lambda)\varepsilon;$$

hence, if $\lambda_0 \in \,]0,1[$ is such that $\lambda f(\bar{x}) - (1-\lambda)\varepsilon < \langle \lambda \bar{x}, x^* \rangle$ for all $\lambda \in \,]0, \lambda_0]$, the last inequality yields

$$\sup_{z^* \in X^* \setminus A_\varepsilon} [\langle \lambda \bar{x}, z^* \rangle - h(z^*)] < \langle \lambda \bar{x}, x^* \rangle \text{ for all } \lambda \in \,]0, \lambda_0].$$

Consequently, by combining this inequality with (5.36), (5.35) gives us $f(\lambda \bar{x}) < \langle \lambda \bar{x}, x^* \rangle$ for all $\lambda \in \,]0, \lambda_0]$; in other words, $f(\lambda \bar{x}) - f(\theta) = f(\lambda \bar{x}) < \langle \lambda \bar{x}, x^* \rangle = \langle \lambda \bar{x} - \theta, x^* \rangle$, showing that $x^* \notin \partial f(\theta)$, as we wanted to prove. ∎

Remark 5 *The same conclusion of Theorem 5.2.2 is valid if, instead of condition (5.10), we assume that*

$$(\mathrm{cl}\, f)(x) = \sup_{t \in T}(\mathrm{cl}\, f_t)(x) \text{ for all } x \in \mathrm{dom}\, f,$$

as the reader will see in Theorem 7.3.2 in section 7.3 (see, also, Exercise 114(i)).

Observe that if $X = \mathbb{R}^n$ in Theorem 5.2.2, then $X \in \mathcal{F}(x)$,

$$N_{\mathrm{dom}\, f}(x) \subset N_{L \cap \mathrm{dom}\, f}(x) \text{ for all } L \in \mathcal{F}(x),$$

and formula (5.26) becomes

$$\partial f(x) = \bigcap_{\varepsilon > 0} \overline{\mathrm{co}} \left(\bigcup_{t \in T_\varepsilon(x)} \partial_\varepsilon f_t(x) + \mathrm{N}_{\mathrm{dom}\, f}(x) \right).$$

The intersection over the L's can also be removed in other situations which are analyzed in section 5.3.

As a consequence of Theorem 5.2.2, the following corollary shows that the functions f_t, $t \in T \setminus T_\varepsilon(x)$, only contribute to building the effective domain of f.

Corollary 5.2.3 *Let $\{f_t, t \in T\}$ be a non-empty family of convex functions, and set $f := \sup_{t \in T} f_t$. Then for every $x \in \mathrm{dom}\, f$ we have that*

$$\partial f(x) = \partial \left(\sup_{t \in T_\varepsilon(x)} f_t + \mathrm{I}_{\mathrm{dom}\, f} \right)(x) \quad \textit{for all } \varepsilon > 0. \qquad (5.37)$$

Consequently, provided that $\mathrm{int}(\mathrm{dom}\, f) \neq \emptyset$ or some of the functions $\sup_{t \in T_\varepsilon(x)} f_t$, $\varepsilon > 0$, is continuous somewhere in $\mathrm{dom}\, f$, there exists some $\varepsilon_0 > 0$ such that

$$\partial f(x) = \mathrm{N}_{\mathrm{dom}\, f}(x) + \partial \left(\sup_{t \in T_\varepsilon(x)} f_t \right)(x) \quad \textit{for all } 0 < \varepsilon \leq \varepsilon_0. \qquad (5.38)$$

Proof. We fix $x \in \mathrm{dom}\, f$ such that $\partial f(x) = \emptyset$; otherwise, if $\partial f(x) = \emptyset$, then the desired formulas hold trivially. First, we assume that the family $\{f_t,\ t \in T\}$ satisfies (5.10). We denote $A_\varepsilon := \cup_{t \in T_\varepsilon(x)} \partial_\varepsilon f_t(x)$ and $g_\varepsilon := \sup_{t \in T_\varepsilon(x)} f_t$. Given $\delta > \varepsilon > 0$, we observe that for all $x^* \in \partial_\varepsilon f_t(x)$ and $t \in T_\varepsilon(x)$ we have that, for all $y \in X$,

$$\langle x^*, y - x \rangle \leq f_t(y) - f_t(x) + \varepsilon \leq g_\varepsilon(y) - f_t(x) + \varepsilon \leq g_\delta(y) - g_\delta(x) + 2\varepsilon;$$

that is, $x^* \in \partial_{2\varepsilon} g_\delta(x)$ and we deduce that $A_\varepsilon \subset \partial_{2\varepsilon} g_\delta(x)$. Hence, by Theorem 5.2.2 and taking into account (4.16), we obtain

5.2. MAIN SUBDIFFERENTIAL FORMULAS

$$\partial f(x) = \bigcap_{L \in \mathcal{F}(x), 0 < \varepsilon < \delta} \overline{\operatorname{co}} \left\{ \bigcup_{t \in T_\varepsilon(x)} \partial_\varepsilon f_t(x) + \mathrm{N}_{L \cap \operatorname{dom} f}(x) \right\}$$

$$\subset \bigcap_{L \in \mathcal{F}(x), 0 < \varepsilon < \delta} \operatorname{cl}(\partial_{2\varepsilon} g_\delta(x) + \mathrm{N}_{L \cap \operatorname{dom} f}(x))$$

$$\subset \bigcap_{L \in \mathcal{F}(x), 0 < \varepsilon < \delta} \partial_{2\varepsilon}(g_\delta + \mathrm{I}_{L \cap \operatorname{dom} f})(x)$$

$$= \bigcap_{0 < \varepsilon < \delta} \partial_{2\varepsilon}(g_\delta + \mathrm{I}_{\operatorname{dom} f})(x) = \partial(g_\delta + \mathrm{I}_{\operatorname{dom} f})(x).$$

Moreover, since $g_\delta \leq f$ and $g_\delta(x) = f(x)$, we also have that $\partial(g_\delta + \mathrm{I}_{\operatorname{dom} f})(x) \subset \partial(f + \mathrm{I}_{\operatorname{dom} f})(x) = \partial f(x)$; that is, by combining the last two inclusions, $\partial f(x) \subset \partial(g_\delta + \mathrm{I}_{\operatorname{dom} f})(x) \subset \partial f(x)$, and $\partial f(x) = \partial(g_\delta + \mathrm{I}_{\operatorname{dom} f})(x)$ for all $\delta > 0$.

To prove (5.37) in the general case, we fix $\varepsilon > 0$ and $L \in \mathcal{F}(x)$. Then

$$\partial f(x) \subset \partial(f + \mathrm{I}_{L \cap \operatorname{dom} f})(x) = \partial \left(\sup_{t \in T}(f_t + \mathrm{I}_{L \cap \operatorname{dom} f}) \right)(x).$$

Moreover, since the family $\{f_t + \mathrm{I}_{L \cap \operatorname{dom} f}, \ t \in T\}$ satisfies (5.10) (Proposition 5.2.4(iv)), by the first part of the proof we infer that $\partial f(x) \subset \partial(\sup_{t \in T_\varepsilon(x)} f_t + \mathrm{I}_{L \cap \operatorname{dom} f})(x)$ and, by intersecting over the L's and using (4.16),

$$\partial f(x) \subset \bigcap_{L \in \mathcal{F}(x)} \partial \left(\sup_{t \in T_\varepsilon(x)} f_t + \mathrm{I}_{L \cap \operatorname{dom} f} \right)(x) = \partial \left(\sup_{t \in T_\varepsilon(x)} f_t + \mathrm{I}_{\operatorname{dom} f} \right)(x).$$

Then we are done as the opposite inclusion is straightforward:

$$\partial \left(\sup_{t \in T_\varepsilon(x)} f_t + \mathrm{I}_{\operatorname{dom} f} \right)(x) \subset \partial(f + \mathrm{I}_{\operatorname{dom} f})(x) = \partial f(x).$$

Assume now that for some $\varepsilon_0 > 0$ the function g_{ε_0} is continuous at $x_0 \in \operatorname{dom} f$. Then, by Proposition 4.1.20 and the first part of the proof, we deduce $\partial f(x) = \partial(g_{\varepsilon_0} + \mathrm{I}_{\operatorname{dom} f})(x) = \partial g_{\varepsilon_0}(x) + \mathrm{N}_{\operatorname{dom} f}(x)$. Also, thanks to Proposition 4.1.20, the same conclusion holds when $\operatorname{int}(\operatorname{dom} f) \neq \emptyset$
∎

The following proposition shows that condition (5.10) is satisfied in a variety of situations.

Proposition 5.2.4 *Let $\{f_t, t \in T\}$ be a non-empty family of convex functions, and set $f := \sup_{t \in T} f_t$. Then condition (5.10) is fulfilled in any of the following situations:*

(i) Each one of the f_t's is continuous somewhere in $\operatorname{dom} f$. This holds if, for instance, the function f is continuous somewhere.

(ii) T is finite and each one of the f_t's, except perhaps one of them, is continuous somewhere in $\operatorname{dom} f$.

(iii) $X = \mathbb{R}^n$ and f is finite at a common point of the sets $\operatorname{ri}(\operatorname{dom} f_t)$, $t \in T$.

(iv) $\operatorname{aff}(\operatorname{dom} f_t) = \operatorname{aff}(\operatorname{dom} f)$, for all $t \in T$, and $f_{|\operatorname{aff}(\operatorname{dom} f)}$ is finite and continuous somewhere in $\operatorname{ri}(\operatorname{dom} f)$.

Proof. Setting $A_t := \operatorname{epi} f_t$ for $t \in T$ and $A := \operatorname{epi} f$, one always has $A = \cap_{t \in T} A_t$, and we have to show that

$$\operatorname{cl}(A) = \bigcap_{t \in T} \operatorname{cl}(A_t).$$

The inclusion "⊂" being obvious, the opposite one remains to be proved in each case.

(i) Take $x_0 \in X$ and $\mu \in \mathbb{R}$ such that $\mu > f(x_0)$. Hence, $f_t(x_0) \le f(x_0) < \mu$ for all $t \in T$, so that $y_0 := (x_0, \mu) \in \cap_{t \in T} \operatorname{int}(A_t)$. Now, if $y \in \cap_{t \in T} \operatorname{cl}(A_t)$, then $(1-\lambda)y + \lambda y_0 \in \cap_{t \in T} \operatorname{int}(A_t) \subset A$ for every $\lambda \in {]0,1[}$, due to (2.15), and so $y \in \operatorname{cl}(A)$.

(ii) We denote $T := \{1, \ldots, k, k+1\}$, and assume that each one of the functions $f_1, \ldots, f_k$ is continuous at some $x_i \in \operatorname{dom} f$, respectively. Set $B := \cap_{1 \le t \le k} A_t$ and $x_0 := \sum_1^k \frac{1}{k} x_i$. Then $x_0 \in \cap_{1 \le t \le k} \operatorname{int}(\operatorname{dom} f_i)$, by (2.15), and so each one of the functions $f_1, \ldots, f_k$ is continuous at x_0. Next, similarly to (i), we can show that $y_0 := (x_0, \mu) \in A_{k+1} \cap (\cap_{1 \le t \le k} \operatorname{int}(A_t)) = A_{k+1} \cap \operatorname{int}(B)$. Hence, using (2.20),

$$\operatorname{cl}\left(\bigcap_{t \in T} A_t\right) = \operatorname{cl}(A_{k+1} \cap B) = \operatorname{cl}(A_{k+1}) \cap \operatorname{cl}(B)$$

$$= \operatorname{cl}(A_{k+1}) \cap \left(\bigcap_{1 \le t \le k} \operatorname{cl}(A_t)\right) = \bigcap_{t \in T} \operatorname{cl}(A_t).$$

(iii) Let x_0 be the given continuity point. If all the f_t's are proper, then the result is known (see the bibliographical notes). Otherwise, given $\alpha \in \mathbb{R}$ such that $\alpha < f(x_0)$, we consider the functions

$$f_{t,\alpha} := \max\{f_t, \alpha\}, \ t \in T;$$

hence, $f_\alpha := \sup_{t \in T} f_{t,\alpha} = \max\{\sup_{t \in T} f_t, \alpha\} = \max\{f, \alpha\}$, and

$$f_{t,\alpha}(x_0) = \max\{f_t(x_0), \alpha\} \le \max\{f(x_0), \alpha\} = f(x_0) < +\infty;$$

5.2. MAIN SUBDIFFERENTIAL FORMULAS

that is, all the $f_{t,\alpha}$'s are proper. Moreover, by assertion (ii) we have that $\operatorname{cl}(f_{t,\alpha}) = \max\{\operatorname{cl}(f_t), \alpha\}$ for all $t \in T$ and, similarly, $\operatorname{cl}(f_\alpha) = \operatorname{cl}(\max\{f, \alpha\}) = \max\{\operatorname{cl} f, \alpha\}$. Since $\operatorname{dom} f = \operatorname{dom} f_\alpha$ and $\operatorname{dom} f_t = \operatorname{dom} f_{t,\alpha}$ for all $t \in T$, and the functions $f_{t,\alpha}$ are proper, we infer that

$$\max\{\operatorname{cl} f, \alpha\} = \operatorname{cl}(f_\alpha) = \sup_{t \in T} \operatorname{cl}(f_{t,\alpha}) = \max\{\sup_{t \in T} \operatorname{cl}(f_t), \alpha\}. \tag{5.39}$$

Thus, (5.10) follows when α goes to $-\infty$. Indeed, we only need to prove that $(\operatorname{cl} f)(x) \leq \sup_{t \in T} \operatorname{cl}(f_t)(x)$ for each $x \in X$. If $(\operatorname{cl} f)(x) = +\infty$, (5.39) yields, for every scalar α,

$$+\infty = \max\{(\operatorname{cl} f)(x), \alpha\} = \max\{\sup_{t \in T}(\operatorname{cl} f_t)(x), \alpha\},$$

and we obtain $\sup_{t \in T}(\operatorname{cl} f_t)(x) = +\infty$. If $(\operatorname{cl} f)(x) \in \mathbb{R}$ and we take $\alpha < (\operatorname{cl} f)(x)$, we get, thanks to (5.39),

$$(\operatorname{cl} f)(x) = \max\{(\operatorname{cl} f)(x), \alpha\} = \max\{\sup_{t \in T}(\operatorname{cl} f_t)(x), \alpha\},$$

and this implies $\sup_{t \in T}(\operatorname{cl} f_t)(x) = (\operatorname{cl} f)(x)$.

Finally, if $(\operatorname{cl} f)(x) = -\infty$, for every scalar α, also by (5.39) we get

$$\alpha = \max\{(\operatorname{cl} f)(x), \alpha\} = \max\{\sup_{t \in T}(\operatorname{cl} f_t)(x), \alpha\},$$

entailing this time $\sup_{t \in T}(\operatorname{cl} f_t)(x) \leq \alpha$, and so $\sup_{t \in T}(\operatorname{cl} f_t)(x) = -\infty$.

(iv) Since the inequality $\sup_{t \in T}(\operatorname{cl} f_t) \leq \operatorname{cl} f$ is always true, we need to prove that $(\operatorname{cl} f)(x) \leq \sup_{t \in T}(\operatorname{cl} f_t)(x)$ for $x \in X$ such that $\sup_{t \in T}(\operatorname{cl} f_t)(x) < +\infty$. Then

$$x \in \operatorname{dom}(\operatorname{cl} f_t) \subset \operatorname{cl}(\operatorname{dom} f_t) \subset \operatorname{aff}(\operatorname{dom} f_t) = \operatorname{aff}(\operatorname{dom} f).$$

We also observe that

$$(\operatorname{cl} f_t)(x) = \operatorname{cl}(f_{t|\operatorname{aff}(\operatorname{dom} f)})(x) \text{ and } (\operatorname{cl} f)(x) = \operatorname{cl}(f_{|\operatorname{aff}(\operatorname{dom} f)})(x).$$

Then the conclusion follows from assertion (i) when applied to the functions $f_{t|\operatorname{aff}(\operatorname{dom} f)}$, $t \in T$, and $f_{|\operatorname{aff}(\operatorname{dom} f)}$. ∎

Relation (5.40) below constitutes an alternative formula to (5.26) in which condition (5.10) is not required. Nonetheless, instead of the original f_t's used in (5.26), formula (5.40) involves the augmented functions $f_t + I_{L \cap \operatorname{dom} f}$, $t \in T$.

Corollary 5.2.5 *Given the convex functions $f_t : X \to \overline{\mathbb{R}}$, $t \in T$, for every $x \in X$ we have*

$$\partial f(x) = \bigcap_{\varepsilon>0,\ L\in\mathcal{F}(x)} \overline{\mathrm{co}}\left(\bigcup_{t\in T_\varepsilon(x)} \partial_\varepsilon(f_t + \mathrm{I}_{L\cap\mathrm{dom}\,f})(x)\right). \qquad (5.40)$$

Proof. We shall consider the non-trivial case $x \in \mathrm{dom}\,f$. To prove the inclusion "$\subset$" in (5.40), we take $L \in \mathcal{F}(x)$ and define the convex functions
$$g_t := f_t + \mathrm{I}_{L\cap\mathrm{dom}\,f},\ t\in T,\ g := \sup_{t\in T} g_t.$$

Observe that $g = f + \mathrm{I}_{L\cap\mathrm{dom}\,f}$ and $\mathrm{dom}\,g_t = (\mathrm{dom}\,f_t) \cap (L \cap \mathrm{dom}\,f) = L \cap (\mathrm{dom}\,f) = \mathrm{dom}\,g$. Then, since $\mathrm{dom}\,g$ is a non-empty finite-dimensional set, the family $\{g_t, t \in T\}$ satisfies condition (5.10) (Proposition 5.2.4(iv)). Thus, due to the following relations, $\partial_\varepsilon g_t(x) + \mathrm{N}_{L\cap\mathrm{dom}\,f}(x) \subset \partial_\varepsilon(f_t + \mathrm{I}_{L\cap\mathrm{dom}\,f})(x)$ for all $t \in T_\varepsilon(x)$, and $\{t \in T : g_t(x) \geq g(x) - \varepsilon\} = T_\varepsilon(x), \varepsilon \geq 0$, Theorem 5.2.2 gives us

$$\partial f(x) \subset \partial(f + \mathrm{I}_L)(x) = \partial g(x) \subset \bigcap_{\varepsilon>0} \overline{\mathrm{co}}\left(\bigcup_{t\in T_\varepsilon(x)} \partial_\varepsilon g_t(x) + \mathrm{N}_{L\cap\mathrm{dom}\,g}(x)\right)$$

$$\subset \bigcap_{\varepsilon>0} \overline{\mathrm{co}}\left(\bigcup_{t\in T_\varepsilon(x)} \partial_\varepsilon(f_t + \mathrm{I}_{L\cap\mathrm{dom}\,f})(x)\right),$$

and the direct inclusion "$\subset$" in (5.40) follows. The proof is finished because the opposite inclusion in (5.40) is straightforward. ∎

In the sequel, we shall need the following representation of $\partial f(x)$ involving augmented lsc convex functions.

Corollary 5.2.6 *Given the lsc convex functions $f_t : X \to \overline{\mathbb{R}}$, $t \in T$, for every $x \in X$ we have*

$$\partial f(x) = \bigcap_{\varepsilon>0,\ L\in\mathcal{F}(x)} \overline{\mathrm{co}}\left(\bigcup_{t\in T_\varepsilon(x)} \partial_\varepsilon(f_t + \mathrm{I}_{\overline{L\cap\mathrm{dom}\,f}})(x)\right). \qquad (5.41)$$

Proof. We shall consider the non-trivial case $x \in \mathrm{dom}\,f$. We apply formula (5.26) to the family of lsc convex functions $\{f_t,\ t \in T\}$ and obtain that

5.2. MAIN SUBDIFFERENTIAL FORMULAS

$$\partial f(x) = \bigcap_{\varepsilon>0,\ L\in\mathcal{F}(x)} \overline{\mathrm{co}}\left(\bigcup_{t\in T_\varepsilon(x)} \partial_\varepsilon f_t(x) + \mathrm{N}_{L\cap\mathrm{dom}\,f}(x)\right)$$

$$= \bigcap_{\varepsilon>0,\ L\in\mathcal{F}(x)} \overline{\mathrm{co}}\left(\bigcup_{t\in T_\varepsilon(x)} \partial_\varepsilon f_t(x) + \mathrm{N}_{\overline{L\cap\mathrm{dom}\,f}}(x)\right)$$

$$\subset \bigcap_{\varepsilon>0,\ L\in\mathcal{F}(x)} \overline{\mathrm{co}}\left(\bigcup_{t\in T_\varepsilon(x)} \partial_\varepsilon(f_t + \mathrm{I}_{\overline{L\cap\mathrm{dom}\,f}})(x)\right),$$

which is the inclusion "⊂" in (5.41). The proof is finished because the opposite inclusion is straightforward. ∎

Remark 6 *If* $\mathrm{dom}\,f = \mathrm{dom}\,f_t$ *for all* $t \in T$, *then the formula in Corollary 5.2.5 reduces to*

$$\partial f(x) = \bigcap_{\varepsilon>0,\ L\in\mathcal{F}(x)} \overline{\mathrm{co}}\left(\bigcup_{t\in T_\varepsilon(x)} \partial_\varepsilon(f_t + \mathrm{I}_L)(x)\right).$$

A Banach version of main formula (5.26) is given next, in Theorems 5.2.7 and 5.2.9, where the subdifferential of f is expressed using the (exact) subdifferentials of the data functions f_t, $t \in T_\varepsilon(x)$, but at nearby points.

Theorem 5.2.7 *Assume that X is a Banach space and let the functions $f_t : X \to \overline{\mathbb{R}}$, $t \in T$, be convex and lsc. Then, for every $x \in X$,*

$$\partial f(x) = \bigcap_{\varepsilon>0,\ L\in\mathcal{F}(x)} \overline{\mathrm{co}}\left\{\bigcup_{t\in T_\varepsilon(x),\, y\in B_t(x,\varepsilon)} \partial f_t(y) \cap S_\varepsilon(y-x) + \mathrm{N}_{L\cap\mathrm{dom}\,f}(x)\right\}, \quad (5.42)$$

where

$$S_\varepsilon(y-x) := \{y^* \in X^* : \langle y^*, y-x \rangle \leq \varepsilon\}, \quad (5.43)$$

and

$$B_t(x,\varepsilon) := \{y \in B_X(x,\varepsilon) : |f_t(y) - f_t(x)| \leq \varepsilon\}. \quad (5.44)$$

Proof. If $x \notin \mathrm{dom}\,f$, then (5.42) holds trivially, thanks to the convention in (2.6), since both sets $\mathrm{N}_{L\cap\mathrm{dom}\,f}(x)$ and $\partial f(x)$ are empty; hence, we suppose that $x \in \mathrm{dom}\,f$.

To prove the inclusion "⊃" we first observe that, given any $\varepsilon > 0$,

$$\partial f_t(y) \cap S_\varepsilon(y-x) \subset \partial_{3\varepsilon} f(x), \quad (5.45)$$

for every $t \in T_\varepsilon(x)$ and $y \in B_t(x, \varepsilon)$. Indeed, if $y^* \in \partial f_t(y) \cap S_\varepsilon(y - x)$, then for every $z \in X$

$$\langle y^*, z - x \rangle = \langle y^*, z - y \rangle + \langle y^*, y - x \rangle$$
$$\leq f_t(z) - f_t(y) + \varepsilon \leq f_t(z) - f_t(x) + 2\varepsilon,$$

and the fact that $t \in T_\varepsilon(x)$ yields $\langle y^*, z - x \rangle \leq f_t(z) - f_t(x) + 2\varepsilon \leq f(z) - f(x) + 3\varepsilon$; that is, $y^* \in \partial_{3\varepsilon} f(x)$. Thus, denoting by E the right-hand side set in (5.42),

$$E \subset \bigcap_{\varepsilon > 0,\ L \in \mathcal{F}(x)} \overline{\text{co}}(\partial_{3\varepsilon} f(x) + N_{L \cap \text{dom } f}(x)) \subset \bigcap_{\varepsilon > 0,\ L \in \mathcal{F}(x)} \partial_{3\varepsilon}(f + I_{L \cap \text{dom } f})(x)$$
$$= \bigcap_{\varepsilon > 0,\ L \in \mathcal{F}(x)} \partial_{3\varepsilon}(f + I_{\text{dom } f} + I_L)(x) = \bigcap_{\varepsilon > 0,\ L \in \mathcal{F}(x)} \partial_{3\varepsilon}(f + I_L)(x),$$

and the inclusion "$\supset$" follows, thanks to (4.15) and (4.16).

To prove the inclusion "$\subset$", we pick $L \in \mathcal{F}(x)$ and $x^* \in \partial_\varepsilon f_t(x)$ (if any), for $\varepsilon > 0$ and $t \in T_\varepsilon(x)$. Since $\partial_\varepsilon f_t(x) \neq \emptyset$, it turns out that f_t is proper. Next, according to Proposition 4.3.7, there are $x_\varepsilon \in B_X(x, \sqrt{\varepsilon})$, $y_\varepsilon^* \in B_{X^*}$, and $\lambda_\varepsilon \in [-1, 1]$ such that

$$x_\varepsilon^* := x^* + \sqrt{\varepsilon}(y_\varepsilon^* + \lambda_\varepsilon x^*) \in \partial f_t(x_\varepsilon), \tag{5.46}$$

$$|\langle x_\varepsilon^*, x_\varepsilon - x \rangle| \leq \varepsilon + \sqrt{\varepsilon},\ |f_t(x_\varepsilon) - f_t(x)| \leq \varepsilon + \sqrt{\varepsilon}, \tag{5.47}$$

entailing that $x_\varepsilon \in B_t(x, \varepsilon + \sqrt{\varepsilon})$ and $x_\varepsilon^* \in \partial f_t(x_\varepsilon) \cap S_{\varepsilon + \sqrt{\varepsilon}}(x_\varepsilon - x)$. From the relation $x^* = (1/(1 + \lambda_\varepsilon \sqrt{\varepsilon}))(x_\varepsilon^* - \sqrt{\varepsilon} y_\varepsilon^*)$ coming from (5.46), we get

$$x^* \in (1/(1 + \lambda_\varepsilon \sqrt{\varepsilon}))\left(\partial f_t(x_\varepsilon) \cap S_{\varepsilon + \sqrt{\varepsilon}}(x_\varepsilon - x) + \sqrt{\varepsilon} B_{X^*}\right)$$
$$\subset (1/(1 + \lambda_\varepsilon \sqrt{\varepsilon}))\left(\partial f_t(x_\varepsilon) \cap S_{\varepsilon + \sqrt{\varepsilon}}(x_\varepsilon - x) + N_{L \cap \text{dom } f}(x) + \sqrt{\varepsilon} B_{X^*}\right)$$
$$\subset (1/(1 + \lambda_\varepsilon \sqrt{\varepsilon}))\,\text{co}\left(A_\varepsilon + \sqrt{\varepsilon} B_{X^*}\right),$$

where

$$A_\varepsilon := \bigcup_{t \in T_\varepsilon(x),\ y \in B_t(x, \varepsilon + \sqrt{\varepsilon})} \partial f_t(y) \cap S_{\varepsilon + \sqrt{\varepsilon}}(y - x) + N_{L \cap \text{dom } f}(x).$$

We take $0 < \varepsilon < 1$, so that $1/(1 + \lambda_\varepsilon \sqrt{\varepsilon}) \in \Lambda_\varepsilon := [1/(1 + \sqrt{\varepsilon}), 1/(1 - \sqrt{\varepsilon})]$, entailing that $x^* \in \Lambda_\varepsilon \,\text{co}\left(A_\varepsilon + \sqrt{\varepsilon} B_{X^*}\right)$. Observe that $0 \notin \Lambda_\varepsilon$ for all $0 < \varepsilon < 1$. Then, according to formula (5.26) in Theorem 5.2.2,

5.2. MAIN SUBDIFFERENTIAL FORMULAS

$$\partial f(x) \subset \bigcap_{\varepsilon>0} \overline{\operatorname{co}}\left(\bigcup_{t\in T_\varepsilon(x)} \partial_\varepsilon f_t(x) + \mathrm{N}_{L\cap\mathrm{dom}\, f}(x)\right) \qquad (5.48)$$

$$\subset \bigcap_{0<\varepsilon<1} \overline{\operatorname{co}}\left(\Lambda_\varepsilon \operatorname{co}(A_\varepsilon + \sqrt{\varepsilon}B_{X^*}) + \mathrm{N}_{L\cap\mathrm{dom}\, f}(x)\right)$$

$$= \bigcap_{0<\varepsilon<1} \overline{\operatorname{co}}\left\{\Lambda_\varepsilon\left(\operatorname{co}(A_\varepsilon + \sqrt{\varepsilon}B_{X^*}) + \mathrm{N}_{L\cap\mathrm{dom}\, f}(x)\right)\right\}$$

$$= \bigcap_{0<\varepsilon<1} \Lambda_\varepsilon\overline{\operatorname{co}}\left\{\operatorname{co}(A_\varepsilon + \sqrt{\varepsilon}B_{X^*}) + \mathrm{N}_{L\cap\mathrm{dom}\, f}(x)\right\}$$

$$= \bigcap_{0<\varepsilon<1} \Lambda_\varepsilon\overline{\operatorname{co}}\left(A_\varepsilon + \mathrm{N}_{L\cap\mathrm{dom}\, f}(x) + \sqrt{\varepsilon}B_{X^*}\right)$$

$$= \bigcap_{0<\varepsilon<1} \Lambda_\varepsilon\overline{\operatorname{co}}\left(A_\varepsilon + \sqrt{\varepsilon}B_{X^*}\right),$$

where the second equality comes from Exercise 11. Moreover, since the families $(\Lambda_\varepsilon)_{\varepsilon>0}$ and $(\overline{\operatorname{co}}\,(A_\varepsilon + \sqrt{\varepsilon}B_{X^*}))_{\varepsilon>0}$ are non-decreasing and $\cap_{\varepsilon>0}\Lambda_\varepsilon = \{1\}$, the last inclusion yields (Exercise 12)

$$\partial f(x) \subset \bigcap_{\varepsilon>0} \overline{\operatorname{co}}\left(A_\varepsilon + \sqrt{\varepsilon}B_{X^*}\right) = \bigcap_{\varepsilon>0} \left(\overline{\operatorname{co}}(A_\varepsilon) + \sqrt{\varepsilon}B_{X^*}\right),$$

and so $\partial f(x) \subset \cap_{\varepsilon>0}\overline{\operatorname{co}}(A_\varepsilon)$ (Exercise 10(ii)). The desired inclusion follows by the arbitrariness of L in $\mathcal{F}(x)$. ∎

The following example draws aside, in general, the possibility of extending Theorem 5.2.7 to non-Banach spaces.

Example 5.2.8 *Assume that there is a proper lsc convex function g, defined on the locally convex space X, which has an empty subdifferential everywhere (such a function exists as it is commented in the bibliographical notes of this chapter). We may suppose that $\theta \in \mathrm{dom}\, g$ and $g(\theta) = 0$. We define the function $f_t \in \Gamma_0(X)$ as*

$$f_t(x) := tg(x), \quad t \in T := \,]0, +\infty[,$$

so that $f := \sup_{t\in T} f_t = \mathrm{I}_{[g\leq 0]}$. Since $\partial f_t \equiv t\partial g \equiv \emptyset$ for all $t \in T$, the set in the right-hand side of (5.42) is empty, whereas $\partial f(\theta) = \mathrm{N}_{[g\leq 0]}(\theta) \neq \emptyset$.

Theorem 5.2.7 can be applied in a general locally convex space X provided that the set $\mathrm{cl}(\mathrm{span}(\cup_{t\in T}\mathrm{dom}\, f_t))$ is a Banach subspace. We have the following result where $\mathcal{P}$ is the family of all continuous seminorms in X.

Theorem 5.2.9 *Let $f_t : X \to \overline{\mathbb{R}}$, $t \in T$, be convex and lsc functions, and denote $f := \sup_{t\in T} f_t$. Assume that $X_0 := \mathrm{cl}(\mathrm{span}(\cup_{t\in T}\mathrm{dom}\, f_t))$ is a Banach linear subspace of X. Then, for every $x \in X$, we have that*

CHAPTER 5. SUPREMUM OF CONVEX FUNCTIONS

$$\partial f(x) = \bigcap_{\varepsilon>0,\, p\in\mathcal{P},\, L\in\mathcal{F}(x)} \overline{\mathrm{co}}\left\{\bigcup_{t\in T_\varepsilon(x),\, y\in B_{p,t}(x,\varepsilon)} \partial f_t(y) \cap S_\varepsilon(y-x) + N_{L\cap\mathrm{dom}\, f}(x)\right\}, \tag{5.49}$$

where

$$B_{p,t}(x,\varepsilon) := \{y \in X : p(y-x) \le \varepsilon,\, |f_t(y) - f_t(x)| \le \varepsilon\}. \tag{5.50}$$

Proof. Fix $x \in \mathrm{dom}\, f$ ($\subset X_0$) and $L \in \mathcal{F}(x)$. Let us first prove that

$$\partial f(x) \subset \bigcap_{\varepsilon>0} \overline{\mathrm{co}}\left\{\bigcup_{t\in T_\varepsilon(x),\, y\in B_{0,t}(x,\varepsilon)} \partial f_t(y) \cap S_\varepsilon(y-x) + N_{L\cap\mathrm{dom}\, f}(x)\right\}, \tag{5.51}$$

where

$$B_{0,t}(x,\varepsilon) := \{y \in X : \|y-x\|_0 \le \varepsilon,\, |f_t(y) - f_t(x)| \le \varepsilon\},$$

with $\|\cdot\|_0$ being the norm in X_0 which generates the induced topology from X. To this aim, fix $\varepsilon > 0$ and denote $X_1 := X_0 + L_0$, where $L_0 := \mathrm{span}\{L \setminus X_0\}$; hence, X_1 is a Banach subspace of X given with the norm

$$\|u+v\|_1 = \|u\|_0 + \|v\|_{L_0}, \quad u \in X_0,\, v \in L_0. \tag{5.52}$$

We define the functions $g_t : X_1 \to \overline{\mathbb{R}}$, $t \in T$, as the restrictions of the corresponding functions f_t to X_1; that is, $g_t(z) = f_t(z)$, for $z \in X_1 \subset X$, together with $g := \sup_{t\in T} g_t$. It is clear that each g_t, $t \in T$, is convex and lsc on L, as for all $z \in X_1$

$$\liminf_{y\to z,\, y\in X_1} g_t(y) = \liminf_{y\to z,\, y\in X_1} f_t(y) \ge \liminf_{y\to z} f_t(y) \ge f_t(z) = g_t(z). \tag{5.53}$$

Thus, since L is also a finite-dimensional subspace of the Banach space X_1, by applying Theorem 5.2.7 in X_1 we get

$$\partial g(x) \subset \overline{\mathrm{co}}(\tilde{A}_\varepsilon + \tilde{N}_{L\cap\mathrm{dom}\, g}(x)), \tag{5.54}$$

where $\tilde{A}_\varepsilon := \bigcup_{t\in\tilde{T}_\varepsilon(x),\, y\in\tilde{B}_{1,t}(x,\varepsilon)} \partial g_t(y) \cap \tilde{S}_\varepsilon(y-x)$, with

$$\tilde{T}_\varepsilon(x) := \{t \in T : g_t(x) \ge g(x) - \varepsilon\} = \{t \in T : f_t(x) \ge f(x) - \varepsilon\} = T_\varepsilon(x),$$

$$\tilde{B}_{1,t}(x,\varepsilon) := \{y \in X_1 : \|y-x\|_1 \le \varepsilon,\, |g_t(y) - g_t(x)| \le \varepsilon\},$$

5.2. MAIN SUBDIFFERENTIAL FORMULAS

and
$$\tilde{S}_\varepsilon(y-x) := \{y^* \in X_1^* : \langle y^*, y-x \rangle \leq \varepsilon\}, \; y \in X_1.$$

More precisely, since $\cup_{t \in T} \operatorname{dom} f_t \subset X_0$, we have that

$$\tilde{B}_{1,t}(x,\varepsilon) = \{y \in X_0 : \|y-x\|_0 \leq \varepsilon, \; |f_t(y) - f_t(x)| \leq \varepsilon\} = B_{0,t}(x,\varepsilon).$$

The closure operation and the normal cone $\tilde{\mathrm{N}}_{L \cap \operatorname{dom} g}(x)$ in (5.54) are given with respect to the subspace X_1. In particular, since $L \cap \operatorname{dom} g \subset X_1$ we verify that (Exercise 55(iv))

$$\tilde{\mathrm{N}}_{L \cap \operatorname{dom} g}(x) = \{u^*_{|X_1} : u^* \in \mathrm{N}_{L \cap \operatorname{dom} g}(x)\} = \{u^*_{|X_1} : u^* \in \mathrm{N}_{L \cap \operatorname{dom} f}(x)\}, \tag{5.55}$$

as $\operatorname{dom} g = \operatorname{dom} f$. Similarly, using again the extension theorem, we verify that

$$\tilde{S}_\varepsilon(y-x) = \{y^*_{|X_1} : y^* \in S_\varepsilon(y-x)\} \text{ for all } y \in X_1. \tag{5.56}$$

Accordingly, each element $y_0^* \in \tilde{A}_\varepsilon$ (if any) satisfies $y_0^* \in \partial g_{t_0}(y_0) \cap \tilde{S}_\varepsilon(y_0 - x)$ for some $t_0 \in \tilde{T}_\varepsilon(x)$ and $y_0 \in \tilde{B}_{1,t_0}(x,\varepsilon)$. Hence, since $\operatorname{dom} f_{t_0} \subset X_0 \subset X_1$, there exists $y^* \in \partial f_{t_0}(y_0)$ such that $y_0^* = y^*_{|X_1}$ (by Exercise 55(i)). Moreover, due to (5.56), we may assume that $y^* \in S_\varepsilon(y_0 - x)$. Thus,

$$y^* \in \bigcup_{t \in T_\varepsilon(x), \, y \in B_{0,t}(x,\varepsilon)} \partial f_t(y) \cap S_\varepsilon(y-x) =: A_\varepsilon.$$

Therefore, given any $x^* \in \partial f(x)$, we have that $x^*_{|X_1} \in \partial g(x)$ and so, by (5.54) and (5.55),

$$x^*_{|X_1} \in \overline{\operatorname{co}}(\tilde{A}_\varepsilon + \tilde{\mathrm{N}}_{L \cap \operatorname{dom} g}(x)) \subset \overline{\operatorname{co}} \left\{ y^*_{|X_1} + u^*_{|X_1} : y^* \in A_\varepsilon, \; u^* \in \mathrm{N}_{L \cap \operatorname{dom} f}(x) \right\}.$$

In other words, for all $v \in X_1$,

$$\langle x^*, v \rangle \leq \sup_{y^* \in A_\varepsilon, \, u^* \in \mathrm{N}_{L \cap \operatorname{dom} f}(x)} \left\langle y^*_{|X_1} + u^*_{|X_1}, v \right\rangle$$
$$= \sup_{y^* \in A_\varepsilon, \, u^* \in \mathrm{N}_{L \cap \operatorname{dom} f}(x)} \langle y^* + u^*, v \rangle$$
$$= \sigma_{A_\varepsilon + \mathrm{N}_{L \cap \operatorname{dom} f}(x)}(v) \leq \sigma_{A_\varepsilon + \mathrm{N}_{L \cap \operatorname{dom} f}(x) + X_1^\perp}(v).$$

Thus, since this last inequality also holds for all $v \notin X_1$ (the term of the right-hand side is equal to $+\infty$ in such a case, due to (3.51)), we deduce that

$$x^* \in \overline{\mathrm{co}}(A_\varepsilon + N_{L \cap \mathrm{dom}\, f}(x) + X_1^\perp)$$
$$\subset \overline{\mathrm{co}}(A_\varepsilon + N_{L \cap \mathrm{dom}\, f}(x) + L^\perp) \subset \overline{\mathrm{co}}(A_\varepsilon + N_{L \cap \mathrm{dom}\, f}(x)),$$

as $N_{L \cap \mathrm{dom}\, f}(x) + L^\perp \subset N_{L \cap \mathrm{dom}\, f}(x)$. Then the inclusion "$\subset$" in (5.51) follows because ε was arbitrarily chosen.

Finally, fix $x^* \in \partial f(x)$. Given $q \in \mathcal{P}$ and $\varepsilon > 0$, we have that

$$B_q(x, \varepsilon) := \{y \in X : q(y - x) \le \varepsilon\}$$
$$\supset \{y \in X_0 : q(y - x) \le \varepsilon\} = \{y \in X_0 : q_{|X_0}(y - x) \le \varepsilon\}.$$

So, since the norm topology in X_0 is equivalently determined by the family $\{p_{|X_0} : p \in \mathcal{P}\}$, we find $0 < \delta < \varepsilon$ such that

$$\{y \in X_0 : \|y - x\|_0 \le \delta\} \subset \{y \in X_0 : q_{|X_0}(y - x) \le \varepsilon\} \subset B_q(x, \varepsilon).$$

Consequently, using (5.51),

$$x^* \in \overline{\mathrm{co}}\left\{\bigcup_{t \in T_\delta(x),\, y \in B_{0,t}(x,\delta)} \partial f_t(y) \cap S_\delta(y - x) + N_{L \cap \mathrm{dom}\, f}(x)\right\}$$
$$\subset \overline{\mathrm{co}}\left\{\bigcup_{t \in T_\varepsilon(x),\, y \in B_{q,t}(x,\varepsilon)} \partial f_t(y) \cap S_\varepsilon(y - x) + N_{L \cap \mathrm{dom}\, f}(x)\right\},$$

and, by the arbitrariness of q and ε, we infer that

$$x^* \in \bigcap_{\varepsilon > 0,\, q \in \mathcal{P}} \overline{\mathrm{co}}\left\{\bigcup_{t \in T_\varepsilon(x),\, y \in B_{q,t}(x,\varepsilon)} \partial f_t(y) \cap S_\varepsilon(y - x) + N_{L \cap \mathrm{dom}\, f}(x)\right\}.$$

So, the inclusion "$\subset$" in (5.49) follows by intersecting over the L's. The proof is complete as the opposite inclusion there is straightforward. ∎

Observe that we can replace $B_{p,t}(x, \varepsilon)$ in formula (5.49) with the larger set

$$B_t(x, \varepsilon) := \{y \in X : |f_t(y) - f_t(x)| \le \varepsilon\},$$

giving rise to a representation of $\partial f(x)$ free of the seminorms. However, the use of seminorms in (5.49) will be crucial in the sequel, namely in Corollary 5.3.6. Moreover, we have the following result.

Corollary 5.2.10 *Under the assumptions of Theorem 5.2.9, we have that*

5.2. MAIN SUBDIFFERENTIAL FORMULAS

$$\partial f(x) = \bigcap_{\varepsilon>0,\, L\in\mathcal{F}(x)} \overline{\mathrm{co}}\left(\bigcup_{t\in T_\varepsilon(x),\, y\in B_t(x,\varepsilon)} \partial f_t(y) \cap S_\varepsilon(y-x) + \mathrm{N}_{L\cap\mathrm{dom}\, f}(x)\right).$$

Proof. The inclusion "⊂" follows from (5.49) as $B_{p,t}(x,\varepsilon) \subset B_t(x,\varepsilon)$ for all $p \in \mathcal{P}$. To verify the opposite inclusion, observe that, for all $t \in T_\varepsilon(x)$ and $y \in B_t(x,\varepsilon)$

$$\partial f_t(y) \cap S_\varepsilon(y-x) \subset \partial_{3\varepsilon} f(x). \tag{5.57}$$

Indeed, if $z^* \in \partial f_t(y) \cap S_\varepsilon(y-x)$, then for all $z \in X$ we get $\langle z^*, z-y\rangle \leq f_t(z) - f_t(y) \leq f(z) - f_t(x) + \varepsilon$, and so,

$$\langle z^*, z-x\rangle \leq f(z) - f_t(x) + \langle z^*, y-x\rangle + \varepsilon$$
$$\leq f(z) - f_t(x) + 2\varepsilon \leq f(z) - f(x) + 3\varepsilon.$$

Consequently, using (4.9),

$$\partial f_t(y) \cap S_\varepsilon(y-x) + \mathrm{N}_{L\cap\mathrm{dom}\,f}(x) \subset \partial_{3\varepsilon} f(x) + \mathrm{N}_{L\cap\mathrm{dom}\,f}(x)$$
$$= \partial_{3\varepsilon}(f + \mathrm{I}_{L\cap\mathrm{dom}\,f})(x) = \partial_{3\varepsilon}(f + \mathrm{I}_L)(x),$$

and (4.15) together with (4.16) yield, intersecting over ε and then over L,

$$\bigcap_{\varepsilon>0,\, L\in\mathcal{F}(x)} \overline{\mathrm{co}}\left(\bigcup_{t\in T_\varepsilon(x),\, y\in B_t(x,\varepsilon)} \partial f_t(y) \cap S_\varepsilon(y-x) + \mathrm{N}_{L\cap\mathrm{dom}\,f}(x)\right)$$
$$\subset \bigcap_{L\in\mathcal{F}(x)}\bigcap_{\varepsilon>0} \partial_{3\varepsilon}(f+\mathrm{I}_L)(x) = \bigcap_{L\in\mathcal{F}(x)} \partial(f+\mathrm{I}_L)(x) = \partial f(x).$$

∎

The proof of the following corollary is similar to the one of Theorem 5.2.7, but uses formula (5.41) instead of (5.26) in Theorem 5.2.2. In Theorem 5.2.12 we will indeed prove that this result holds in any locally convex space.

Corollary 5.2.11 *Assume that X is a Banach space and let the functions $f_t : X \to \overline{\mathbb{R}}$, $t \in T$, be convex and lsc. Then, for every $x \in X$, we have that*

$$\partial f(x) = \bigcap_{\varepsilon>0,\, L\in\mathcal{F}(x)} \overline{\mathrm{co}}\left\{\bigcup_{t\in T_\varepsilon(x),\, y\in B_t(x,\varepsilon)} \partial(f_t + \mathrm{I}_{\overline{L\cap\mathrm{dom}\,f}})(y) \cap S_\varepsilon(y-x)\right\}.$$

We can place ourselves in the same setting as that of Theorem 5.2.9 by using the augmented data functions $f_t + \mathrm{I}_{\overline{L\cap\mathrm{dom}\,f}}$, $t \in T$. Additionally, in this case, the normal cone $N_{L\cap \mathrm{dom}\, f}(x)$ is dropped out from formula (5.49) as we show next.

Theorem 5.2.12 *Given lsc convex functions $f_t : X \to \overline{\mathbb{R}}$, $t \in T$, we denote $f := \sup_{t \in T} f_t$. Then, for every $x \in X$, we have that*

$$\partial f(x) = \bigcap_{\varepsilon > 0,\ p \in \mathcal{P},\ L \in \mathcal{F}(x)} \overline{\mathrm{co}} \left(\bigcup_{t \in T_\varepsilon(x),\ y \in B_{p,t}(x,\varepsilon)} \partial(f_t + \mathrm{I}_{\overline{L \cap \mathrm{dom}\, f}})(y) \cap S_\varepsilon(y - x) \right). \tag{5.58}$$

Remark 7 (before the proof) *Taking into account that, for every $t \in T_\varepsilon(x)$ and $y \in B_{p,t}(x,\varepsilon)$,*

$$\partial(f_t + \mathrm{I}_{\overline{L \cap \mathrm{dom}\, f}})(y) \cap S_\varepsilon(y - x) \subset \partial_{3\varepsilon}(f + \mathrm{I}_L)(x), \text{ for every } L \subset X, \tag{5.59}$$

formula (5.58) is equivalently written as

$$\partial f(x) = \bigcap_{\substack{\varepsilon > 0,\ p \in \mathcal{P} \\ L \in \mathcal{F}(x)}} \overline{\mathrm{co}} \left(\bigcup_{\substack{t \in T_\varepsilon(x) \\ y \in B_{p,t}(x,\varepsilon)}} \partial(f_t + \mathrm{I}_{\overline{L \cap \mathrm{dom}\, f}})(y) \cap S_\varepsilon(y - x) \cap \partial_\varepsilon(f + \mathrm{I}_L)(x) \right). \tag{5.60}$$

Proof. First we suppose that X is Banach. The proof in this case is similar to the one in Theorem 5.2.7, and so we only give a sketch of the proof of the inclusion "$\subset$". Fix $0 < \varepsilon < 1$ and $L \in \mathcal{F}(x)$, and consider the lsc convex functions $g_t : X \to \overline{\mathbb{R}}$, $t \in T$, defined as

$$g_t := f_t + \mathrm{I}_{\overline{L \cap \mathrm{dom}\, f}}. \tag{5.61}$$

So, by Corollary 5.2.6,

$$\partial f(x) \subset \overline{\mathrm{co}} \left(\bigcup_{t \in T_\varepsilon(x)} \partial_\varepsilon g_t(x) \right). \tag{5.62}$$

Then, by Proposition 4.3.7, associated with each element $x^* \in \partial_\varepsilon g_t(x)$, $t \in T_\varepsilon(x)$, there are $x_\varepsilon \in B_X(x, \sqrt{\varepsilon}) \cap (\overline{L \cap \mathrm{dom}\, f})$, $y_\varepsilon^* \in B_{X^*}$, and $\lambda_\varepsilon \in [-1, 1]$ such that

$$x_\varepsilon^* := x^* + \sqrt{\varepsilon}(y_\varepsilon^* + \lambda_\varepsilon x^*) \in \partial g_t(x_\varepsilon), \tag{5.63}$$

$$|\langle x_\varepsilon^*, x_\varepsilon - x \rangle| \le \varepsilon + \sqrt{\varepsilon},\ |f_t(x_\varepsilon) - f_t(x)| \le \varepsilon + \sqrt{\varepsilon},$$

entailing that $x_\varepsilon^* \in \partial g_t(x_\varepsilon) \cap S_{\varepsilon + \sqrt{\varepsilon}}(x_\varepsilon - x)$. Hence, since $x^* = (1/(1 + \lambda_\varepsilon \sqrt{\varepsilon}))(x_\varepsilon^* - \sqrt{\varepsilon} y_\varepsilon^*)$, coming from (5.63), we get

5.2. MAIN SUBDIFFERENTIAL FORMULAS

$$x^* \in \left(1/(1+\lambda_\varepsilon\sqrt{\varepsilon})\right)\left(\partial g_t(x_\varepsilon) \cap S_{\varepsilon+\sqrt{\varepsilon}}(x_\varepsilon - x) + \sqrt{\varepsilon}B_{X^*}\right)$$
$$\subset \left(1/(1+\lambda_\varepsilon\sqrt{\varepsilon})\right) \operatorname{co}\left(A_\varepsilon + \sqrt{\varepsilon}B_{X^*}\right),$$

where $\Lambda_\varepsilon := [1/(1+\sqrt{\varepsilon}), 1/(1-\sqrt{\varepsilon})]$ and

$$A_\varepsilon := \bigcup_{t \in T_\varepsilon(x),\, y \in B_t(x,\varepsilon+\sqrt{\varepsilon})} \partial g_t(y) \cap S_{\varepsilon+\sqrt{\varepsilon}}(y-x).$$

Consequently, (5.62) entails (Exercise 12)

$$\partial f(x) \subset \bigcap_{0<\varepsilon<1,\, L\in\mathcal{F}(x)} \Lambda_\varepsilon \overline{\operatorname{co}}\left(\bigcup_{t\in T_\varepsilon(x)} A_\varepsilon\right) = \bigcap_{0<\varepsilon<1,\, L\in\mathcal{F}(x)} \overline{\operatorname{co}}\left(\bigcup_{t\in T_\varepsilon(x)} A_\varepsilon\right).$$

More generally, we suppose that X is not necessarily Banach. Given $x \in \operatorname{dom} f$, $x^* \in \partial f(x)$, and $L \in \mathcal{F}(x)$, we introduce the lsc convex functions $g_t : L \to \overline{\mathbb{R}}$, $t \in T$, defined as

$$g_t(z) := f_t(z) + I_{\overline{L \cap \operatorname{dom} f}}(z), \text{ for } z \in L,$$

together with the associated supremum $g := \sup_{t \in T} g_t$; hence, g is the restriction of $f + I_{\overline{L \cap \operatorname{dom} f}}$ to L whose effective domain is given by

$$\operatorname{dom} g = L \cap (\operatorname{dom} f \cap \overline{L \cap \operatorname{dom} f}) = L \cap \operatorname{dom} f.$$

Fix $\varepsilon > 0$ and $p \in \mathcal{P}$. Then, since $x_0^* := x_{|L}^* \in \partial g(x)$, by applying Corollary 5.2.11 to the family $\{g_t,\, t \in T\}$ we get

$$x_0^* \subset \overline{\operatorname{co}}\left(\bigcup_{t \in \tilde{T}_\varepsilon(x),\, y \in \tilde{B}_{p,t}(x,\varepsilon)} \partial g_t(y) \cap \tilde{S}_\varepsilon(y-x)\right), \qquad (5.64)$$

where $\tilde{T}_\varepsilon(x) := \{t \in T : g_t(x) \geq g(x) - \varepsilon\} = T_\varepsilon(x)$,

$$\tilde{B}_{p,t}(x,\varepsilon) := \{y \in \overline{L \cap \operatorname{dom} f} : p_{|L}(y-x) \leq \varepsilon,\, |f_t(y) - f_t(x)| \leq \varepsilon\} \subset B_{p,t}(x,\varepsilon),$$

and $\tilde{S}_\varepsilon(y-x) := \{y \in L^* : \langle y^*, y-x \rangle \leq \varepsilon\}$. Observe that for all $t \in \tilde{T}_\varepsilon(x)$ and $y \in \tilde{B}_{p,t}(x,\varepsilon)$ we have that (Exercise 55(i))

$$\partial g_t(y) \cap \tilde{S}_\varepsilon(y-x) = \{y_{|L}^* \in L^* : y^* \in \partial(f_t + I_{\overline{L \cap \operatorname{dom} f}})(y) \cap S_\varepsilon(y-x)\}.$$

Therefore, (5.64) entails

$$x^* \subset \overline{\mathrm{co}}\left(\bigcup_{t \in \tilde{T}_\varepsilon(x),\, y \in \tilde{B}_{p,t}(x,\varepsilon)} \partial(f_t + \mathrm{I}_{\overline{L \cap \mathrm{dom}\, f}})(y) \cap S_\varepsilon(y-x)\right) + L^\perp$$

$$\subset \overline{\mathrm{co}}\left(\bigcup_{t \in T_\varepsilon(x),\, y \in \tilde{B}_{p,t}(x,\varepsilon)} \partial(f_t + \mathrm{I}_{\overline{L \cap \mathrm{dom}\, f}})(y) \cap S_\varepsilon(y-x)\right) + L^\perp,$$

and so, by taking the intersection over L and after that over ε and p (Exercise 10(*iii*)), we get the inclusion "$\subset$" in (5.58). The opposite inclusion is straightforward. ∎

The lower semicontinuity assumption cannot be dropped from Theorem 5.2.7, as we show in the following example, which imitates the construction in Example 5.2.8. Alternative formulas for the non-lsc case can be done for functions satisfying the closure condition (5.10) (Exercise 87).

Example 5.2.13 *Let X be any infinite-dimensional Banach space, let g be a non-continuous linear mapping, and define the functions $f_t : X \to \mathbb{R}$ as*

$$f_t(x) := tg(x),\ t \in T := \,]0,+\infty[.$$

Also now $f := \sup_{t \in T} f_t = \mathrm{I}_{[g \leq 0]}$ and $\partial f(\theta) = \mathrm{N}_{[g \leq 0]}(\theta) \neq \emptyset$. Simultaneously, $\partial f_t \equiv t \partial g \equiv \emptyset$ for all $t \in T$, and the conclusion of Theorem 5.2.7 fails.

We continue with Example 5.2.1 to give the corresponding formula of ∂f, using Theorem 5.2.7.

Example 5.2.14 (Example 5.2.1, revisited) *Let $f_1, f_2 : \mathbb{R} \to \mathbb{R}_\infty$ be defined as*

$$f_1(x) = -\sqrt{|x|} + \mathrm{I}_{[0,+\infty[}\ \text{and}\ f_2(x) = f_1(-x).$$

Then $f := \max\{f_1, f_2\} = \mathrm{I}_{\{0\}}$ and, so, $\partial f(0) = \mathbb{R}$. This can be confirmed by Theorem 5.2.7. Indeed, denote

$$A_i := \{y^* \in \partial f_i(y) : y \in \varepsilon B_\mathbb{R},\ |f_i(y)| \leq \varepsilon,\ y^* y \leq \varepsilon\},\ i = 1, 2,$$

where $0 < \varepsilon < 1$ is fixed. Then we have that

$$A_1 = \{-1/(2\sqrt{y}) : 0 < y \leq \varepsilon^2\} = \,]-\infty, -1/(2\varepsilon)],$$

and, similarly, $A_2 = [1/(2\varepsilon), +\infty[$. Therefore, since $T_\varepsilon(0) = \{1,2\}$ and $\text{dom } f = \{0\}$, Theorem 5.2.7 yields $\partial f(0) = \cap_{\varepsilon>0} \overline{\text{co}} \{(A_1 \cup A_2) + \mathbb{R}\} = \mathbb{R}$.

5.3 The role of continuity assumptions

In this part, we derive other characterizations of the subdifferential of f when the data functions satisfy additional continuity assumptions. The first result allows us to remove the finite-dimensional sections of $\text{dom } f$ used in the general characterization of the subdifferential of f given in Theorem 5.2.2.

Proposition 5.3.1 *Let $x \in \text{dom } f$ be such that either $\text{ri}(\text{cone}(\text{dom } f - x)) \neq \emptyset$ or $\text{cone}(\text{dom } f - x)$ is closed. Then, provided that (5.10) holds, we have that*

$$\partial f(x) = \bigcap_{\varepsilon>0} \overline{\text{co}} \left(\bigcup_{t \in T_\varepsilon(x)} \partial_\varepsilon f_t(x) + N_{\text{dom } f}(x) \right). \quad (5.65)$$

Proof. The inclusion "$\supset$" always holds, due to Theorem 5.2.2. To prove the inclusion "$\subset$", we only consider the case when $\partial f(x) \neq \emptyset$; hence, we may suppose that $x = \theta$ and $f(\theta) = 0$ (Exercise 78).

Assume first that $\text{ri}(\text{cone}(\text{dom } f)) \neq \emptyset$. We fix $\varepsilon > 0$, a θ-neighborhood $V \subset X^*$, and choose $L \in \mathcal{F}(\theta)$ such that $L^\perp \subset (1/2)V$ and $L \cap \text{ri}(\text{cone}(\text{dom } f)) \neq \emptyset$. We also denote $A_\varepsilon := \cup_{t \in T_\varepsilon(\theta)} \partial_\varepsilon f_t(\theta)$. Then, by (2.15), we have that $\text{cl}(L \cap (\text{cone}(\text{dom } f))) = L \cap \text{cl}(\text{cone}(\text{dom } f))$, which leads us to

$$N_{L \cap \text{dom } f}(\theta) = (L \cap \text{cl}(\text{cone}(\text{dom } f)))^-$$
$$= \text{cl}(L^- + (\text{cone}(\text{dom } f))^-) = \text{cl}\left(L^\perp + N_{\text{dom } f}(\theta)\right).$$

Thus, according to Theorem 5.2.2,

$$\partial f(\theta) \subset \overline{\text{co}}\left(A_\varepsilon + N_{L \cap \text{dom } f}(\theta)\right) = \overline{\text{co}}\left(A_\varepsilon + L^\perp + N_{\text{dom } f}(\theta)\right) \subset \text{co}(A_\varepsilon) + N_{\text{dom } f}(\theta) + V,$$

and the desired inclusion follows then by intersecting over V and $\varepsilon > 0$.

Now, we assume that $\text{cone}(\text{dom } f)$ is closed. Then, for every $L \in \mathcal{F}(\theta)$, we have

$$N_{L \cap \text{dom } f}(\theta) = N_{L \cap (\text{cone}(\text{dom } f))}(\theta) = \partial(I_L + I_{\text{cone}(\text{dom } f)})(\theta).$$

Then, using (4.45) together with the relations $\partial_\varepsilon I_L(\theta) = L^\perp$ and

$$\partial_\varepsilon I_{\text{cone}(\text{dom } f)}(\theta) = \partial I_{\text{cone}(\text{dom } f)}(\theta) = N_{\text{dom } f}(\theta),$$

we obtain

$$N_{L \cap \text{dom } f}(\theta) = \bigcap_{\varepsilon > 0} \text{cl}(\partial_\varepsilon I_L(\theta) + \partial_\varepsilon I_{\text{cone}(\text{dom } f)}(\theta)) = \text{cl}(L^\perp + N_{\text{dom } f}(\theta)).$$

Consequently, by Theorem 5.2.2 and Exercise 10,

$$\partial f(\theta) = \bigcap_{\varepsilon > 0,\ L \in \mathcal{F}(\theta)} \text{cl}\left(\text{co}(A_\varepsilon) + N_{\text{dom } f}(\theta) + L^\perp\right) = \bigcap_{\varepsilon > 0} \overline{\text{co}}\left(A_\varepsilon + N_{\text{dom } f}(\theta)\right),$$

as we wanted to prove. ∎

Remark 8 *The conditions in Proposition (5.3.1) guarantee that (see Exercise 5(i))*

$$\bigcup_{L \in \mathcal{F}_X} \text{cl}(L \cap (\text{cone}(\text{dom } f - x))) = \text{cl}(\text{cone}(\text{dom } f - x)),$$

where $\mathcal{F}_X = \mathcal{F}_X(\theta)$. At the same time, the second part in that exercise gives an example where the last equality is not fulfilled.

Corollary 5.3.2 *Assume that (5.10) holds, and let $x \in \text{dom } f$ be such that the set $\text{int}(\text{cone}(\text{dom } f - x))$ is non-empty. Then we have*

$$\partial f(x) = N_{\text{dom } f}(x) + \bigcap_{\varepsilon > 0} \overline{\text{co}}\left\{\bigcup_{t \in T_\varepsilon(x)} \partial_\varepsilon f_t(x)\right\}. \tag{5.66}$$

Proof. We may assume that $\partial f(x) \neq \emptyset$. By Proposition 5.3.1 we have that

$$\partial f(x) = \bigcap_{\varepsilon > 0} \overline{\text{co}}\left\{\bigcup_{t \in T_\varepsilon(x)} \partial_\varepsilon f_t(x) + N_{\text{dom } f}(x)\right\},$$

and the conclusion follows by applying Proposition 4.1.28 to the non-decreasing family of non-empty convex sets $A_\varepsilon = \text{co}\left\{\bigcup_{t \in T_\varepsilon(x)} \partial_\varepsilon f_t(x)\right\}$, $\varepsilon > 0$ (non-empty because we are assuming that $\partial f(x) \neq \emptyset$). ∎

Corollary 5.3.3 *Assume that condition (5.10) holds. Fix $x \in \text{dom } f$ and let $\varepsilon_0 > 0$ such that the function $f_{\varepsilon_0} := \sup_{t \in T_{\varepsilon_0}(x)} f_t$ is finite and continuous at some point in $\text{dom } f$. Then we have that*

5.3. THE ROLE OF CONTINUITY ASSUMPTIONS

$$\partial f(x) = \mathrm{N}_{\mathrm{dom}\, f}(x) + \bigcap_{\varepsilon>0} \overline{\mathrm{co}} \left\{ \bigcup_{t \in T_\varepsilon(x)} \partial_\varepsilon f_t(x) \right\}.$$

Proof. Theorem 5.2.2 together with Proposition 4.1.20 yields, due to Corollary 5.2.3,

$$\partial f(x) = \partial(f_{\varepsilon_0} + \mathrm{I}_{\mathrm{dom}\, f})(x) = \partial f_{\varepsilon_0}(x) + \mathrm{N}_{\mathrm{dom}\, f}(x). \qquad (5.67)$$

Moreover, since the supremum function f_{ε_0} is continuous somewhere, the family $\{f_t,\ t \in T_{\varepsilon_0}(x)\}$ satisfies condition (5.10) (Exercise 67(i)). Thus, Corollary 5.3.2 gives rise to

$$\partial f_{\varepsilon_0}(x) = \mathrm{N}_{\mathrm{dom}\, f_{\varepsilon_0}}(x) + \bigcap_{\varepsilon>0} \overline{\mathrm{co}} \left\{ \bigcup_{t \in \tilde{T}_\varepsilon(x)} \partial_\varepsilon f_t(x) \right\}$$
$$= \mathrm{N}_{\mathrm{dom}\, f_{\varepsilon_0}}(x) + \bigcap_{0<\varepsilon<\varepsilon_0} \overline{\mathrm{co}} \left\{ \bigcup_{t \in \tilde{T}_\varepsilon(x)} \partial_\varepsilon f_t(x) \right\},$$

where $\tilde{T}_\varepsilon(x) := \{t \in T_{\varepsilon_0}(x) : f_t(x) \geq f_{\varepsilon_0}(x) - \varepsilon\} = \{t \in T_{\varepsilon_0}(x) : f_t(x) \geq f(x) - \varepsilon\}$. Hence, $\tilde{T}_\varepsilon(x) = T_\varepsilon(x)$ for all $\varepsilon \in]0, \varepsilon_0[$, and the relation above, together with (5.67), entails

$$\partial f(x) = \mathrm{N}_{\mathrm{dom}\, f}(x) + \mathrm{N}_{\mathrm{dom}\, f_{\varepsilon_0}}(x) + \bigcap_{0<\varepsilon<\varepsilon_0} \overline{\mathrm{co}} \left\{ \bigcup_{t \in T_\varepsilon(x)} \partial_\varepsilon f_t(x) \right\}. \qquad (5.68)$$

More precisely, again by Proposition 4.1.20, we have that $\mathrm{N}_{\mathrm{dom}\, f}(x) + \mathrm{N}_{\mathrm{dom}\, f_{\varepsilon_0}}(x) = \mathrm{N}_{\mathrm{dom}\, f \cap \mathrm{dom}\, f_{\varepsilon_0}}(x) = \mathrm{N}_{\mathrm{dom}\, f}(x)$, and (5.68) implies the desired formula. ∎

In the following result, the continuity assumption on the f_t's allows us to avoid the assumption that the whole space X is Banach in Theorem 5.2.7.

Theorem 5.3.4 *Let the functions f_t, $t \in T$, be convex and lsc, and denote $f := \sup_{t \in T} f_t$. Assume that there exists a Banach linear subspace X_0 such that each one of the f_t's has a point of continuity in $X_0 \cap f_t^{-1}(\mathbb{R})$. Then for every $x \in X$ we have that*

$$\partial f(x) = \bigcap_{\varepsilon>0,\ p \in \mathcal{P},\ L \in \mathcal{F}(x)} \overline{\mathrm{co}} \left\{ \bigcup_{t \in T_\varepsilon(x),\ y \in B_{p,t}(x,\varepsilon)} \partial f_t(y) \cap S_\varepsilon(y-x) + \mathrm{N}_{L \cap \mathrm{dom}\, f}(x) \right\}. \qquad (5.69)$$

Remark 9 (before the proof) *The hypothesis above is fulfilled if, for instance, all the f_t's have a common continuity point $x_0 \in X$; in such a case, we can take $X_0 \equiv \mathbb{R}\{x_0\}$.*

Proof. Fix $x \in \text{dom } f$, $\varepsilon > 0$, and $p \in \mathcal{P}$. Given $L \in \mathcal{F}(x)$, we denote $L_0 := \text{span}\{L \setminus X_0\}$ and $X_1 := X_0 + L_0$, so that $L \subset X_1$. Hence, X_1 is a Banach space and $\text{cl}(\text{span}(\cup_{t \in T} \text{dom}(f_t + \text{I}_{X_1}))) \subset X_1$. So, by applying Theorem 5.2.9 to the supremum function, $f + \text{I}_{X_1}$, of the family of lsc convex functions $\{f_t + \text{I}_{X_1}, t \in T\}$ we obtain that

$$\partial f(x) \subset \partial(f + \text{I}_{X_1})(x) \subset \overline{\text{co}}\left\{\tilde{A}_{\varepsilon,p} + \text{N}_{L \cap \text{dom } f \cap X_1}(x)\right\} = \overline{\text{co}}\left\{\tilde{A}_{\varepsilon,p} + \text{N}_{L \cap \text{dom } f}(x)\right\}, \tag{5.70}$$

where

$$\tilde{A}_{\varepsilon,p} := \bigcup_{t \in T_\varepsilon(x), \, y \in \tilde{B}_{p,t}(x,\varepsilon)} \partial(f_t + \text{I}_{X_1})(y) \cap S_\varepsilon(y - x),$$

and

$$\tilde{B}_{p,t}(x,\varepsilon) := \{y \in X_1 : p(y - x) \le \varepsilon, \ |f_t(y) - f_t(x)| \le \varepsilon\} \subset B_{p,t}(x,\varepsilon).$$

Observe that $\partial(f_t + \text{I}_{X_1})(y) = \partial f_t(y) + X_1^\perp$, due to Proposition 4.1.20. Thus, because $S_\varepsilon(y - x) + X_1^\perp \subset S_\varepsilon(y - x)$ for all $y \in \tilde{B}_{p,t}(x,\varepsilon) \subset X_1$, we can easily verify that

$$\partial(f_t + \text{I}_{X_1})(y) \cap S_\varepsilon(y - x) \subset (\partial f_t(y) \cap S_\varepsilon(y - x)) + X_1^\perp \subset (\partial f_t(y) \cap S_\varepsilon(y - x)) + L^\perp.$$

Thus, (5.70) entails

$$\partial f(x) \subset \overline{\text{co}}\left\{\tilde{A}_{\varepsilon,p} + \text{N}_{L \cap \text{dom } f}(x)\right\}$$

$$\subset \overline{\text{co}}\left\{\bigcup_{t \in T_\varepsilon(x), \, y \in B_{p,t}(x,\varepsilon)} \partial f_t(y) \cap S_\varepsilon(y - x) + \text{N}_{L \cap \text{dom } f}(x) + L^\perp\right\}$$

$$= \overline{\text{co}}\left\{\bigcup_{t \in T_\varepsilon(x), \, y \in B_{p,t}(x,\varepsilon)} \partial f_t(y) \cap S_\varepsilon(y - x) + \text{N}_{L \cap \text{dom } f}(x)\right\}.$$

Hence, the inclusion "$\subset$" holds by intersecting over $\varepsilon > 0$, $p \in \mathcal{P}$, and $L \in \mathcal{F}(x)$, and we are done as the opposite inclusion is straightforward. ∎

The previous formulas of $\partial f(x)$ are significantly simplified under the continuity of the supremum function.

Theorem 5.3.5 *Let the functions f_t, $t \in T$, be convex, and assume that $f := \sup_{t \in T} f_t$ is finite and continuous somewhere. Then, for every $x \in X$, we have that*

5.3. THE ROLE OF CONTINUITY ASSUMPTIONS

$$\partial f(x) = \mathrm{N}_{\mathrm{dom}\, f}(x) + \bigcap_{\varepsilon>0} \overline{\mathrm{co}} \left\{ \bigcup_{t \in T_\varepsilon(x)} \partial_\varepsilon f_t(x) \right\} \tag{5.71}$$

and, provided that the f_t's are lsc,

$$\partial f(x) = \mathrm{N}_{\mathrm{dom}\, f}(x) + \bigcap_{\varepsilon>0,\ p \in \mathcal{P}} \overline{\mathrm{co}} \left\{ \bigcup_{t \in T_\varepsilon(x), y \in B_{p,t}(x,\varepsilon)} \partial f_t(y) \cap S_\varepsilon(y - x) \right\}. \tag{5.72}$$

Remark 10 (before the proof) *Observe that formula (5.71) uses approximate subdifferentials of the f_t's at the reference point x, like in (5.26), whereas (5.72) is given in terms of exact subdifferentials at nearby points, as in (5.58).*

Proof. Formula (5.71) follows from Corollary 5.3.2, as condition (5.10) automatically holds in the current setting (Proposition 5.2.4(i)). To prove formula (5.72) we denote

$$A_{\varepsilon,p} := \bigcup_{t \in T_\varepsilon(x),\, y \in B_{p,t}(x,\varepsilon)} \partial f_t(y) \cap S_\varepsilon(y - x).$$

Since $A_{\varepsilon,p} \subset \partial_{3\varepsilon} f(x)$, coming from (5.57), we have that

$$\mathrm{N}_{\mathrm{dom}\, f}(x) + \bigcap_{\varepsilon>0,\, p \in \mathcal{P}} \overline{\mathrm{co}}(A_{\varepsilon,p}) \subset \mathrm{N}_{\mathrm{dom}\, f}(x) + \bigcap_{\varepsilon>0} \partial_{3\varepsilon} f(x)$$
$$= \mathrm{N}_{\mathrm{dom}\, f}(x) + \partial f(x) = \partial f(x),$$

showing that the inclusion "$\supset$" in (5.72) holds. Hence, (5.72) holds whenever $\partial f(x) = \emptyset$, and we only need to establish the inclusion "$\subset$" in (5.72) when $\partial f(x) \neq \emptyset$.

To this aim, we assume that all the f_t's are proper; the general case is treated in Exercise 89. Then, due to Proposition 2.2.6, the current continuity hypothesis implies that all the f_t's are finite and continuous at some point $x_0 \in \mathrm{dom}\, f$. Therefore, Theorem 5.3.4 applies and yields

$$\partial f(x) = \bigcap_{\varepsilon>0,\, p \in \mathcal{P},\, L \in \mathcal{F}(x)} \overline{\mathrm{co}}\left(\mathrm{N}_{L \cap \mathrm{dom}\, f}(x) + A_{\varepsilon,p} \right).$$

Observe that the same relation holds if $\mathcal{F}(x)$ is replaced with the family $\mathcal{F}(x, x_0) := \{L \in \mathcal{F}(x) : x_0 \in L\}$ (Exercise 82). Notice also that, due to Proposition 4.1.20, $\mathrm{N}_{L \cap \mathrm{dom}\, f}(x) = \mathrm{N}_{\mathrm{dom}\, f}(x) + L^\perp$ for all $L \in \mathcal{F}(x, x_0)$, and the relation above yields (Exercise 10(i))

$$\partial f(x) = \bigcap_{\varepsilon > 0,\, p \in \mathcal{P},\, L \in \mathcal{F}(x,x_0)} \overline{\mathrm{co}}\left(\mathrm{N}_{\mathrm{dom}\, f}(x) + A_{\varepsilon,p} + L^{\perp}\right)$$

$$= \bigcap_{\varepsilon > 0,\, p \in \mathcal{P}} \overline{\mathrm{co}}\left(\mathrm{N}_{\mathrm{dom}\, f}(x) + A_{\varepsilon,p}\right) = \bigcap_{\varepsilon > 0,\, p \in \mathcal{P}} \mathrm{cl}\left(\mathrm{N}_{\mathrm{dom}\, f}(x) + \mathrm{co}\,(A_{\varepsilon,p})\right). \quad (5.73)$$

Now we consider the following partial order in the set $]0, +\infty[\times \mathcal{P}$,

$$(\varepsilon_1, p_1) \preccurlyeq (\varepsilon_2, p_2) \iff \varepsilon_1 \leq \varepsilon_2 \text{ and } p_1 \geq p_2,$$

so that $(]0, +\infty[\times \mathcal{P}, \preccurlyeq)$ becomes a directed set, and the net $(\mathrm{co}\, A_{\varepsilon,p})_{(\varepsilon,p)}$ is non-decreasing. Moreover, using again the relation $A_{\varepsilon,p} \subset \partial_{3\varepsilon} f(x)$, for every $z \in \mathrm{dom}\, f$, $\varepsilon > 0$, and $p \in \mathcal{P}$ we have that

$$\sigma_{\mathrm{co}\, A_{\varepsilon,p}}(z - x) = \sigma_{A_{\varepsilon,p}}(z - x) \leq \sigma_{\partial_{3\varepsilon} f(x)}(z - x) \leq f(z) - f(x) + 3\varepsilon < +\infty;$$

that is, $\mathrm{dom}\, f - x \subset \bigcap_{\varepsilon > 0,\, p \in \mathcal{P}} \mathrm{dom}\, \sigma_{A_{\varepsilon,p}}$ and, obviously, $\mathrm{dom}\, f - x \subset \bigcup_{\varepsilon > 0,\, p \in \mathcal{P}} \mathrm{dom}\, \sigma_{A_{\varepsilon,p}}$. Therefore, Proposition 4.1.28 applies and (5.73) produces the desired result. ■

Corollary 5.3.6 *Let $f_t : X \to \mathbb{R}_{\infty}$, $t \in T$, be convex functions and assume that $f := \sup_{t \in T} f_t$ is finite and continuous at $x \in X$. Then we have that*

$$\partial f(x) = \bigcap_{\varepsilon > 0} \overline{\mathrm{co}} \left\{ \bigcup_{t \in T_{\varepsilon}(x)} \partial_{\varepsilon} f_t(x) \right\} \quad (5.74)$$

$$= \bigcap_{\varepsilon > 0,\, p \in \mathcal{P}} \overline{\mathrm{co}} \left\{ \bigcup_{t \in T_{\varepsilon}(x),\, y \in B_p(x,\varepsilon)} \partial f_t(y) \right\}, \quad (5.75)$$

where $B_p(x, \varepsilon) := \{y \in X : p(y - x) \leq \varepsilon\}$.

Proof. Without any loss of generality, we may suppose that all the f_t's are proper. Otherwise, according to Corollary 5.2.3, and since

$$\partial f(x) = \partial \left(\sup_{t \in T_{\varepsilon}(x)} f_t \right)(x) + \mathrm{N}_{\mathrm{dom}\, f}(x) \text{ for all } \varepsilon > 0, \quad (5.76)$$

we can deal with the subfamily $\{f_t,\, t \in T_{\varepsilon}(x)\}$ of functions, which are continuous at x and, therefore, proper. We may assume that $x = \theta$ and $f(\theta) = 0$ (Exercise 78). First, formula (5.74) is straightforward from (5.71) as $\mathrm{N}_{\mathrm{dom}\, f}(\theta) = \{\theta\}$, whereas formula (5.72) implies that

$$\partial f(\theta) = \bigcap_{\varepsilon > 0,\, p \in \mathcal{P}} \overline{\mathrm{co}}(A_{\varepsilon,p}) \subset \bigcap_{\varepsilon > 0,\, p \in \mathcal{P}} \overline{\mathrm{co}} \left\{ \bigcup_{t \in T_{\varepsilon}(\theta),\, y \in B_p(\theta,\varepsilon)} \partial f_t(y) \right\},$$

5.3. THE ROLE OF CONTINUITY ASSUMPTIONS

where we remember that

$$A_{\varepsilon,p} := \bigcup_{t \in T_\varepsilon(\theta), y \in B_{p,t}(\theta,\varepsilon)} \partial f_t(y) \cap S_\varepsilon(y).$$

This yields the inclusion "$\subset$" in (5.75). To prove the other inclusion in (5.75), we only need to verify that

$$C := \bigcap_{\varepsilon>0, p \in \mathcal{P}} \overline{\mathrm{co}}(C_{\varepsilon,p}) \subset \bigcap_{\varepsilon>0, p \in \mathcal{P}} \overline{\mathrm{co}}(A_{\varepsilon,p}),$$

where $C_{\varepsilon,p} := \bigcup_{t \in T_\varepsilon(\theta), y \in B_p(\theta,\varepsilon)} \partial f_t(y)$. To this aim, we take $\varepsilon_0 \in \,]0,1[$ and $U \in \mathcal{N}_X$ such that (see Proposition 2.2.6)

$$|f_t(y) - f_t(z)| \le p_U(y-z) \text{ for all } y, z \in U,\, t \in T_{\varepsilon_0}(\theta), \tag{5.77}$$

where p_U is the gauge function of U. Next, take $\varepsilon \in \,]0, \varepsilon_0/2[$, $p \in \mathcal{P}$ and consider $\tilde{p} := \max\{p, p_U\}$. Then, given $y \in B_{\tilde{p}}(\theta, \varepsilon)$ ($\subset B_{p_U}(\theta, \varepsilon) = \varepsilon[p_U \le 1] = \varepsilon U$), $y^* \in \partial f_t(y)$, and $t \in T_\varepsilon(\theta) \subset T_{\varepsilon_0}(\theta)$, relation (5.77) gives rise to

$$\langle y^*, z - y \rangle \le f_t(z) - f_t(y) \le |f_t(y) - f_t(z)| \le p_U(y-z) \text{ for all } z \in U.$$

In particular, taking $z = \theta$ we obtain that $|f_t(y) - f_t(\theta)| \le p_U(y) \le \varepsilon$, while $z = 2y \in 2\varepsilon U \subset U$ gives $\langle y^*, y \rangle \le p_U(y) \le \varepsilon$, entailing that $y^* \in \partial f_t(y) \cap S_\varepsilon(y)$ and $y \in B_{\tilde{p},t}(\theta, \varepsilon)$; in other words, $C_{\varepsilon,\tilde{p}} \subset A_{\varepsilon,\tilde{p}}$. Therefore, choosing $p_0 \in \mathcal{P}$ such that $\tilde{p} \le p_0$,

$$C \subset \bigcap_{0<\varepsilon<\varepsilon_0/2} \overline{\mathrm{co}}(C_{\varepsilon,p_0}) \subset \bigcap_{0<\varepsilon<\varepsilon_0/2} \overline{\mathrm{co}}(C_{\varepsilon,\tilde{p}})$$
$$\subset \bigcap_{0<\varepsilon<\varepsilon_0/2} \overline{\mathrm{co}}(A_{\varepsilon,\tilde{p}}) \subset \bigcap_{0<\varepsilon<\varepsilon_0/2} \overline{\mathrm{co}}(A_{\varepsilon,p}).$$

Hence, $C \subset \bigcap_{0<\varepsilon<\varepsilon_0/2,\, p \in \mathcal{P}} \overline{\mathrm{co}}(A_{\varepsilon,p}) = \bigcap_{\varepsilon>0,\, p \in \mathcal{P}} \overline{\mathrm{co}}(A_{\varepsilon,p})$ and the proof is complete. ∎

Remark 11 *Taking into account (5.76), the proof above shows that formulas (5.74) and (5.75) also hold if, instead of the continuity of f at x, we assume that the function $\sup_{t \in T_\varepsilon(x)} f_t$, for some $\varepsilon > 0$, is continuous at x and that $x \in \mathrm{int}(\mathrm{dom}\, f)$.*

The following corollary deals with a non-convex situation, where Theorem 5.2.2 can also be applied.

Corollary 5.3.7 *Given non-necessarily convex functions $f_t : X \to \overline{\mathbb{R}}$, $t \in T$, and $f := \sup_{t \in T} f_t$, formula (5.26) is still valid under the condition*

$$\overline{\mathrm{co}} f = \sup_{t \in T} (\overline{\mathrm{co}} f_t). \tag{5.78}$$

Proof. We only need to prove the inclusion "$\subset$". We fix $x \in \mathrm{dom}\, f$ such that $\partial f(x) \neq \emptyset$; hence, $(\overline{\mathrm{co}} f)(x) = f(x) \in \mathbb{R}$ and $\partial f(x) = \partial(\overline{\mathrm{co}} f)(x)$ (Exercise 78). Then, by applying Theorem 5.2.2 to the family of lsc convex functions $\{\overline{\mathrm{co}} f_t, t \in T\}$, we obtain that

$$\partial f(x) = \bigcap_{\varepsilon > 0,\, L \in \mathcal{F}(x)} \overline{\mathrm{co}} \left(\bigcup_{t \in \widetilde{T}_\varepsilon(x)} \partial_\varepsilon (\overline{\mathrm{co}} f_t)(x) + \mathrm{N}_{L \cap \mathrm{dom}(\overline{\mathrm{co}} f)}(x) \right), \tag{5.79}$$

where $\widetilde{T}_\varepsilon(x) := \{t \in T : (\overline{\mathrm{co}} f_t)(x) \geq f(x) - \varepsilon\} \subset T_\varepsilon(x)$. Observe that for all $t \in \widetilde{T}_\varepsilon(x)$ we have $(\overline{\mathrm{co}} f_t)(x) \geq f(x) - \varepsilon \geq f_t(x) - \varepsilon$, giving rise to $\partial_\varepsilon (\overline{\mathrm{co}} f_t)(x) \subset \partial_{2\varepsilon} f_t(x)$ (Exercise 62). In fact, if $x^* \in \partial_\varepsilon (\overline{\mathrm{co}} f_t)(x)$, then for all $y \in X$

$$\langle x^*, y - x \rangle \leq (\overline{\mathrm{co}} f_t)(y) - (\overline{\mathrm{co}} f_t)(x) + \varepsilon \leq f_t(y) - f_t(x) + 2\varepsilon.$$

Consequently, taking into account that

$$\widetilde{T}_\varepsilon(x) \subset T_\varepsilon(x) \text{ and } \mathrm{N}_{L \cap \mathrm{dom}(\overline{\mathrm{co}} f)}(x) \subset \mathrm{N}_{L \cap \mathrm{dom}\, f}(x),$$

Exercise 62 entails

$$\partial f(x) \subset \bigcap_{\varepsilon > 0,\, L \in \mathcal{F}(x)} \overline{\mathrm{co}} \left(\bigcup_{t \in T_\varepsilon(x)} \partial_{2\varepsilon} f_t(x) + \mathrm{N}_{L \cap \mathrm{dom}\, f}(x) \right),$$

and we are done. ∎

As an illustration of the previous results, in the following example we provide some formulas for the subdifferential of the supremum of affine functions.

Example 5.3.8 *Assume that*

$$f(x) := \sup \{\langle a_t, x \rangle - b_t : t \in T\},$$

with $(a_t, b_t) \in X^* \times \mathbb{R}$. Then, according to Theorem 5.2.2, for every fixed $x \in \mathrm{dom}\, f$ we have that

5.3. THE ROLE OF CONTINUITY ASSUMPTIONS

$$\partial f(x) = \bigcap_{\varepsilon>0,\ L\in\mathcal{F}(x)} \overline{\mathrm{co}}(A_\varepsilon + N_{L\cap\mathrm{dom}\,f}(x)), \tag{5.80}$$

where $A_\varepsilon := \{a_t \in A : \langle a_t, x\rangle - b_t \geq f(x) - \varepsilon\}$.

Let us suppose that $X = \mathbb{R}$ and $a_t > 0$ for all $t \in T$. We also assume that $x = 0$, so that $\bar{b} := \sup\{-b_t : t \in T\} < +\infty$. Then, by Proposition 5.3.1, the intersection over the L's is removed from (5.80), which is now written in two forms:

1) When $1 \in [\overline{\mathrm{co}}\{a_t,\ t \in T\}]_\infty$. In this case, we have $\sup_{t\in T} a_t = +\infty$, and so $\mathrm{dom}\,f =]-\infty, c]$ for some $c \in \mathbb{R}$. If $c = 0$, then $N_{\mathrm{dom}\,f}(0) = \mathbb{R}_+$ and (5.80) simplifies to $\partial f(0) = \bigcap_{\varepsilon>0}[\inf_{a_t\in A_\varepsilon} a_t, +\infty[$. If $c > 0$, then $N_{\mathrm{dom}\,f}(0) = \{0\}$ and (5.80) reads

$$\partial f(0) = \bigcap_{\varepsilon>0} \overline{\mathrm{co}}\{a_t \in A : -b_t \geq \bar{b} - \varepsilon\}.$$

2) When $1 \notin [\overline{\mathrm{co}}\{a_t,\ t \in T\}]_\infty$. In this case, the set $\{a_t : t \in T\}$ is bounded, and so f is finite everywhere; that is, $\mathrm{dom}\,f = \mathbb{R}$. Hence, the last formula holds in this case too.

We shall discuss a pair of particular cases:

i) $(a_t, b_t) = (t, 1/t)$, $t > 0$. We have $\mathrm{dom}\,f =]-\infty, 0]$ so that $0 \in \mathrm{bd}(\mathrm{dom}\,f)$, and the formula in 1) entails $\partial f(0) = \bigcap_{\varepsilon>0}[1/\varepsilon, +\infty[= \emptyset$; i.e., f has no subgradient at 0. In fact, we have in this case $f(x) = -2\sqrt{-x}$, for $x \leq 0$.

ii) $(a_t, b_t) = (t|\sin(t)|, 1/t)$, $t > 0$. Again $\mathrm{dom}\,f =]-\infty, 0]$ and the formula in 1) also applies because

$$(1,0) = \lim_{k\to\infty} \frac{2}{(2k+1)\pi}\left(\frac{(2k+1)\pi}{2}|\sin((2k+1)\pi/2)|, \frac{2}{(2k+1)\pi}\right).$$

Thus $\partial f(0) = \bigcap_{\varepsilon>0}[0, +\infty[= \mathbb{R}_+$.

In the case of the support function, let us say $f = \sigma_A$ for a nonempty set $A \subset X^*$, the formula in Theorem 5.2.2 is different from the following classical characterization, which can be derived from the Fenchel inequality (see (5.1)).

Example 5.3.9 *Let $T \subset \mathbb{R}^3$ be the set given by*

$$T := \{(1, \alpha, \beta) : \alpha \geq 0,\ \beta \in \mathbb{R}\} \cup \{(0, \gamma, -\log\gamma) : 0 < \gamma \leq 1\},$$

so that $f_t := \langle t, \cdot\rangle$, $t \in T$, and $f := \sup_{t\in T} f_t$ is the support function of the set T. For $x := (-1, -1, 0)$ we have $f(x) = 0$ and, with $\varepsilon \leq 1$ fixed,

$$T_\varepsilon(x) := \{t \in T : \langle t, x\rangle \geq f(x) - \varepsilon\} = \{t \in T : \langle t, x\rangle = -t_1 - t_2 \geq -\varepsilon\}$$
$$= \{(0, \gamma, -\log\gamma) : 0 < \gamma \leq \varepsilon\},$$

and

$$\operatorname{co}\left\{\bigcup_{t\in T_\varepsilon(x)} \partial_\varepsilon f_t(x)\right\} = \{(0,\gamma,\delta) : 0 < \gamma \leq \varepsilon;\ -\log\varepsilon \leq \delta \leq -\log\gamma\}.$$

Moreover, $N_{\operatorname{dom} f}(x) = \mathbb{R}(0,0,1)$ and

$$\operatorname{co}\left\{\bigcup_{t\in T_\varepsilon(x)} \partial_\varepsilon f_t(x)\right\} + N_{\operatorname{dom} f}(x) = (\overline{\operatorname{co}}T_\varepsilon(x)) + \mathbb{R}(0,0,1)$$
$$= \{(0,\gamma,\delta) : 0 < \gamma \leq \varepsilon,\ \delta \in \mathbb{R}\},$$

which obviously is not closed. Then

$$\partial f(x) = \bigcap_{\varepsilon > 0} \overline{\operatorname{co}}\left\{\bigcup_{t\in T_\varepsilon(x)} \partial_\varepsilon f_t(x) + N_{\operatorname{dom} f}(x)\right\} = \mathbb{R}\{(0,0,1)\}.$$

However, for every $\varepsilon > 0$ the Fenchel equality yields

$$\partial_\varepsilon f(x) = \{t \in \overline{\operatorname{co}}T : \langle t, x\rangle \geq -\varepsilon\}$$
$$= \{t \in \mathbb{R}^3 : 0 \leq t_1 \leq 1,\ t_2 \geq 0,\ t_1 + t_2 \leq \varepsilon\}.$$

This shows that, for every $\varepsilon > 0$,

$$\overline{\operatorname{co}}\left\{\bigcup_{t\in T_\varepsilon(x)} \partial_\varepsilon f_t(x) + N_{\operatorname{dom} f}(x)\right\} = \operatorname{cl}\left(\operatorname{co}\left\{\bigcup_{t\in T_\varepsilon(x)} \partial_\varepsilon f_t(x)\right\} + N_{\operatorname{dom} f}(x)\right) \subsetneq \partial_\varepsilon f(x).$$

The above example also shows that even in finite-dimensional spaces, one may have that

$$\bigcap_{\varepsilon > 0}\left(\overline{\operatorname{co}}\left\{\bigcup_{t\in T_\varepsilon(x)} \partial_\varepsilon f_t(x)\right\} + N_{\operatorname{dom} f}(x)\right) \subsetneq \partial f(x).$$

The following example shows that, in general, the set $N_{\operatorname{dom} f}(x)$ in formula (5.26) cannot be removed.

Example 5.3.10 Consider the family of linear functions $f_t : \mathbb{R}^2 \to \mathbb{R}_\infty$, $t \in T := \{0\} \cup\,]1, +\infty[$, given by

$$f_t(u,v) := \begin{cases} -v, & \text{for } t = 0, \\ tu + \frac{v}{t-1}, & \text{for } t > 1, \end{cases}$$

and the associated supremum function

$$f(u,v) := \max\{-v, \ \sup\{tu + v/(t-1) : t > 1\}\}.$$

Observe that f is the support function of the set $\{(u,v) \in \mathbb{R}^2 : u \geq 0, v \geq -1\}$. On the one hand, for $x := (-1, 0)$ we easily check that $f(x) = 0$ and $\partial f(x) = \{0\} \times [-1, +\infty[$. On the other hand, for all $\varepsilon < 1$ we have that $T_\varepsilon(x) = \{t \in T : f_t(x) \geq -\varepsilon\} = \{0\}$, and

$$\bigcap_{\varepsilon > 0} \overline{\mathrm{co}} \left\{ \bigcup_{t \in T_\varepsilon(x)} \partial_\varepsilon f_t(x) \right\} = \bigcap_{\varepsilon \in]0,1[} \partial_\varepsilon f_0(x) = \partial f_0(x) = \{(0, -1)\}.$$

Thus, $\bigcap_{\varepsilon > 0} \overline{\mathrm{co}} \left\{ \bigcup_{t \in T_\varepsilon(x)} \partial_\varepsilon f_t(x) \right\} \subsetneq \partial f(x)$. At the same time, we have

$$\mathrm{dom}\, f =]-\infty, 0] \times]-\infty, 0].$$

Then $\mathrm{N}_{\mathrm{dom}\, f}(x) = \{0\} \times [0, +\infty[$, and formula (5.71) reads

$$\partial f(x) = \mathrm{N}_{\mathrm{dom}\, f}(x) + \bigcap_{\varepsilon > 0} \overline{\mathrm{co}} \left\{ \bigcup_{t \in T_\varepsilon(x)} \partial_\varepsilon f_t(x) \right\}$$
$$= \{0\} \times [0, +\infty[+ \{(0, -1)\} = \{0\} \times [-1, +\infty[.$$

5.4 Exercises

Exercise 67 *Let $\{f_t, t \in T\}$ be a non-empty family of proper convex functions, and set $f := \sup_{t \in T} f_t$. Assume that $\{f_t, t \in T\}$ satisfies condition (5.10).*

(i) Given $T_i \subset T$ such that $\cup_i T_i = T$, we set $g_i := \sup_{t \in T_i} f_t$. Prove that the family $\{g_i\}$ satisfies (5.10). Prove in addition that (5.26) holds if there exists some $\varepsilon_0 > 0$ such that the following condition holds:

$$\mathrm{cl}\, f = \sup \left\{ \mathrm{cl}\, f_t, \ t \in T_{\varepsilon_0}(x); \ \mathrm{cl} \left(\sup_{t \in T \setminus T_{\varepsilon_0}(x)} f_t \right) \right\}.$$

(ii) Assume that $\{\mathrm{cl}\, f_t : t \in T\} \subset \Gamma_0(X)$. Prove that, for every convex set $A \supset \mathrm{dom}\, f$, the family $\{f_t + \mathrm{I}_A : t \in T\}$ satisfies (5.10).

Exercise 68 Let $A \subset X^*$ be a non-empty set. Prove that, for every $x \in X$ and $\varepsilon \geq 0$,

$$\partial_\varepsilon \sigma_A(x) = \bigcap_{\delta > \varepsilon} \operatorname{cl} \{x^* \in \operatorname{co} A : \langle x^*, x \rangle \geq \sigma_A(x) - \delta\}. \tag{5.81}$$

Exercise 69 Let $A := \{(a_t, b_t) \in X^* \times \mathbb{R}, \, t \in T\}$ be a non-empty set, and denote $f := \sup_{t \in T}(\langle a_t, \cdot \rangle - b)$. Prove that, for every $x \in X$ and $\varepsilon \geq 0$,

$$\partial_\varepsilon f(x) = \bigcap_{\delta > \varepsilon} \operatorname{cl} \{x^* \in X^* : (x^*, \alpha^*) \in \operatorname{co} A, \, \langle x^*, x \rangle - \alpha^* \geq f(x) - \delta\}.$$

Exercise 70 Let $C := \{(a_t, b_t) \in X^* \times \mathbb{R}, \, t \in T\}$ be a non-empty set, and denote $f := \sup_{t \in T}(\langle a_t, \cdot \rangle - b)$. Let the mappings $A, A_0 : X \to X \times \mathbb{R}$ be such that $A_0 x := (x, \theta)$, $Ax := A_0 x + (\theta, -1)$. Prove the following statements:
 (i) $f = \sigma_C \circ A$.
 (ii) For every $x \in X$ and $\varepsilon > 0$,

$$\partial_\varepsilon f(x) = \operatorname{cl}(A_0^*(\partial_\varepsilon \sigma_C(x, -1)))$$
$$= \operatorname{cl}(\{x^* \in X^* : (x^*, \alpha^*) \in \partial_\varepsilon \sigma_C(x, -1)\}).$$

Exercise 71 We have proved in Theorem 5.1.4 that, when $\{f_t, \, t \in T\} \subset \Gamma_0(X)$ is closed for convex combinations and $f := \sup_{t \in T} f_t$, for all $x \in X$ and $\varepsilon > 0$ we have

$$\partial_\varepsilon f(x) = \operatorname{cl}\left\{\bigcup_{t \in T} \partial_{(\varepsilon + f_t(x) - f(x))} f_t(x)\right\}. \tag{5.82}$$

The purpose of this exercise is to show, under the assumptions that the functions f_t, $t \in T$, are not necessarily in $\Gamma_0(X)$ but satisfy (5.10), that for all $x \in X$ and $\varepsilon > 0$

$$\partial_\varepsilon f(x) = \operatorname{cl}\left\{\bigcup_{t \in S_0} \partial_{(\varepsilon + f_t(x) - f(x))} (f_t + \mathrm{I}_D)(x)\right\}, \tag{5.83}$$

where $D \subset X$ is as in (5.11). To this aim, proceed by proving the following facts.

 (i) $\bar{f} = \sup_{t \in S_0} g_t$, where $g_t := \bar{f}_t + \mathrm{I}_{\operatorname{cl}(D)}$.
 (ii) $\{g_t, \, t \in S_0\} \subset \Gamma_0(X)$ is closed for convex combinations.
 (iii) Formula (5.83) holds.

5.4. EXERCISES

Exercise 72 *Assume that X is a reflexive Banach space. Prove that formula (5.16) also holds if the closure is taken with respect to the norm topology.*

Exercise 73 *Prove that formula (5.22) also holds if the closure is taken with respect to the norm topology when X is a reflexive Banach space.*

Exercise 74 *Give a proof of Example 5.1.6 by using Proposition 3.2.8.*

Exercise 75 *Given any function $f : X \to \mathbb{R}_\infty$ with a proper conjugate, prove that, for all $x^* \in X^*$ and $\varepsilon > 0$,*

$$\partial_\varepsilon f^*(x^*) = \mathrm{cl}\left\{ \sum_{x \in \mathrm{dom}\, f} \lambda_x (\partial_{\varepsilon_x} f)^{-1}(x^*) : \lambda \in \Delta(\mathrm{dom}\, f),\, \varepsilon_x \geq 0,\, \sum_{x \in \mathrm{dom}\, f} \lambda_x \varepsilon_x \leq \varepsilon \right\}. \tag{5.84}$$

Exercise 76 *Let $(f_n)_n \subset \Gamma_0(X)$ be a countable family, and denote $f := \sup_{n \geq 1} f_n$. Using Corollary 5.1.9 and Example 5.1.6, prove that, for all $x \in \mathrm{dom}\, f$ and $\varepsilon \geq 0$,*

$$\partial_\varepsilon f(x) = \bigcap_{\delta > 0} \mathrm{cl} \left(\bigcup_{\lambda \in \Delta_n,\, n \geq 1} \partial_{\varepsilon + \delta + \Sigma_{1 \leq i \leq n} \lambda_i f_i(x) - f(x)} \left(\sum_{1 \leq i \leq n} \lambda_i f_i \right)(x) \right).$$

Exercise 77 *Assume that X is a reflexive Banach space. Let $\{f_t, t \in T\}$ be a family of convex functions satisfying (5.10), denote $f := \sup_{t \in T} f_t$, and let $D \subset X$ be convex set satisfying (5.11). Prove the following assertions, for every $x \in \mathrm{dom}\, f$ and $\varepsilon \geq 0$:*

(i) $\partial_\varepsilon f(x) = \bigcup\limits_{J \subset S_0,\, J \text{ countable}} \partial_{(\varepsilon + f_J(x) - f(x))} (f_J + \mathrm{I}_D)(x).$

(ii) $\partial_\varepsilon f(x) = \bigcup\limits_{J \subset T,\, J \text{ countable}} \partial_{(\varepsilon + f_J(x) - f(x))} f_J(x).$

Exercise 78 *Given convex functions f_t, $t \in T$, $f := \sup_{t \in T} f_t$, and $x \in f^{-1}(\mathbb{R})$, we introduce the functions*

$$\tilde{f}_t := f_t(\cdot + x) - f(x),\, t \in T,$$

and the associated supremum function

$$\tilde{f} := \sup_{t \in T} \tilde{f}_t = f(\cdot + x) - f(x).$$

(i) Prove that if the family $\{f_t,\ t \in T\}$ satisfies condition (5.10), then the family $\{\tilde{f}_t,\ t \in T\}$ also does.

(ii) Use the $\tilde{f}_t$'s to prove that, in order to establish (5.26) and related formulas when $\partial f(x) \neq \emptyset$, it suffices to suppose that $x = \theta$, $\partial f(\theta) = \partial (\operatorname{cl} f)(\theta) \neq \emptyset$, and $f(\theta) = (\operatorname{cl} f)(\theta) = 0$.

Exercise 79 Let C be a convex set in X, and let $\{A_\varepsilon\}_{\varepsilon > 0}$ be a family of convex subsets of X^*, which is non-increasing as $\varepsilon \downarrow 0$. Given $x \in C$, we assume that either $\mathbb{R}_+(C - x)$ is closed or $\operatorname{ri}(C - x) \cap \operatorname{dom} \sigma_{A_{\varepsilon_0}} \neq \emptyset$ for some $\varepsilon_0 > 0$. Prove that

$$\bigcap_{\varepsilon > 0} \operatorname{cl}\{A_\varepsilon + \mathrm{N}_C^\varepsilon(x)\} = \bigcap_{\varepsilon > 0} \operatorname{cl}\{A_\varepsilon + \mathrm{N}_C(x)\}. \tag{5.85}$$

Exercise 80 Complete the proof of Theorem 5.2.2 where, instead of supposing that $\{f_t,\ t \in T\} \subset \Gamma_0(X)$, we assume that the functions f_t are lsc.

Exercise 81 Complete the proof of Theorem 5.2.2 when the convex functions $f_t,\ t \in T$, are not necessarily lsc (as in Exercise 80) but satisfy condition (5.10).

Exercise 82 Let the functions $f_t,\ t \in T$, be convex and lsc, and denote $f := \sup_{t \in T} f_t$. Let $A \subset X$ be a finite-dimensional subset. Prove that in formulas (5.26), (5.42), (5.49), (5.69), etc., $\mathcal{F}(x)$ can be replaced with the family $\mathcal{F}(x, A) := \{L \in \mathcal{F}(x) : A \subset L\}$.

Exercise 83 (i) Prove that if X is a reflexive Banach space, then formula (5.26) holds when the convex closure is taken with respect to the norm topology.

(ii) Prove that if X is a Banach space, then the formula in Corollary 5.2.5 is also true when the convex closure is taken with respect to the norm topology.

Exercise 84 Given a non-empty set $A \subset \mathbb{R}^p$, $p \geq 1$, and the associated support function σ_A, prove that for every $x \in \operatorname{dom} \sigma_A$ we have

$$\partial \sigma_A(x) = \bigcap_{\varepsilon > 0} \operatorname{cl}\left((\operatorname{co}\{a \in A : \langle a, x \rangle \geq \sigma_A(x) - \varepsilon\}) + A(x)\right), \tag{5.86}$$

where

$$A(x) := [\overline{\operatorname{co}} A]_\infty \cap \{x\}^\perp. \tag{5.87}$$

In the particular case when $x \in \operatorname{ri}(\operatorname{dom} \sigma_A)$ prove that

$$\partial \sigma_A(x) = \bigcap_{\varepsilon > 0} \mathrm{cl}\left((\mathrm{co}\{a \in A : \langle a, x \rangle \geq \sigma_A(x) - \varepsilon\}) + \mathrm{lin}(\overline{\mathrm{co}}A) \right),$$

where $\mathrm{lin}(\overline{\mathrm{co}}A)$ is the lineality space of $\overline{\mathrm{co}}A$ (see (2.21)). Moreover, when $x \in \mathrm{int}(\mathrm{dom}\,\sigma_A)$ prove that

$$\partial \sigma_A(x) = \bigcap_{\varepsilon > 0} \overline{\mathrm{co}}\{a \in A : \langle a, x \rangle \geq \sigma_A(x) - \varepsilon\}.$$

Exercise 85 *Given a non-empty set $A := \{(a_t, b_t) : t \in T\} \subset \mathbb{R}^{p+1}$, $p \geq 1$, and the supremum function $f(x) := \sup\{\langle a_t, x \rangle - b_t : t \in T\}$, prove that, for every $x \in \mathrm{dom}\,f$,*

$$\partial f(x) = \bigcap_{\varepsilon > 0} \mathrm{cl}\left(\mathrm{co}\{a_t : t \in T_\varepsilon(x)\} + B(x) \right), \quad (5.88)$$

where $B(x) := \{v \in \mathbb{R}^p : (v, \langle v, x \rangle) \in [\overline{\mathrm{co}}(A)]_\infty\}$. In particular, if $x \in \mathrm{ri}(\mathrm{dom}\,f)$, one has

$$\partial f(x) = \bigcap_{\varepsilon > 0} \mathrm{cl}\left(\mathrm{co}\{a_t : t \in T_\varepsilon(x)\} + C(x) \right), \quad (5.89)$$

where $C(x) := \{v \in \mathbb{R}^p : (v, \langle v, x \rangle) \in \mathrm{lin}(\overline{\mathrm{co}}\{A\})\}$, and if $x \in \mathrm{int}(\mathrm{dom}\,f)$

$$\partial f(x) = \bigcap_{\varepsilon > 0} \overline{\mathrm{co}}\{a_t : t \in T_\varepsilon(x)\}. \quad (5.90)$$

Exercise 86 *Let $T \neq \emptyset$ and $\{f_t,\ t \in T\} \subset \Gamma_0(\mathbb{R}^n)$, and set $f := \sup_{t \in T} f_t$. Prove that, for every $x \in X$, we have*

$$\partial f(x) = \bigcap_{\varepsilon > 0} \overline{\mathrm{co}}\left\{ A + \bigcup_{t \in T_\varepsilon(z)} \partial_\varepsilon f_t(x) \right\},$$

where

$$A := \left\{ v^* \in X^* : (v^*, \langle v^*, x \rangle) \in \left[\overline{\mathrm{co}}\left\{ \bigcup_{t \in T} \mathrm{epi}\,f_t^* \right\} \right]_\infty \right\}.$$

Exercise 87 *Given convex functions $f_t,\ t \in T$, we denote $f := \sup_{t \in T} f_t$ and assume that condition (5.10) holds. Prove that, for all $x \in X$,*

$$\partial f(x) = \bigcap_{\substack{\varepsilon>0 \\ L\in\mathcal{F}(x),\, p\in\mathcal{P}}} \overline{\mathrm{co}}\left(\bigcup_{\substack{t\in\tilde{T}_\varepsilon(x) \\ y\in B_{p,t}(x,\varepsilon)}} \partial((\mathrm{cl}\, f_t) + \mathrm{I}_{\overline{L\cap\mathrm{dom}(\mathrm{cl}\, f)}})(y) \cap S_\varepsilon(y-x)\right),$$

where $\tilde{T}_\varepsilon(x) := \{t \in T : (\mathrm{cl}\, f_t)(x) \geq f(x) - \varepsilon\}$ and

$$B_{p,t}(x,\varepsilon) := \{y \in X : p(y-x) \leq \varepsilon, |(\mathrm{cl}\, f_t)(y) - f_t(x)| \leq \varepsilon\}.$$

Exercise 88 Let $f_t : \mathbb{R}^n \to \overline{\mathbb{R}}$, $t \in T$, be convex and lsc, and denote $f := \sup_{t\in T} f_t$. Prove that, for every $x \in \mathbb{R}^n$,

(i)

$$\partial f(x) = \bigcap_{\varepsilon>0} \overline{\mathrm{co}}\left\{\bigcup_{t\in T_\varepsilon(x),\, y\in B_t(x,\varepsilon)} \partial f_t(y) \cap S_\varepsilon(y-x) + \mathrm{N}_{\mathrm{dom}\, f}(x)\right\}, \tag{5.91}$$

where S_ε and $B_t(x,\varepsilon)$ are defined in (5.43) and (5.44), respectively.

(ii)

$$\partial f(x) = \bigcap_{\varepsilon>0} \overline{\mathrm{co}}\left(\bigcup_{t\in T_\varepsilon(x),\, y\in B_t(x,\varepsilon)} \partial(f_t + \mathrm{I}_{\overline{\mathrm{dom}\, f}})(y) \cap S_\varepsilon(y-x) \cap \partial_\varepsilon f(x)\right). \tag{5.92}$$

(iii) In addition, if $\mathrm{dom}\, f_t = \mathrm{dom}\, f$ for all $t \in T$, then

$$\partial f(x) = \bigcap_{\varepsilon>0} \overline{\mathrm{co}}\left\{\bigcup_{t\in T_\varepsilon(x),\, y\in B_t(x,\varepsilon)} \partial f_t(y) \cap S_\varepsilon(y-x)\right\}. \tag{5.93}$$

Exercise 89 Complete the proof of Theorem 5.3.5 when the properness assumption is dropped out.

Exercise 90 Let $f_t : X \to \overline{\mathbb{R}}$, $t \in T$, be convex, and suppose that $f := \sup_{t\in T} f_t$ is finite and continuous at some point. Prove that, for every $x \in X$,

$$\partial f(x) = \mathrm{N}_{\mathrm{dom}\, f}(x) + \bigcap_{\varepsilon>0,\, p\in\mathcal{P}} \overline{\mathrm{co}}\left\{\bigcup_{t\in T_\varepsilon(x),\, y\in \tilde{B}_{p,t}(x,\varepsilon)} \partial(\mathrm{cl}\, f_t)(y) \cap S_\varepsilon(y-x)\right\}, \tag{5.94}$$

where $S_\varepsilon(y-x) := \{y^* \in X^* : \langle y^*, y-x\rangle \leq \varepsilon\}$ and

$$\tilde{B}_{p,t}(x,\varepsilon) := \{y \in X : p(y-x) \leq \varepsilon, \ |(\mathrm{cl}\, f_t)(y) - f_t(x)| \leq \varepsilon\}.$$

(Observe that when the f_t's are lsc, $\tilde{B}_{p,t}(x,\varepsilon)$ reduces to $B_{p,t}(x,\varepsilon)$ introduced in (5.50).)

Exercise 91 *Let the functions f_t, $t \in T$, be convex, and let $x \in \mathrm{dom}\, f$. Assume the existence of some $\varepsilon_0 > 0$ such that $f_{\varepsilon_0} := \sup_{t \in T_{\varepsilon_0}(x)} f_t$ is finite and continuous somewhere in $\mathrm{dom}\, f$. Prove that*

$$\partial f(x) = \mathrm{N}_{\mathrm{dom}\, f}(x) + \bigcap_{\varepsilon > 0} \overline{\mathrm{co}} \left\{ \bigcup_{t \in T_\varepsilon(x)} \partial_\varepsilon f_t(x) \right\}, \quad (5.95)$$

and, provided that the f_t's are lsc,

$$\partial f(x) = \mathrm{N}_{\mathrm{dom}\, f}(x) + \bigcap_{\varepsilon > 0,\, p \in \mathcal{P}} \overline{\mathrm{co}} \left\{ \bigcup_{t \in T_\varepsilon(x),\, y \in B_{p,t}(x,\varepsilon)} \partial f_t(y) \cap S_\varepsilon(y - x) \right\}. \quad (5.96)$$

5.5 Bibliographical notes

The chapter presents extensions of the following historical results: Dubovitskii and Milyutin (see [115]) for a finite number of continuous functions, Levin [131] for infinitely many finite-valued convex functions, Rockafellar [177, Theorem 4] and Volle [196] under weaker continuity assumptions with finitely many functions, Brøndsted [29] who used the concept of approximate subdifferentials, Valadier [191] in normed spaces, assuming the continuity of the supremum function and resorting to the exact subdifferential at nearby points around it, and Volle [195] who obtained another characterization in terms of approximate subgradients at the nominal point. The material in this chapter has been mainly extracted from [50], [51], [52], [53], [99], [100], and [103].

The large number of references allows us to emphasize the importance of this subject. Let us quote a paragraph on the second page of [107]: "One of the most specific constructions in convex or nonsmooth analysis is certainly taking the supremum of a (possibly infinite) collection of functions". In the years 1965–1970, various calculus rules concerning the subdifferential of supremum functions started to emerge. There is extensive literature dealing with subdifferential calculus rules for the supremum of convex functions including, among many others, [64, 113, 114, 122, 134, 139, 141, 142, 185, 191, 196, 198, 201], etc. Pioneering works for the subdifferential/directional derivative of the supremum function are attributed to A. Y. Dubovitskii and

A. A. Milyutin ([76]), and J. M. Danskin ([66, 67]) (and references therein). The first two authors dealt with the supremum of finitely many continuous convex functions, whereas the results in [66, 67] concern finite families of $\mathcal{C}^1$-functions. These results were extended by Valadier ([191]) to arbitrary families provided that the data functions are continuous with respect to both the state variable and the index parameter. In such a case, and assuming the compactness of the index set and the upper semicontinuity of the index mappings, the subdifferential of the supremum function is completely characterized by means of the subdifferentials of active functions at the nominal point. In the general case, maintaining only the continuity of the supremum with respect to the state variable, the same author provided other characterizations using the subdifferentials at nearby points of the ε-active functions. The last result was rewritten by M. Volle ([195, Theorem A]) replacing exact subdifferentials with ε-subdifferentials of the ε-active functions f_t, evaluated at the reference point instead of nearby ones. Therefore, the mathematical interest of this topic, which is the main subject of this book, has been widely recognized by such prestigious authors since the very beginning of convex analysis history. Although we are mainly concerned with the convex case, the general case is also of great interest. For example, besides the aforementioned works by Danskin, dealing with the maximum of a finite family of $\mathcal{C}^1$-functions, we also cite [169] and [40] for further extensions. The last one deals with the maximum of Lipschitz functions and appeals to the concept of the generalized gradient introduced by the same author (see, also, [130]). More recently, [157] considered the supremum of an arbitrary family of non-convex Lipschitz functions and established some formulas via different notions of nonsmooth subdifferentials.

Corollary 5.1.9 and Example 5.1.10 are well-known; see, for instance, [201, Corollary 2.8.11 and Example 2.8.1, respectively]. Example 5.1.12 is given in [105, Corollary 7] for the case of lsc convex functions (see, also, [32] and references therein). For results related to Corollary 5.1.14 we refer to [6]. Example 5.2.1 can also be found in [110]. Theorem 5.1.7 is in the line of some results in [105], [142], and [167, Proposition 3.1].

Theorem 5.2.2 is established in [103], but the first attempts to obtain it have been made in [100] and [99] in the finite-dimensional setting. Namely, the approach used in [100] starts from the analysis of the subdifferential of the support function, which is extended, in the second step, to the supremum of affine functions. Then, the general case of arbitrary convex functions is obtained based on the Fenchel–Moreau–Rockafellar theorem (Theorem 3.2.2). The lower semicontinu-

5.5. BIBLIOGRAPHICAL NOTES

ity assumption (5.10) is an intrinsic condition for developing subdifferential calculus rules for convex functions. A variant of (5.10) was shown in [134, Theorem 3.1] to be necessary for the validity of formula (5.26) in the Banach setting. For other variants of [103], see [134, Theorem 3.1] and [113]. Proposition 5.2.4(iii) can be found in [174, Theorem 9.4] in the case where all the functions are proper. In Banach spaces, [139] gives a formula using the exact subdifferentials of the data functions but at points which are close to the reference point. The locally convex version of this result is investigated in [52]. We also quote here [185], which deals with the directional derivative of the supremum function, under certain conditions on the index set. The first non-convex counterpart to Theorem 5.2.2, given in Corollary 5.3.7, has been proposed in [141]. The condition relying on the local closedness of the set cone(dom $f - x$) in Proposition 5.3.1 has been used in [141] and [113]. Corollary 6.4.4, established in [55], extends to the compact setting the well-known Brøndsted formula ([29]), given for the supremum of a finite number of proper lsc convex functions, all of them being active at the reference point. Other extensions of the Brøndsted formula were provided in [100, Proposition 6.3] for the finite-dimensional setting. Theorem 5.2.7 is given in [51, Theorem X] (see, also, [139]). The existence of proper lsc convex functions with empty subdifferential mapping, such as those considered in Example 5.2.8, can be consulted in [172] (for some Fréchet spaces) and in [21] (for some non-complete normed spaces; actually, certain subspaces of $\ell_2(\mathbb{N})$). Exercise 66 is [43, Theorem 5.3]. Exercise 84 is [100, Proposition 2.1]. Exercise 90 extends formula (5.72) to convex (non-necessarily lsc) functions.

Chapter 6

The supremum in specific contexts

This chapter aims to provide formulas for the subdifferential of supremum functions in specific contexts. In these scenarios, an additional structure is available so that the proposed characterizations can adopt particular formats, generally of greater simplicity. As in the previous chapter, X is an lcs with a topological dual X^*. Unless otherwise stated, X^* is endowed with a compatible topology, in particular, the topologies $\sigma(X^*, X)$ and $\tau(X^*, X)$, or the dual norm topology when X is a reflexive Banach space. The associated bilinear form is represented by $\langle \cdot, \cdot \rangle$.

6.1 The compact-continuous setting

In this section, we provide precise formulas for the subdifferential of the supremum function $f := \sup_{t \in T} f_t$, which only involve the subdifferentials of the active data functions at the reference point. We focus on the compact-continuous setting where the following standard hypothesis holds:

T is compact and the mapping $t \mapsto f_t(z)$ is usc, for every $z \in \operatorname{dom} f$;
$$(6.1)$$

the topological space T needs not to be Hausdorff, and so its compact subsets are not necessarily closed. As in the previous chapters, given $x \in X$ and $\varepsilon \geq 0$, we denote $\mathcal{F}(x) := \{L$ is a finite-dimensional linear subspace of X containing $x\}$ and $T_\varepsilon(x) := \{t \in T : f_t(x) \geq f(x) - \varepsilon\}$, with $T(x) := T_0(x)$.

Below, in Theorems 6.1.4 and 6.1.5, we establish the main result of this section. Formula (6.12) in Theorem 6.1.4 characterizes $\partial f(x)$ by means of exact subdifferentials of the active enlarged data functions $f_t + I_{L \cap \mathrm{dom}\, f}$, $t \in T(x)$. The proof of this theorem appeals to the following technical lemmas.

Lemma 6.1.1 *Given the lsc convex functions f_t, $t \in T$, and $f := \sup_{t \in T} f_t$, we suppose that $f(\theta) = 0$ and $L \subset X$ is a finite-dimensional linear subspace. We denote $D := L \cap \mathrm{dom}\, f$ and $Z := \mathrm{span}\, D$, and introduce the lsc convex functions*

$$g_t := (f_t + I_{\mathrm{cl}\, D})_{|Z} \text{ and } g := \sup_{t \in T} g_t.$$

Then $\mathrm{dom}\, g = D$ and

$$\partial f_{|Z}(\theta) \subset \partial g(\theta) = \bigcap_{\varepsilon > 0} \overline{\mathrm{co}} \left\{ \bigcup_{t \in T_\varepsilon(\theta),\ z \in B_t(\theta, \varepsilon)} A(t, z, \varepsilon) \right\}, \quad (6.2)$$

where $T_\varepsilon(\theta) := \{t \in T : g_t(\theta) \geq -\varepsilon\}$,

$$B_t(\theta, \varepsilon) := \{z \in Z : \|z\|_Z \leq \varepsilon,\ |g_t(z) - g_t(\theta)| \leq \varepsilon\},$$

and

$$A(t, z, \varepsilon) := \partial g_t(z) \cap S_\varepsilon(z) \cap \partial_\varepsilon g(\theta), \text{ with } t \in T,\ z \in Z.$$

Proof. Observe that $g = (f + I_{\mathrm{cl}\, D})_{|Z}$, and since $\mathrm{cl}\, D \subset L$, we have the following relation in X:

$$\mathrm{dom}\, g = Z \cap \mathrm{dom}(f + I_{\mathrm{cl}\, D}) = (\mathrm{cl}\, D) \cap \mathrm{dom}\, f = D. \quad (6.3)$$

Consequently,

$$\partial f_{|Z}(\theta) \subset \partial\left((f + I_{\mathrm{cl}\, D})_{|Z}\right)(\theta) = \partial g(\theta). \quad (6.4)$$

To prove the second inclusion in (6.2), we apply formula (5.60) (or, more precisely, formula (5.92)) to the family of lsc convex functions $\{g_t,\ t \in T\}$. Indeed, thanks to (5.92), we get

6.1. THE COMPACT-CONTINUOUS SETTING

$$\partial g(\theta) = \bigcap_{\varepsilon > 0} \overline{\mathrm{co}} \left(\bigcup_{t \in T_\varepsilon(\theta),\ z \in B_t(\theta,\varepsilon)} \partial g_t(z) \cap S_\varepsilon(z) \cap \partial_\varepsilon g(\theta) \right). \quad (6.5)$$

∎

The following lemma gives more information on the convex combinations involved in (6.2).

Lemma 6.1.2 *With the same assumptions and notations as Lemma 6.1.1, for every $z^* \in \partial f_{|Z}(\theta)$ there exist $\rho > 0$ and $d \geq 1$ such that, for each $\varepsilon \in\,]0, 1]$, there are $\lambda \in \Delta_{d+1}$, $u^* \in B_{Z^*}$ and some*

$$t_i \in T_\varepsilon(\theta),\ z_i \in B_{t_i}(\theta,\varepsilon) \cap \mathrm{cl}\, D,\ \text{and}\ z_i^* \in A(t_i, z_i, \varepsilon),\ 1 \leq i \leq d+1,$$

such that $\lambda_i z_i^* \in \rho B_{Z^*}$ and $z^* = \sum_{1 \leq i \leq d+1} \lambda_i z_i^* + \varepsilon u^*$.

Proof. First, we pick $z_0 \in \mathrm{ri}\, D$. Then $z_0 \in \mathrm{ri}\, D = \mathrm{ri}(\mathrm{dom}\, g)$, by Lemma 6.1.1, and the continuity of g at z_0 entails the existence of $m \geq 0$ and $\delta > 0$ such that $z_0 + \delta B_Z \subset D$ and

$$g(z_0 + z) \leq m \text{ for all } z \in \delta B_Z. \quad (6.6)$$

Now, we fix $\varepsilon \in\,]0,1]$. According to (6.2), we have that

$$z^* \in \partial f_{|Z}(\theta) \subset \partial g(\theta) \subset \overline{\mathrm{co}} \left\{ \bigcup_{t \in T_\varepsilon(\theta),\ z \in B_t(\theta,\varepsilon)} A(t, z, \varepsilon) \right\} \subset Z^*.$$

Hence, taking into account the Carathéodory theorem, and denoting $d := \dim Z$, we find $\lambda \in \Delta_{d+1}$, $u^* \in B_{Z^*}$, $t_i \in T_\varepsilon(\theta)$, $z_i \in B_{t_i}(\theta,\varepsilon)$, and $z_i^* \in A(t_i, z_i, \varepsilon)$, $1 \leq i \leq d+1$, such that

$$z^* = \sum_{1 \leq i \leq d+1} \lambda_i z_i^* + \varepsilon u^*. \quad (6.7)$$

Notice also that $z_i \in B_{t_i}(\theta,\varepsilon) \subset \mathrm{dom}\, g_{t_i} \subset \mathrm{cl}\, D$ for all $1 \leq i \leq d+1$. Next, for each fixed $1 \leq i \leq d+1$, the relation $z_i^* \in A(t_i, z_i, \varepsilon) \subset \partial_\varepsilon g(\theta)$ together with (6.6) implies that

$$\langle z_i^*, z_0 + z \rangle \leq g(z_0 + z) - g(\theta) + \varepsilon \leq m + 1 \text{ for all } z \in \delta B_Z;$$

that is, taking $z = \theta$ and multiplying by λ_i, we get $\langle \lambda_i z_i^*, z_0 \rangle \leq \lambda_i(m+1)$ and

$$\langle \lambda_i z_i^*, z \rangle \leq -\langle \lambda_i z_i^*, z_0 \rangle + m + 1 \text{ for all } z \in \delta B_Z. \quad (6.8)$$

Hence, (6.7) yields

$$\langle \lambda_i z_i^*, z_0 \rangle = \langle z^*, z_0 \rangle - \sum_{1 \le j \le d+1, j \ne i} \langle \lambda_j z_j^*, z_0 \rangle - \varepsilon \langle u^*, z_0 \rangle$$
$$\ge \langle z^*, z_0 \rangle - m - 1 - \|z_0\|_Z;$$

that is, $|\langle \lambda_i z_i^*, z_0 \rangle| \le M$, where $M \ge \max\{m+1, m+1+\|z_0\|_Z - \langle z^*, z_0 \rangle\}$. Therefore, (6.8) implies that $\langle \lambda_i z_i^*, z \rangle \le M + m + 1$ for all $z \in \delta B_Z$, showing that $\lambda_i z_i^* \in (M+m+1)\delta^{-1} B_{Z^*}$, and the conclusion holds by taking $\rho := (M+m+1)\delta^{-1} > 0$. ∎

Lemma 6.1.3 *Given the convex functions f_t, $t \in T$, and $f := \sup_{t \in T} f_t$, we suppose that $f(\theta) = 0$, $\partial f(\theta) \ne \emptyset$, and $L \in \mathcal{F}(\theta)$. We denote $D := L \cap \operatorname{dom} f$ and $Z := \operatorname{span} D$, and introduce the lsc convex functions*

$$h_t := \operatorname{cl}(f_t + I_D), \ t \in T, \text{ and } h := \sup_{t \in T} h_t. \quad (6.9)$$

Then $h = \operatorname{cl}(f + I_D)$, $h(\theta) = 0$,

$$\operatorname{dom} h \subset \operatorname{cl}(D), \ \operatorname{cl}(\operatorname{dom} h) = \operatorname{cl}(D), \ Z = \operatorname{span}(L \cap \operatorname{dom} h), \quad (6.10)$$

and

$$\partial f(\theta) \subset \partial h(\theta). \quad (6.11)$$

Proof. Notice that, for all $t \in T$,

$$\operatorname{dom}(f_t + I_D) = D \cap \operatorname{dom} f_t = L \cap \operatorname{dom} f = \operatorname{dom}(f + I_D) = D,$$

so that

$$\operatorname{aff}(\operatorname{dom}(f_t + I_D)) = \operatorname{aff}(\operatorname{dom}(f + I_D)) = \operatorname{aff}(D) \text{ and } (f + I_D)_{|\operatorname{aff} D}$$

is finite and continuous on ri D, because D is finite-dimensional. Hence, since $f + I_D = \sup_{t \in T}(f_t + I_D)$, Proposition 5.2.4(iv) applies and we get

$$\operatorname{cl}(f + I_D) = \sup_{t \in T} \operatorname{cl}(f_t + I_D) = h,$$

which also implies that

$$\operatorname{dom} h = \operatorname{dom}(\operatorname{cl}(f + I_D)) \subset \operatorname{cl}(D \cap \operatorname{dom} f)) = \operatorname{cl}(D) \subset L$$

6.1. THE COMPACT-CONTINUOUS SETTING

and, using (2.35),

$$\mathrm{cl}(\mathrm{dom}\, h) = \mathrm{cl}(\mathrm{dom}(f + \mathrm{I}_D)) = \mathrm{cl}(D).$$

Therefore, $Z = \mathrm{aff}(\mathrm{dom}\, h) = \mathrm{aff}(L \cap \mathrm{dom}\, h)$,

$$\emptyset \neq \partial f(\theta) \subset \partial(f + \mathrm{I}_D)(\theta) = \partial(\mathrm{cl}(f + \mathrm{I}_D))(\theta) = \partial h(\theta),$$

and $0 = (f + \mathrm{I}_D)(\theta) = (\mathrm{cl}(f + \mathrm{I}_D))(\theta) = h(\theta)$. ∎

Theorem 6.1.4 *Given the convex functions f_t, $t \in T$, we denote $f := \sup_{t \in T} f_t$. Under hypothesis (6.1), for every $x \in \mathrm{dom}\, f$, we have*

$$\partial f(x) = \bigcap_{L \in \mathcal{F}(x)} \mathrm{co}\left\{\bigcup_{t \in T(x)} \partial(f_t + \mathrm{I}_{L \cap \mathrm{dom}\, f})(x)\right\}. \quad (6.12)$$

Remark 12 *(before the proof) Actually, we prove the following formula:*

$$\partial f(x) = \bigcap_{L \in \mathcal{F}(x)} \mathrm{co}\left\{\bigcup_{t \in T(x)} \partial(f_t + \mathrm{I}_{\mathrm{cl}(L \cap \mathrm{dom}\, f)})(x)\right\}, \quad (6.13)$$

which will be used in the proof of Theorem 6.1.5 and elsewhere. This formula easily leads us to the inclusion "$\subset$" in (6.12), as $\partial(f_t + \mathrm{I}_{\mathrm{cl}(L \cap \mathrm{dom}\, f)})(x) \subset \partial(f_t + \mathrm{I}_{L \cap \mathrm{dom}\, f})(x)$, for all $t \in T$ and $x \in L \cap \mathrm{dom}\, f$, while the inclusion "$\supset$" in (6.12) is straightforward (Exercise 92).

Proof. We only need to prove the inclusion "$\subset$" in (6.13) when $x = \theta$, $\partial f(\theta) \neq \emptyset$ and $f(\theta) = 0$, by using Lemmas 6.1.1, 6.1.2, 6.1.3, and the notations used there. Given a fixed $L \in \mathcal{F}(\theta)$, we consider the lsc convex functions h_t and h defined in (6.9), together with the sets $D = L \cap \mathrm{dom}\, f$ and $Z := \mathrm{span}\, D$; hence $\mathrm{cl}\, D = \mathrm{cl}(\mathrm{dom}\, h) = \mathrm{cl}(L \cap \mathrm{dom}\, h)$ due to (6.10).

Fix $x^* \in \partial f(\theta) \subset \partial h(\theta)$, so that $z^* := x^*_{|Z} \in \partial h_{|Z}(\theta)$. By applying Lemma 6.1.2 to the family $\{h_t,\, t \in T\}$, we find some $\rho > 0$ and $d \geq 1$ such that, for each $k \geq 1$, there are $\lambda_k \in \Delta_{d+1}$, $u^*_k \in B_{Z^*}$, $t_{i,k} \in T^h_{1/k}(\theta) := \{t \in T : h_t(\theta) \geq -1/k\}$,

$$z_{i,k} \in B^h_{t_{i,k}}(\theta, 1/k) := \{z \in Z : \|z\|_Z \leq 1/k, |h_{t_{i,k}}(z) - h_{t_{i,k}}(\theta)| \leq 1/k\},$$

and

$$z_{i,k}^* \in \partial(h_{t_{i,k}} + \mathrm{I}_{\mathrm{cl}\,D})_{|Z}(z_{i,k}) \cap S_{1/k}(z_{i,k}) \cap \partial_{1/k}(h + \mathrm{I}_{\mathrm{cl}\,D})_{|Z}(\theta)$$

with $(\lambda_{i,k} z_{i,k}^*)_k \subset \rho B_{Z^*}$, $1 \leq i \leq d+1$, and

$$z^* = \sum_{1 \leq i \leq d+1} \lambda_{i,k} z_{i,k}^* + (1/k) u_k^*. \tag{6.14}$$

By taking subnets if necessary, we may suppose that $t_{i,k} \to t_i \in T$ (T is compact by (6.1)) and $\lambda_{i,k} \to \lambda_i \in \Delta_{d+1}$, $1 \leq i \leq d+1$. At the same time, by the Weierstrass theorem, we also may suppose that $(\lambda_{i,k} z_{i,k}^*)_k$ converges to some $z_i^* \in Z^*$, $1 \leq i \leq d+1$. Consequently, since $z_{i,k}^* \in \partial(h_{t_{i,k}} + \mathrm{I}_{\mathrm{cl}\,D})_{|Z}(z_{i,k})$ and $z_{i,k} \in B_{t_{i,k}}^h(\theta, 1/k) \cap (\mathrm{cl}\,D)$, for all $z \in \mathrm{cl}\,D$ we obtain

$$\langle z_i^*, z \rangle = \lim_k \langle \lambda_{i,k} z_{i,k}^*, z \rangle = \lim_k (\lambda_{i,k} \langle z_{i,k}^*, z - z_{i,k} \rangle + \langle \lambda_{i,k} z_{i,k}^*, z_{i,k} \rangle)$$
$$\leq \liminf_k (\lambda_{i,k} (h_{t_{i,k}}(z) - h_{t_{i,k}}(z_{i,k})) + (1/k) \| \lambda_{i,k} z_{i,k}^* \|_{Z^*})$$
$$= \liminf_k \lambda_{i,k} (h_{t_{i,k}}(z) - h_{t_{i,k}}(z_{i,k})).$$

More precisely, because $z_{i,k} \in B_{t_{i,k}}^h(\theta, 1/k)$ and $t_{i,k} \in T_{1/k}^h(\theta)$, the last inequality yields for all $z \in \mathrm{cl}\,D$

$$\langle z_i^*, z \rangle \leq \liminf_k \lambda_{i,k} (h_{t_{i,k}}(z) - h_{t_{i,k}}(\theta) + 1/k)$$
$$\leq \liminf_k \lambda_{i,k} (h_{t_{i,k}}(z) + 2/k) = \liminf_k \lambda_{i,k} h_{t_{i,k}}(z).$$

In particular, using the upper semicontinuity assumption in (6.1), for every $z \in D$ ($\subset \mathrm{dom}\, f_{t_i}$) the last inequality gives rise to

$$\langle z_i^*, z \rangle \leq \liminf_k \lambda_{i,k} f_{t_{i,k}}(z) \leq \lambda_i f_{t_i}(z). \tag{6.15}$$

Therefore, if i is such that $\lambda_i = 0$ (if any), then $\langle y_i^*, z \rangle \leq 0$ for all $z \in D$, and so $z_i^* \in \tilde{N}_D(\theta)$ (the normal cone to D in Z). Otherwise, if $\lambda_i > 0$, then (6.15) yields $0 = \langle z_i^*, \theta \rangle \leq \lambda_i f_{t_i}(\theta) \leq \lambda_i f(\theta) = 0$ (that is, $t_i \in T(\theta)$) and $\langle \lambda_i^{-1} z_i^*, z \rangle \leq f_{t_i}(z) = f_{t_i}(z) - f_{t_i}(\theta)$ for all $z \in D$. Moreover, as $(\mathrm{ri}\,D) \cap (\mathrm{dom}\, f_{t_i}) = \mathrm{ri}\,D \neq \emptyset$, the last inequality also implies that $\langle \lambda_i^{-1} z_i^*, z \rangle \leq f_{t_i}(z) - f_{t_i}(\theta)$ for all $z \in \mathrm{cl}\,D$ (Exercise 14 applied in Z), and we get $z_i^* \in \lambda_i \partial(f_{t_i} + \mathrm{I}_{\mathrm{cl}\,D})_{|Z}(\theta)$. Thus, by taking limits in (6.14), and remembering that each one of $(\lambda_{i,k} z_{i,k}^*)_k$, $1 \leq i \leq d+1$, converges, we deduce

6.1. THE COMPACT-CONTINUOUS SETTING

$$x^*_{|Z} = z^* = \lim_k \left(\sum_{1\leq i\leq d+1,\, \lambda_i=0} \lambda_{i,k} z^*_{i,k} + \sum_{1\leq i\leq d+1,\, \lambda_i>0} \lambda_{i,k} z^*_{i,k} \right)$$

$$\in \tilde{N}_D(\theta) + \sum_{1\leq i\leq d+1,\, \lambda_i>0} \lambda_i \partial(f_{t_i} + I_{\operatorname{cl} D})_{|Z}(\theta).$$

Hence, because (see Exercise 55(iv))

$$\tilde{N}_D(\theta) = \{u^*_{|Z} : u^* \in N_D(\theta)\}, \ \partial(f_{t_i} + I_{\operatorname{cl} D})(\theta) = \{u^*_{|Z} : u^* \in \partial(f_{t_i} + I_{\operatorname{cl} D})(\theta)\},$$

and $\{1 \leq i \leq d+1 : \lambda_i > 0\} \neq \emptyset$, the relation above yields

$$x^* \in N_D(\theta) + \sum_{1\leq i\leq d+1,\, \lambda_i>0} \lambda_i \partial(f_{t_i} + I_{\operatorname{cl} D})(\theta) + Z^\perp$$

$$\subset \sum_{1\leq i\leq d+1,\, \lambda_i>0} \lambda_i \partial(f_{t_i} + I_{\operatorname{cl} D} + I_Z)(\theta) = \operatorname{co}\left\{ \bigcup_{t\in T(\theta)} \partial(f_t + I_{\operatorname{cl} D})(\theta) \right\}.$$

Finally, (6.13) follows by the arbitrariness of the L's in $\mathcal{F}(\theta)$. ∎

Formula (6.16) below gives another variant of (5.26) involving the approximate subdifferentials of the active data functions f_t, $t \in T(x)$.

Theorem 6.1.5 *Given the lsc convex functions f_t, $t \in T$, and $f := \sup_{t\in T} f_t$, we assume that hypothesis (6.1) fulfills. Then, for every $x \in \operatorname{dom} f$, we have*

$$\partial f(x) = \bigcap_{\varepsilon>0,\ L\in\mathcal{F}(x)} \overline{\operatorname{co}}\left\{ \bigcup_{t\in T(x)} \partial_\varepsilon f_t(x) + N_{L\cap \operatorname{dom} f}(x) \right\}. \tag{6.16}$$

Proof. As in the proof of Theorem 6.1.4, only the inclusion "$\subset$" in (6.16) needs to be verified, as the opposite one comes easily from Theorem 5.2.2. Again, we suppose that $\partial f(x) \neq \emptyset$, $x = \theta$, and $f(\theta) = (\operatorname{cl} f)(\theta) = 0$, consider a fixed finite-dimensional linear subspace $L \subset X$, and denote $D = L \cap \operatorname{dom} f$. Then, by (6.13), we get

$$\partial f(\theta) \subset \operatorname{co}\left\{ \bigcup_{t\in T(\theta)} \partial(f_t + I_{\operatorname{cl} D})(\theta) \right\}. \tag{6.17}$$

Thus, since each f_t for $t \in T(\theta)$ is proper and lsc, Proposition 4.1.16 entails

$$\partial f(\theta) \subset \bigcap_{\varepsilon>0} \mathrm{co}\left\{\bigcup_{t \in T(\theta)} \mathrm{cl}\left(\partial_\varepsilon f_t(\theta) + \mathrm{N}_{\mathrm{cl}\,D}^\varepsilon(\theta)\right)\right\}$$

$$\subset \bigcap_{\varepsilon>0} \mathrm{co}\left\{\mathrm{cl}\left(\bigcup_{t \in T(\theta)} \partial_\varepsilon f_t(\theta) + \mathrm{N}_{\mathrm{cl}\,D}^\varepsilon(\theta)\right)\right\}$$

$$\subset \bigcap_{\varepsilon>0} \mathrm{cl}(A_\varepsilon + \mathrm{N}_D^\varepsilon(\theta)),$$

where $A_\varepsilon := \mathrm{co}\left\{\bigcup_{t \in T(\theta)} \partial_\varepsilon f_t(\theta)\right\}$. But we have $\emptyset \ne \mathrm{ri}\, D \subset \mathrm{dom}\, \sigma_{A_\varepsilon}$ for all $\varepsilon > 0$, as the following inequality holds for all $x_0 \in \mathrm{ri}\, D$,

$$\sigma_{A_\varepsilon}(x_0) = \sup_{t \in T(\theta)} \sigma_{\partial_\varepsilon f_t(\theta)}(x_0) \le \sup_{t \in T(\theta)} \sigma_{\partial_\varepsilon f(\theta)}(x_0) \le f(x_0) + \varepsilon < +\infty,$$

and so Exercise 79 applies and yields the desired inclusion. ∎

Remark 13 *Due to the relation (see Corollary 5.2.3)*

$$\partial f(x) = \partial\left(\sup_{t \in T_\varepsilon(x)} (f_t + \mathrm{I}_{\mathrm{dom}\, f})\right)(x) \text{ for all } \varepsilon > 0,$$

it suffices to apply formulas (6.12) and (6.16) to the smaller family of functions $\{f_t + \mathrm{I}_{\mathrm{dom}\, f},\, t \in T_\varepsilon(x)\}$. In other words, when calculating $\partial f(x)$, the compactness of the index set used in hypothesis (6.1) can be relaxed by taking, instead of the whole set T, the subset $T_\varepsilon(x)$ for sufficiently small $\varepsilon > 0$.

As the following corollary shows, $\partial f(x)$ is based only on the active functions at the reference point, while the rest of the functions are beyond the construction of $\mathrm{dom}\, f$.

Corollary 6.1.6 *Given the convex functions $f_t : X \to \overline{\mathbb{R}},\, t \in T$, and $f := \sup_{t \in T} f_t$, suppose that hypothesis (6.1) fulfills. Then, for every $x \in \mathrm{dom}\, f$, we have that*

$$\partial f(x) = \partial\left(\sup_{t \in T(x)} f_t + \mathrm{I}_{\mathrm{dom}\, f}\right)(x).$$

Consequently, provided that $\mathrm{int}(\mathrm{dom}\, f) \ne \emptyset$ or $\sup_{t \in T(x)} f_t$ is continuous somewhere in $\mathrm{dom}\, f$, we get

$$\partial f(x) = \partial\left(\sup_{t \in T(x)} f_t\right)(x) + \mathrm{N}_{\mathrm{dom}\, f}(x).$$

6.1. THE COMPACT-CONTINUOUS SETTING

Proof. It suffices to prove the inclusion "$\subset$". By Theorem 6.1.4, we have

$$\partial f(x) = \bigcap_{L \in \mathcal{F}(x)} \overline{\text{co}} \left\{ \bigcup_{t \in T(x)} \partial(f_t + I_{L \cap \text{dom} f})(x) \right\}.$$

Observe that, for all $t \in T(x)$ and $L \in \mathcal{F}(x)$, one has $f_t + I_{L \cap \text{dom} f} \leq \sup_{s \in T(x)} f_s + I_{L \cap \text{dom} f}$, and both functions coincide at x, entailing $\partial(f_t + I_{L \cap \text{dom} f})(x) \subset \partial(\sup_{s \in T(x)} f_s + I_{L \cap \text{dom} f})(x)$. Consequently, using (4.16),

$$\partial f(x) \subset \bigcap_{L \in \mathcal{F}(x)} \partial \left(\sup_{t \in T(x)} f_t + I_{L \cap \text{dom} f} \right)(x) = \partial \left(\sup_{t \in T(x)} f_t + I_{\text{dom} f} \right)(x),$$

and the proof of the first statement is complete since the opposite inclusion is straightforward. The last statement comes by applying Proposition 4.1.20. ■

We can relax the upper semicontinuity assumption in Theorems 6.1.4 and 6.1.5, but maintaining the compactness of the index set T. To this aim, we introduce the usc regularizations of the functions f_t, $\tilde{f}_t : X \to \overline{\mathbb{R}}$, $t \in T$, defined by

$$\tilde{f}_t(z) := \limsup_{s \to t} f_s(z). \tag{6.18}$$

The following lemma gives some properties of the functions $\tilde{f}_t$, $t \in T$, which are exploited in the sequel.

Lemma 6.1.7 *Let $f_t : X \to \overline{\mathbb{R}}$, $t \in T$, be convex and let $\tilde{f}_t$ be defined as in (6.18). Then the following statements hold true:*
(i) The $\tilde{f}_t$'s are convex and satisfy $f = \sup_{t \in T} \tilde{f}_t$.
(ii) The family $\{\tilde{f}_t, t \in T\}$ satisfies condition (5.10) if the original family $\{f_t, t \in T\}$ does.
(iii) The mappings $s \in T \mapsto \tilde{f}_s(z)$, $z \in \text{dom} f$, are usc on T.

Proof. (i) Given $t \in T$, for every $x, y \in X$ and $\lambda \in]0, 1[$, we have

$$\tilde{f}_t(\lambda x + (1-\lambda)y) \leq \limsup_{s \to t} (\lambda f_s(x) + (1-\lambda)f_s(y))$$

$$\leq \limsup_{s \to t} \lambda f_s(x) + \limsup_{s \to t}(1-\lambda)f_s(y) = \lambda \tilde{f}_t(x) + (1-\lambda)\tilde{f}_t(y),$$

entailing the convexity of $\tilde{f}_t$. To verify that $f = \sup_{t \in T} \tilde{f}_t$, we observe that for every $z \in X$

$$\sup_{t\in T} \tilde{f}_t(z) = \sup_{t\in T}\left(\limsup_{s\to t} f_s(z)\right) \leq f(z).$$

The opposite inclusion follows as $\limsup_{s\to t} f_s(z) \geq f_t(z)$ for all $z \in X$.

(*ii*) Assume that the family $\{f_t,\ t \in T\}$ satisfies condition (5.10). We observe that $\operatorname{cl} f_t \leq f_t \leq \tilde{f}_t \leq f$, for all $t \in T$. So, $\operatorname{cl} f_t \leq \operatorname{cl} \tilde{f}_t \leq \operatorname{cl} f$ and, taking the supremum over $t \in T$ and using the equality $f = \sup_{t\in T} \tilde{f}_t$ coming from (*i*),

$$\operatorname{cl} f = \sup_{t\in T}(\operatorname{cl} f_t) \leq \sup_{t\in T}(\operatorname{cl} \tilde{f}_t) \leq \operatorname{cl} f = \operatorname{cl}(\sup_{t\in T}\tilde{f}_t);$$

that is, $\{\tilde{f}_t,\ t \in T\}$ satisfies condition (5.10) too.

(*iii*) Take any $t \in T$ and $z \in \operatorname{dom} f\ (= \operatorname{dom}(\sup_{t\in T}\tilde{f}_t))$, and consider a net $(t_i)_i \subset T$ such that $t_i \to t$ and $\limsup_{s\to t} \tilde{f}_s(z) = \lim_i \tilde{f}_{t_i}(z)$. Similarly, for each i, there exists a net $(t_{i,j})_j \subset T$ such that $t_{i,j} \to_j t_i$ and $\tilde{f}_{t_i}(z) = \lim_j f_{t_{i,j}}(z)$. Next, we can find a diagonal net $(t_{i,j_i})_i$ such that $t_{i,j_i} \to_i t$ and $\lim_i f_{t_{i,j_i}}(z) = \limsup_{s\to t} \tilde{f}_s(z)$; that is,

$$\limsup_{s\to t} \tilde{f}_s(z) = \lim_i f_{t_{i,j_i}}(z) \leq \limsup_{s\to t} f_t(z) = \tilde{f}_t(z),$$

showing the desired property. ∎

Corollary 6.1.8 *Given the convex functions $f_t : X \to \overline{\mathbb{R}},\ t \in T$, and $f := \sup_{t\in T} f_t$, assume that T is compact. Then, for every $x \in \operatorname{dom} f$, we have*

$$\partial f(x) = \bigcap_{L\in\mathcal{F}(x)} \operatorname{co}\left\{\bigcup_{t\in \tilde{T}(x)} \partial(\tilde{f}_t + \mathrm{I}_{L\cap\operatorname{dom} f})(x)\right\}, \qquad (6.19)$$

where $\tilde{f}_t$ is defined in (6.18) and $\tilde{T}(x) := \{t \in T : \tilde{f}_t(x) = f(x)\}$.

Proof. According to Lemma 6.1.7, the family $\{\tilde{f}_t,\ t \in T\}$ satisfies (6.1) and its supremum is f. Therefore, formula (6.19) follows by applying formula (6.12) to the family $\{\tilde{f}_t,\ t \in T\}$. ∎

Example 6.1.9 *Consider the set $T = [0,1]$ and define the affine functions $f_t : \mathbb{R} \to \mathbb{R},\ t \in T$, by*

$$f_t(x) := \frac{x}{t} - t,\ t \in\ ]0,1],\ \text{and}\ f_0(x) := x - 1.$$

Hence,

6.1. THE COMPACT-CONTINUOUS SETTING

$$f(x) := \sup_{t \in T} f_t(x) = \begin{cases} +\infty, & \text{if } x > 0, \\ -2\sqrt{-x}, & \text{if } -1 \leq x \leq 0, \\ x - 1, & \text{if } x < -1. \end{cases}$$

We have that $T(0) = \emptyset$, $\tilde{f}_t := \limsup_{s \to t} f_s = f_t$ for all $t \in \,]0,1]$, and $\tilde{f}_0(x) := \limsup_{s \to 0} f_s(x) = \max\{x - 1, \limsup_{t \downarrow 0} \frac{x}{t}\}$; that is,

$$\tilde{f}_0(x) = \begin{cases} +\infty, & \text{if } x > 0, \\ 0, & \text{if } x = 0, \\ x - 1, & \text{if } x < 0. \end{cases}$$

Hence, $\tilde{T}(0) = \{t \in [0,1] : \tilde{f}_t(0) = f(0) = 0\} = \{0\}$ and formula (6.19) entails

$$\partial f(0) = \partial(\tilde{f}_0 + \mathrm{I}_{]-\infty,0]})(0) = \partial \tilde{f}_0(0) = \emptyset.$$

Let us show how this conclusion can be achieved using formula (5.26). Observe that, for every $0 < \varepsilon < 1$,

$$T_\varepsilon(0) := \{t \in [0,1] : f_t(0) \geq -\varepsilon\} = \{t \in \,]0,1] : t \leq \varepsilon\} = \,]0,\varepsilon],$$

and $\partial_\varepsilon f_t(0) = \{1/t\}$ for all $t \in \,]0,\varepsilon]$. Then formula (5.26) (or better its variant given in (5.65)) yields

$$\partial f(0) = \bigcap_{\varepsilon > 0} \overline{\mathrm{co}} \left(\bigcup_{t \in T_\varepsilon(0)} \partial_\varepsilon f_t(0) + \mathrm{N}_{\mathrm{dom}\, f}(0) \right)$$

$$= \bigcap_{0 < \varepsilon < 1} \overline{\mathrm{co}} \left(\bigcup_{t \in \,]0,\varepsilon]} \{1/t\} + \mathbb{R}_+ \right) = \bigcap_{0 < \varepsilon < 1} [1/\varepsilon, +\infty[= \emptyset.$$

Remark 14 *The L's are naturally dropped out from Theorems 6.1.4 and 6.1.5 if the underlying space is finite-dimensional. For instance, Theorem 6.1.4 simplifies to*

$$\partial f(x) = \mathrm{co} \left\{ \bigcup_{t \in T(x)} \partial(f_t + \mathrm{I}_{\mathrm{dom}\, f})(x) \right\}. \tag{6.20}$$

The following corollary establishes a slight generalization of this last relation.

Corollary 6.1.10 *Given convex functions $f_t : X \to \overline{\mathbb{R}}$, $t \in T$, satisfying hypothesis (6.1), assume that $\mathrm{ri}(\mathrm{dom}\, f) \neq \emptyset$ and that the restriction of $f := \sup_{t \in T} f_t$ to $\mathrm{aff}(\mathrm{dom}\, f)$ is continuous on $\mathrm{ri}(\mathrm{dom}\, f)$. Then, for* *every $x \in X$, we have*

$$\partial f(x) = \overline{\mathrm{co}}\left\{\bigcup_{t \in T(x)} \partial(f_t + \mathrm{I}_{\mathrm{dom}\, f})(x)\right\}.$$

Proof. We may assume that $x = \theta$ and $\partial f(\theta) \neq \emptyset$. Only the inclusion "⊂" needs to be proved. The key idea of the proof consists of simplifying the formula provided in Theorem 6.1.4, by means of Corollary 4.1.27 applied in the subspace $Y := \mathrm{aff}(\mathrm{dom}\, f)$, which is closed because $\mathrm{ri}(\mathrm{dom}\, f) \neq \emptyset$.

Fix $x_0 \in \mathrm{ri}(\mathrm{dom}\, f)$, $L \in \mathcal{F}(\theta)$ and denote $L_0 := \mathrm{span}\{L, x_0\}$. Then, for every $t \in T(\theta)$, Corollary 4.1.27 gives rise to

$$\partial(f_t + \mathrm{I}_{L_0 \cap \mathrm{dom}\, f})(\theta) = \partial(f_t + \mathrm{I}_{\mathrm{dom}\, f} + \mathrm{I}_{L_0})(\theta) = \partial(f_t + \mathrm{I}_{\mathrm{dom}\, f})(\theta) + \partial(\mathrm{I}_{L_0} + \mathrm{I}_Y)(\theta),$$

and Proposition 4.1.16 (applied to the functions I_{L_0}, I_Y) yields

$$\partial(f_t + \mathrm{I}_{L_0 \cap \mathrm{dom}\, f})(\theta) = \partial(f_t + \mathrm{I}_{\mathrm{dom}\, f})(\theta) + \mathrm{cl}(L_0^\perp + Y^\perp).$$

Therefore, Theorem 6.1.4 implies that

$$\partial f(\theta) \subset \mathrm{co}\left\{\bigcup_{t \in T(\theta)} \partial(f_t + \mathrm{I}_{L_0 \cap \mathrm{dom}\, f})(x)\right\}$$

$$= \mathrm{co}\left\{\bigcup_{t \in T(\theta)} \partial(f_t + \mathrm{I}_{\mathrm{dom}\, f})(\theta) + \mathrm{cl}(L_0^\perp + Y^\perp)\right\}$$

$$\subset \overline{\mathrm{co}}\left\{\bigcup_{t \in T(\theta)} \partial(f_t + \mathrm{I}_{\mathrm{dom}\, f})(\theta) + L_0^\perp + Y^\perp\right\}.$$

Notice that $Y^\perp = \partial \mathrm{I}_Y(\theta)$, so that $\partial(f_t + \mathrm{I}_{\mathrm{dom}\, f})(\theta) + Y^\perp \subset \partial(f_t + \mathrm{I}_{\mathrm{dom}\, f} + \mathrm{I}_Y)(\theta) = \partial(f_t + \mathrm{I}_{\mathrm{dom}\, f})(\theta)$. Then, since $L_0^\perp \subset L^\perp$, the last inclusion gives

$$\partial f(\theta) \subset \overline{\mathrm{co}}\left\{\bigcup_{t \in T(\theta)} \partial(f_t + \mathrm{I}_{\mathrm{dom}\, f})(\theta) + L^\perp\right\},$$

and the desired conclusion follows by intersecting over the L's (Exercise 10(i)). ∎

Theorem 6.1.11 *Given the lsc convex functions $f_t : X \to \overline{\mathbb{R}}$, $t \in T$, and $f := \sup_{t \in T} f_t$, suppose that hypothesis (6.1) fulfills. Then, provided that $\mathrm{ri}(\mathrm{dom}\, f) \neq \emptyset$, for every $x \in \mathrm{dom}\, f$, we have that*

6.1. THE COMPACT-CONTINUOUS SETTING

$$\partial f(x) = \bigcap_{\varepsilon > 0} \overline{\mathrm{co}} \left\{ \bigcup_{t \in T(x)} \partial_\varepsilon f_t(x) + \mathrm{N}_{\mathrm{dom}\, f}(x) \right\}. \quad (6.21)$$

In particular, when $\mathrm{int}(\mathrm{dom}\, f) \neq \emptyset$, *we have*

$$\partial f(x) = \mathrm{N}_{\mathrm{dom}\, f}(x) + \bigcap_{\varepsilon > 0} \overline{\mathrm{co}} \left\{ \bigcup_{t \in T(x)} \partial_\varepsilon f_t(x) \right\}. \quad (6.22)$$

Proof. We only prove the inclusions "⊂" in (6.21) and (6.22), assuming again that $x = \theta$, $f(\theta) = 0$ and $\partial f(\theta) \neq \emptyset$. Indeed, the converse inclusions "⊃" follow easily from Theorem 5.2.2 and Corollary 5.3.2, respectively.

Fix $x_0 \in \mathrm{ri}(\mathrm{dom}\, f)$, $L \in \mathcal{F}(\theta)$, $\varepsilon > 0$, and consider $L_0 := \mathrm{span}\{L, x_0\}$ and $Y := \mathrm{aff}(\mathrm{dom}\, f)$, which is closed as $\mathrm{ri}(\mathrm{dom}\, f) \neq \emptyset$. Then, according to Corollary 4.1.27(ii), we have that $\mathrm{N}_{L_0 \cap \mathrm{dom}\, f}(\theta) = \mathrm{N}_{\mathrm{dom}\, f}(\theta) + \mathrm{cl}(L_0^\perp + Y^\perp)$. Hence, denoting $A_\varepsilon := \cup_{t \in T(\theta)} \partial_\varepsilon f_t(\theta)$, by (6.16), (2.9), and the identity $\mathrm{N}_{\mathrm{dom}\, f}(\theta) + Y^\perp = \mathrm{N}_{\mathrm{dom}\, f}(\theta)$, we obtain that

$$\partial f(\theta) \subset \overline{\mathrm{co}} \left\{ \bigcup_{t \in T(\theta)} \partial_\varepsilon f_t(\theta) + \mathrm{N}_{L_0 \cap \mathrm{dom}\, f}(\theta) \right\}$$
$$= \overline{\mathrm{co}} \left\{ A_\varepsilon + \mathrm{N}_{\mathrm{dom}\, f}(\theta) + \mathrm{cl}(L_0^\perp + Y^\perp) \right\} = \overline{\mathrm{co}} \left\{ A_\varepsilon + \mathrm{N}_{\mathrm{dom}\, f}(\theta) + L_0^\perp + Y^\perp \right\}$$
$$= \overline{\mathrm{co}} \left\{ A_\varepsilon + \mathrm{N}_{\mathrm{dom}\, f}(\theta) + L_0^\perp \right\},$$

and the first conclusion follows from Exercise 10, by intersecting first over the L's, and then over the positive ε's.

Now, we proceed with the proof of inclusion "⊂" in (6.22) when $\mathrm{int}(\mathrm{dom}\, f) \neq \emptyset$. By (6.21), we have

$$\emptyset \neq \partial f(\theta) = \bigcap_{\varepsilon > 0} \mathrm{cl}\left(\mathrm{co}(A_\varepsilon) + \mathrm{N}_{\mathrm{dom}\, f}(\theta)\right),$$

and so all the sets $\mathrm{co}(A_\varepsilon)$, $\varepsilon > 0$, are non-empty. Notice that $\mathrm{co}(A_\varepsilon) \subset \partial_\varepsilon f(\theta)$ and the family $\{\mathrm{co}(A_\varepsilon), \varepsilon > 0\}$ is non-decreasing. Therefore, Proposition 4.1.28 leads us to $\partial f(\theta) = \mathrm{N}_{\mathrm{dom}\, f}(\theta) + \cap_{\varepsilon > 0} \overline{\mathrm{co}}(A_\varepsilon)$. ∎

Next we derive simpler formulas for $\partial f(x)$, which involve only the exact subdifferential of active functions, adopting the assumption that f is continuous somewhere. This assumption is possibly stronger than the condition $\mathrm{int}(\mathrm{dom}\, f) \neq \emptyset$, namely out of Banach spaces. In Corollary 6.5.3, formula (6.23) is shown to be valid with $\overline{\mathrm{co}}$ replaced with co, provided that T is finite and each function f_t, except perhaps one of them, is continuous somewhere in $\mathrm{dom}\, f$.

Corollary 6.1.12 *Given convex functions f_t, $t \in T$, we suppose that (6.1) fulfills. If the function $f := \sup_{t \in T} f_t$ is continuous somewhere, then for every $x \in \mathrm{dom}\, f$, we have*

$$\partial f(x) = \mathrm{N}_{\mathrm{dom}\, f}(x) + \overline{\mathrm{co}} \left\{ \bigcup_{t \in T(x)} \partial f_t(x) \right\} \tag{6.23}$$

$$= \mathrm{N}_{\mathrm{dom}\, f}(x) + \mathrm{co} \left\{ \bigcup_{t \in T(x)} \partial f_t(x) \right\} \; (\textit{if } X = \mathbb{R}^n).$$

Proof. Only the inclusion "$\subset$" will be verified, and so we may assume that $x = \theta$, $f(\theta) = 0$ and $\partial f(\theta) \neq \emptyset$. Applying Proposition 4.1.20, Corollary 6.1.10 entails

$$\partial f(\theta) = \overline{\mathrm{co}} \left\{ \bigcup_{t \in T(\theta)} \partial (f_t + \mathrm{I}_{\mathrm{dom}\, f})(\theta) \right\} = \mathrm{cl}\,(A + \mathrm{N}_{\mathrm{dom}\, f}(\theta)),$$

where $A := \mathrm{co}\left\{ \bigcup_{t \in T(\theta)} \partial f_t(\theta) \right\}$. Notice that $\emptyset \neq A \subset \partial f(\theta)$, and applying Proposition 4.1.28 with the constant family $A_\varepsilon := A$, $\varepsilon > 0$, we get the first formula of the corollary.

Finally, by combining formula (6.20) and Proposition 4.1.20, we get

$$\partial f(\theta) = \mathrm{co} \left\{ \bigcup_{t \in T(\theta)} \partial (f_t + \mathrm{I}_{\mathrm{dom}\, f})(\theta) \right\} = \mathrm{co} \left\{ \bigcup_{t \in T(\theta)} \partial f_t(\theta) + \mathrm{N}_{\mathrm{dom}\, f}(\theta) \right\},$$

and the second formula follows. ∎

The following result is straightforward from Corollary 6.1.12.

Corollary 6.1.13 *Given convex functions $f_t : X \to \overline{\mathbb{R}}$, $t \in T$, we suppose that hypothesis (6.1) fulfills. If the function $f := \sup_{t \in T} f_t$ is continuous at $x \in X$, then we have*

$$\partial f(x) = \overline{\mathrm{co}} \left\{ \bigcup_{t \in T(x)} \partial f_t(x) \right\} \tag{6.24}$$

$$= \mathrm{co} \left\{ \bigcup_{t \in T(x)} \partial f_t(x) \right\} \; (\textit{when } X = \mathbb{R}^n). \tag{6.25}$$

Corollary 6.1.14 *Given convex functions $f_t : X \to \overline{\mathbb{R}}$, $t \in T$, and $f := \sup_{t \in T} f_t$, suppose that hypothesis (6.1) holds. Assume the f_t's are*

6.2. COMPACTIFICATION APPROACH

finite and continuous on an open set $U \subset X$. Then formulas (6.24) and (6.25) hold on U.

Proof. As in Corollary 6.1.12, we only prove the inclusion "⊂", assuming that $x = \theta \in U$, $f(\theta) = 0$ and $\partial f(\theta) \neq \emptyset$. Take $x^* \in \partial f(\theta)$. Given an $L \in \mathcal{F}(\theta)$, we introduce the convex functions $g_t : L \to \overline{\mathbb{R}}$, $t \in T$, defined for $z \in L$ by

$$g_t(z) := f_t(z) \text{ and } g(z) := \sup_{t \in T} g_t(z) = f(z).$$

Since the family $\{g_t, t \in T\}$ satisfies condition (6.1) and the g_t's are finite and continuous on the open set $U \cap L$, the function g is finite and continuous on $U \cap L$ (Exercise 96). Hence, formula (6.25) entails $z^* := x^*_{|L} \in \partial g(\theta) = \text{co}\left\{\cup_{t \in T(\theta)} \partial g_t(\theta)\right\}$, where $T(\theta) = \{t \in T : f_t(\theta) = 0\}$. Consequently, taking into account Proposition 4.1.20, we obtain that (Exercise 55(i))

$$x^* \in \text{co}\left\{\bigcup_{t \in T(\theta)} \partial (f_t + I_L)(\theta)\right\} = \text{co}\left\{\bigcup_{t \in T(\theta)} \partial f_t(\theta) + L^\perp\right\},$$

and the inclusion "⊂" in (6.25) follows (when $X = \mathbb{R}^n$). The inclusion "⊂" in formula (6.24) comes from the relation above by intersecting over $L \in \mathcal{F}(\theta)$ (Exercise 10(i)). ∎

Remark 15 *When the space X is Banach, the continuity assumption in Corollary 6.1.14 implies the continuity of the function f on U (Exercise 96), and Corollary 6.1.13 comes into play.*

6.2 Compactification approach

The analysis of the compact framework developed in section 6.1 is applied in this section to more general families, where the sets of indices and the associated index mappings are not necessarily compact and upper semicontinuous, respectively. To this aim, we propose a compactification procedure for the index set and an upper semicontinuous regularization of the index mappings. Doing so, we generate new indices and enlarge the family of the data functions, but maintaining the same supremum function. In a further step, and applying the results of the compact-continuous setting of section 6.1, we obtain new formulas for the subdifferential of the supremum which involve these new objects. In this way, we provide a unified analysis for the compact-continuous and the non-compact non-continuous settings.

We consider a family of extended real-valued convex functions $f_t : X \to \overline{\mathbb{R}}$, $t \in T$, and its supremum function $f := \sup_{t \in T} f_t$, defined on the locally convex space X, which are indexed by an arbitrary topological space T. In particular, if no topology is known on T, we shall consider the discrete topology, which is *completely regular* and Hausdorff.

The first step toward the compactification process is the identification of the index set T as a subset of the compact *product space*

$$\mathbb{S} := [0,1]^{\mathcal{C}(T,[0,1])} \equiv \{\gamma : \mathcal{C}(T,[0,1]) \to [0,1]\},$$

where $\mathcal{C}(T,[0,1])$ is the set of continuous functions from T to $[0,1]$. The product space $\mathbb{S}$ is endowed with the product topology, so that $\mathbb{S}$ is Hausdorff for being the product of infinite copies of the Hausdorff space $[0,1]$. Hence, a given net $(\gamma_i)_i \subset \mathbb{S}$ converges to $\gamma \in \mathbb{S}$, written $\gamma_i \to \gamma$, if and only if

$$\gamma_i(\varphi) \to \gamma(\varphi) \text{ for all } \varphi \in \mathcal{C}(T,[0,1]). \tag{6.26}$$

The identification of T as a subset of $\mathbb{S}$ is made possible thanks to the mapping $\mathfrak{w} : T \to \mathbb{S}$, defined as

$$\mathfrak{w}(t) \equiv \gamma_t, \tag{6.27}$$

where γ_t, $t \in T$, is the *evaluation function* defined as

$$\gamma_t(\varphi) := \varphi(t), \; \varphi \in \mathcal{C}(T,[0,1]). \tag{6.28}$$

By abuse of language and notation, the closure of $\mathfrak{w}(T)$ with respect to the product topology, denoted by

$$\beta T := \mathrm{cl}(\mathfrak{w}(T)), \tag{6.29}$$

is called *Stone-Čech compactification* (or just *Stone-Čech compact extension*) of T. Recall that the formal definition of the Stone-Čech compactification additionally requires that T and $\mathfrak{w}(T)$ be homeomorphic; indeed, this fact occurs if and only if T is completely regular and Hausdorff (i.e., Tychonoff). In our analysis, we don't need this additional condition since our approach exclusively relies on the properties

6.2. COMPACTIFICATION APPROACH

of the associated regularized family, in particular, on the fact that the new family maintains the same supremum (see Proposition 6.2.3).

The proof of the following lemma, gathering standard properties of the compact extension βT, is given in Exercise 95.

Lemma 6.2.1 *The following assertions hold true:*

(i) The set βT is a Hausdorff compact subset of $\mathbb{S}$.

(ii) The mapping $\mathfrak{w}$ is continuous. So, if T is compact, then $\beta T = \mathfrak{w}(T)$.

(iii) If T is Tychonoff, then the mapping $\mathfrak{w}$ is a homeomorphism between T and $\mathfrak{w}(T)$; that is,

$$\gamma_{t_i} \to \gamma_t \text{ if and only if } t_i \to t \text{ in } T,$$

for every $t \in T$ and net $(t_i)_i \subset T$.

Next, we enlarge the original family $\{f_t, t \in T\}$ for the aim of agreeing with the upper semicontinuity condition of the index mappings in the compact framework (6.1).

Definition 6.2.2 *Given $\gamma \in \beta T$, we define the function $f_\gamma : X \to \overline{\mathbb{R}}$ as*

$$f_\gamma(x) := \limsup_{\gamma_t \to \gamma,\, t \in T} f_t(x). \tag{6.30}$$

Remark 16 *The family $\{f_\gamma, \gamma \in \beta T\}$ includes all the elements of the form $f_{\gamma_t}, t \in T$, defined as*

$$f_{\gamma_t}(x) = \limsup_{\gamma_s \to \gamma_t,\, s \in T} f_s(x);$$

hence, provided that T is Tychonoff, Lemma 6.2.1(iii) yields

$$f_{\gamma_t}(x) = \limsup_{\gamma_s \to \gamma_t,\, s \in T} f_s(x) = \limsup_{s \to t,\, s \in T} f_s(x),$$

which is the usc regularization defined in (6.18). Let us point out that these functions may not belong to the original family $\{f_t, t \in T\}$ as Example 6.1.9 shows.

Let us emphasize that the changes we made to the original functions do not modify the value of the supremum function f. Other properties of these new functions f_γ are given in the following proposition.

Proposition 6.2.3 *The following statements hold true:*
(i) The functions f_γ, $\gamma \in \beta T$, are convex and satisfy

$$f = \sup\nolimits_{t \in T} f_t = \max\nolimits_{\gamma \in \beta T} f_\gamma.$$

(ii) For all $t \in T$,

$$f_{\gamma_t} \geq f_t. \tag{6.31}$$

(iii) The mappings $\gamma \mapsto f_\gamma(z)$, $z \in \mathrm{dom}\, f$, are usc.

Proof. (i) The convexity of the f_γ's follows easily from the convexity of the f_t's (see (2.51)). Next, for each $\gamma \in \beta T$ and $x \in X$, we have $f_\gamma(x) = \limsup_{\gamma_s \to \gamma} f_s(x) \leq f(x)$, entailing that $\sup_{\gamma \in \beta T} f_\gamma \leq f$. In addition, given $x \in X$, if the sequence $(t_n)_n \subset T$ is such that $f(x) = \lim_n f_{t_n}(x)$, then the compactness of βT gives rise to the existence of some subnet $(t_i)_i$ of $(t_n)_n$ and $\gamma_0 \in \beta T$ such that $\gamma_{t_i} \to \gamma_0$. Hence, we obtain

$$f(x) \geq f_{\gamma_0}(x) \geq \limsup\nolimits_i f_{t_i}(x) = \lim\nolimits_n f_{t_n}(x) = f(x),$$

and $f_{\gamma_0}(x) = f(x)$.

(ii) Since the mapping $\mathfrak{w}$ is continuous by Lemma 6.2.1(ii), for all $t \in T$ and $x \in X$, we have

$$f_{\gamma_t}(x) = \limsup_{\gamma_s \to \gamma_t} f_s(x) \geq \limsup_{s \to t} f_s(x) \geq f_t(x).$$

(iii) We prove that $\limsup_{\gamma \to \gamma_0} f_\gamma(z) \leq f_{\gamma_0}(z)$, for every fixed $\gamma_0 \in \beta T$. Let $(\gamma_i)_i \subset \beta T$ such that $\gamma_i \to \gamma_0$ and

$$\lim\nolimits_i f_{\gamma_i}(z) = \limsup\nolimits_{\gamma \to \gamma_0} f_\gamma(z); \tag{6.32}$$

hence, by (6.26), we have $\gamma_i(\varphi) \to \gamma_0(\varphi)$ for all $\varphi \in \mathcal{C}(T, [0,1])$. Next, for each i, there exists a net $(t_{i,j})_j \subset T$ such that $\gamma_i = \lim_j \gamma_{t_{i,j}}$ and $f_{\gamma_i}(z) = \lim_j f_{t_{i,j}}(z)$. So, applying a diagonal process to the scheme

$$(f_{t_{i,j}}(z), \gamma_{t_{i,j}}) \to_j (\gamma_i, f_{\gamma_i}(z)) \to_i (\gamma_0, \lim\nolimits_i f_{\gamma_i}(z)),$$

we find a net $(t_{i,j_i})_i \subset T$ such that $\gamma_{t,j_i} \to_i \gamma_0$ and $\lim_i f_{t,j_i}(z) = \lim_i f_{\gamma_i}(z)$. Hence, by (6.32) and the definition of f_{γ_0} in (6.30),

$$\limsup\nolimits_{\gamma \to \gamma_0} f_\gamma(z) = \lim\nolimits_i f_{\gamma_i}(z) = \lim\nolimits_i f_{t,j_i}(z) \leq \limsup\nolimits_{\gamma_t \to \gamma_0} f_t(z) = f_{\gamma_0}(z),$$

6.2. COMPACTIFICATION APPROACH

and we are done. ∎

Now, given $x \in f^{-1}(\mathbb{R})$ and $\varepsilon \geq 0$, we introduce the *extended ε-active index set* of f at x defined by

$$\widehat{T}_\varepsilon(x) := \{\gamma \in \beta T : f_\gamma(x) \geq f(x) - \varepsilon\}, \tag{6.33}$$

and the *extended active index set* of f at x, which is

$$\widehat{T}(x) := \widehat{T}_0(x). \tag{6.34}$$

Obviously, $\widehat{T}(x) \subset \widehat{T}_\varepsilon(x)$ for all $\varepsilon \geq 0$. Moreover, the following proposition highlights the structure and properties of the extended active index set $\widehat{T}_\varepsilon(x)$ and its relationships with $T_\varepsilon(x)$.

Proposition 6.2.4 *Given $x \in f^{-1}(\mathbb{R})$ and $\varepsilon \geq 0$, the following statements hold:*

(i) $\widehat{T}_\varepsilon(x)$ *is non-empty and compact.*

(ii) $\mathfrak{w}(T(x)) \subset \widehat{T}(x)$.

(iii) $\widehat{T}(x) = \bigcap_{\varepsilon > 0} \mathrm{cl}\,(\mathfrak{w}(T_\varepsilon(x)))$.

(iv) $\widehat{T}(x) = \mathfrak{w}(T(x)) = \bigcap_{\varepsilon > 0} \mathfrak{w}(T_\varepsilon(x))$, *provided that (6.1) holds.*

Proof. (i) First, the non-emptiness of the set $\widehat{T}(x)$ comes from Proposition 6.2.3, entailing that $\widehat{T}_\varepsilon(x)$ is also non-empty, for all $\varepsilon > 0$. To prove the compactness of $\widehat{T}_\varepsilon(x)$, $\varepsilon \geq 0$, we only need to verify that it is closed in the Hausdorff compact space βT. Indeed, given a net $(\gamma_i)_i \subset \widehat{T}_\varepsilon(x)$ that converges to γ ($\in \beta T$) and, by the definition of each f_{γ_i}, there exists a net $(t_{i,j})_j \subset T$ such that $\gamma_{t_{i,j}} \to_j \gamma_i$ and $f(x) - \varepsilon \leq f_{\gamma_i}(x) = \lim_j f_{t_{i,j}}(x) \leq f(x)$.

We may assume that $f_{\gamma_i}(x) \to \alpha \in [f(x) - \varepsilon, f(x)]$. Thus, there exists a diagonal net $(\gamma_{t_{i,j_i}}, f_{t_{i,j_i}}(x))_i \subset (\beta T) \times \mathbb{R}$ such that $\gamma_{t_{i,j_i}} \to_i \gamma$ and $f_{t_{i,j_i}}(x) \to_i \alpha$, and we get $f_\gamma(x) \geq \limsup_i f_{t_{i,j_i}}(x) = \alpha \geq f(x) - \varepsilon$, implying that $\gamma \in \widehat{T}_\varepsilon(x)$. Hence, we are done.

(ii) The inclusion obviously holds if $T(x) = \emptyset$; otherwise, take $t \in T(x)$. So, by Proposition 6.2.3 and (6.31), $f(x) \geq f_{\gamma_t}(x) \geq f_t(x) = f(x)$ and we get $\mathfrak{w}(t) = \gamma_t \in \widehat{T}(x)$.

(iii) We take $\gamma \in \widehat{T}(x)$. Then there exists a net $(t_i)_i \subset T$ such that $\gamma_{t_i} \to \gamma$ and $f(x) = f_\gamma(x) = \lim_i f_{t_i}(x)$. Hence, for each $\varepsilon > 0$, we have $t_i \in T_\varepsilon(x)$ eventually, and so $\gamma_{t_i} \in \mathfrak{w}(T_\varepsilon(x))$, entailing that $\gamma \in \mathrm{cl}\,(\mathfrak{w}(T_\varepsilon(x)))$.

Conversely, take $\gamma \in \bigcap_{\varepsilon>0} \mathrm{cl}\,(\mathfrak{w}(T_\varepsilon(x)))$. Then, using a diagonal process, we can find nets $(t_i)_i$ and $\varepsilon_i \downarrow 0$ such that $\gamma = \lim_i \gamma_{t_i}$ and $t_i \in T_{\varepsilon_i}(x)$ for all i. Hence, $f(x) - \varepsilon_i \le f_{t_i}(x) \le f_{\gamma_{t_i}}(x)$ by (6.31), and we obtain $f(x) \le \limsup_i f_{\gamma_{t_i}}(x) \le f_\gamma(x) \le f(x)$; that is, $\gamma \in \widehat{T}(x)$ and we are done.

(iv) Under condition (6.1), the sets $T_\varepsilon(x)$, $\varepsilon > 0$, are compact by Exercise 94. Thus, using the continuity of the mapping $\mathfrak{w}$, from assertion (iii) and the fact that βT is Hausdorff compact, we obtain

$$\widehat{T}(x) = \bigcap_{\varepsilon>0} \mathrm{cl}\,(\mathfrak{w}(T_\varepsilon(x))) = \bigcap_{\varepsilon>0} \mathfrak{w}(T_\varepsilon(x)).$$

Take $\gamma \in \widehat{T}(x)$. Then, for every $k \ge 1$, the last relation yields some $t_k \in T_{1/k}(x)$ such that $\gamma = \gamma_{t_k}$. Since T is compact, we find a subnet $(t_{k_i})_i$ of $(t_k)_k$ and $t \in T$ such that $t_{k_i} \to t$. Hence, (6.1) leads us to $f_t(x) \ge \limsup_i f_{t_{k_i}}(x) \ge \limsup_i (f(x) - 1/k_i) = f(x)$, that is, $t \in T(x)$. Moreover, the continuity of the mapping $\mathfrak{w}$ implies that $\gamma = \gamma_{t_{k_i}} = \mathfrak{w}(t_{k_i}) \to \mathfrak{w}(t)$, and so $\gamma = \mathfrak{w}(t) \in \mathfrak{w}(T(x))$.

Conversely, if $t \in T(x)$, then $\mathfrak{w}(t) \in \mathfrak{w}(T(x)) \subset \mathfrak{w}(T_\varepsilon(x))$ for all $\varepsilon > 0$, and so the desired inclusion follows by (iii). ∎

We characterize in (6.35) below the subdifferential set $\partial f(x)$ in terms of (exact) subdifferentials involving the new functions f_γ, $\gamma \in \widehat{T}(x)$.

Theorem 6.2.5 *Given the convex functions $f_t : X \to \overline{\mathbb{R}}$, $t \in T$, and $f := \sup_{t \in T} f_t$, where T is a topological space, for every $x \in \mathrm{dom}\, f$, we have*

$$\partial f(x) = \bigcap_{L \in \mathcal{F}(x)} \mathrm{co}\left\{ \bigcup_{\gamma \in \widehat{T}(x)} \partial(f_\gamma + \mathrm{I}_{L \cap \mathrm{dom}\, f})(x) \right\}. \tag{6.35}$$

Proof. By Proposition 6.2.3(i), the functions f_γ, $\gamma \in \beta T$, are convex and their supremum is f. Therefore, since the set βT is compact in $\mathbb{S}$ by Lemma 6.2.1, and the mappings $\gamma \mapsto f_\gamma(z)$, $z \in \mathrm{dom}\, f$, are usc thanks to Proposition 6.2.3(iii), the desired formula comes from Theorem 6.1.4. ∎

The following result provides an explicit reformulation of Theorem 6.2.5, in which the elements coming from the set $[0,1]^{\mathcal{C}(T,[0,1])}$ are now replaced with nets in the index set T. The limits $\lim_i (f_{t_i} + \mathrm{I}_{L \cap \mathrm{dom}\, f})$ involved in (6.36) are defined locally around the reference point x with values in $\mathbb{R}_\infty$; in other words, $\partial(\lim_i (f_{t_i} + \mathrm{I}_{L \cap \mathrm{dom}\, f}))(x)$ refers to the

6.2. COMPACTIFICATION APPROACH

subdifferential of the function $\lim_i (f_{t_i} + I_{U \cap L \cap \operatorname{dom} f})$, where $U \subset X$ is some convex neighborhood of x in which the limit $\lim_i (f_{t_i} + I_{L \cap \operatorname{dom} f})$ exists in $\overline{\mathbb{R}}_\infty$.

Theorem 6.2.6 *Given the convex functions $f_t : X \to \overline{\mathbb{R}}$, $t \in T$, and $f := \sup_{t \in T} f_t$, for every $x \in \operatorname{dom} f$, we have*

$$\partial f(x) = \bigcap_{L \in \mathcal{F}(x)} \operatorname{co} \left\{ \bigcup_{(t_i)_i \in \mathcal{T}(x)} \partial(\lim_i (f_{t_i} + I_{L \cap \operatorname{dom} f}))(x) \right\}, \qquad (6.36)$$

where each limit $\lim_i (f_{t_i} + I_{L \cap \operatorname{dom} f})$ exists in $\overline{\mathbb{R}}_\infty$ in a convex neighborhood of x, and

$$\mathcal{T}(x) := \{(t_i)_i \subset T : \lim_i f_{t_i}(x) = f(x)\}. \qquad (6.37)$$

Proof. We may assume that $x = \theta$ and $f(\theta) = 0$. We introduce the family of convex functions $g_t : X \to \overline{\mathbb{R}}_\infty$, $t \in T$, defined as

$$g_t := \max\{f_t, -1\},$$

together with their associated supremum function $g := \sup_{t \in T} g_t$. Then we obtain $g = \max\{f, -1\}$, $g(\theta) = 0$, and $\partial f(\theta) = \partial g(\theta)$ (Exercise 93). Next we endow T with the discrete topology and apply (6.35) to get

$$\partial f(\theta) = \partial g(\theta) = \bigcap_{L \in \mathcal{F}(x)} \operatorname{co} \left\{ \bigcup_{\gamma \in \widehat{T}(\theta)} \partial(g_\gamma + I_{L \cap \operatorname{dom} f})(\theta) \right\}, \qquad (6.38)$$

where $g_\gamma := \limsup_{\gamma_t \to \gamma} g_t$ (see (6.30)) and

$$\widehat{T}(\theta) := \{\gamma \in \beta T : g_\gamma(\theta) = 0\} = \{\gamma \in \beta T : f_\gamma(\theta) = 0\}.$$

Let us introduce the real-valued functions φ_z, $z \in \operatorname{dom} f$, defined on T by

$$\varphi_z(t) := (\max\{f(z) + 1, 1\})^{-1}(g_t(z) + 1). \qquad (6.39)$$

Then $\varphi_z \in \mathcal{C}(T, [0, 1])$, since T has been endowed with the discrete topology, and whenever the convergence $\gamma_{t_i} \to \gamma$ occurs, we have $\varphi_z(t_i) \to_i \gamma(\varphi_z)$ $(\in [0, 1])$ for all $z \in \operatorname{dom} f$; hence, thanks to (6.39),

$$g_{t_i}(z) \to_i \gamma(\varphi_z) \max\{f(z) + 1, 1\} - 1 \in \mathbb{R}. \qquad (6.40)$$

In other words, for all $z \in \operatorname{dom} f$ and all net $(t_i)_i$ satisfying $\gamma_{t_i} \to_i \gamma$, we have

$$g_\gamma(z) = \limsup_{\gamma_t \to \gamma} g_t(z) = \lim_{\gamma_t \to \gamma} g_t(z) = \lim_i g_{t_i}(z) = \gamma(\varphi_z) \max\{f(z)+1, 1\} - 1 \in \mathbb{R}. \tag{6.41}$$

Now, given $L \in \mathcal{F}(\theta)$ and $\gamma \in \widehat{T}(\theta)$ such that $\partial(g_\gamma + I_{L \cap \operatorname{dom} f})(\theta) \neq \emptyset$, we choose a net $(t_i)_i \subset T$ such that $\gamma_{t_i} \to \gamma$ and

$$\lim_i g_{t_i}(\theta) = \lim_i f_{t_i}(\theta) = 0; \tag{6.42}$$

that is, $g_\gamma(\theta) = \lim_i g_{t_i}(\theta) = 0$. Hence, for any $z \in L \cap \operatorname{dom} f$, the equality $g_\gamma(z) = \lim_i g_{t_i}(z)$, coming from (6.41) and the definition of the g_t's, leads us to $g_\gamma(z) = \lim_i g_{t_i}(z) = \lim_i \max\{f_{t_i}(z), -1\}$. In particular, for a subnet of $(t_i)_i$ that realizes the upper limit of $(f_{t_i}(z))_i$; that is, $\lim_j f_{t_j}(z) = \limsup_i f_{t_i}(z)$, we get

$$g_\gamma(z) = \lim_j g_{t_j}(z) = \lim_j \max\left\{f_{t_{i_j}}(z), -1\right\}$$
$$= \max\left\{\lim_j f_{t_{i_j}}(z), -1\right\} = \max\left\{\limsup_i f_{t_i}(z), -1\right\}.$$

By applying the same argument to the lower limit $\liminf_i f_{t_i}(z)$, we show that

$$g_\gamma(z) = \max\{\liminf_i f_{t_i}(z), -1\};$$

that is, for all $z \in L \cap \operatorname{dom} f$ such that $g_\gamma(z) > -1$, we have

$$g_\gamma(z) = \liminf_i f_{t_i}(z) = \limsup_i f_{t_i}(z);$$

in other words, for all $z \in X$ such that $(g_\gamma + I_{L \cap \operatorname{dom} f})(z) > -1$ we have

$$(g_\gamma + I_{L \cap \operatorname{dom} f})(z) = \lim_i (f_{t_i} + I_{L \cap \operatorname{dom} f})(z). \tag{6.43}$$

But we have assumed that the function $g_\gamma + I_{L \cap \operatorname{dom} f}$ is subdifferentiable at θ, and so it is proper, and lsc at θ. Thus, we find (convex) $U \in \mathcal{N}_X$ such that

$$(g_\gamma + I_{L \cap \operatorname{dom} f})(z) > (g_\gamma + I_{L \cap \operatorname{dom} f})(\theta) - 1 = -1 \text{ for all } z \in U,$$

and, *a fortiori*, (6.43) yields

6.2. COMPACTIFICATION APPROACH

$$(g_\gamma + I_{L \cap \text{dom} f})(z) = \lim_i (f_{t_i} + I_{U \cap L \cap \text{dom} f})(z) \in \mathbb{R}_\infty \text{ for all } z \in X,$$

entailing that

$$\partial(g_\gamma + I_{L \cap \text{dom} f})(\theta) = \partial \left(\lim_i (f_{t_i} + I_{U \cap L \cap \text{dom} f}) \right)(\theta) = \partial \left(\lim_i (f_{t_i} + I_{L \cap \text{dom} f}) \right)(\theta),$$

because of the local coincidence of the involved convex functions. This yields the inclusion "$\subset$" in (6.36), due to (6.38). The proof is over because the opposite inclusion in (6.36) is straightforward. ∎

Example 6.2.7 *Assume that T is finite and $T = T(x)$. Therefore, T is compact for the discrete topology and Theorem 6.1.4 applies and gives*

$$\partial f(x) = \bigcap_{L \in \mathcal{F}(x)} \text{co} \left\{ \bigcup_{t \in T(x)} \partial(f_t + I_{L \cap \text{dom} f})(x) \right\}. \tag{6.44}$$

Now we see how to recover this formula starting from (6.36), asserting that

$$\partial f(x) = \bigcap_{L \in \mathcal{F}(x)} \text{co} \left\{ \bigcup_{(t_i)_i \in \mathcal{T}(x)} \partial(\lim_i (f_{t_i} + I_{L \cap \text{dom} f}))(x) \right\}. \tag{6.45}$$

Take $(t_i)_i \in \mathcal{T}(x)$, so that $\lim_i f_{t_i}(x) = f(x)$ and $\lim_i(f_{t_i} + I_{L \cap \text{dom} f}) \in \mathbb{R}_\infty$ locally around x. Then there exists some $S \subset T(x)$ such that each element of S is visited infinitely many times by the net $(t_i)_i$. Pick $t_0 \in S$ and consider the constant subnet $t_{i_j} = t_0$ of $(t_i)_i$. Hence, for all z close enough to x,

$$\lim_i (f_{t_i} + I_{L \cap \text{dom} f})(z) = \lim_j (f_{t_{i_j}} + I_{L \cap \text{dom} f})(z) = (f_{t_0} + I_{L \cap \text{dom} f})(z),$$

so that

$$\partial(\lim_i (f_{t_i} + I_{L \cap \text{dom} f}))(x) = \partial(f_{t_0} + I_{L \cap \text{dom} f})(x)$$
$$\subset \bigcup_{t \in S} \partial(f_t + I_{L \cap \text{dom} f})(x) \subset \bigcup_{t \in T(x)} \partial(f_t + I_{L \cap \text{dom} f})(x).$$

Hence, (6.45) yields the inclusion "$\subset$" in (6.44), whereas the opposite inclusion is easily verified.

We show below that the results of Theorem 6.2.6, based on the Stone-Čech compactification, also gives rise to formulas for $\partial f(x)$ involving the *one-point compact extension* of the index set T. Given a topology $\mathfrak{T}$ on T, we choose an element $\omega \notin T$ and consider the compact topological space $T \cup \{\omega\}$, where the topology is given by

$$\mathfrak{T}_\omega := \mathfrak{T} \cup \{\{\omega\} \cup (T \setminus C) : C \in \mathfrak{C}(T)\}, \qquad (6.46)$$

and

$$\mathfrak{C}(T) := \{C \subset T : C \text{ is compact and closed}\};$$

remember that T and $T \cup \{\omega\}$ are not required to be Hausdorff. Let us recall that the space $(T \cup \{\omega\}, \mathfrak{T}_\omega)$ is Hausdorff if and only if $(T, \mathfrak{T})$ is Hausdorff and locally compact.

In the following result, $\tilde{f}_\omega : X \to \overline{\mathbb{R}}$ denotes the convex function defined by

$$\tilde{f}_\omega := \limsup_{s \to \omega,\, s \in T} f_s; \qquad (6.47)$$

that is, denoting by $\mathcal{V}(\omega)$ the family of neighborhoods of ω,

$$\tilde{f}_\omega = \inf_{V \in \mathcal{V}(\omega)} \left(\sup_{s \in V \setminus \{\omega\}} f_s \right) = \inf_{C \in \mathfrak{C}(T)} \left(\sup_{s \in T \setminus C} f_s \right).$$

Observe that

$$\tilde{f}_\omega \leq f. \qquad (6.48)$$

Remember that $\tilde{f}_t : X \to \overline{\mathbb{R}}$, $t \in T$, denote the usc regularizations of the f_t's defined in (6.18); that is,

$$\tilde{f}_t := \limsup_{s \to t} f_s, \qquad (6.49)$$

and

$$\tilde{T}(x) := \{t \in T : \tilde{f}_t(x) = f(x)\}.$$

Theorem 6.2.8 *Assume that T is a topological space. Given the convex functions $f_t : X \to \overline{\mathbb{R}}$, $t \in T$, and $f := \sup_{t \in T} f_t$, for every $x \in \operatorname{dom} f$, we have*

$$\partial f(x) = \bigcap_{L \in \mathcal{F}(x)} \operatorname{co} \left\{ \bigcup_{s \in \tilde{T}^\omega(x)} \partial (\tilde{f}_s + \mathrm{I}_{L \cap \operatorname{dom} f})(x) \right\}, \qquad (6.50)$$

6.2. COMPACTIFICATION APPROACH

where
$$T^\omega(x) := \begin{cases} \tilde{T}(x), & \text{if } \tilde{f}_\omega(x) < f(x), \\ \tilde{T}(x) \cup \{\omega\}, & \text{if } \tilde{f}_\omega(x) = f(x). \end{cases}$$

Proof. Fix $L \in \mathcal{F}(x)$ and denote

$$\mathcal{T}_1(x) := \left\{ (t_i)_i \in \mathcal{T}(x) : \begin{array}{l} \lim_i (f_{t_i} + \mathrm{I}_{L \cap \mathrm{dom}\, f}) \in \mathbb{R}_\infty \text{ locally around } x, \\ \text{and } (t_i)_i \text{ has a cluster point} \end{array} \right\},$$

and

$$\mathcal{T}_2(x) := \{(t_i)_i \in \mathcal{T}(x) \setminus \mathcal{T}_1(x) : \lim_i (f_{t_i} + \mathrm{I}_{L \cap \mathrm{dom}\, f}) \in \mathbb{R}_\infty \text{ locally around } x\}.$$

Pick $(t_i)_i \in \mathcal{T}_1(x)$ and choose a convergent subnet $(t_{i_j})_j$ such that $t_{i_j} \to t_0 \in T$. Then $f(x) = \lim_i f_{t_i}(x) = \lim_j f_{t_{i_j}}(x) = \tilde{f}_{t_0}(x)$ and the following relation holds locally around the point x,

$$\lim_i (f_{t_i} + \mathrm{I}_{L \cap \mathrm{dom}\, f}) = \tilde{f}_{t_0} + \mathrm{I}_{L \cap \mathrm{dom}\, f}. \tag{6.51}$$

Then $t_0 \in \tilde{T}(x)$ and

$$\partial \left(\lim_i (f_{t_i} + \mathrm{I}_{L \cap \mathrm{dom}\, f}) \right)(x) = \partial \left(\tilde{f}_{t_0} + \mathrm{I}_{L \cap \mathrm{dom}\, f} \right)(x) \subset \bigcup_{t \in \tilde{T}(x)} \partial(\tilde{f}_t + \mathrm{I}_{L \cap \mathrm{dom}\, f})(x). \tag{6.52}$$

Now take $(t_i)_i \in \mathcal{T}_2(x)$. Then, for any $C \in \mathfrak{C}(T)$, we have $t_i \in T \setminus C$ eventually, and the inequality $\lim_i (f_{t_i} + \mathrm{I}_{L \cap \mathrm{dom}\, f}) \leq \sup_{t \in T \setminus C} f_t + \mathrm{I}_{L \cap \mathrm{dom}\, f}$ holds locally around x; that is, taking the infimum over the sets $C \in \mathfrak{C}(T)$,

$$\lim_i (f_{t_i} + \mathrm{I}_{L \cap \mathrm{dom}\, f}) \leq \tilde{f}_\omega + \mathrm{I}_{L \cap \mathrm{dom}\, f}. \tag{6.53}$$

Thus, taking into account (6.48), since

$$f(x) = \lim_i (f_{t_i} + \mathrm{I}_{L \cap \mathrm{dom}\, f})(x) \leq (\tilde{f}_\omega + \mathrm{I}_{L \cap \mathrm{dom}\, f})(x) = \tilde{f}_\omega(x), \tag{6.54}$$

relation (6.53) gives rise to

$$\partial \left(\lim_i (f_{t_i} + \mathrm{I}_{L \cap \mathrm{dom}\, f}) \right)(x) \subset \partial(\tilde{f}_\omega + \mathrm{I}_{L \cap \mathrm{dom}\, f})(x). \tag{6.55}$$

Consequently, when $\tilde{f}_\omega(x) < f(x)$, relations (6.54) and (6.55) imply that

$$\{(t_i)_i \in \mathcal{T}(x) : \lim_i (f_{t_i} + \mathrm{I}_{L \cap \mathrm{dom}\, f}) \in \mathbb{R}_\infty \text{ locally around } x\} = \mathcal{T}_1(x),$$

and combining (6.36) and (6.52), we infer that

$$\partial f(x) \subset \mathrm{co}\left\{\bigcup_{t \in \tilde{\mathcal{T}}(x)} \partial(\tilde{f}_t + \mathrm{I}_{L \cap \mathrm{dom}\, f})(x)\right\}.$$

Hence, by intersecting over the L's, the inclusion "$\subset$" in (6.50) follows, and we are done as the opposite inclusion is straightforward. In the other case, where $\tilde{f}_\omega(x) = f(x)$, (6.50) follows by combining (6.36), (6.52), and (6.55). ∎

The special case of $T = \mathbb{N}$ easily comes from Theorem 6.2.8. In this case, the function $\tilde{f}_\omega$ defined in (6.47) is

$$\tilde{f}_\infty := \limsup_{n \to +\infty} f_n.$$

Corollary 6.2.9 *Given the convex functions* $f_n : X \to \overline{\mathbb{R}}$, $n \geq 1$, *and* $f := \sup_{n \geq 1} f_t$, *for every* $x \in \mathrm{dom}\, f$, *we have*

$$\partial f(x) = \bigcap_{L \in \mathcal{F}(x)} \mathrm{co}\left\{\bigcup_{s \in T^\infty(x)} \partial(f_s + \mathrm{I}_{L \cap \mathrm{dom}\, f})(x)\right\},$$

where

$$T^\infty(x) := \begin{cases} T(x), & \text{if } \tilde{f}_\infty(x) < f(x), \\ T(x) \cup \{\infty\}, & \text{if } \tilde{f}_\infty(x) = f(x). \end{cases}$$

Proof. We are going to apply Theorem 6.2.8 to $\mathbb{N}$, endowed with the discrete topology, whose one-point compact extension is denoted $\mathbb{N} \cup \{\infty\}$. In the current case, we have $\tilde{f}_n := \limsup_{m \to n} f_m = f_n$ for all $n \geq 1$, and

$$\tilde{f}_\omega := \inf_{C \in \mathfrak{C}(\mathbb{N})}\left(\sup_{n \in \mathbb{N}\setminus C} f_n\right) \equiv \tilde{f}_\infty,$$

where $\mathfrak{C}(\mathbb{N}) = \{C \subset \mathbb{N} : C \text{ finite}\}$. Hence, the conclusion follows from Theorem 6.2.8. ∎

The following result gives rise to a slight extension of Corollary 6.1.8, stated in a kind of compact-like framework.

Corollary 6.2.10 *Given the convex functions* $f_t : X \to \overline{\mathbb{R}}$, $t \in T$, *the supremum* $f := \sup_{t \in T} f_t$, *and* $x \in \mathrm{dom}\, f$, *let us assume that, for each*

6.2. COMPACTIFICATION APPROACH

net $(t_i)_i \in \mathcal{T}(x)$ (see (6.37)), there exist a subnet $(t_{i_j})_j \subset T$ and $t \in T$ such that
$$\limsup_j f_{t_{i_j}}(z) \leq f_t(z) \text{ for all } z \in \text{dom } f.$$

Then we have

$$\partial f(x) = \bigcap_{L \in \mathcal{F}(x)} \text{co} \left\{ \bigcup_{t \in \mathcal{T}(x)} \partial(f_t + \text{I}_{L \cap \text{dom } f})(x) \right\}. \tag{6.56}$$

Proof. Fix $L \in \mathcal{F}(x)$ and take $(t_i)_i \in \mathcal{T}(x)$ such that the limit $\lim_i (f_{t_i} + \text{I}_{L \cap \text{dom } f})$ exists in $\mathbb{R}_\infty$, locally around x. By the current assumption, there exist a subnet $(t_{i_j})_j \subset T$ and $t \in T$ such that $\limsup_j f_{t_{i_j}} + \text{I}_{L \cap \text{dom } f} \leq f_t + \text{I}_{L \cap \text{dom } f}$; that is, locally around x,

$$\lim_i (f_{t_i} + \text{I}_{L \cap \text{dom } f}) = \lim_j \left(f_{t_{i_j}} + \text{I}_{L \cap \text{dom } f} \right) \leq f_t + \text{I}_{L \cap \text{dom } f}.$$

But these functions take the same value at x, because

$$f(x) = \lim_i (f_{t_i} + \text{I}_{L \cap \text{dom } f})(x) \leq (f_t + \text{I}_{L \cap \text{dom } f})(x) \leq f(x),$$

so that $t \in \mathcal{T}(x)$ and $\partial(\lim_i (f_{t_i} + \text{I}_{L \cap \text{dom } f}))(x) \subset \partial(f_t + \text{I}_{L \cap \text{dom } f})(x)$; hence the inclusion "$\subset$" in (6.56) comes from (6.36). The other inclusion is straightforward. ∎

Additional continuity assumptions allow simple characterizations of $\partial f(x)$. The following corollary simplifies the formula of Theorem 6.2.6. Actually, we can apply the same analysis to simplify the formula of Theorem 6.2.8.

Corollary 6.2.11 *Given convex functions $f_t : X \to \overline{\mathbb{R}}, t \in T$, and $f := \sup_{t \in T} f_t$, the following assertions hold true for every $x \in \text{dom } f$:*
(i) If $f_{|\text{aff}(\text{dom } f)}$ is continuous at some point in $\text{ri}(\text{dom } f)$, then

$$\partial f(x) = \overline{\text{co}} \left\{ \bigcup_{(t_i)_i \in \mathcal{T}(x)} \partial(\limsup_j f_{t_{i_j}} + \text{I}_{\text{dom } f})(x) \right\}, \tag{6.57}$$

where $(t_{i_j})_j$ denotes any particular subnet of the net $(t_i)_i$.
(ii) If f is continuous somewhere, then

$$\partial f(x) = \mathrm{N}_{\mathrm{dom}\, f}(x) + \overline{\mathrm{co}} \left\{ \bigcup_{(t_i)_i \in \mathcal{T}(x)} \partial (\limsup_j f_{t_{i_j}})(x) \right\} \tag{6.58}$$

$$= \mathrm{N}_{\mathrm{dom}\, f}(x) + \mathrm{co} \left\{ \bigcup_{(t_i)_i \in \mathcal{T}(x)} \partial (\limsup_j f_{t_{i_j}})(x) \right\} \ (if\ X = \mathbb{R}^n), \tag{6.59}$$

where $(t_{i_j})_j$ is as in (i).

Proof. We may assume that $x = \theta$, $f(\theta) = 0$ and $\partial f(\theta) \neq \emptyset$. Fix $L \in \mathcal{F}(\theta)$ such that $L \cap \mathrm{ri}(\mathrm{dom}\, f) \neq \emptyset$, and let $(t_i)_i \in \mathcal{T}(\theta)$ such that the limit function $\lim_i (f_{t_i} + \mathrm{I}_{L \cap \mathrm{dom}\, f})$ exists in $\mathbb{R}_\infty$ around θ. Then, for every subnet $(t_{i_j})_j$ of $(t_i)_i$, locally around θ, we have

$$\lim_i (f_{t_i} + \mathrm{I}_{L \cap \mathrm{dom}\, f}) = \lim_j (f_{t_{i_j}} + \mathrm{I}_{L \cap \mathrm{dom}\, f})$$

$$= \limsup_j (f_{t_{i_j}} + \mathrm{I}_{L \cap \mathrm{dom}\, f}) \leq \left(\limsup_j f_{t_{i_j}} \right) + \mathrm{I}_{L \cap \mathrm{dom}\, f}.$$

Thus, since these two convex functions take the same value at θ, we get

$$\partial (\lim_i f_{t_i} + \mathrm{I}_{L \cap \mathrm{dom}\, f})(\theta) \subset \partial (\limsup_j f_{t_{i_j}} + \mathrm{I}_{L \cap \mathrm{dom}\, f})(\theta). \tag{6.60}$$

Moreover, applying Corollary 4.1.27(i) to the convex functions $(f, \limsup_j f_{t_{i_j}}, \mathrm{I}_L)$, we obtain

$$\partial (\lim_i f_{t_i} + \mathrm{I}_{L \cap \mathrm{dom}\, f})(\theta) \subset \partial (\limsup_j f_{t_{i_j}} + \mathrm{I}_{\mathrm{dom}\, f})(\theta) + \partial \mathrm{I}_{L \cap (\mathrm{aff}\, \mathrm{dom}\, f)}(\theta).$$

Since $\partial \mathrm{I}_{L \cap (\mathrm{aff}\, \mathrm{dom}\, f)}(\theta) = \mathrm{cl}(L^\perp + (\mathrm{aff}\, \mathrm{dom}\, f)^\perp)$, due to Proposition 4.1.16, from (6.60) we get

$$\partial (\lim_i f_{t_i} + \mathrm{I}_{L \cap \mathrm{dom}\, f})(\theta) \subset \partial (\limsup_j f_{t_{i_j}} + \mathrm{I}_{\mathrm{dom}\, f})(\theta) + \mathrm{cl}(L^\perp + (\mathrm{aff}\, \mathrm{dom}\, f)^\perp)$$

$$\subset \mathrm{cl}(\partial (\limsup_j f_{t_{i_j}} + \mathrm{I}_{\mathrm{dom}\, f})(\theta) + L^\perp + \partial \mathrm{I}_{\mathrm{aff}\, \mathrm{dom}\, f}(\theta)),$$

and Proposition 4.1.6(iii) yields

$$\partial (\lim_i f_{t_i} + \mathrm{I}_{L \cap \mathrm{dom}\, f})(\theta) \subset \mathrm{cl}(\partial (\limsup_j f_{t_{i_j}} + \mathrm{I}_{\mathrm{dom}\, f} + \mathrm{I}_{\mathrm{aff}\, \mathrm{dom}\, f})(\theta) + L^\perp)$$

$$= \mathrm{cl}(\partial (\limsup_j f_{t_{i_j}} + \mathrm{I}_{\mathrm{dom}\, f})(\theta) + L^\perp).$$

Therefore, formula (6.36) leads us to

6.2. COMPACTIFICATION APPROACH

$$\partial f(\theta) \subset \mathrm{co}\left\{\bigcup_{(t_i)_i \in \mathcal{T}(\theta)} \partial(\lim_i f_{t_i} + \mathrm{I}_{L \cap \mathrm{dom}\, f})(\theta)\right\}$$

$$\subset \mathrm{co}\left\{\bigcup_{(t_i)_i \in \mathcal{T}(\theta)} \partial(\limsup_j f_{t_{i_j}} + \mathrm{I}_{\mathrm{dom}\, f})(\theta) + L^\perp\right\}, \quad (6.61)$$

and the inclusion "$\subset$" in (6.57) follows by intersecting over the L's (Exercise 10(i)). We are done with assertion (i) since the inclusion "$\supset$" in (6.57) is easily verified.

Now we establish formula (6.58). According to Proposition 4.1.20, formula (6.57) simplifies to

$$\partial f(\theta) = \overline{\mathrm{co}}\left\{\bigcup_{(t_i)_i \in \mathcal{T}(\theta)} \partial(\limsup_j f_{t_{i_j}} + \mathrm{I}_{\mathrm{dom}\, f})(\theta)\right\}$$

$$= \overline{\mathrm{co}}\left\{\bigcup_{(t_i)_i \in \mathcal{T}(\theta)} \partial(\limsup_j f_{t_{i_j}})(\theta) + \mathrm{N}_{\mathrm{dom}\, f}(\theta)\right\} = \mathrm{cl}\left(A + \mathrm{N}_{\mathrm{dom}\, f}(\theta)\right), \quad (6.62)$$

where A is the non-empty convex set (because we have assumed that $\partial f(\theta) \neq \emptyset$)

$$A := \mathrm{co}\left\{\bigcup_{(t_i)_i \in \mathcal{T}(\theta)} \partial(\limsup_j f_{t_{i_j}})(\theta)\right\}.$$

With the same arguments as in the proof of Theorem 6.1.11, we prove that $\mathrm{dom}\, f \subset \mathrm{dom}\, \sigma_A$. In fact, observe that, for all $z \in \mathrm{dom}\, f$,

$$\sigma_A(z) = \sup_{(t_i)_i \in \mathcal{T}(\theta)} \sigma_{\partial(\limsup_j f_{t_{i_j}})(\theta)}(z)$$

$$\leq \sup_{(t_i)_i \in \mathcal{T}(\theta)} ((\limsup_j f_{t_{i_j}})(z) - (\limsup_j f_{t_{i_j}})(\theta))$$

$$\leq f(z) - f(\theta) = f(z) < +\infty.$$

Therefore, Proposition 4.1.28 ensures that $\mathrm{cl}(A + \mathrm{N}_{\mathrm{dom}\, f}(\theta)) = (\mathrm{cl}\, A) + \mathrm{N}_{\mathrm{dom}\, f}(\theta)$, and (6.62) yields the desired formula,

$$\partial f(\theta) = \overline{\mathrm{co}}\left\{\bigcup_{(t_i)_i \in \mathcal{T}(\theta)} \partial(\limsup_j f_{t_{i_j}})\right\} + \mathrm{N}_{\mathrm{dom}\, f}(\theta).$$

Finally, to prove formula (6.59), we suppose $X = \mathbb{R}^n$. Then, taking into account Proposition 4.1.20, (6.61) with $L = \mathbb{R}^n$ yields

$$\partial f(\theta) \subset \mathrm{co}\left\{\bigcup_{(t_i)_i \in \mathcal{T}(\theta)} \partial(\limsup_j f_{t_{i_j}} + \mathrm{I}_{\mathrm{dom}\, f})(\theta)\right\}$$

$$= \mathrm{N}_{\mathrm{dom}\, f}(\theta) + \mathrm{co}\left\{\bigcup_{(t_i)_i \in \mathcal{T}(\theta)} \partial(\limsup_j f_{t_{i_j}})(\theta)\right\},$$

and we conclude the proof as the opposite inclusion easily comes from (6.58). ∎

A good choice of the subnet $(t_{i_j})_j$ in formulas (6.57) and (6.58) would lead to operative representations of $\partial f(x)$, whereas inappropriate choices make these formulas useless. This fact is illustrated in the following example.

Example 6.2.12 *Take $T := \{1, \ldots, m\}$ and $f := \max_{t \in T} f_t$. Let $x \in X$ such that $T(x) = T$ and f is continuous at x. We are going to show that if we choose as a subnet $(t_{i_j})_j$ the same net $(t_i)_i$ in formula (6.58); that is,*

$$\partial f(x) = \mathrm{co}\left\{\bigcup_{(t_i)_i \in \mathcal{T}(x)} \partial(\limsup_i f_{t_i})(x)\right\}, \qquad (6.63)$$

may be useless. Notice that

$$\mathcal{T}(x) = \{(t_i)_i \subset T : f_{t_i}(x) = f(x),\ \text{eventually}\} = \{(t_i)_i \subset T(x),\ \text{eventually}\}; \quad (6.64)$$

in other words, associated with each net $(t_i)_i$, there exists a subset $S \subset T(x)$ such that each element of S is visited by $(t_i)_i$ infinitely many times. Therefore, $\limsup_i f_{t_i}(z) = \max_{t \in S} f_t(z)$ for all $z \in X$. Conversely, given any set $S \subset T(x)$, we can easily construct a net (even a sequence) such that all the elements of S are visited infinitely many times. Consequently, formula (6.63) yields

$$\partial f(x) = \mathrm{co}\left\{\bigcup_{S \subset T(x)} \partial(\max_{t \in S} f_t)(x)\right\}.$$

This representation is useless because the set in the right-hand side could contain the subdifferential of the same supremum function f; this is the case when $S = T(x) = T$. Alternatively, given our net $(t_i)_i$, if we choose a constant subnet $(t_{i_j})_j$ such that $t_{i_j} = t_0 \in T(x)$ for all j, then $\limsup_j f_{t_{i_j}} = f_{t_0}$, and formula (6.63) reads

6.2. COMPACTIFICATION APPROACH

$$\partial f(x) = \mathrm{co}\left\{\bigcup_{t \in T(x)} \partial f_t(x)\right\},$$

which is nothing else but formula (6.24).

We give the following example to illustrate the difference between Theorems 6.2.8 and 6.2.6 (or, more precisely, Corollary 6.2.11), which are based on the one-point and the Stone-Čech compactifications, respectively.

Example 6.2.13 Consider the family of convex functions $g_{2n+1}, h_{2n} : X \to \overline{\mathbb{R}}, n \in \mathbb{N}$, defined on $\mathbb{R}$ as

$$g_{2n+1}(x) := \max\left\{\frac{nx}{n+1}, 0\right\}, \quad h_{2n}(x) := \max\left\{\frac{-nx}{n+1}, 0\right\}.$$

We introduce the family $\{f_n, n \in \mathbb{N}\}$ such that $f_{2n+1} := g_{2n+1}$ and $f_{2n} := h_{2n}$, together with the supremum function $f := \sup_{n \in \mathbb{N}} f_n = \sup_{n \in \mathbb{N}} \{g_{2n+1}, h_{2n}\}$. Obviously, $f(x) = |x|$,

$$T(x) = \begin{cases} \mathbb{N}, & \text{if } x = 0, \\ \emptyset, & \text{if } x \neq 0, \end{cases} \quad \text{and } \partial f(x) = \begin{cases} [-1,1], & \text{if } x = 0, \\ \{1\}, & \text{if } x > 0, \\ \{-1\}, & \text{if } x < 0. \end{cases}$$

If we apply the formula established in Theorem 6.1.4, then we attain a false conclusion since, obviously, $\mathbb{N}$ is not compact: for every $x \in \mathbb{R}$

$$\partial f(x) = \begin{cases}]-1,1[, & \text{if } x = 0, \\ \emptyset, & \text{if } x \neq 0. \end{cases}$$

Next, we apply the formulas obtained in Corollaries 6.2.9 and 6.2.11.

(i) *One-point compact extension (Corollary 6.2.9)*: the function $\tilde{f}_\infty$ defined in (6.47) is given in the current case by $\tilde{f}_\infty(x) = \limsup_{n \to \infty} f_n(x) = |x| = f(x)$, and Corollary 6.2.9 is useless.

(ii) *Stone-Čech compact extension (finite-dimensional formula (6.59) in Corollary 6.2.11(ii))*: take $x \in \mathbb{R}$ and $(n_i)_i \in \mathcal{T}(x)$, so that $\lim_i f_{n_i}(x) = |x|$. Here $\mathcal{T}(x)$ is (see (6.64))

$$\mathcal{T}(x) = \{(n_i)_i \subset \mathbb{N} : f_{n_i}(x) = f(x), \text{ eventually}\} = \{(n_i)_i \subset T(x), \text{ eventually}\}.$$

If $x > 0$, then n_i must be odd eventually, so that

$$\limsup_i f_{n_i}(z) = \limsup_i g_{2n_i+1}(z) = \max\{z, 0\} =: g_{\tilde{\gamma}}(z).$$

Thus, taking for $(n_{i_j})_j$ the own net $(n_i)_i$ in Corollary 6.2.11(ii),

$$\partial f(x) = \overline{co}\left\{\bigcup_{(n_i)_i \in \mathcal{T}(x),\, n_i \text{ odd}} \partial(\limsup_i f_{n_i})(x)\right\} = \partial g_{\bar{\gamma}}(x) = \{1\}.$$

Similarly, if $x < 0$, then n_i must be even eventually, so that

$$\limsup_i f_{n_i}(z) = \limsup_i h_{2n_i}(z) = \max\{-z, 0\} =: h_{\bar{\gamma}}(z),$$

and Corollary 6.2.11(ii) again yields $\partial f(x) = \partial h_{\bar{\gamma}}(x) = \{-1\}$.

Assume now that $x = 0$. In this case, given $(n_i)_i \in \mathcal{T}(0)$, we choose a subnet $(n_{i_j})_j$ of $(n_i)_i$ that is composed uniquely of odd or even numbers. Since

$$\limsup_j f_{n_{i_j}} = \limsup_j g_{2n_{i_j}+1} \le \limsup_{n\to\infty} g_{2n+1} = g_{\bar{\gamma}}, \text{ if } n_{i_j} \text{ is odd},$$

$$\limsup_j f_{n_{i_j}} = \limsup_j h_{2n_{i_j}} \le \limsup_{n\to\infty} h_{2n} = h_{\bar{\gamma}}, \text{ if } n_{i_j} \text{ is even},$$

and all these functions are equal at 0, we derive that

$$\partial(\limsup_j f_{n_{i_j}})(0) \subset \partial g_{\bar{\gamma}}(0), \text{ if } n_{i_j} \text{ is odd for all } j,$$

$$\partial(\limsup_j f_{n_{i_j}})(0) \subset \partial h_{\bar{\gamma}}(0), \text{ if } n_{i_j} \text{ is even for all } j;$$

that is, $\partial(\limsup_j f_{n_{i_j}})(0) \subset \partial g_{\bar{\gamma}}(0) \cup \partial h_{\bar{\gamma}}(0)$. So, by formula (6.59),

$$\partial f(0) = co\left\{\bigcup_{n\ge 1}(\partial g_{2n+1}(0) \cup \partial h_{2n}(0))\bigcup \partial g_{\bar{\gamma}}(0) \bigcup \partial h_{\bar{\gamma}}(0)\right\}$$

$$= co\left\{\bigcup_{n\ge 1}\left([0, \frac{n}{n+1}] \cup [\frac{-n}{n+1}, 0]\right) \cup [0, 1] \cup [-1, 0]\right\}$$

$$= [0, 1[\cup]-1, 0] \cup [0, 1] \cup [-1, 0] = [-1, 1],$$

and we recover the whole set $\partial f(0)$.

6.3 Main subdifferential formula revisited

In this section, we again consider a family $f_t : X \to \overline{\mathbb{R}}$, $t \in T$, of convex functions defined on the locally convex space X, together with the supremum function $f := \sup_{t \in T} f_t$. Our goal is to derive new characterizations for ∂f, starting from the compactification processes developed in section 6.2. To do this, we will write the subdifferential of the regularizing functions f_γ, $\gamma \in \beta T$, introduced in the Stone-Čech compactification, in terms of the original data f_t, $t \in T$. Theorems 6.1.4 and 6.2.6, and their consequences, will allow us to refine the results of chapter 5. We will need the following technical proposition.

Proposition 6.3.1 *Assume that the f_t's are proper and lsc such that $\theta \in f^{-1}(0)$ and $f_{|\mathrm{span}(\mathrm{dom}\, f)}$ is continuous at some point $x_0 \in \mathrm{ri}(\mathrm{dom}\, f)$. We consider a net $(z_i^*)_i \subset X^*$ such that*

$$\lim_i \left(\inf_{t \in T} f_t^* \right)(z_i^*) = 0 \text{ and } \limsup_i \langle z_i^*, x_0 \rangle > -\infty. \tag{6.65}$$

Then there exist a subnet $(z_{i_j}^*)_j$ of $(z_i^*)_i$ and $z^* \in X^*$ such that $\langle z_{i_j}^* - z^*, z \rangle \to_j 0$, for all $z \in \mathrm{span}(\mathrm{dom}\, f)$, and

$$z^* \in \bigcap_{\varepsilon > 0} \mathrm{cl} \left(\bigcup_{t \in T_\varepsilon(\theta)} \partial_\varepsilon f_t(\theta) + (\mathrm{dom}\, f)^\perp \right). \tag{6.66}$$

Proof. We denote $E := \mathrm{span}(\mathrm{dom}\, f)$ with E^* standing for the dual of E; hence E is a closed subspace because $\mathrm{ri}(\mathrm{dom}\, f) \neq \emptyset$. We also denote $h := \inf_{t \in T} f_t^*$ so that, by (3.10) and Theorem 3.2.2,

$$h^* = \left(\inf_{t \in T} f_t^* \right)^* = \sup_{t \in T} f_t^{**} = \sup_{t \in T} f_t - f, \tag{6.67}$$

and (6.65) gives rise to

$$h^*(\theta) + h(z_i^*) = f(\theta) + h(z_i^*) = (\inf_{t \in T} f_t^*)(z_i^*) \to 0. \tag{6.68}$$

Hence, for every fixed $\varepsilon > 0$, eventually on i, we have

$$h^*(\theta) + h^{**}(z_i^*) \leq h^*(\theta) + h(z_i^*) = h(z_i^*) < \varepsilon, \tag{6.69}$$

and (6.67) implies, also eventually on i,

$$z_i^* \in \partial_\varepsilon h^*(\theta) = \partial_\varepsilon f(\theta). \tag{6.70}$$

Without loss of generality, we may assume that (6.70) holds for all i. Moreover, using (6.69), for each i there exists $t_i \in T$ such that $f_{t_i}^*(z_i^*) < \varepsilon$, and so

$$f_{t_i}(\theta) + f_{t_i}^*(z_i^*) \leq f(\theta) + f_{t_i}^*(z_i^*) = f_{t_i}^*(z_i^*) < \varepsilon.$$

This entails that $z_i^* \in \partial_\varepsilon f_{t_i}(\theta)$ and $-f_{t_i}(\theta) = \langle z_i^*, \theta \rangle - f_{t_i}(\theta) \leq f_{t_i}^*(z_i^*) < \varepsilon$; that is, $t_i \in T_\varepsilon(\theta)$ and $(z_i^*)_i \in \partial_\varepsilon f_{t_i}(\theta) \subset \cup_{t \in T_\varepsilon(\theta)} \partial_\varepsilon f_t(\theta)$. Now, thanks to the continuity assumption of f at $x_0 \in \operatorname{dom} f$, we choose $U \in \mathcal{N}_X$ and $r \geq 0$ such that

$$f(x_0 + y) \leq r \text{ for all } y \in U \cap E. \tag{6.71}$$

Also, due to (6.65), there exists some $M \geq 0$ such that $\inf_i \langle z_i^*, x_0 \rangle \geq -M$ for all i (without loss of generality). Therefore, taking into account (6.70), (6.71) yields the existence of some $m \geq r + M + \varepsilon$ such that, for all $y \in U \cap E$ and i,

$$\langle z_i^*, y \rangle \leq f(x_0 + y) + \varepsilon - \inf_i \langle z_i^*, x_0 \rangle \leq m, \tag{6.72}$$

showing that $(z_i^*)_i \subset (U \cap E)^\circ$. Consequently, using the w^*-compactness of $(U \cap E)^\circ$ and the fact that the dual E^* is isomorphic to the quotient space $X^*/E^\perp$, Exercise 106 yields the existence of a subnet $(z_{i_j}^*)_j$ and $z^* \in \operatorname{cl}(\cup_{t \in T_\varepsilon(\theta)} \partial_\varepsilon f_t(\theta) + E^\perp)$ such that $\langle z_{i_j}^* - z^*, u \rangle \to_j 0$ for all $u \in E$. The conclusion follows by intersecting over V and then over $\varepsilon > 0$. ∎

We proceed by giving the main result of this section. The difference between formula (6.73) below and main formula (5.26) in Theorem 5.2.2 can be particularly appreciated in section 8.3, where we obtained expressions of the optimal sets of relaxed-convex optimization problems. While formula (5.26) produces characterizations by means of approximate solutions of the original problem (Theorem 8.3.2), formula (6.73) below gives rise to characterizations in terms of exact rather than approximate solutions (Theorem 8.3.3). In Exercise 97, a non-convex version of Theorem 6.3.2 is given.

Theorem 6.3.2 *Let convex functions* $f_t : X \to \overline{\mathbb{R}}$, $t \in T$, *and* $f := \sup_{t \in T} f_t$ *satisfy the same assumption as in Theorem 5.2.2; that is,* $\operatorname{cl} f = \sup_{t \in T} (\operatorname{cl} f_t)$. *Then, for every* $x \in X$, *we have*

6.3. MAIN SUBDIFFERENTIAL FORMULA ...

$$\partial f(x) = \bigcap_{L \in \mathcal{F}(x)} \operatorname{co} \left\{ \bigcap_{\varepsilon > 0} \operatorname{cl} \left(\bigcup_{t \in T_\varepsilon(x)} \partial_\varepsilon f_t(x) + \mathrm{N}_{L \cap \operatorname{dom} f}(x) \right) \right\}. \quad (6.73)$$

Proof. The inclusion "⊃" follows easily from formula (5.26). To prove the inclusion "⊂" we assume, without loss of generality, that $x = \theta$, $f(\theta) = 0$ and $\partial f(\theta) \neq \emptyset$; hence, $\partial(\operatorname{cl} f)(\theta) = \partial f(\theta)$ and $f(\theta) = (\operatorname{cl} f)(\theta) = 0$. We give the proof only for the case $\{f_t, t \in T\} \subset \Gamma_0(X)$, the general case is treated in Exercise 105. We fix $L \in \mathcal{F}(\theta)$ and introduce the proper lsc convex functions $\tilde{f}_t : X \to \overline{\mathbb{R}}$, $t \in T$, defined by

$$\tilde{f}_t := f_t + \mathrm{I}_L, \ t \in T, \quad (6.74)$$

together with the function $h : X \to \overline{\mathbb{R}}$ given by

$$h := \inf_{t \in T} \tilde{f}_t^*.$$

Hence, using Theorem 3.2.2,

$$f + \mathrm{I}_L = \sup_{t \in T} \tilde{f}_t^{**} = \left(\inf_{t \in T} \tilde{f}_t^* \right)^* = h^* = \sup_{z^* \in X^*} h_{z^*}, \quad (6.75)$$

where $h_{z^*} := \langle z^*, \cdot \rangle - h(z^*)$. Then, applying Theorem 6.2.6 to the function $f + \mathrm{I}_L$ and the index set $T = X^*$, we get

$$\partial f(\theta) \subset \partial (f + \mathrm{I}_L)(\theta) \subset \operatorname{co} \left\{ \bigcup_{(z_i^*)_i \subset X^*, \, h_{z_i^*}(\theta) \to 0} \partial \left(\lim_i (h_{z_i^*} + \mathrm{I}_{L \cap \operatorname{dom} f}) \right)(\theta) \right\}, \quad (6.76)$$

where each limit $\lim_i (h_{z_i^*} + \mathrm{I}_{L \cap \operatorname{dom} f})$ exists in $\mathbb{R}_\infty$ locally around θ, say, in some (convex) neighborhood $U \in \mathcal{N}_X$. Let $(z_i^*)_i \subset X^*$ be any of the nets involved in (6.76). Hence, since $U \cap \operatorname{ri}(L \cap \operatorname{dom} f) \neq \emptyset$ by (2.15), for every $x_0 \in U \cap \operatorname{ri}(L \cap \operatorname{dom} f)$ we have

$$\limsup_i \langle z_i^*, x_0 \rangle = \lim_i \left(\langle z_i^*, \cdot \rangle + \mathrm{I}_{L \cap \operatorname{dom} f} \right)(x_0) > -\infty.$$

Moreover, because $h_{z_i^*}(\theta) \to 0$, (6.75) implies that $h(z_i^*) = \langle z_i^*, \theta \rangle - h_{z^*}(\theta) \to 0$, and Proposition 6.3.1 yields the existence of $z^* \in X^*$ such that

$$z^* \in \bigcap_{\varepsilon>0} \mathrm{cl}\left(\bigcup_{t \in T_\varepsilon(\theta)} \partial_\varepsilon (f_t + \mathrm{I}_L)(\theta) + (L \cap \mathrm{dom}\, f)^\perp\right), \qquad (6.77)$$

and a subnet $(z^*_{i_j})_j$ such that $\langle z^*_{i_j} - z^*, z\rangle \to_j 0$ for all $z \in \mathrm{span}(L \cap \mathrm{dom}\, f)$. Equivalently, for all $z \in X$,

$$\begin{aligned}\lim_i (h_{z^*_i} + \mathrm{I}_{U \cap L \cap \mathrm{dom}\, f})(z) &= \lim_j \left(\langle z^*_{i_j}, z\rangle - h(z^*_{i_j}) + \mathrm{I}_{U \cap L \cap \mathrm{dom}\, f}(z)\right) \\ &= \lim_j \langle z^*_{i_j}, z\rangle + \mathrm{I}_{U \cap L \cap \mathrm{dom}\, f}(z) = \langle z^*, z\rangle + \mathrm{I}_{U \cap L \cap \mathrm{dom}\, f}(z).\end{aligned} \qquad (6.78)$$

In other words, taking into account (6.77) and the fact that $(L \cap \mathrm{dom}\, f)^\perp + \mathrm{N}_{L \cap \mathrm{dom}\, f}(\theta) \subset \mathrm{N}_{L \cap \mathrm{dom}\, f}(\theta)$, we obtain

$$\partial\left(\lim_i (h_{z^*_i} + \mathrm{I}_{U \cap L \cap \mathrm{dom}\, f})\right)(\theta) \subset \bigcap_{\varepsilon>0} \mathrm{cl}\left(\bigcup_{t \in T_\varepsilon(\theta)} \partial_\varepsilon(f_t + \mathrm{I}_L)(\theta) + \mathrm{N}_{L \cap \mathrm{dom}\, f}(\theta)\right).$$

Finally, using Proposition 4.1.16, we have $\partial_\varepsilon (f_t + \mathrm{I}_L)(\theta) \subset \mathrm{cl}(\partial_\varepsilon f_t(\theta) + L^\perp)$, and the last inclusion gives rise to

$$\partial\left(\lim_i (h_{z^*_i} + \mathrm{I}_{U \cap L \cap \mathrm{dom}\, f})\right)(\theta) \in \bigcap_{\varepsilon>0} \mathrm{cl}\left(\bigcup_{t \in T_\varepsilon(\theta)} \partial_\varepsilon f_t(\theta) + \mathrm{N}_{L \cap \mathrm{dom}\, f}(\theta)\right).$$

The desired inclusion "$\subset$" comes from (6.76) by intersecting over $L \in \mathcal{F}(\theta)$. ∎

In the following corollary, we remove the closedness condition (5.10).

Corollary 6.3.3 *Given convex functions $f_t : X \to \overline{\mathbb{R}}$, $t \in T$, and $f := \sup_{t \in T} f_t$, for every $x \in X$ we have*

$$\partial f(x) = \bigcap_{L \in \mathcal{F}(x)} \mathrm{co}\left\{\bigcap_{\varepsilon>0} \mathrm{cl}\left(\bigcup_{t \in T_\varepsilon(x)} \partial_\varepsilon(f_t + \mathrm{I}_{L \cap \mathrm{dom}\, f})(x)\right)\right\}. \qquad (6.79)$$

Proof. Given any $x \in \mathrm{dom}\, f$ and $L \in \mathcal{F}(x)$, the family of convex functions $\{f_t + \mathrm{I}_{L \cap \mathrm{dom}\, f},\, t \in T\}$ satisfies condition (5.10) (Proposition 5.2.4(iv)). Therefore, since $\sup_{t \in T}(f_t + \mathrm{I}_{L \cap \mathrm{dom}\, f}) = f + \mathrm{I}_{L \cap \mathrm{dom}\, f}$, Theorem 6.3.2 entails

6.3. MAIN SUBDIFFERENTIAL FORMULA ... 263

$$\partial f(x) \subset \operatorname{co} \left\{ \bigcap_{\varepsilon > 0} \operatorname{cl} \left(\bigcup_{t \in T_\varepsilon(x)} \partial_\varepsilon (f_t + \mathrm{I}_{L \cap \operatorname{dom} f})(x) + \mathrm{N}_{L \cap \operatorname{dom} f}(x) \right) \right\}$$

$$\subset \operatorname{co} \left\{ \bigcap_{\varepsilon > 0} \operatorname{cl} \left(\bigcup_{t \in T_\varepsilon(x)} \partial_\varepsilon (f_t + \mathrm{I}_{L \cap \operatorname{dom} f})(x) \right) \right\}.$$

Then the inclusion "$\subset$" follows by intersecting over L. The proof is complete since the opposite inclusion is straightforward. ∎

It is clear that Theorem 6.3.2 easily implies Theorem 5.2.2, while the converse implication is more complicated. To better appreciate the difference between the scope of both results we give the following example. It shows how to retrieve Theorem 6.1.4 from formula (6.73) in the case of a finite number of functions. Notice that this analysis cannot be done, at least directly, from formula (5.26).

Example 6.3.4 *Assume that T is finite. Then formula (6.79) easily implies, for any given $x \in \operatorname{dom} f$, that*

$$\partial f(x) = \bigcap_{L \in \mathcal{F}(x)} \operatorname{co} \left\{ \bigcup_{t \in T(x)} \partial (f_t + \mathrm{I}_{L \cap \operatorname{dom} f})(x) \right\}, \qquad (6.80)$$

which is formula (6.12), whereas formula (5.26) only gives (by combining Corollary 5.2.5 and Exercise 107)

$$\partial f(x) = \bigcap_{L \in \mathcal{F}(x),\ \varepsilon > 0} \operatorname{co} \left\{ \bigcup_{t \in T(x)} \partial_\varepsilon (f_t + \mathrm{I}_{L \cap \operatorname{dom} f})(x) \right\}. \qquad (6.81)$$

In fact, to obtain (6.80), we observe that $T_\varepsilon(x) = T(x)$, for all small enough $\varepsilon > 0$, so formula (6.79) yields

$$\partial f(x) = \bigcap_{L \in \mathcal{F}(x)} \operatorname{co} \left(\bigcap_{\varepsilon > 0} \bigcup_{t \in T(x)} \partial_\varepsilon (f_t + \mathrm{I}_{L \cap \operatorname{dom} f})(x) \right). \qquad (6.82)$$

Take $x^* \in \bigcap_{\varepsilon > 0} \bigcup_{t \in T(x)} \partial_\varepsilon (f_t + \mathrm{I}_{L \cap \operatorname{dom} f})(x)$. Then, for each $n \geq 1$, we find $t_n \in T(x)$ such that $x^* \in \partial_{1/n}(f_{t_n} + \mathrm{I}_{L \cap \operatorname{dom} f})(x)$. Since $T(x)$ is finite, by taking a subsequence if needed, we may assume that $t_n \equiv t_0 \in T(x)$ for all $n \geq 1$, and (4.15) implies $x^* \in \partial (f_{t_0} + \mathrm{I}_{L \cap \operatorname{dom} f})(x)$. Consequently, (6.82) gives rise to

$$\partial f(x) \subset \bigcap_{L \in \mathcal{F}(x)} \mathrm{co}\left(\bigcup_{t \in T(x)} \partial(f_t + \mathrm{I}_{L \cap \mathrm{dom}\,f})(x)\right),$$

and (6.80) follows, since the converse inclusion is straightforward.

If $X = \mathbb{R}^n$ in Theorem 6.3.2, then formula (6.73) reads

$$\partial f(x) = \mathrm{co}\left\{\bigcap_{\varepsilon > 0} \mathrm{cl}\left(\bigcup_{t \in T_\varepsilon(x)} \partial_\varepsilon f_t(x) + \mathrm{N}_{\mathrm{dom}\,f}(x)\right)\right\}. \tag{6.83}$$

More generally, we have the following result.

Corollary 6.3.5 *Given convex functions $f_t : X \to \overline{\mathbb{R}}$, $t \in T$, and $f := \sup_{t \in T} f_t$, we assume that (5.10) holds. Provided that $\mathrm{ri}(\mathrm{dom}\,f) \neq \emptyset$ and $f_{|\mathrm{aff}(\mathrm{dom}\,f)}$ is continuous on $\mathrm{ri}(\mathrm{dom}\,f)$, for every $x \in X$, we have*

$$\partial f(x) = \overline{\mathrm{co}}\left\{\bigcap_{\varepsilon > 0} \mathrm{cl}\left(\bigcup_{t \in T_\varepsilon(x)} \partial_\varepsilon f_t(x) + \mathrm{N}_{\mathrm{dom}\,f}(x)\right)\right\}.$$

Proof. The proof is similar to that of Theorem 6.3.2, except that we use formula (6.57) instead of Theorem 6.2.6. For the sake of completeness, we give the proof of the non-trivial inclusion "⊂" when $\{f_t, t \in T\} \subset \Gamma_0(X)$, $x = \theta$ and $\partial(\mathrm{cl}\,f)(\theta) = \partial f(\theta) \neq \emptyset$, $f(\theta) = (\mathrm{cl}\,f)(\theta) = 0$. We consider the function $h : X \to \overline{\mathbb{R}}$ defined by $h := \inf_{t \in T} f_t^*$, so that

$$f = h^* = \sup_{z^* \in X^*} h_{z^*}(z),$$

with $h_{z^*} := \langle z^*, \cdot \rangle - h(z^*)$. Moreover, (6.57) yields

$$\emptyset \neq \partial f(\theta) = \overline{\mathrm{co}}\left\{\bigcup_{(z_i^*)_i \subset X^*,\, h(z_i^*) \to 0} \partial\left(\limsup_j h_{z_{i_j}^*} + \mathrm{I}_{\mathrm{dom}\,f}\right)(\theta)\right\}, \tag{6.84}$$

where $(z_{i_j}^*)_j$ denotes any particular subnet of $(z_i^*)_i$. Let nets $(z_i^*)_i$, $(z_{i_j}^*)_j \subset X^*$ be as in (6.84) such that $h(z_i^*) = \langle z_i^*, \theta \rangle - h_{z^*}(\theta) \to 0$ and $\partial(\limsup_j h_{z_{i_j}^*} + \mathrm{I}_{\mathrm{dom}\,f})(\theta) \neq \emptyset$. Observe that

$$\limsup_j h_{z_{i_j}^*} \leq \limsup_i h_{z_i^*}$$

and

6.3. MAIN SUBDIFFERENTIAL FORMULA ...

$$0 = \limsup_j h_{z^*_{i_j}}(\theta) \le \limsup_i h_{z^*_i}(\theta) \le h^*(\theta) = f(\theta) = 0,$$

so that $\emptyset \ne \partial(\limsup_j h_{z^*_{i_j}} + \mathrm{I}_{\mathrm{dom}\,f})(\theta) \subset \partial(\limsup_i h_{z^*_i} + \mathrm{I}_{\mathrm{dom}\,f})(\theta)$ and the convex function $\limsup_i (h_{z^*_i} + \mathrm{I}_{\mathrm{dom}\,f})$ is proper. Hence, $\limsup_i \langle z^*_i, x_0 \rangle = \limsup_i (h_{z^*_i}(x_0) + \mathrm{I}_{\mathrm{dom}\,f}(x_0)) > -\infty$ for all $x_0 \in \mathrm{dom}\,f$ ($\supset \mathrm{ri}(\mathrm{dom}\,f)$). Consequently, Proposition 6.3.1 applies and yields some $z^* \in X^*$ such that

$$z^* \in \bigcap_{\varepsilon>0} \mathrm{cl}\left(\bigcup_{t \in T_\varepsilon(\theta)} \partial_\varepsilon f_t(\theta) + (\mathrm{dom}\,f)^\perp \right), \tag{6.85}$$

together with the existence of a subnet $(z^*_{i_k})_k$ of $(z^*_i)_i$ such that $\langle z^*_{i_k} - z^*, z \rangle \to_k 0$ for all $z \in \mathrm{span}(\mathrm{dom}\,f)$; that is, for all $z \in X$

$$\limsup_k (h_{z^*_{i_k}} + \mathrm{I}_{\mathrm{dom}\,f})(z) = \lim_k \left(\langle z^*_{i_k}, z \rangle + \mathrm{I}_{\mathrm{dom}\,f}(z) \right) = \mathrm{I}_{\mathrm{dom}\,f}(z) + \langle z^*, z \rangle.$$

Therefore, by (6.85),

$$\partial\left(\limsup_k (h_{z^*_{i_k}} + \mathrm{I}_{\mathrm{dom}\,f}) \right)(\theta) \subset \mathrm{N}_{\mathrm{dom}\,f}(\theta) + \bigcap_{\varepsilon>0} \mathrm{cl}\left(\bigcup_{t \in T_\varepsilon(\theta)} \partial_\varepsilon f_t(\theta) + (\mathrm{dom}\,f)^\perp \right)$$
$$\subset \bigcap_{\varepsilon>0} \mathrm{cl}\left(\bigcup_{t \in T_\varepsilon(\theta)} \partial_\varepsilon f_t(\theta) + (\mathrm{dom}\,f)^\perp + \mathrm{N}_{\mathrm{dom}\,f}(\theta) \right),$$

and the relation

$$(\mathrm{dom}\,f)^\perp + \mathrm{N}_{\mathrm{dom}\,f}(\theta) = \partial \mathrm{I}_{\mathrm{span}\,f}(\theta) + \mathrm{N}_{\mathrm{dom}\,f}(\theta) \subset \partial(\mathrm{I}_{\mathrm{dom}\,f} + \mathrm{I}_{\mathrm{span}\,f})(\theta) = \mathrm{N}_{\mathrm{dom}\,f}(\theta)$$

leads us to

$$\partial\left(\limsup_k (h_{z^*_{i_k}} + \mathrm{I}_{\mathrm{dom}\,f}) \right)(\theta) \subset \bigcap_{\varepsilon>0} \mathrm{cl}\left(\bigcup_{t \in T_\varepsilon(\theta)} \partial_\varepsilon f_t(\theta) + \mathrm{N}_{\mathrm{dom}\,f}(\theta) \right).$$

Then the inclusion "$\subset$" follows from (6.84). ∎

Corollary 6.3.6 *Given convex functions $f_t : X \to \overline{\mathbb{R}}, t \in T$, we assume that $f := \sup_{t \in T} f_t$ is continuous somewhere. Then, for every $x \in X$,*

$$\partial f(x) = \mathrm{N}_{\mathrm{dom}\, f}(x) + \overline{\mathrm{co}} \left\{ \bigcap_{\varepsilon > 0} \mathrm{cl} \left(\bigcup_{t \in T_\varepsilon(x)} \partial_\varepsilon f_t(x) \right) \right\} \tag{6.86}$$

$$= \mathrm{N}_{\mathrm{dom}\, f}(x) + \mathrm{co} \left\{ \bigcap_{\varepsilon > 0} \mathrm{cl} \left(\bigcup_{t \in T_\varepsilon(x)} \partial_\varepsilon f_t(x) \right) \right\} \quad (if\ X = \mathbb{R}^n). \tag{6.87}$$

Proof. First, the current continuity assumption on f implies condition (5.10) (Proposition 5.2.4(i)). We proceed as in the proof of Corollary 6.3.5, using formula (6.58) and outlining only the proof of the inclusion "$\subset$" when $\{f_t,\ t \in T\} \subset \Gamma_0(X)$, $x = \theta$ and $\partial(\mathrm{cl}\, f)(\theta) = \partial f(\theta) \neq \emptyset$, $f(\theta) = (\mathrm{cl}\, f)(\theta) = 0$. Again we consider the function $h : X \to \overline{\mathbb{R}}$ defined by $h := \inf_{t \in T} f_t^*$, so that $f = h^* = \sup_{z^* \in X^*} h_{z^*}(z)$, where $h_{z^*} := \langle z^*, \cdot \rangle - h(z^*)$. So, (6.58) yields

$$\emptyset \neq \partial f(\theta) = \mathrm{N}_{\mathrm{dom}\, f}(\theta) + \overline{\mathrm{co}} \left\{ \bigcup_{(z_i^*)_i \subset X^*,\, h(z_i^*) \to 0} \partial \left(\limsup_j h_{z_{i_j}^*} \right)(\theta) \right\}, \tag{6.88}$$

where $(z_{i_j}^*)_j$ denotes any particular subnet of $(z_i^*)_i$. As in the proof of Corollary 6.3.5, we consider those nets $(z_i^*)_i \subset X^*$ involved in (6.88), which satisfy $h(z_i^*) \to 0$ and $\limsup_i \langle z_i^*, x_0 \rangle > -\infty$ for all $x_0 \in \mathrm{int}(\mathrm{dom}\, f)$. Consequently, since $(\mathrm{dom}\, f)^\perp = \{\theta\}$ due to the current assumption, Proposition 6.3.1 yields the existence of $z^* \in X^*$ such that

$$z^* \in \bigcap_{\varepsilon > 0} \mathrm{cl} \left(\bigcup_{t \in T_\varepsilon(\theta)} \partial_\varepsilon f_t(\theta) \right), \tag{6.89}$$

and a subnet $(z_{i_j}^*)_j$ such that $\langle z_{i_j}^* - z^*, z \rangle \to_j 0$ for all $z \in X$; that is, $\limsup_j h_{z_{i_j}^*}(z) = \lim_j \langle z_{i_j}^*, z \rangle = \langle z^*, z \rangle$ for all $z \in X$. Thus, the inclusion "$\subset$" in (6.86) comes by combining (6.88) and (6.89).

The proof of (6.87) is similar, except that we use formula (6.59) instead of (6.58). ∎

We end this section by illustrating the previous formulas with the support function.

Example 6.3.7 Consider the support function $f(x) := \sigma_A(x) = \sup\{\langle a, x \rangle : a \in A\}$, where $A \subset X \equiv \mathbb{R}^n$ is a non-empty set. Let us show that, for every given $x \in \mathrm{dom}\, f$,

6.4. HOMOGENEOUS FORMULAS

$$\partial f(x) = \mathrm{co}\left\{\bigcap_{\varepsilon>0} \mathrm{cl}\left(A_\varepsilon(x) + (\overline{\mathrm{co}}A)_\infty \cap \{x\}^\perp\right)\right\}, \qquad (6.90)$$

where $A_\varepsilon(x) := \{a \in A : \langle a, x\rangle \geq f(x) - \varepsilon\}$ (and $A(x) := \{a \in A : \langle a, x\rangle = f(x)\}$). Moreover, if A is closed and $\mathrm{int}\,([(\overline{\mathrm{co}}A)_\infty]^-) \neq \emptyset$, then

$$\partial f(x) = (\overline{\mathrm{co}}A)_\infty \cap \{x\}^\perp + \mathrm{co}\,(A(x)). \qquad (6.91)$$

Indeed, since $\mathrm{N}_{\mathrm{dom}\,f}(x) = (\overline{\mathrm{co}}A)_\infty \cap \{x\}^\perp$ (see Exercise 54), (6.90) follows by (6.73) (or, more precisely, its finite-dimensional version given in formula (6.83)).

Assume now that A is closed and $\mathrm{int}\,([(\overline{\mathrm{co}}A)_\infty]^-) \neq \emptyset$, so that $A_\varepsilon(x)$ is closed and we obtain

$$\bigcap_{\varepsilon>0} \mathrm{cl}\,(A_\varepsilon(x)) = \bigcap_{\varepsilon>0} A_\varepsilon(x) = A(x).$$

Therefore, thanks to (3.52), we have $\mathrm{int}(\mathrm{dom}\,f) = \mathrm{int}\,([(\overline{\mathrm{co}}A)_\infty]^-) \neq \emptyset$, and f is continuous on $\mathrm{int}(\mathrm{dom}\,f)$. So, using again the relation $\mathrm{N}_{\mathrm{dom}\,f}(x) = (\overline{\mathrm{co}}A)_\infty \cap \{x\}^\perp$, formula (6.87) entails (6.91). If, in addition $x \in \mathrm{int}\,([(\overline{\mathrm{co}}A)_\infty]^-)$, then

$$\partial f(x) = \mathrm{co}\,(A(x)). \qquad (6.92)$$

Consequently, f is (Fréchet-) differentiable at $x \in \mathrm{int}\,([(\overline{\mathrm{co}}A)_\infty]^-)$ if and only if the set $A(x)$ is a singleton; in this case, we have $A(x) = \{\nabla f(x)\}$.

6.4 Homogeneous formulas

In this section, we represent the subdifferential of the supremum function $f := \sup_{t \in T} f_t$, using exclusively the data functions f_t, $t \in T$, belonging to $\Gamma_0(X)$. If the (almost) active functions are still present in the final formulas, the normal cone $\mathrm{N}_{\mathrm{dom}\,f}(x)$ is now removed and replaced with the approximate subdifferential of non-active functions, but affected by appropriate weights. The resulting formulas are called homogeneous because they do not depend, at least explicitly, on the normal cone to the effective domain of f (or to finite-dimensional sections of it).

To this aim, we first characterize the normal cone $N_{\text{dom} f}(x)$ by means of the data functions f_t, $t \in T$. The following result deals with the so-called compact-continuous case.

Proposition 6.4.1 *Assume that $\{f_t, t \in T\} \subset \Gamma_0(X)$ with T compact and that the mappings $t \mapsto f_t(z)$, $z \in X$, are usc. Then, for every $x \in$ dom f, such that*
$$\inf_{t \in T} f_t(x) > -\infty, \tag{6.93}$$

we have
$$N_{\text{dom} f}(x) = \left[\overline{\text{co}}\left(\bigcup_{t \in T} \partial_\varepsilon f_t(x)\right)\right]_\infty \quad \text{for all } \varepsilon > 0. \tag{6.94}$$

Proof. We fix $\varepsilon > 0$ and denote
$$E_\varepsilon := \bigcup_{t \in T} \partial_\varepsilon f_t(x);$$

observe that E_ε is non-empty by Proposition 4.1.10.

To establish the inclusion "$\supset$" in (6.94), we take $x^* \in [\overline{\text{co}}(E_\varepsilon)]_\infty$ and fix $x_0^* \in E_\varepsilon$. Then for every $\alpha > 0$, we have $x_0^* + \alpha x^* \in \overline{\text{co}}(E_\varepsilon)$, and so there are nets $(\lambda_{j,1}, \ldots, \lambda_{j,k_j}) \in \Delta_{k_j}^*$, $t_{j,1}, \ldots, t_{j,k_j} \in T$, and $x_{j,1}^* \in \partial_\varepsilon f_{t_{j,1}}(x), \ldots, x_{j,k_j}^* \in \partial_\varepsilon f_{t_{j,k_j}}(x)$ such that $x_0^* + \alpha x^* = \lim_j (\lambda_{j,1} x_{j,1}^* + \ldots + \lambda_{j,k_j} x_{j,k_j}^*)$. Hence, for every fixed $y \in \text{dom } f$,

$$\langle x_0^* + \alpha x^*, y - x \rangle = \lim_j \left\langle \lambda_{j,1} x_{j,1}^* + \ldots + \lambda_{j,k_j} x_{j,k_j}^*, y - x \right\rangle$$
$$\leq \limsup_j \left(\sum_{i=1,\ldots,k_j} \lambda_{j,i}(f_{t_{j,i}}(y) - f_{t_{j,i}}(x) + \varepsilon)\right)$$
$$\leq \limsup_j \left(\sum_{i=1,\ldots,k_j} \lambda_{j,i}(f^+(y) - f_{t_{j,i}}(x) + \varepsilon)\right)$$
$$\leq f^+(y) - \inf\{f_t(x), t \in T\} + \varepsilon,$$

where f^+ is the positive part of f. Next, dividing by α and making $\alpha \uparrow +\infty$, condition (6.93) ensures that $\langle x^*, y - x \rangle \leq 0$ for all $y \in \text{dom } f = \text{dom } f^+$; that is, $x^* \in N_{\text{dom} f}(x)$, as we wanted to prove.

Now, we prove the inclusion
$$([\overline{\text{co}}(E_\varepsilon)]_\infty)^- \subset (N_{\text{dom} f}(x))^-, \tag{6.95}$$

6.4. HOMOGENEOUS FORMULAS

or equivalently, according to (3.52) and the fact that $(N_{\text{dom }f}(x))^- = \text{cl}(\mathbb{R}_+(\text{dom }f - x))$,

$$\text{cl}(\text{dom }\sigma_{E_\varepsilon}) \subset \text{cl}(\mathbb{R}_+(\text{dom }f - x)). \qquad (6.96)$$

Take $z \in \text{dom }\sigma_{E_\varepsilon} = \text{dom}\left(\sup_{t\in T} \sigma_{\partial_\varepsilon f_t(x)}\right) = \text{dom}\left(\sup_{t\in T}(f_t)'_\varepsilon(x;\cdot)\right)$, where the last equality comes from Proposition 4.1.12. Then, by (4.1),

$$z \in \bigcap_{t\in T} \text{dom}((f_t)'_\varepsilon(x;\cdot)) = \bigcap_{t\in T} \mathbb{R}_+(\text{dom }f_t - x),$$

and (see Exercise 9)

$$z \in \bigcap_{t\in T} \mathbb{R}_+(\text{dom }f_t - x) = \mathbb{R}_+(\text{dom }f - x) \subset \text{cl}(\mathbb{R}_+(\text{dom }f - x)).$$

Hence, (6.96) holds and (6.95) follows. Finally, the inclusion "⊂" in (6.94) follows from (6.95) by the bipolar theorem (3.51). ■

The following result provides the non-compact counterpart to Proposition 6.4.1. Now, instead of the individual functions f_t, $t \in T$, we use the following max-type functions (introduced in (5.21))

$$f_J := \max\{f_t, t \in J\}, \ J \in \mathcal{T} := \{J \subset T : |J| < +\infty\}. \qquad (6.97)$$

The proof of this result uses a countable reduction argument given in Exercise 100 whose proof is based on strictly topological arguments.

Proposition 6.4.2 *Consider the family* $\{f_t, t \in T\} \subset \Gamma_0(X)$, *the function* $f := \sup_{t\in T} f_t$, *and let* $x \in \text{dom }f$ *such that condition (6.93) holds; that is,* $\inf_{t\in T} f_t(x) > -\infty$. *Then, for every* $\varepsilon > 0$,

$$N_{\text{dom }f}(x) = \left[\overline{\text{co}}\left(\bigcup_{J\in\mathcal{T}} \partial_\varepsilon f_J(x)\right)\right]_\infty. \qquad (6.98)$$

Proof. Take $u^* \in N_{\text{dom }f}(x)$ and $\varepsilon > 0$. By Exercise 100, for every fixed $L \in \mathcal{F}(x)$ there exists a sequence $(t_n)_n \subset T$ such that $u^* \in N_{\text{dom}(\sup_{n\geq 1} f_{t_n})\cap L}(x)$. We denote $J_n := \{t_1,\ldots,t_n\}$, $n \geq 1$, and introduce the convex functions $\hat{f}_n := f_{J_n} + I_L$, $n \geq 1$ (see (6.97)). So, $(\hat{f}_n)_n$ is non-decreasing and satisfies $\sup_{n\geq 1}(f_{t_n} + I_L) = \sup_{n\geq 1} \hat{f}_n$ and $\text{dom}\left(\sup_{n\geq 1} f_{t_n}\right) \cap L = \text{dom}\left(\sup_{n\geq 1} \hat{f}_n\right)$. In addition, according to Example 5.1.6 (formula (5.15)), we have

$$\partial_{\varepsilon/2}\left(\sup_{n\geq 1}\hat{f}_n\right)(x) = \bigcap_{\delta>0}\mathrm{cl}\left(\bigcup_{k\geq 1}\bigcap_{n\geq k}\partial_{\delta+\varepsilon/2}\hat{f}_n(x)\right) \subset \mathrm{cl}\left(\bigcup_{k\geq 1}\partial_\varepsilon(f_{J_k}+\mathrm{I}_L)(x)\right),$$

taking $\delta = \varepsilon/2$ and $n = k$. Next, using formula (4.46) in Proposition 4.1.16, we get

$$\partial_{\varepsilon/2}\left(\sup_{n\geq 1}\hat{f}_n\right)(x) \subset \mathrm{cl}\left(\bigcup_{k\geq 1}\mathrm{cl}\left(\bigcup_{\substack{\varepsilon_1+\varepsilon_2=\varepsilon\\ \varepsilon_1,\varepsilon_2\geq 0}}(\partial_{\varepsilon_1}f_{J_k}(x) + \partial_{\varepsilon_2}\mathrm{I}_L(x))\right)\right)$$

$$\subset \overline{\mathrm{co}}\left(\bigcup_{k\geq 1}\partial_\varepsilon f_{J_k}(x) + L^\perp\right).$$

Therefore, applying (4.9) and Proposition 4.1.10,

$$u^* \in \mathrm{N}_{\mathrm{dom}(\sup_{n\geq 1}f_{t_n})\cap L}(x) = \left[\partial_{\varepsilon/2}\left(\sup_{n\geq 1}\hat{f}_n\right)(x)\right]_\infty$$

$$\subset \left[\overline{\mathrm{co}}\left(\bigcup_{k\geq 1}\partial_\varepsilon f_{J_k}(x) + L^\perp\right)\right]_\infty \subset \left[\overline{\mathrm{co}}\left(\bigcup_{J\in\mathcal{T}}\partial_\varepsilon f_J(x) + L^\perp\right)\right]_\infty,$$

and so, by (2.23),

$$u^* \in \bigcap_{L\in\mathcal{F}(x)}\left[\overline{\mathrm{co}}\left(\bigcup_{J\in\mathcal{T}}\partial_\varepsilon f_J(x) + L^\perp\right)\right]_\infty = \left[\bigcap_{L\in\mathcal{F}(x)}\overline{\mathrm{co}}\left(\bigcup_{J\in\mathcal{T}}\partial_\varepsilon f_J(x) + L^\perp\right)\right]_\infty,$$

and we obtain that $u^* \in [\overline{\mathrm{co}}(\cup_{J\in\mathcal{T}}\partial_\varepsilon f_J(x))]_\infty$ (see Exercise 10(iv)). Hence, the inclusion "⊂" in (6.98) follows, and we are done since the opposite inclusion follows as in the proof of Proposition 6.4.1. ∎

Next, we give the first homogeneous formula of $\partial f(x)$, corresponding to the so-called compact-continuous framework.

Theorem 6.4.3 Assume that $\{f_t,\ t\in T\} \subset \Gamma_0(X)$ with T compact and that the mappings $t \mapsto f_t(z)$, $z \in X$, are usc. Denote $f := \sup_{t\in T} f_t$ and let $x \in \mathrm{dom}\, f$ such that condition (6.93) holds; that is, $\inf_{t\in T} f_t(x) > -\infty$. Then we have

$$\partial f(x) = \bigcap_{\varepsilon>0}\overline{\mathrm{co}}\left(\left(\bigcup_{t\in T(x)}\partial_\varepsilon f_t(x)\right) + \left(\bigcup_{t\in T\setminus T(x)}\{0,\varepsilon\}\partial_\varepsilon f_t(x)\right)\right).$$

(6.99)

6.4. HOMOGENEOUS FORMULAS

Proof. We may assume that $f(x) = 0$ (without loss of generality). Fix $\varepsilon > 0$, $U \in \mathcal{N}_{X^*}$, and pick $L \in \mathcal{F}(x)$ such that $L^\perp \subset U$. An isolated index is assigned to the function I_L so that the family $\{f_t, t \in T; I_L\} \subset \Gamma_0(X)$ also satisfies the same compactness and upper semicontinuity assumptions. Therefore, applying Proposition 6.4.1 to the family $\{f_t, t \in T; I_L\}$, we obtain

$$N_{L \cap \operatorname{dom} f}(x) = \left[\overline{\operatorname{co}}\left(\left(\bigcup_{t \in T(x)} \partial_\varepsilon f_t(x)\right) \cup \left(\bigcup_{t \in T \setminus T(x)} \partial_\varepsilon f_t(x)\right) \cup L^\perp\right)\right]_\infty, \quad (6.100)$$

and so (see Exercises 22 and 23)

$$N_{L \cap \operatorname{dom} f}(x) = \left[\overline{\operatorname{co}}\left(\left(\bigcup_{t \in T(x)} \partial_\varepsilon f_t(x)\right) \cup \left(\bigcup_{t \in T \setminus T(x)} \varepsilon \partial_\varepsilon f_t(x)\right) \cup L^\perp\right)\right]_\infty$$

$$= \left[\overline{\operatorname{co}}\left(\left(\bigcup_{t \in T(x)} \partial_\varepsilon f_t(x)\right) + \left(\bigcup_{t \in T \setminus T(x)} \varepsilon \partial_\varepsilon f_t(x)\right) + L^\perp\right)\right]_\infty$$

$$\subset \left[\overline{\operatorname{co}}\left(A_\varepsilon + L^\perp\right)\right]_\infty,$$

where we denote

$$A_\varepsilon := \left(\bigcup_{t \in T(x)} \partial_\varepsilon f_t(x)\right) + \left(\bigcup_{t \in T \setminus T(x)} \{0, \varepsilon\} \partial_\varepsilon f_t(x)\right).$$

Observe that, due to the lower semicontinuity of the f_t's, the sets $\partial_\varepsilon f_t(x)$ are non-empty and we have $\theta \in \bigcup_{t \in T \setminus T(x)} \{0, \varepsilon\} \partial_\varepsilon f_t(x) + L^\perp$. Thus, using (6.16),

$$\partial f(x) \subset \overline{\operatorname{co}}\left(\bigcup_{t \in T(x)} \partial_\varepsilon f_t(x) + N_{L \cap \operatorname{dom} f}(x)\right)$$

$$\subset \overline{\operatorname{co}}\left(\bigcup_{t \in T(x)} \partial_\varepsilon f_t(x) + \left[\overline{\operatorname{co}}\left(A_\varepsilon + L^\perp\right)\right]_\infty\right),$$

and, taking into account that $\bigcup_{t \in T(x)} \partial_\varepsilon f_t(x) \subset A_\varepsilon$, we have

$$\partial f(x) \subset \overline{\operatorname{co}}\left(A_\varepsilon + \left[\overline{\operatorname{co}}\left(A_\varepsilon + L^\perp\right)\right]_\infty\right)$$
$$\subset \overline{\operatorname{co}}\left(A_\varepsilon + L^\perp\right) \subset \operatorname{co}\left(A_\varepsilon + L^\perp\right) + U = \operatorname{co}(A_\varepsilon) + L^\perp + U \subset \operatorname{co}(A_\varepsilon) + 2U.$$

Therefore, the first inclusion "$\subset$" in (6.99) follows by intersecting over $U \in \mathcal{N}_{X^*}$ and then over $\varepsilon > 0$. Conversely, to show the inclusion "$\supset$"

in (6.99) we fix $\varepsilon > 0$. Then we have $\partial_\varepsilon f_t(x) \subset \partial_{\varepsilon-M} f(x)$, for all $\varepsilon > 0$ and $t \in T$, where $M := \inf_{t \in T} f_t(x)$. Thus, given any x^* in the right-hand side set of (6.99), we obtain

$$x^* \in \overline{\operatorname{co}}\left(\left(\bigcup_{t \in T(x)} \partial_\varepsilon f_t(x)\right) + \left(\bigcup_{t \in T \setminus T(x)} \{0, \varepsilon\} \partial_{\varepsilon-M} f(x)\right)\right)$$
$$\subset \operatorname{cl}\left(\partial_\varepsilon f(x) + \operatorname{co}\left(\{0, \varepsilon\} \partial_{\varepsilon-M} f(x)\right)\right) = \operatorname{cl}\left(\partial_\varepsilon f(x) + [0, \varepsilon] \partial_{\varepsilon-M} f(x)\right),$$

where in the last inclusion we used the fact that $\partial_\varepsilon f_t(x) \subset \partial_\varepsilon f(x)$ for all $t \in T(x)$. Let us consider nets $(y_i^*)_i \subset \partial_\varepsilon f(x)$, $(\varepsilon_i)_i \subset [0, \varepsilon]$ and $(z_i^*)_i \subset \partial_{\varepsilon-M} f(x)$ such that $x^* = \lim_i (y_i^* + \varepsilon_i z_i^*)$. Then, for each $y \in \operatorname{dom} f$,

$$\langle x^*, y - x \rangle = \lim_i \langle y_i^* + \varepsilon_i z_i^*, y - x \rangle$$
$$\leq \limsup_i \left((f(y) - f(x) + \varepsilon) + \varepsilon_i (f(y) - f(x) - M + \varepsilon)\right)$$
$$\leq (f(y) - f(x) + \varepsilon) + \varepsilon (f(y) - f(x) - M + \varepsilon)^+;$$

that is,

$$\langle x^*, y - x \rangle \leq (f(y) - f(x) + \varepsilon) + \varepsilon (f(y) - f(x) - M + \varepsilon)^+ \text{ for all } \varepsilon > 0,$$

and we deduce that $x^* \in \partial f(x)$ as $\varepsilon \downarrow 0$. ■

The following corollary deals with the particular case in which all the functions are active; it is an immediate consequence of Theorem 6.4.3.

Corollary 6.4.4 *Assume in Theorem 6.4.3 that $T(x) = T$. Then*

$$\partial f(x) = \bigcap_{\varepsilon > 0} \overline{\operatorname{co}}\left(\bigcup_{t \in T} \partial_\varepsilon f_t(x)\right).$$

Next, we give a non-compact counterpart to Theorem 6.4.3.

Theorem 6.4.5 *Consider the family $\{f_t, t \in T\} \subset \Gamma_0(X)$ and $f := \sup_{t \in T} f_t$. Then, for every $x \in \operatorname{dom} f$, we have*

$$\partial f(x) = \bigcap_{\varepsilon > 0} \overline{\operatorname{co}}\left(\left(\bigcup_{t \in T_\varepsilon(x)} \partial_\varepsilon f_t(x)\right) + \{0, \varepsilon\} \left(\bigcup_{J \in \mathcal{T}_\varepsilon(x)} \partial_\varepsilon f_J(x)\right)\right), \quad (6.101)$$

where f_J and $\mathcal{T}$ are defined in (6.97), and

6.4. HOMOGENEOUS FORMULAS

$$\mathcal{T}_\varepsilon(x) := \{J \in \mathcal{T} : f_J(x) \geq f(x) - \varepsilon\}. \tag{6.102}$$

Proof. Fix $x \in \operatorname{dom} f$ and $\varepsilon > 0$. So, by Theorem 5.2.2, for each $L \in \mathcal{F}(x)$ one has

$$\partial f(x) \subset \overline{\operatorname{co}}\left(\bigcup_{t \in \mathcal{T}_\varepsilon(x)} \partial_\varepsilon f_t(x) + \mathrm{N}_{L \cap \operatorname{dom} f}(x)\right). \tag{6.103}$$

Now, given a fixed $t_0 \in \mathcal{T}_\varepsilon(x)$, we introduce the functions

$$g_t := \max\{f_t, f_{t_0}\} + \mathrm{I}_L, \ t \in T;$$

hence $\sup_{t \in T} g_t = f + \mathrm{I}_L$, and for all $J \in \mathcal{T}$,

$$g_J := \max_{t \in J} g_t = f_{J \cup \{t_0\}} + \mathrm{I}_L. \tag{6.104}$$

Observe that $g_J(x) \geq f_{t_0}(x) \geq f(x) - \varepsilon$; that is, $J' := J \cup \{t_0\} \in \mathcal{T}_\varepsilon(x)$. Therefore, since $\{g_t, t \in T\} \subset \Gamma_0(X)$ and satisfies condition (6.93); that is, $\inf_{t \in T} g_t(x) \geq f_{t_0}(x) > -\infty$, Proposition 6.4.2 entails

$$\mathrm{N}_{L \cap \operatorname{dom} f}(x) = \mathrm{N}_{\operatorname{dom}(\sup_{t \in T} g_t)}(x) = \left[\overline{\operatorname{co}}\left(\bigcup_{J \in \mathcal{T}} \partial_\varepsilon g_J(x)\right)\right]_\infty$$

$$= \left[\overline{\operatorname{co}}\left(\bigcup_{J \in \mathcal{T}} \partial_\varepsilon (f_{J \cup \{t_0\}} + \mathrm{I}_L)(x)\right)\right]_\infty \subset \left[\overline{\operatorname{co}}\left(\bigcup_{J' \in \mathcal{T}_\varepsilon(x)} \partial_\varepsilon (f_{J'} + \mathrm{I}_L)(x)\right)\right]_\infty.$$

Thus, using formula (4.46) in Proposition 4.1.16,

$$\mathrm{N}_{L \cap \operatorname{dom} f}(x) \subset \left[\overline{\operatorname{co}}\left(\bigcup_{J \in \mathcal{T}_\varepsilon(x)} \partial_\varepsilon f_J(x) + L^\perp\right)\right]_\infty$$

$$\subset \left[\overline{\operatorname{co}}\left(\left(\bigcup_{t \in \mathcal{T}_\varepsilon(x)} \partial_\varepsilon f_t(x) + L^\perp\right) \cup \left(\bigcup_{J \in \mathcal{T}_\varepsilon(x)} \partial_\varepsilon f_J(x) + L^\perp\right)\right)\right]_\infty,$$

and we deduce that $\mathrm{N}_{L \cap \operatorname{dom} f}(x) \subset \left[\overline{\operatorname{co}}\left(E_\varepsilon + L^\perp\right)\right]_\infty$ (see Exercises 22 and 23), where

$$E_\varepsilon := \left(\bigcup_{t \in \mathcal{T}_\varepsilon(x)} \partial_\varepsilon f_t(x)\right) + \{0, \varepsilon\} \left(\bigcup_{J \in \mathcal{T}_\varepsilon(x)} \partial_\varepsilon f_J(x)\right).$$

Consequently, (6.103) gives rise to

$$\partial f(x) \subset \overline{\operatorname{co}} \left(\bigcup_{t \in T_\varepsilon(x)} \partial_\varepsilon f_t(x) + \left[\overline{\operatorname{co}} \left(E_\varepsilon + L^\perp \right) \right]_\infty \right)$$

$$\subset \overline{\operatorname{co}} \left(E_\varepsilon + L^\perp + \left[\overline{\operatorname{co}} \left(E_\varepsilon + L^\perp \right) \right]_\infty \right) = \overline{\operatorname{co}} \left(E_\varepsilon + L^\perp \right)$$

$$\subset \operatorname{co} \left(E_\varepsilon + L^\perp \right) + U = \operatorname{co} (E_\varepsilon) + L^\perp + U \subset \operatorname{co} (E_\varepsilon) + 2U,$$

and the desired inclusion "⊂" follows once we intersect over all $U \in \mathcal{N}_{X^*}$ and after over $\varepsilon > 0$. To verify the opposite inclusion, we easily observe that, for all $\varepsilon > 0$,

$$\bigcup_{t \in T_\varepsilon(x)} \partial_\varepsilon f_t(x) \subset \partial_{2\varepsilon} f(x) \text{ and } \bigcup_{J \in \mathcal{T}_\varepsilon(x)} \partial_\varepsilon f_J(x) \subset \partial_{2\varepsilon} f(x),$$

and so

$$\bigcap_{\varepsilon > 0} \overline{\operatorname{co}} \left(E_\varepsilon \right) \subset \bigcap_{\varepsilon > 0} \overline{\operatorname{co}} \left(\partial_{2\varepsilon} f(x) + \{0, \varepsilon\} \partial_{2\varepsilon} f(x) \right)$$

$$\subset \bigcap_{\varepsilon > 0} [1, 1 + \varepsilon] \partial_{2\varepsilon} f(x) = \partial f(x).$$

■

6.5 Qualification conditions

This section addresses the conditions on the data functions, f_t, $t \in T$, which allow us to derive formulas for the subdifferential of the supremum function $f := \sup_{t \in T} f_t$ that are simpler than those given previously. These conditions are of a geometric and topological nature and depend on the way in which the functions involved are related; that is, when their effective domains overlap sufficiently. The term qualification applies here because the conditions we give are of the same type as those often used to derive optimality conditions for convex optimization problems (see chapter 8). More precisely, some constraints are qualified and lead to simple expressions of the subdifferential of f by means of convex combinations of the subdifferentials of the qualified f_t's. This approach avoids the use of limiting processes.

The first result in this section is based on a finite-dimensional qualification. In contrast to Corollary 6.1.12, which assumes the continuity of the supremum function f, the condition given here is only based on the effective domains of the f_t's.

6.5. QUALIFICATION CONDITIONS

Theorem 6.5.1 *Given convex functions $f_t : \mathbb{R}^n \to \mathbb{R}_\infty$, $t \in T$, and $f := \sup_{t \in T} f_t$, take $x \in \operatorname{dom} f$ and assume that the standard compactness hypothesis (6.1) holds. If, additionally, we assume that*

$$\operatorname{ri}(\operatorname{dom} f_t) \cap \operatorname{dom} f \neq \emptyset \text{ for all } t \in T(x), \qquad (6.105)$$

then we have

$$\partial f(x) = \mathrm{N}_{\operatorname{dom} f}(x) + \operatorname{co}\left\{\bigcup_{t \in T(x)} \partial f_t(x)\right\}. \qquad (6.106)$$

Proof. Fix $t \in T(x)$ and, by (6.105), take $x_0 \in \operatorname{ri}(\operatorname{dom} f_t) \cap \operatorname{dom} f$. Then, by (2.15), for any $x_1 \in \operatorname{ri}(\operatorname{dom} f) \subset \operatorname{dom} f_t$ and $\lambda \in\ ]0,1[$, we have

$$\lambda x_1 + (1-\lambda) x_0 \in \operatorname{ri}(\operatorname{dom} f_t) \cap \operatorname{ri}(\operatorname{dom} f).$$

Thus, according Proposition 4.1.26, $\partial (f_t + \mathrm{I}_{\operatorname{dom} f})(x) = \partial f_t(x) + \mathrm{N}_{\operatorname{dom} f}(x)$, and consequently, (6.106) comes by combining Corollary 6.1.10 and Exercise 107. ∎

Remark 17 *Theorem 6.5.1 remains true if, instead of condition (6.105), we assume that $\cap_{t \in T} \operatorname{ri}(\operatorname{dom} f_t) \neq \emptyset$. Indeed, this condition ensures that $\operatorname{ri}(\operatorname{dom} f) = \cap_{t \in T} \operatorname{ri}(\operatorname{dom} f_t)$ (see Exercise 13), which in turn yields condition (6.105).*

Theorem 6.5.2 below gives an infinite-dimensional counterpart to Theorem 6.5.1. It uses the continuity somewhere in $\operatorname{dom} f$ of all but one of the active functions. This condition is much weaker than the continuity of the supremum function f imposed in Corollary 6.1.12.

Theorem 6.5.2 *Given convex functions $f_t : X \to \mathbb{R}_\infty$, $t \in T$, we denote $f := \sup_{t \in T} f_t$ and assume the standard compactness hypothesis (6.1). Let $x \in \operatorname{dom} f$ such that $T(x) := \{1, ..., m+1\} \subset T$, $m \geq 0$, and assume that each one of the functions f_i, $i \in T(x)$, except perhaps one of them, say f_{m+1}, is continuous somewhere in $\operatorname{dom} f$. Then we have*

$$\partial f(x) = \operatorname{co}\left\{\bigcup_{1 \leq i \leq m} \partial f_i(x) \cup \partial (f_{m+1} + \mathrm{I}_{\operatorname{dom} f})(x)\right\} + \mathrm{N}_{\operatorname{dom} f}(x).$$

Proof. The inclusion "⊃" is straightforward. To prove "⊂" we may assume that $\partial f(x) \neq \emptyset$; hence $f(x) = (\operatorname{cl} f)(x) \in \mathbb{R}$ and $\partial f(x) = \partial (\operatorname{cl} f)(x)$. We also assume that $x = \theta$ and $f(\theta) = (\operatorname{cl} f)(\theta) = 0$. Observe that if $m = 0$, so that $T(\theta) = \{1\}$, then Corollary 6.1.6 yields

$$\partial f(\theta) = \partial \left(\max_{k \in T(\theta)} f_k + I_{\operatorname{dom} f} \right)(\theta) \equiv \partial \left(f_1 + I_{\operatorname{dom} f} \right)(\theta),$$

and the desired formula follows as (see (4.9))

$$N_{\operatorname{dom} f}(\theta) = N_{\operatorname{dom} f \cap \operatorname{dom} f_1}(\theta) = (\partial \left(f_1 + I_{\operatorname{dom} f} \right)(\theta))_\infty.$$

Now we consider the case $m > 0$. By the current assumption, for each $i \in \{1, \ldots, m\}$ there exists some $x_i \in \operatorname{dom} f$ such that f_i is continuous at x_i. More precisely, (2.15) (see also Exercise 13) ensures that

$$x_0 := \frac{1}{m} \sum_{1 \leq i \leq m} x_i \in \operatorname{dom} f \cap \operatorname{int}(\operatorname{dom} f_1) \cap \ldots \cap \operatorname{int}(\operatorname{dom} f_m),$$

and the functions $f_1, \ldots, f_m$ are continuous at $x_0 \in \operatorname{dom} f$. Consequently, using Proposition 5.2.4(ii), we have

$$\operatorname{cl}\left(\max\left\{ \max_{1 \leq i \leq m} f_i, f_{m+1} + I_{\operatorname{dom} f} \right\} \right) = \max\left\{ \max_{1 \leq i \leq m} \operatorname{cl} f_i, \operatorname{cl}(f_{m+1} + I_{\operatorname{dom} f}) \right\}.$$

Therefore, taking into account Corollary 6.1.6 and observing that

$$\max_{i \in T(\theta)} f_i + I_{\operatorname{dom} f} = \max\{f_i, 1 \leq i \leq m; f_{m+1} + I_{\operatorname{dom} f}\},$$

by applying Corollary 5.1.9, we get

$$\partial f(\theta) = \partial \left(\max_{t \in T(\theta)} f_t + I_{\operatorname{dom} f} \right)(\theta) = \bigcup_{\lambda \in S(\theta)} \partial \left(\sum_{1 \leq k \leq m} \lambda_k f_k + (\lambda_{m+1} f_{m+1} + I_{\operatorname{dom} f}) \right)(\theta),$$

where $\lambda_k f_k = I_{\operatorname{dom} f_k}$ if $\lambda_k = 0$, and

$$S(\theta) := \left\{ \lambda \in \Delta_{m+1} : \sum_{1 \leq k \leq m+1} \lambda_k f_k(\theta) = 0 \right\}.$$

Hence, since the convex functions $\lambda_1 f_1, \ldots, \lambda_m f_m$ are also continuous at $x_0 \in \operatorname{dom} f$ ($\subset \operatorname{dom}(\lambda_{m+1} f_{m+1} + I_{\operatorname{dom} f})$), by applying Proposition 4.1.20, we get

$$\partial f(\theta) = \bigcup_{\lambda \in S(\theta)} \sum_{1 \leq k \leq m} \partial(\lambda_k f_k)(\theta) + \partial(\lambda_{m+1} f_{m+1} + I_{\operatorname{dom} f})(\theta)$$

$$= \bigcup_{\lambda \in S(\theta)} \sum_{\substack{\lambda_k > 0 \\ 1 \leq k \leq m}} \lambda_k \partial f_k(\theta) + \sum_{\substack{\lambda_k = 0 \\ 1 \leq k \leq m+1}} N_{\operatorname{dom} f_k}(\theta) + \partial(\lambda_{m+1} f_{m+1} + I_{\operatorname{dom} f})(\theta)$$

6.5. QUALIFICATION CONDITIONS

$$\subset \bigcup_{\substack{\lambda \in S(\theta) \\ \lambda_k > 0 \\ 1 \leq k \leq m}} \sum \lambda_k \partial f_k(\theta) + \partial \left(\lambda_{m+1} f_{m+1} + I_{\operatorname{dom} f} + \sum_{\substack{\lambda_k = 0 \\ 1 \leq k \leq m+1}} I_{\operatorname{dom} f_k} \right)(\theta)$$

$$= \bigcup_{\substack{\lambda \in S(\theta) \\ \lambda_k > 0 \\ 1 \leq k \leq m}} \sum \lambda_k \partial f_k(\theta) + \partial \left(\lambda_{m+1} f_{m+1} + I_{\operatorname{dom} f} \right)(\theta), \qquad (6.107)$$

where the last inclusion holds because $\operatorname{dom} f \subset \operatorname{dom} f_k$ for all $1 \leq k \leq m+1$. Finally, given any $x^* \in \partial f(\theta)$, by (6.107) there exists some $\lambda \in \Delta_{m+1}$ such that

$$x^* \in \sum_{\lambda_k > 0, 1 \leq k \leq m} \lambda_k \partial f_k(\theta) + \partial(\lambda_{m+1} f_{m+1} + I_{\operatorname{dom} f})(\theta).$$

Denote $A := \cup_{k \in T(\theta)} \partial f_k(\theta)$. If $\lambda_{m+1} = 0$, then

$$x^* \in \sum_{\lambda_k > 0, 1 \leq k \leq m} \lambda_k \partial f_k(\theta) + N_{\operatorname{dom} f}(\theta)$$
$$\subset \operatorname{co} A + N_{\operatorname{dom} f}(\theta) \subset \operatorname{co} \{A \cup \partial(f_{m+1} + I_{\operatorname{dom} f})(\theta)\} + N_{\operatorname{dom} f}(\theta),$$

and otherwise, i.e., when $\lambda_{m+1} > 0$,

$$x^* \in \sum_{\lambda_k > 0, 1 \leq k \leq m} \lambda_k \partial f_k(\theta) + \lambda_{m+1} \partial(f_{m+1} + I_{\operatorname{dom} f})(\theta)$$
$$\subset \operatorname{co} \{A \cup \partial(f_{m+1} + I_{\operatorname{dom} f})(\theta)\} \subset \operatorname{co} \{A \cup \partial(f_{m+1} + I_{\operatorname{dom} f})(\theta)\} + N_{\operatorname{dom} f}(\theta).$$

The proof of the inclusion "$\subset$" is complete. ■

As a consequence of Theorem 6.5.2, we obtain the following result for the maximum of a finite family of convex functions. This model constitutes a relevant particular case of the compactly indexed setting studied in section 6.1.

Corollary 6.5.3 *Given a finite family of convex functions, $f_k : X \to \mathbb{R}_\infty$, $k \in T := \{1, ..., p\}$, $p \geq 1$, we denote $f := \max_{k \in T} f_k$. Assume that each one of the functions f_k, $k \in T$, except perhaps one of them, say f_p, is continuous somewhere in $\operatorname{dom} f$. Then, for every $x \in X$, we have*

$$\partial f(x) = \operatorname{co} \left\{ \bigcup_{k \in T(x)} \partial f_k(x) \right\} + \sum_{k \in T} N_{\operatorname{dom} f_k}(x). \qquad (6.108)$$

Proof. First, if $p = 1$, then (6.108) trivially holds as it reduces to $\partial f(x) = \partial f(x) + N_{\operatorname{dom} f}(x)$. Assume that $p > 1$ and fix $x \in X$. If $p \notin T(x)$, so that all the active functions at x are continuous somewhere in $\operatorname{dom} f$, then Theorem 6.5.2 yields

$$\partial f(x) = \operatorname{co}\left\{\bigcup_{k \in T(x)} \partial f_k(x)\right\} + \operatorname{N}_{\operatorname{dom} f}(x).$$

Thus, (6.108) follows by applying Proposition 4.1.20:

$$\operatorname{N}_{\operatorname{dom} f}(x) = \sum_{k \in T} \operatorname{N}_{\operatorname{dom} f_k}(x). \tag{6.109}$$

Next we assume that $p \in T(x)$. By Corollary 6.1.6, we have

$$\partial f(x) = \partial\left(\max_{k \in T(x)} f_k + \operatorname{I}_{\operatorname{dom} f}\right)(x) = \partial g(x), \tag{6.110}$$

where $g = \max\{f_k,\ k \in T(x) \setminus \{p\};\ f_p + \operatorname{I}_{\operatorname{dom} f}\}$. Since $f_p + \operatorname{I}_{\operatorname{dom} f} \le g \le f$, we observe that $\operatorname{dom} f = \operatorname{dom} g$. Then, since all the functions f_k, $k \in T(x) \setminus \{p\}$, are continuous somewhere in $\operatorname{dom} f = \operatorname{dom}(f_p + \operatorname{I}_{\operatorname{dom} f})$, by applying again Theorem 6.5.2, we obtain

$$\partial g(x) = \operatorname{co}\left\{\bigcup_{k \in T(x) \setminus \{p\}} \partial f_k(x) \bigcup \partial(f_p + \operatorname{I}_{\operatorname{dom} f})(x)\right\} + \operatorname{N}_{\operatorname{dom} f}(x). \tag{6.111}$$

In addition, we have that

$$f_p + \operatorname{I}_{\operatorname{dom} f} = f_p + \operatorname{I}_{\bigcap_{k \in T \setminus \{p\}} \operatorname{dom} f_k} = f_p + \sum_{k \in T \setminus \{p\}} \operatorname{I}_{\operatorname{dom} f_k}$$

and the functions $\operatorname{I}_{\operatorname{dom} f_k}$, $k \in T \setminus \{p\}$, are continuous at a common point in $\operatorname{dom} f \subset \operatorname{dom} f_p$ (by (2.15)). Then Proposition 4.1.20 yields

$$\partial(f_p + \operatorname{I}_{\operatorname{dom} f})(x) = \partial f_p(x) + \sum_{k \in T \setminus \{p\}} \operatorname{N}_{\operatorname{dom} f_k}(x),$$

and (6.110), (6.111) and (6.109) entail

$$\partial f(x) = \operatorname{co}\left\{\bigcup_{k \in T(x) \setminus \{p\}} \partial f_k(x) \bigcup \left(\partial f_p(x) + \sum_{k \in T \setminus \{p\}} \operatorname{N}_{\operatorname{dom} f_k}(x)\right)\right\} + \sum_{k \in T} \operatorname{N}_{\operatorname{dom} f_k}(x)$$

$$\subset \operatorname{co}\left\{\bigcup_{k \in T(x)} \partial f_k(x)\right\} + \sum_{k \in T} \operatorname{N}_{\operatorname{dom} f_k}(x).$$

Thus, we are done since the opposite inclusion is straightforward. ■

Corollary 6.5.4 *In addition to the assumption of Corollary 6.5.3, suppose that $T = T(x)$ and $\partial f_k(x) \ne \emptyset$ for all $k \in T$. Then*

$$\partial f(x) = \overline{co}\left(\bigcup_{k\in T}\partial f_k(x)\right).$$

Proof. By taking into account (4.9), Corollary 6.5.3 yields

$$\partial f(x) = co\left\{\bigcup_{k\in T}\partial f_k(x)\right\} + \sum_{k\in T} N_{dom\, f_k}(x) = co\left\{\bigcup_{k\in T}\partial f_k(x)\right\} + \sum_{k\in T}(\partial f_k(x))_\infty$$

$$\subset co\left\{\bigcup_{k\in T}\partial f_k(x)\right\} + \left(\overline{co}\left(\bigcup_{k\in T}\partial f_k(x)\right)\right)_\infty \subset \overline{co}\left(\bigcup_{k\in T}\partial f_k(x)\right).$$

The desired formula follows because the opposite inclusion clearly holds. ∎

6.6 Exercises

Exercise 92 *Prove the inclusion "⊃" in (6.12) and (6.16).*

Exercise 93 *(Exercise 40 revisited) Let f be a proper convex function and take $x \in \mathrm{dom}\, f$.*
(a) If $g := \max\{f, f(x) - 1\}$, prove that $\partial f(x) = \partial g(x)$ (Hint: Apply Theorem 6.1.4).
(b) If $\{f_t : X \to \overline{\mathbb{R}}, t \in T\}$ is a family of convex functions and $f = \sup_{t\in T} f_t$ is proper, prove that

$$\partial f(x) = \partial\left(\sup_{t\in T}(\max\{f_t, f(x) - 1\})\right).$$

Exercise 94 *Under hypothesis (6.1), prove that the sets $T_\varepsilon(x)$, $x \in f^{-1}(\mathbb{R})$, $\varepsilon \geq 0$, are non-empty, closed and compact (remember that T is not necessarily Hausdorff).*

Exercise 95 *Prove Lemma 6.2.1.*

Exercise 96 *Assume that X is Banach, and let the convex functions $f_t : X \to \overline{\mathbb{R}}$, $t \in T$, be such that (6.1) holds. If the f_t's are finite and continuous on some open set $U \subset X$, prove that the function $f := \sup_{t\in T} f_t$ is continuous on U.*

Exercise 97 *Given the functions $f_t : X \to \overline{\mathbb{R}}$, $t \in T$, and $f := \sup_{t\in T} f_t$, we assume that $f^{**} = \sup_{t\in T} f_t^{**}$. Prove that, for every $x \in X$,*

$$\partial f(x) = \bigcap_{L \in \mathcal{F}(x)} \mathrm{co}\left\{\bigcap_{\varepsilon > 0} \mathrm{cl}\left(\bigcup_{t \in T_\varepsilon(x)} \partial_\varepsilon f_t(x) + \mathrm{N}_{L \cap \mathrm{dom}\, f}(x)\right)\right\}.$$

Exercise 98 *Assume that $\{f_t,\, t \in T\} \subset \Gamma_0(X)$ with T compact and that the mappings $t \mapsto f_t(z)$, $z \in X$, are usc. Take $x \in \mathrm{dom}\, f$ such that $\inf_{t \in T} f_t(x) > -\infty$ and, given $t_0 \in T(x)$ and $0 < \mu_t < 1$, we define*

$$\tilde{f}_t := \begin{cases} f_t, & \text{if } t \in T(x), \\ \mu_t f_t + (1 - \mu_t) f_{t_0}, & \text{if } t \in T \setminus T(x), \end{cases}$$

and $\tilde{f} := \sup_{t \in T} \tilde{f}_t$. Prove that, for every $\varepsilon > 0$,

$$\mathrm{N}_{\mathrm{dom}\, f}(x) = \mathrm{N}_{\mathrm{dom}\, \tilde{f}}(x) = \left[\overline{\mathrm{co}}\left(\bigcup_{t \in T} \partial_\varepsilon \tilde{f}_t(x)\right)\right]_\infty. \tag{6.112}$$

Exercise 99 *Let $\{f_t,\, t \in T\}$ and T be as in Exercise 98. Given $t_0 \in T(x)$ and $\varepsilon > 0$, we define*

$$\tilde{f}_{t,\varepsilon} := \begin{cases} f_t, & \text{if } t \in T(x), \\ \mu_{t,\varepsilon} f_t + (1 - \mu_{t,\varepsilon}) f_{t_0}, & \text{if } t \in T \setminus T(x), \end{cases}$$

where $\mu_{t,\varepsilon} := \varepsilon(2f(x) - 2f_t(x) + \varepsilon)^{-1}$, $t \in T \setminus T(x)$, $\varepsilon > 0$. Prove that

$$\partial f(x) = \bigcap_{\varepsilon > 0} \overline{\mathrm{co}}\left(\bigcup_{t \in T} \partial_\varepsilon \tilde{f}_{t,\varepsilon}(x)\right), \tag{6.113}$$

and so, provided that $T(x) = T$,

$$\partial f(x) = \bigcap_{\varepsilon > 0} \overline{\mathrm{co}}\left(\bigcup_{t \in T} \partial_\varepsilon f_{t,\varepsilon}(x)\right). \tag{6.114}$$

Exercise 100 *Consider a family $\{f_t,\, t \in T\} \subset \Gamma_0(X)$ and $f := \sup_{t \in T} f_t$. Given $x \in \mathrm{dom}\, f$, $L \in \mathcal{F}(x)$, and $u^* \in \mathrm{N}_{\mathrm{dom}\, f}(x)$, prove the existence of a sequence $(t_n)_n \subset T$ such that $u^* \in \mathrm{N}_{\mathrm{dom}(\sup_{n \geq 1} f_{t_n}) \cap L}(x)$.*

Exercise 101 *Let f be a proper convex function and take $x \in \mathrm{dom}\, f$. Prove that*

$$\partial f(x) = \bigcap_{0 < \varepsilon \leq \varepsilon_0} \overline{\mathrm{co}}\left(\partial_\varepsilon f(x) + \{0, \varepsilon\} \partial_{\varepsilon + \delta} f(x)\right),$$

for every $\delta \geq 0$ and $\varepsilon_0 > 0$.

6.6. EXERCISES

Exercise 102 *Consider a convex function f that attains its minimum at $x \in \operatorname{dom} f$. Prove that, for every $\delta \geq 0$ and $\varepsilon_0 > 0$,*

$$\partial f(x) = \bigcap_{0 < \varepsilon \leq \varepsilon_0} \overline{\operatorname{co}} \left(\partial_\varepsilon f(x) \cup \varepsilon \partial_{\varepsilon + \delta} f(x) \right). \tag{6.115}$$

Exercise 103 *Assume that $\{f_t, t \in T\} \subset \Gamma_0(X)$ with T compact and that the mappings $t \mapsto f_t(z)$, $z \in X$, are usc. Denote $f := \sup_{t \in T} f_t$.*
(i) Prove that

$$\partial f(x) \subset \bigcap_{\varepsilon > 0} \overline{\operatorname{co}} \left(\left(\bigcup_{t \in T(x)} \partial_\varepsilon f_t(x) \right) \cup \left(\bigcup_{t \in T \setminus T(x)} \varepsilon \partial_\varepsilon f_t(x) \right) \right), \tag{6.116}$$

and that (6.116) becomes an equality when $\inf_{t \in T} f_t(x) > -\infty$ and x is a minimum of f.

(ii) If we consider the family $\{f_t, t \in T; h\}$, where h is the constant function $h \equiv f(x) - 1$ with $x \in \operatorname{dom} f$ not being a minimum of f, prove that the inclusion in (6.116) may be strict for this new family.

Exercise 104 *Given a family of convex functions, $f_t : X \to \mathbb{R}_\infty$, $t \in T$, and $f := \sup_{t \in T} f_t$, we suppose that condition (6.1) holds. Given $x \in \operatorname{dom} f$, we suppose that $T_c := \{t \in T(x) : f_t \text{ is continuous somewhere in } \operatorname{dom} f\}$ is finite. Prove that, for every $x \in \operatorname{dom} f$, we have*

$$\partial f(x) = \overline{\operatorname{co}} \left\{ \bigcup_{t \in T_c} \partial f_t(x) \bigcup \partial (\tilde{f} + \mathrm{I}_{\operatorname{dom} f})(x) \right\} + \mathrm{N}_{\operatorname{dom} f}(x),$$

where $\tilde{f} := \sup_{t \in T(x) \setminus T_c} f_t$.

Exercise 105 *Complete the proof of Theorem 6.3.2 where, instead of being in $\Gamma_0(X)$, the f_t's satisfy condition (5.10).*

Exercise 106 *Let E be a closed linear subspace of X. Given a nonempty set $A \subset X^*$ and a net $(z_i^*)_i \subset A$, we assume that $(z_i^*)_i \subset (U \cap E)^\circ$ for some $U \in \mathcal{N}_X$. Prove the existence of a subnet $(z_{i_j}^*)_j$ and $z^* \in \operatorname{cl}(A + E^\perp)$ such that $\left\langle z_{i_j}^* - z^*, u \right\rangle \to_j 0$ for all $u \in E$.*

Exercise 107 *Given the convex functions f_t, $t \in T$, we assume that T is compact and that the mappings $t \mapsto f_t(z)$, $z \in X$, are usc. Denote $f := \sup_{t \in T} f_t$ and, for $x \in X$, $\varepsilon_1, \varepsilon_2 \geq 0$ and $L \in \mathcal{F}(x)$, consider the set*

$$E_L := \operatorname{co}\left\{\bigcup_{t\in T_{\varepsilon_2}(x)} \partial_{\varepsilon_1}(f_t + \mathrm{I}_{L\cap\operatorname{dom} f})(x)\right\}.$$

Prove that E_L is closed.

6.7 Bibliographical notes

An early version of Theorem 6.1.4 was established in [53, Theorem 3.6] under slightly weaker conditions of continuity and compactness. The current version of this result is [54, Proposition 1] and allows us to use co instead of $\overline{\operatorname{co}}$. Formula (6.16) is given in [53, Theorem 3.8], whereas (6.21) corresponds to [53, Corollary 3.10]. Corollary 6.1.12, which is [53, Theorem 3.12], gives rise to the well-known characterization (6.24), originally from [191]. Formula (6.25) was also proved in [53, Theorem 3.12].

In section 6.2, the compactification process of the index set is performed using the Stone-Čech and one-point compact extensions; meanwhile, the regularization of the family of data functions is done by enlarging it by means of the associated usc hulls. The main result in this section is Theorem 6.2.5; some consequences of it were given in [54, Theorem 1]. Theorem 6.2.8 is based on the one-point compact extension. Formula (6.73) in Theorem 6.3.2 is [56, Theorem 12, formula (59)], whereas (6.92) can also be found in [204, Proposition 8]. The main results of section 6.4 are Proposition 6.4.1 and Theorem 6.4.3, and are stated in [55, Theorems 5 and 10, respectively]. The first result in section 6.5 is Theorem 6.5.1, given in [52, Theorem 3(ii)]. Theorem 6.5.2 represents an improvement of [52, Theorem 9]. Corollary 6.5.3 in [196] goes back to [177] when the supremum function is continuous at the reference point. Exercise 98 can be found in [102, Lemma 1].

Chapter 7
Other subdifferential calculus rules

The main objective of this chapter is to establish new formulas for the subdifferential of the sum under conditions that are at an intermediate level of generality between those in Proposition 4.1.16, which hold for lsc proper convex functions, and Proposition 4.1.20, which requires additional continuity assumptions. Alternative sufficient conditions for the validity of Proposition 4.1.16 will also be supplied, including the case of polyhedral functions. Throughout this chapter, X stands for an lcs with a dual X^* endowed with a compatible topology (unless otherwise stated). The associated bilinear form is represented by $\langle \cdot, \cdot \rangle$.

7.1 Subdifferential of the sum

Throughout this section, we consider two convex functions $f : Y \to \overline{\mathbb{R}}$ and $g : X \to \overline{\mathbb{R}}$, where Y is another lcs, and a continuous linear mapping $A : X \to Y$ with continuous adjoint A^*. We show that Theorem 5.2.2, which provides a characterization of the subdifferential of suprema, also leads to other calculus rules for functions that can be written as a supremum. This includes the sum of functions and the composition with affine mappings.

First, we give an alternative supremum-based proof to formula (4.45) in Proposition 4.1.16 (for $\varepsilon = 0$). We actually assume a hypothesis that is weaker than the lower semicontinuity assumption required there.

Theorem 7.1.1 *Consider two convex functions $f : Y \to \overline{\mathbb{R}}$ and $g : X \to \overline{\mathbb{R}}$, and a continuous linear mapping $A : X \to Y$ such that*

$$\operatorname{cl}(g + f \circ A) = (\operatorname{cl} g) + (\operatorname{cl} f) \circ A. \tag{7.1}$$

Then, for every $x \in X$, we have

$$\partial(g + f \circ A)(x) = \bigcap_{\varepsilon > 0} \operatorname{cl}\left(\partial_\varepsilon g(x) + A^* \partial_\varepsilon f(Ax)\right). \tag{7.2}$$

Proof. Let us denote $\varphi := g + f \circ A$ and $\psi := (\operatorname{cl} g) + (\operatorname{cl} f) \circ A$. The inclusion "$\supset$" always holds as $\partial_\varepsilon g(x) + A^* \partial_\varepsilon f(Ax) \subset \partial_{2\varepsilon}(g + f \circ A)(x)$ for all $x \in X$ and $\varepsilon > 0$. Consequently, it suffices to establish the opposite inclusion "$\subset$" when $\partial \varphi(x) \neq \emptyset$. In such a case, by (7.1), we have that

$$(\operatorname{cl} g)(x) + (\operatorname{cl} f)(Ax) = (\operatorname{cl} \varphi)(x) = \varphi(x) = g(x) + f(Ax) \in \mathbb{R}, \tag{7.3}$$

and (by Exercise 62)

$$\partial \varphi(x) = \partial (\operatorname{cl} \varphi)(x) = \partial ((\operatorname{cl} g) + (\operatorname{cl} f) \circ A)(x) = \partial \psi(x). \tag{7.4}$$

Since $(\operatorname{cl} g)(x) \leq g(x)$ and $(\operatorname{cl} f)(Ax) \leq f(Ax)$, from (7.3) we get $(\operatorname{cl} g)(x) = g(x) \in \mathbb{R}$ and $(\operatorname{cl} f)(Ax) = f(Ax) \in \mathbb{R}$; therefore, $\operatorname{cl} f \in \Gamma_0(Y)$ and $\operatorname{cl} g \in \Gamma_0(X)$. Furthermore, for every $\varepsilon \geq 0$, one has (again by Exercise 62)

$$\partial_\varepsilon (\operatorname{cl} g)(x) = \partial_\varepsilon g(x) \text{ and } \partial_\varepsilon (\operatorname{cl} f)(Ax) = \partial_\varepsilon f(Ax). \tag{7.5}$$

Now, taking into account that $\operatorname{cl} f \in \Gamma_0(Y)$, Theorem 3.2.2 yields, for every $x \in X$

$$\psi(x) = (\operatorname{cl} g)(x) + f^{**}(Ax) = \sup_{y^* \in \operatorname{dom} f^*} \{(\operatorname{cl} g)(x) + \langle A^* y^*, x \rangle - f^*(y^*)\}.$$

Hence, since the functions $(\operatorname{cl} g)(\cdot) + \langle A^* y^*, \cdot \rangle - f^*(y^*)$, $y^* \in \operatorname{dom} f^*$, are obviously lsc, condition (5.10) trivially holds, and Theorem 5.2.2 gives rise to

7.1. SUBDIFFERENTIAL OF THE SUM

$$\partial \psi(x) = \bigcap_{L \in \mathcal{F}(x), \varepsilon > 0} \mathrm{cl} \left(\mathrm{co} \left(\bigcup_{y^* \in T_\varepsilon(x)} (\partial_\varepsilon(\mathrm{cl}\, g)(x) + A^* y^*) \right) + \mathrm{N}_{L \cap \mathrm{dom}\, \psi}(x) \right),$$

where we also have applied Proposition 4.1.6(i). Now, remembering that $\psi(x) = (\mathrm{cl}\, g)(x) + (\mathrm{cl}\, f)(Ax)$, we write

$$\begin{aligned}
T_\varepsilon(x) &= \{y^* \in Y^* : (\mathrm{cl}\, g)(x) + \langle A^* y^*, x \rangle - f^*(y^*) \geq \psi(x) - \varepsilon\} \\
&= \{y^* \in Y^* : (\psi(x) - (\mathrm{cl}\, f)(Ax)) + \langle A^* y^*, x \rangle - f^*(y^*) \geq \psi(x) - \varepsilon\} \\
&= \{y^* \in Y^* : (\mathrm{cl}\, f)(Ax) + f^*(y^*) \leq \langle y^*, Ax \rangle + \varepsilon\} \\
&= \partial_\varepsilon(\mathrm{cl}\, f)(Ax) = \partial_\varepsilon f(Ax).
\end{aligned}$$

Therefore,

$$\partial \psi(x) = \bigcap_{L \in \mathcal{F}(x), \varepsilon > 0} \mathrm{cl}\left(\partial_\varepsilon(\mathrm{cl}\, g)(x) + A^* \partial_\varepsilon f(Ax) + \mathrm{N}_{L \cap \mathrm{dom}\, \psi}(x) \right). \tag{7.6}$$

Now, take $V \in \mathcal{N}_{X^*}$, $\varepsilon > 0$ and let $L \in \mathcal{F}(x)$ such that $L^\perp \subset (1/2)V$. Then the normal cone $\mathrm{N}_{L \cap \mathrm{dom}\, \psi}(x)$ is written (Exercise 41)

$$\mathrm{N}_{L \cap \mathrm{dom}\, \psi}(x) = \mathrm{N}_{\mathrm{dom}((\mathrm{cl}\, g) + I_L + (\mathrm{cl}\, f) \circ A)}(x) = \left[\mathrm{cl}\left(\partial_\varepsilon(\mathrm{cl}\, g)(x) + L^\perp + A^* \partial_\varepsilon(\mathrm{cl}\, f)(Ax) \right) \right]_\infty,$$

and (7.6), (7.4) and (7.5) lead us to

$$\begin{aligned}
\partial \varphi(x) &= \partial \psi(x) \\
&\subset \mathrm{cl}\left(\partial_\varepsilon(\mathrm{cl}\, g)(x) + A^* \partial_\varepsilon f(Ax) + \left[\mathrm{cl}\left(\partial_\varepsilon(\mathrm{cl}\, g)(x) + L^\perp + A^* \partial_\varepsilon(\mathrm{cl}\, f)(Ax) \right) \right]_\infty \right) \\
&\subset \mathrm{cl}(\partial_\varepsilon(\mathrm{cl}\, g)(x) + A^* \partial_\varepsilon(\mathrm{cl}\, f)(Ax) + L^\perp) \\
&\subset \partial_\varepsilon(\mathrm{cl}\, g)(x) + A^* \partial_\varepsilon(\mathrm{cl}\, f)(Ax) + L^\perp + (1/2)V \\
&\subset \partial_\varepsilon(\mathrm{cl}\, g)(x) + A^* \partial_\varepsilon(\mathrm{cl}\, f)(Ax) + V \\
&= \partial_\varepsilon g(x) + A^* \partial_\varepsilon f(Ax) + V.
\end{aligned}$$

Consequently,

$$\partial \varphi(x) \subset \bigcap_{\varepsilon > 0} \bigcap_{V \in \mathcal{N}_{X^*}} (\partial_\varepsilon g(x) + A^* \partial_\varepsilon f(Ax) + V) = \bigcap_{\varepsilon > 0} \mathrm{cl}\left(\partial_\varepsilon g(x) + A^* \partial_\varepsilon f(Ax) \right),$$

and the proof is complete. ■

Remark 18 *The main motivation of Theorem 7.1.1 is to unify the sum and the supremum calculus rules. An alternative proof, using Proposition 4.1.16, is proposed in Exercises 108 and 109.*

Again using our supremum-based methodology, we provide another proof for formula (4.58) (when $\varepsilon = 0$). First we need the following lemma which shows the effect of continuity conditions on the topological properties of the sets $\partial_\varepsilon g(x) + A^*\partial_\varepsilon f(Ax)$, $\varepsilon \geq 0$.

Lemma 7.1.2 *Let $f : Y \to \overline{\mathbb{R}}$ and $g : X \to \overline{\mathbb{R}}$ be convex functions and $A : X \to Y$ be a continuous linear mapping. Assume that f is continuous at some point in $A(\operatorname{dom} g)$. Then, for every $x \in \operatorname{dom} g \cap A^{-1}(\operatorname{dom} f)$ and $\varepsilon \geq 0$, the set $\partial_\varepsilon g(x) + A^*\partial_\varepsilon f(Ax)$ is closed.*

Proof. Fix $x \in \operatorname{dom} g \cap A^{-1}(\operatorname{dom} f)$ and $\varepsilon \geq 0$. Then it suffices to prove that the convex set $E := \partial_\varepsilon g(x) + A^*\partial_\varepsilon f(Ax)$ is w^*-closed, because we are considering a compatible topology in X^*. Take $x^* \in \operatorname{cl}(E) = \operatorname{cl}^{w^*}(E)$. Let $x_0 \in \operatorname{dom} g$ such that f is continuous at Ax_0. Given nets $(u_i^*)_i \subset \partial_\varepsilon g(x)$ and $(v_i^*)_i \subset \partial_\varepsilon f(Ax)$ such that $u_i^* + A^*v_i^* \to^{w^*} x^*$, we may assume that $\langle u_i^* + A^*v_i^*, x - x_0 \rangle \leq \langle x^*, x - x_0 \rangle + 1$ for every i. Moreover, since $u_i^* \in \partial_\varepsilon g(x)$, we have

$$\langle v_i^*, Ax - Ax_0 \rangle \leq \langle u_i^*, x_0 - x \rangle + \langle x^*, x - x_0 \rangle + 1$$
$$\leq g(x_0) - g(x) + \langle x^*, x - x_0 \rangle + \varepsilon + 1.$$

Now, due to the continuity of f at Ax_0, we can choose $U \in \mathcal{N}_Y$ such that $\sup_{y \in U} f(y + Ax_0) \leq f(Ax_0) + 1$. Hence, by the last inequality above, we get for all $y \in U$

$$\langle v_i^*, y \rangle = \langle v_i^*, Ax - Ax_0 \rangle + \langle v_i^*, y + Ax_0 - Ax \rangle$$
$$\leq \langle v_i^*, Ax - Ax_0 \rangle + f(y + Ax_0) - f(Ax) + \varepsilon$$
$$\leq g(x_0) - g(x) + \langle x^*, x - x_0 \rangle + f(Ax_0) - f(Ax) + 2\varepsilon + 2 \leq \mu,$$

for some $\mu > 0$ (since the expression in the last inequality depends only on x and x_0). In other words, $(v_i^*)_i \subset (\mu^{-1}U)^\circ$ and Theorem 2.1.9 ensures that $(v_i^*)_i$ is w^*-convergent (without loss of generality) to some $v^* \in \partial_\varepsilon f(Ax) \cap (\mu^{-1}U)^\circ$. Moreover, since $u_i^* + A^*v_i^* \to^{w^*} x^*$, the corresponding net $(u_i^*)_i$ also w^*-converges to $u^* := x^* - A^*v^* \in \partial_\varepsilon g(x)$; that is, $x^* = u^* + A^*v^* \in \partial_\varepsilon g(x) + A^*\partial_\varepsilon f(Ax)$, and we are done. ∎

We recover the rules in Proposition 4.1.20 (when $\varepsilon = 0$).

Corollary 7.1.3 *Under the assumptions of Lemma 7.1.2, we have*

$$\partial(g + f \circ A)(x) = \partial g(x) + A^*\partial f(Ax) \text{ for every } x \in X.$$

Proof. Since the inclusion "⊃" trivially holds, we only need to prove "⊂" when $\partial(g + f \circ A)(x) \neq \emptyset$; hence, $g(x), f(Ax) \in \mathbb{R}$. Moreover, using Proposition 2.2.11, the current continuity assumption

7.2. SYMMETRIC VERSUS ASYMMETRIC ...

guarantees that $\operatorname{cl}(g + f \circ A) = (\operatorname{cl} g) + (\operatorname{cl} f) \circ A$, and Theorem 7.1.1 together with Lemma 7.1.2 leads us to

$$\partial(g + f \circ A)(x) = \bigcap_{\varepsilon > 0} \operatorname{cl}\left(\partial_\varepsilon g(x) + A^*\partial_\varepsilon f(Ax)\right) = \bigcap_{\varepsilon > 0} \left(\partial_\varepsilon g(x) + A^*\partial_\varepsilon f(Ax)\right).$$

Thus, the proof will be over once we show that

$$\bigcap_{\varepsilon > 0} \left(\partial_\varepsilon g(x) + A^*\partial_\varepsilon f(Ax)\right) \subset \partial g(x) + A^*\partial f(Ax). \tag{7.7}$$

Take $x^* \in \bigcap_{\varepsilon > 0} \left(\partial_\varepsilon g(x) + A^*\partial_\varepsilon f(Ax)\right)$. Then, for each $r = 1, 2, ...$, there exist $u_r^* \in \partial_{1/r} g(x)$ and $v_r^* \in \partial_{1/r} f(Ax)$ such that $x^* = u_r^* + A^* v_r^*$. Proceeding as in the proof of Lemma 7.1.2, we choose $x_0 \in \operatorname{dom} g$ and $U \in \mathcal{N}_Y$ such that $\sup_{y \in U} f(y + Ax_0) \leq f(Ax_0) + 1$. Hence, for all $y \in U$,

$$\begin{aligned}
\langle v_r^*, y \rangle &= \langle v_r^*, Ax - Ax_0 \rangle + \langle v_r^*, y + Ax_0 - Ax \rangle \\
&\leq \langle x^*, x - x_0 \rangle + \langle u_r^*, x_0 - x \rangle + f(y + Ax_0) - f(Ax) + 1/r \\
&\leq \langle x^*, x - x_0 \rangle + g(x_0) - g(x) + f(Ax_0) - f(Ax) + 1 + 2/r \leq \mu,
\end{aligned}$$

where $\mu > 0$. Then $(v_r^*)_r \subset (\mu^{-1} U)^\circ$ and, without loss of generality, we may assume that $(v_r^*)_r$ and $(u_r^*)_r$ are w^*-convergent to some $v^* \in X^*$ and $x^* - A^* v^*$, respectively. Since $u_r^* \in \partial_{1/r} g(x)$ and $v_r^* \in \partial_{1/r} f(Ax)$ for all $r \geq 1$, by Proposition 4.1.6(vi), we deduce that $u^* \in \partial g(x)$ and $v^* \in \partial f(Ax)$. Thus, $x^* = u^* + A^* v^* \in \partial g(x) + A^* \partial f(Ax)$ and (7.7) follows. ∎

7.2 Symmetric versus asymmetric conditions

As in the previous section, we consider two convex functions $f : Y \to \overline{\mathbb{R}}$ and $g : X \to \overline{\mathbb{R}}$, where Y is another lcs, and a continuous linear mapping $A : X \to Y$ with continuous adjoint A^*. We establish a subdifferential sum rule that is given in terms of the exact subdifferential of one function and the approximate subdifferential of the other. The given formula requires less restrictive conditions than those of Theorem 7.1.1 and Corollary 7.1.3. More precisely, Corollary 7.1.3 requires the continuity of, say, g at some point in $\operatorname{dom} f$, while its finite-dimensional counterpart in Proposition 4.1.26 appeals to the condition

$\operatorname{ri}(\operatorname{dom} f) \cap \operatorname{ri}(\operatorname{dom} g) \neq \emptyset$. If one of the functions, say, g is polyhedral, it is enough to assume that $\operatorname{ri}(\operatorname{dom} f) \cap \operatorname{dom} g \neq \emptyset$. Also, Theorem 7.1.1 uses condition (7.1); that is, $\operatorname{cl}(g + f) = (\operatorname{cl} g) + (\operatorname{cl} f)$. In the following theorem, $X = Y$ and A is the identity mapping.

Theorem 7.2.1 *Consider two functions $f, g \in \Gamma_0(X)$, and $x \in \operatorname{dom} f \cap \operatorname{dom} g$. We assume that at least one of the following conditions holds:*
  (i) $\mathbb{R}_+(\operatorname{epi} g - (x, g(x)))$ is closed,
  (ii) $\operatorname{dom} f \cap \operatorname{ri}(\operatorname{dom} g) \neq \emptyset$ and $g_{|\operatorname{aff}(\operatorname{dom} g)}$ is continuous on $\operatorname{ri}(\operatorname{dom} g)$.
Then we have

$$\partial(f + g)(x) = \bigcap_{\varepsilon > 0} \operatorname{cl}(\partial_\varepsilon f(x) + \partial g(x)). \tag{7.8}$$

Proof. According to Theorem 7.1.1 (see, also, Proposition 4.1.16), we have
$$\partial(f + g)(x) = \bigcap_{\varepsilon > 0} \operatorname{cl}(\partial_\varepsilon f(x) + \partial_\varepsilon g(x)). \tag{7.9}$$

Hence, the inclusion "$\supset$" in (7.8) follows immediately. To show "$\subset$" we pick x^* in $\partial(f + g)(x)$ (if $\partial(f + g)(x)$ is empty, the inclusion is trivial). Then, since $\partial_\varepsilon g(x) \times \{-1\} \subset N^\varepsilon_{\operatorname{epi} g}(x, g(x))$, (by Proposition 4.1.6(vii)), (7.9) yields,

$$(x^*, -1) \in \bigcap_{\varepsilon > 0} \operatorname{cl}\left(\partial_\varepsilon f(x) \times \{0\} + \partial_\varepsilon g(x) \times \{-1\}\right)$$
$$\subset \bigcap_{\varepsilon > 0} \operatorname{cl}\left(\partial_\varepsilon f(x) \times \{0\} + N^\varepsilon_{\operatorname{epi} g}(x, g(x))\right). \tag{7.10}$$

We appeal now to Exercise 79. If (i) holds, then it follows from the last inclusion that

$$(x^*, -1) \in \bigcap_{\varepsilon > 0} \operatorname{cl}\left(\partial_\varepsilon f(x) \times \{0\} + N_{\operatorname{epi} g}(x, g(x))\right). \tag{7.11}$$

Otherwise, if (ii) holds, then we can check that (see Exercise 19)

$$(\operatorname{ri}(\operatorname{epi} g - (x, g(x)))) \cap ((\operatorname{dom} f - x) \times \mathbb{R}) \neq \emptyset,$$

and so, since

$$(\operatorname{dom} f - x) \times \mathbb{R} \subset \left(\operatorname{dom} \sigma_{\partial_\varepsilon f(x)}\right) \times \mathbb{R} = \operatorname{dom} \sigma_{\partial_\varepsilon f(x) \times \{0\}},$$

7.2. SYMMETRIC VERSUS ASYMMETRIC...

we conclude that

$$(\text{ri}(\text{epi}\, g - (x, g(x)))) \cap \text{dom}\, \sigma_{\partial_\varepsilon f(x) \times \{0\}} \neq \emptyset.$$

Therefore, (7.11) also follows in the present case by Exercise 79.

Now we claim that, for each $\varepsilon > 0$ and $u \in X$,

$$\sigma_{\partial(f+g)(x)}(u) \leq \sigma_{\partial_\varepsilon f(x) + \partial g(x)}(u), \tag{7.12}$$

as this clearly entails the inclusion "$\subset$" in (7.8). To prove (7.12), we take $\alpha < \sigma_{\partial(f+g)(x)}(u)$. Let $u^* \in \partial(f+g)(x)$ such that $\alpha < \langle u^*, u \rangle$. Then, using (7.11), we find nets $(y_i^*)_i \subset \partial_\varepsilon f(x)$, $(z_i^*)_i \subset X^*$ and $(\beta_i)_i \subset \mathbb{R}_+$ such that $((z_i^*, -\beta_i))_i \subset N_{\text{epi}\, g}(x, g(x))$, $\beta_i \to 1$, and $u^* = \lim_i (y_i^* + z_i^*)$. Since $\beta_i \to 1$, we can suppose that $\beta_i > 0$, and then $\beta_i^{-1} z_i^* \in \partial g(x)$ (thanks again to Proposition 4.1.6(vii)). Consequently, $\partial g(x) \neq \emptyset$ and $\sigma_{\partial_\varepsilon f(x) + \partial g(x)}(u) = \sigma_{\partial_\varepsilon f(x)}(u) + \sigma_{\partial g(x)}(u)$. Then, writing

$$\sigma_{\partial_\varepsilon f(x)}(u) + \beta_i \sigma_{\partial g(x)}(u) \geq \langle y_i^*, u \rangle + \beta_i \langle \beta_i^{-1} z_i^*, u \rangle = \langle y_i^* + z_i^*, u \rangle,$$

and taking limits, we get

$$\sigma_{\partial_\varepsilon f(x) + \partial g(x)}(u) = \lim_i (\sigma_{\partial_\varepsilon f(x)}(u) + \beta_i \sigma_{\partial g(x)}(u)) \geq \lim_i \langle y_i^* + z_i^*, u \rangle = \langle u^*, u \rangle > \alpha.$$

Therefore, (7.12) follows when α approaches the value $\sigma_{\partial(f+g)(x)}(u)$. ∎

The following result produces a nearly exact subdifferential rule for convex functions whose domains or epigraphs overlap sufficiently. Statement (iii) below has been proven in Corollary 4.1.27(ii) in the case of indicator functions.

Theorem 7.2.2 *Let functions $f, g \in \Gamma_0(X)$. Given $x \in \text{dom}\, f \cap \text{dom}\, g$, we assume that one of the following conditions holds:*

(i) $\mathbb{R}_+(\text{epi}\, f - (x, f(x)))$ is closed, $\text{dom}\, f \cap \text{ri}(\text{dom}\, g) \neq \emptyset$ and $g_{|\text{aff}(\text{dom}\, g)}$ is continuous on $\text{ri}(\text{dom}\, g)$.

(ii) $\mathbb{R}_+(\text{epi}\, f - (x, f(x)))$ and $\mathbb{R}_+(\text{epi}\, g - (x, g(x)))$ are closed.

(iii) $\text{ri}(\text{dom}\, f) \cap \text{ri}(\text{dom}\, g) \neq \emptyset$ and $f_{|\text{aff}(\text{dom}\, f)}$ and $g_{|\text{aff}(\text{dom}\, g)}$ are continuous on $\text{ri}(\text{dom}\, f)$ and $\text{ri}(\text{dom}\, g)$, respectively.

Then we have

$$\partial(f+g)(x) = \text{cl}\,(\partial f(x) + \partial g(x)). \tag{7.13}$$

Proof. We only need to prove that, for every $x \in X$ such that $\partial(f+g)(x) \neq \emptyset$,

$$\partial(f+g)(x) \subset \mathrm{cl}(\partial f(x) + \partial g(x)), \tag{7.14}$$

since the converse inclusion is straightforward. We take $x^* \in \partial(f+g)(x)$. So, arguing as in the proof of Theorem 7.2.1 (statement (7.11)), we show that

$$(x^*, -1) \in \bigcap_{\varepsilon > 0} \mathrm{cl}\left(\partial_\varepsilon g(x) \times \{0\} + \mathrm{N}_{\mathrm{epi}\, f}(x, g(x))\right), \tag{7.15}$$

under conditions (i) and (ii), and also under (iii), thanks to

$$\mathrm{ri}(\mathrm{epi}\, f - (x, f(x))) \cap \mathrm{dom}\, \sigma_{\partial_\varepsilon g(x) \times \{0\}} \neq \emptyset.$$

Similarly, (7.15) implies that $\sigma_{\partial(f+g)(x)}(u) \leq \sigma_{\partial f(x) + \partial_\varepsilon g(x)}(u)$, for all $\varepsilon > 0$ and $u \in X$, which in turn leads us to $x^* \in \bigcap_{\varepsilon > 0} \mathrm{cl}(\partial f(x) + \partial_\varepsilon g(x))$. In order to remove ε from the last expression, we repeat the same arguments as in the previous theorem to obtain that

$$(x^*, -1) \in \bigcap_{\varepsilon > 0} \mathrm{cl}\left(\partial f(x) \times \{0\} + \mathrm{N}_{\mathrm{epi}\, g}(x, g(x))\right), \tag{7.16}$$

leading us to the desired inclusion. ∎

Remark 19 *If in Theorem 7.2.2 f, g are the indicators functions of two closed linear subspaces C, $D \subset X$, respectively, then (7.13) yields the following formula (see, also, Exercise 53 for a standard proof based on the bipolar theorem),*

$$(C \cap D)^\perp = \partial(\mathrm{I}_C + \mathrm{I}_D)(\theta) = \mathrm{cl}\left(C^\perp + D^\perp\right).$$

Note that the closure here can be removed when X is finite-dimensional.

Theorem 7.2.1 is applied below to derive a calculus rule for the subdifferential of the composition with a linear mapping. This result will be used in Theorem 7.2.5.

Theorem 7.2.3 *Consider $f \in \Gamma_0(Y)$ and a continuous linear mapping $A: X \to Y$. Given $x \in A^{-1}(\mathrm{dom}\, f)$, we assume that at least one of the following conditions holds:*
  (i) $\mathbb{R}_+(\mathrm{epi}\, f - (Ax, f(Ax)))$ is closed.
  (ii) $A(X) \cap \mathrm{ri}(\mathrm{dom}\, f) \neq \emptyset$ and $f_{|\mathrm{aff}(\mathrm{dom}\, f)}$ is continuous at some point in $\mathrm{ri}(\mathrm{dom}\, f)$.

7.2. SYMMETRIC VERSUS ASYMMETRIC ...

Then we have that

$$\partial(f \circ A)(x) = \mathrm{cl}(A^*\partial f(Ax)). \tag{7.17}$$

Proof. We consider the convex functions $g, h : X \times Y \to \mathbb{R}_\infty$ defined by

$$g(x, y) := f(y) \text{ and } h(x, y) := I_{\mathrm{gph}\, A}(x, y),$$

which satisfy (Exercise 46)

$$\partial(f \circ A)(x) \times \{\theta\} = \partial(g + h)(x, Ax) \cap (X^* \times \{\theta\}). \tag{7.18}$$

Obviously, $\mathrm{dom}\, g = X \times \mathrm{dom}\, f$ and $\mathrm{dom}\, h = \mathrm{gph}\, A$, and we have that $g, h \in \Gamma_0(X \times Y)$. Then, on the one hand, from the relation

$$\mathbb{R}_+(\mathrm{epi}\, h - ((x, Ax), 0)) = \mathbb{R}_+(((\mathrm{gph}\, A) \times \mathbb{R}_+) - ((x, Ax), 0)) = (\mathrm{gph}\, A) \times \mathbb{R}_+,$$

it follows that $\mathbb{R}_+(\mathrm{epi}\, h - ((x, Ax), 0))$ is a closed set. On the other hand, we have

$$\mathbb{R}_+(\mathrm{epi}\, g - ((x, Ax), f(Ax))) = \mathbb{R}_+((X \times \mathrm{epi}\, f) - ((x, Ax), f(Ax)))$$
$$= X \times (\mathbb{R}_+(\mathrm{epi}\, f - (Ax, f(Ax)))).$$

If (i) holds, then $\mathbb{R}_+(\mathrm{epi}\, g - ((x, Ax), f(Ax)))$ is closed, and so, by Theorem 7.2.2(ii),

$$\partial(g+h)(x, Ax) = \mathrm{cl}(\partial g(x, Ax) + \partial I_{\mathrm{gph}\, A}(x, Ax)) = \mathrm{cl}(\{\theta\} \times \partial f(Ax) + N_{\mathrm{gph}\, A}(x, Ax)),$$

where
$$N_{\mathrm{gph}\, A}(x, Ax) = \{(u^*, v^*) \in X^* \times Y^* : \langle u^*, z - x \rangle + \langle v^*, A(z-x) \rangle \le 0 \text{ for all } z \in X\}$$
$$= \{(u^*, v^*) \in X^* \times Y^* : \langle u^* + A^*v^*, z - x \rangle \le 0 \text{ for all } z \in X\}$$
$$= \{(A^*v^*, -v^*) : v^* \in Y^*\}.$$

Thus, using (7.18),

$$\partial(f \circ A)(x) \times \{\theta\} = \mathrm{cl}\{(A^*v^*, y^* - v^*) : y^* \in \partial f(Ax), v^* \in Y^*\} \cap (X^* \times \{\theta\}). \tag{7.19}$$

In other words, if $x^* \in \partial(f \circ A)(x)$, then we find nets $(y_i^*)_i \subset \partial f(Ax)$ and $(v_i^*)_i \subset Y^*$ such that $x^* = \lim_i A^*v_i^*$ and $\lim_i(y_i^* - v_i^*) = \theta$. Since A^* is continuous, we deduce that

$$x^* = \lim_i A^*v_i^* = \lim_i A^*y_i^* \in \mathrm{cl}(A^*\partial f(Ax)),$$

and the inclusion $\partial(f \circ A)(x) \subset \mathrm{cl}\,(A^*\partial f(Ax))$ holds. The opposite inclusion easily follows.

If (ii) is fulfilled and $y_0 := Ax_0 \in \mathrm{ri}(\mathrm{dom}\,f)$ for some $x_0 \in X$, then

$$(x_0, y_0) \in (\mathrm{gph}\,A) \cap (X \times \mathrm{ri}(\mathrm{dom}\,f)) = (\mathrm{gph}\,A) \cap \mathrm{ri}(X \times \mathrm{dom}\,f) = \mathrm{dom}\,h \cap \mathrm{ri}(\mathrm{dom}\,g),$$

and $g_{|\mathrm{aff}(\mathrm{dom}\,g)}$ is continuous relative to $\mathrm{ri}(\mathrm{dom}\,g) = X \times \mathrm{ri}(\mathrm{dom}\,f)$. Thus, we conclude as in the first part of the proof, but now applying Theorem 7.2.2(i). ∎

Conditions (iii) in Theorem 7.2.2 and (ii) in Theorem 7.2.3 are clearly weaker than the continuity assumption used in Proposition 4.1.20. However, the conclusion of the latter is stronger since it produces formulas without the closure. The following corollary shows that the closure in formula (7.13) of Theorem 7.2.2 can be sometimes ruled out.

Corollary 7.2.4 *Let f and g be the same as in Theorem 7.2.2(iii). If $x \in X$ is such that $\partial f(x)$ or $\partial g(x)$ is locally compact, then*

$$\partial(f+g)(x) = \partial f(x) + \partial g(x).$$

Proof. We may assume that $f(x), g(x) \in \mathbb{R}$. According to Theorem 7.2.2(iii), we have that $\partial(f+g)(x) = \mathrm{cl}(\partial f(x) + \partial g(x))$. If $\partial f(x)$ or $\partial g(x)$ is empty, then we are done. So, we may assume that $\partial f(x)$, $\partial g(x)$, and $\partial(f+g)(x)$ are non-empty; hence, $x \in \mathrm{dom}\,f \cap \mathrm{dom}\,g$. Assuming that $\theta \in \mathrm{ri}(\mathrm{dom}\,f) \cap \mathrm{ri}(\mathrm{dom}\,g)$, without loss of generality, we choose $U \in \mathcal{N}_X$ such that

$$U_1 := U \cap \mathrm{span}(\mathrm{dom}\,f) \subset \mathrm{dom}\,f, U_2 := U \cap \mathrm{span}(\mathrm{dom}\,g) \subset \mathrm{dom}\,g;$$

hence U_1 and U_2 are absorbing, balanced, and convex neighborhoods of θ in $\mathrm{span}(\mathrm{dom}\,f)$ and $\mathrm{span}(\mathrm{dom}\,g)$, respectively. Take $v^* \in (\partial f(x))_\infty \cap (-(\partial g(x))_\infty) = \mathrm{N}_{\mathrm{dom}\,f}(x) \cap (-\mathrm{N}_{\mathrm{dom}\,g}(x))$, by (4.9). Then, for every $u_1 \in U_1$ and $u_2 \in U_2$, we obtain

$$\langle v^*, u_1 - x \rangle \leq 0 \text{ and } \langle -v^*, u_2 - x \rangle \leq 0, \qquad (7.20)$$

and by summing up, $\langle v^*, u_1 - u_2 \rangle \leq 0$. Moreover, since $-u_1 \in U_1$ and $-u_2 \in U_2$, we also have $\langle v^*, u_2 - u_1 \rangle \leq 0$ for all $u_1 \in U_1$ and $u_2 \in U_2$; that is, $v^* \in (\mathrm{span}(\mathrm{dom}\,f - \mathrm{dom}\,g))^-$. Then

$$\mathrm{N}_{\mathrm{dom}\,f}(x) \cap (-\mathrm{N}_{\mathrm{dom}\,g}(x)) \subset (\mathrm{span}(\mathrm{dom}\,f - \mathrm{dom}\,g))^-,$$

7.2. SYMMETRIC VERSUS ASYMMETRIC ...

and observing that

$$(\text{span}(\text{dom } f - \text{dom } g))^- \subset (\text{dom } f - x)^- \cap \left(-\left[(\text{dom } g - x)^-\right]\right) = N_{\text{dom } f}(x) \cap (-N_{\text{dom } g}(x)),$$

we deduce that

$$N_{\text{dom } f}(x) \cap (-N_{\text{dom } g}(x)) = (\text{span}(\text{dom } f - \text{dom } g))^-.$$

Therefore, the closedness of $\partial f(x) + \partial g(x)$ follows by Theorem 2.1.8. ∎

The following theorem deals with a typical situation in which finite and infinite-dimensional settings coexist (as in Remark 2). The resulting formula only involves, once again, the exact subdifferential of the qualified function as in Theorem 7.2.1.

Theorem 7.2.5 *Consider $f \in \Gamma_0(\mathbb{R}^n)$, $g \in \Gamma_0(X)$, and a continuous linear mapping $A : X \to \mathbb{R}^n$. Given $x \in \text{dom } g \cap A^{-1}(\text{dom } f)$, we assume that one of the following conditions holds:*
(i) $\mathbb{R}_+(\text{epi } f - (Ax, f(Ax)))$ is closed in $\mathbb{R}^{n+1}$.
(ii) $(A(\text{dom } g)) \cap \text{ri}(\text{dom } f) \neq \emptyset$.
Then we have

$$\partial(g + f \circ A)(x) = \bigcap_{\varepsilon > 0} \text{cl}(\partial_\varepsilon g(x) + A^* \partial f(Ax)).$$

Proof. We suppose that $\partial(g + f \circ A)(x) \neq \emptyset$, so that the conditions of Theorem 7.2.3 are fulfilled, and we obtain

$$\partial(f \circ A)(x) = \text{cl}(A^* \partial f(Ax)). \tag{7.21}$$

Hence, we only need to check that the lsc convex functions $f \circ A$ and g satisfy the assumptions of Theorem 7.2.1, with $f \circ A$ playing the role of the qualified function; that is, $f \circ A$ satisfies

$$\mathbb{R}_+(\text{epi}(f \circ A) - (x, (f(Ax)))) \text{ is closed}, \tag{7.22}$$

and

$$\begin{cases} \text{dom } g \cap \text{ri}(\text{dom}(f \circ A)) \neq \emptyset \text{ and} \\ (f \circ A)_{|\text{aff}(\text{dom}(f \circ A))} \text{ is continuous on } \text{ri}(\text{dom}(f \circ A)). \end{cases} \tag{7.23}$$

First, we assume that (i) holds. To show (7.22) we take

$$(u, \mu) \in \text{cl}(\mathbb{R}_+(\text{epi}(f \circ A) - (x, (f(Ax))))),$$

and consider nets $(\alpha_i)_i \subset \mathbb{R}_+$ and $(x_i, \lambda_i)_i \subset \operatorname{epi}(f \circ A)$ such that $\alpha_i((x_i, \lambda_i) - (x, f(Ax))) \to (u, \mu)$; hence,
$$(Ax_i, \lambda_i) \in \operatorname{epi} f \text{ and } \alpha_i((Ax_i, \lambda_i) - (Ax, f(Ax))) \to (Au, \mu).$$

Thus, condition (i) leads us to $(Au, \mu) \in \mathbb{R}_+(\operatorname{epi} f - (Ax, f(Ax)))$, and there are $\alpha \in \mathbb{R}_+$ and $(y, \lambda) \in \operatorname{epi} f$ such that $(Au, \mu) = \alpha((y, \lambda) - (Ax, f(Ax)))$; that is, $Au = \alpha(y - Ax)$ and $\mu = \alpha(\lambda - f(Ax))$. Next, thanks to the convexity of f, for all $\gamma > 0$ small enough to satisfy $\alpha\gamma < 1$, we obtain

$$\begin{aligned}(f \circ A)(\gamma u + x) &= f(A(\gamma u + x)) = f(\gamma\alpha(y - Ax) + Ax) \\ &\leq \alpha\gamma f(y) + (1 - \gamma\alpha)f(Ax) \leq \alpha\gamma\lambda + (1 - \gamma\alpha)f(Ax) \\ &= \gamma(\mu + \alpha f(Ax)) + (1 - \gamma\alpha)f(Ax) = f(Ax) + \gamma\mu,\end{aligned}$$

and so
$$(u, \mu) \in \gamma^{-1}(\operatorname{epi}(f \circ A) - (x, f(Ax))),$$

showing that (7.22) holds.

Now, assuming (ii), we prove that (7.23) holds. We choose $x_0 \in \operatorname{dom} g$ such that $y_0 := Ax_0 \in \operatorname{ri}(\operatorname{dom} f)$. Thus, taking into account Corollary 2.2.9, there are some $m \geq 0$ and $V \in \mathcal{N}_{\mathbb{R}^n}$ (i.e., a ball centered at 0_n) such that

$$f(y_0 + y) \leq m \text{ for all } y \in V \cap \operatorname{aff}(\operatorname{dom} f).$$

Let $U \in \mathcal{N}_X$ such that $A(U) \subset V$. Then, for every $z \in (x_0 + U) \cap \operatorname{aff}(\operatorname{dom}(f \circ A))$, we have

$$(f \circ A)(z) \leq \sup_{u \in U} f(y_0 + Au) \leq \sup_{y \in V} f(y_0 + y) \leq m,$$

and, particularly, $x_0 \in \operatorname{ri}(\operatorname{dom}(f \circ A))$; that is,

$$x_0 \in \operatorname{dom} g \cap \operatorname{ri}(\operatorname{dom}(f \circ A)).$$

Therefore, (7.23) follows. Finally, taking into account (7.21), Theorem 7.2.1 entails

$$\begin{aligned}\partial(g + f \circ A)(x) &= \bigcap_{\varepsilon > 0} \operatorname{cl}(\partial_\varepsilon g(x) + \partial(f \circ A)(x)) \\ &= \bigcap_{\varepsilon > 0} \operatorname{cl}(\partial_\varepsilon g(x) + \operatorname{cl}(A^* \partial f(Ax))) = \bigcap_{\varepsilon > 0} \operatorname{cl}(\partial_\varepsilon g(x) + A^* \partial f(Ax)),\end{aligned}$$

yielding the conclusion of the theorem. ■

7.2. SYMMETRIC VERSUS ASYMMETRIC ...

The proof of the following result is similar to that of Theorem 7.2.5, but its proof is based on Theorem 7.2.2 instead of Theorem 7.2.1 (see Exercise 111 for the proof).

Corollary 7.2.6 *With the notation of Theorem 7.2.5, assume that the sets $\mathbb{R}_+(\text{epi } f - (Ax, f(Ax)))$ and $\mathbb{R}_+(\text{epi } g - (x, g(x)))$ are closed. Then we have that*

$$\partial(g + f \circ A)(x) = \text{cl}(\partial g(x) + A^*(\partial f(Ax))).$$

The following corollary establishes a sequential rule. It only uses exact subdifferentials at the reference point of the qualified function (the one whose relative interior or epigraph is involved in the assumption), while the subdifferential of the other function is taken at nearby points. This result is stated in Banach spaces because it appeals to Proposition 4.3.8 (or 4.3.7). The reflexivity assumption comes to justify the use of sequences instead of nets.

Corollary 7.2.7 *Assume that X is a reflexive Banach space, and let $f, g \in \Gamma_0(X)$. Given $x \in \text{dom } f \cap \text{dom } g$, we assume that at least one of conditions (i) and (ii) in Theorem 7.2.1 holds. Then, $x^* \in \partial(f + g)(x)$ if and only if there are sequences $(x_n)_n \subset X$ and $(x_n^*)_n \subset X$, $(y_n^*)_n \subset \partial g(x)$ such that $x_n^* \in \partial f(x_n)$, for all $n \geq 1$, and*

$$x_n \to x, \; f(x_n) + \langle x_n^*, x - x_n \rangle \to f(x), \; \text{and } x_n^* + y_n^* \to x^*.$$

Proof. Fix $x^* \in \partial(f + g)(x)$. Then, using Theorem 7.2.1 and the reflexivity assumption, we have

$$\partial(f + g)(x) = \bigcap_{\varepsilon > 0} \text{cl}^{\|\cdot\|}(\partial_\varepsilon f(x) + \partial g(x)),$$

where $\|\cdot\|$ refers to the dual norm in X^*. So, for each integer $n \geq 1$, there are $z_n^* \in \partial_{1/n^2} f(x)$ and $y_n^* \in \partial g(x)$ such that $x^* \in z_n^* + y_n^* + (1/n)B_{X^*}$. Now, appealing to Proposition 4.3.8, we find $x_n \in x + (1/n)B_X$ and $x_n^* \in \partial f(x_n)$ such that $\|x_n^* - z_n^*\| \leq 1/n$ and $|f(x_n) - f(x) + \langle z_n^*, x - x_n \rangle| \leq 2/n^2$. Thus, using the Cauchy–Schwarz inequality,

$$|f(x_n) - f(x) + \langle x_n^*, x - x_n \rangle| \leq |f(x_n) - f(x) + \langle z_n^*, x - x_n \rangle| + |\langle z_n^* - x_n^*, x - x_n \rangle|$$

$$\leq \frac{2}{n^2} + \|x_n^* - z_n^*\| \|x - x_n\| \leq \frac{3}{n^2}.$$

Consequently, $x^* \in x_n^* + y_n^* + (2/n)B_{X^*}$, and so $\lim_n (x_n^* + y_n^*) = x^*$.

To prove the opposite implication, we take sequences $(x_n)_n \subset X$ and $(x_n^*)_n \subset X^*$ such that $x_n^* \in \partial f(x_n)$, $(y_n^*)_n \subset \partial g(x)$, $x_n \to x$, $f(x_n) + \langle x_n^*, x - x_n \rangle \to f(x)$ and $x_n^* + y_n^* \to x^*$. Hence, for all $z \in X$,

$$\begin{aligned}\langle x_n^* + y_n^*, z - x \rangle &= \langle x_n^*, z - x_n \rangle + \langle y_n^*, z - x \rangle + \langle x_n^*, x_n - x \rangle \\ &\leq f(z) - f(x_n) + \langle x_n^*, x_n - x \rangle + g(z) - g(x) \\ &= f(z) - f(x) + g(z) - g(x) + f(x) - f(x_n) + \langle x_n^*, x_n - x \rangle,\end{aligned}$$

and, taking the limits when $n \to \infty$, we get

$$\langle x^*, z - x \rangle \leq f(z) - f(x) + g(z) - g(x) + f(x),$$

showing that $x^* \in \partial(f + g)(x)$. ∎

7.3 Supremum-sum subdifferential calculus

In this section, we deal jointly with a non-empty family of convex functions $f_t : X \to \overline{\mathbb{R}}$, $t \in T$, together with a distinguished convex function $g : X \to \overline{\mathbb{R}}$. If $f := \sup_{t \in T} f_t$, the main purpose of this section is to characterize the subdifferential of the sum $f + g$ by means, exclusively, of the ε-subdifferential of the functions f_t, $t \in T$, and the (exact) subdifferential of g. It is worth observing that, given that $f + g = \sup_{t \in T}(f_t + g)$, we can apply Theorem 5.2.2 to obtain the following formula

$$\partial(f+g)(x) = \bigcap_{\varepsilon > 0,\, L \in \mathcal{F}(x)} \overline{\operatorname{co}}\left\{\bigcup_{t \in T_\varepsilon(x)} \partial_\varepsilon f_t(x) + \partial_\varepsilon g(x) + \mathrm{N}_{L \cap \operatorname{dom} f \cap \operatorname{dom} g}(x)\right\},$$

where $T_\varepsilon(x) = \{t \in T : f_t(x) \geq f(x) - \varepsilon\}$. This expression uses the ε-subdifferential of f_t and g (see details in Exercise 113). However, our purpose here is to provide formulas involving the (exact) subdifferential of the functions $g + \mathrm{I}_{L \cap \operatorname{dom} f}$, $L \in \mathcal{F}(x)$, rather than the ε-subdifferential of g.

Theorem 7.3.2 below states the main result of this section, constituting the desired extension of (5.26). In its proof, we use the following result, which involves the family $\mathcal{F}$ of finite-dimensional linear subspaces of X.

Proposition 7.3.1 *Let $f_t, g : X \to \mathbb{R}_\infty$, $t \in T$, be proper convex functions and consider $f := \sup_{t \in T} f_t$. Assume that*

7.3. SUPREMUM-SUM SUBDIFFERENTIAL ...

$$\operatorname{cl}(f+g)(x) = \sup_{t \in T}(\operatorname{cl} f_t)(x) + g(x) \ \text{for all} \ x \in \operatorname{dom} f \cap \operatorname{dom} g.$$
(7.24)

Then we have

$$\operatorname{cl}(f+g) = \operatorname{cl}\left(\sup_{t \in T}(\operatorname{cl} f_t) + \inf_{L \in \mathcal{F}} \operatorname{cl}(g + \mathrm{I}_{L \cap \operatorname{dom} f})\right). \quad (7.25)$$

Proof. Let us denote

$$\tilde{f} := \sup_{t \in T}(\operatorname{cl} f_t) \ \text{and} \ g_L := g + \mathrm{I}_{L \cap \operatorname{dom} f}, L \in \mathcal{F}. \quad (7.26)$$

Observe that, for every $F \in \mathcal{F}$,

$$\tilde{f} + \inf_{L \in \mathcal{F}}(\operatorname{cl} g_L) \leq \tilde{f} + \operatorname{cl} g_F \leq \tilde{f} + g_F \leq f + g + \mathrm{I}_F,$$

and by taking the infimum over $F \in \mathcal{F}$, we get

$$\tilde{f} + \inf_{L \in \mathcal{F}}(\operatorname{cl} g_L) \leq \inf_{F \in \mathcal{F}}(f + g + \mathrm{I}_F) = f + g + \mathrm{I}_{\cup\{F : F \in \mathcal{F}\}} = f + g.$$

Thus, passing to the closure in each side, $\operatorname{cl}\left(\tilde{f} + \inf_{L \in \mathcal{F}}(\operatorname{cl} g_L)\right) \leq \operatorname{cl}(f+g)$, and the inequality "$\geq$" in (7.25) follows. To establish the inequality "$\leq$" in (7.25) we fix $L \in \mathcal{F}$ and take $z \in \operatorname{cl}(L \cap \operatorname{dom} f \cap \operatorname{dom} g)$. Since $\operatorname{dom}(g + \mathrm{I}_{L \cap \operatorname{dom} f}) \subset L$, we pick $z_0 \in \operatorname{ri}(L \cap \operatorname{dom} f \cap \operatorname{dom} g)$. Then, for every $\lambda \in \,]0, 1[$, (2.15) implies that

$$z_\lambda := \lambda z_0 + (1 - \lambda)z \in \operatorname{ri}(L \cap \operatorname{dom} f \cap \operatorname{dom} g).$$

Moreover, since $\operatorname{dom}(f + g + \mathrm{I}_L) = L \cap \operatorname{dom} f \cap \operatorname{dom} g = \operatorname{dom} g_L$, we have $z_\lambda \in \operatorname{ri}(\operatorname{dom} g_L)$ and $g_L(z_\lambda) = (\operatorname{cl} g_L)(z_\lambda)$, due to Corollary 2.2.9. Thus, the current assumption and the convexity of the data functions entail

$$\operatorname{cl}(f+g)(z_\lambda) = \tilde{f}(z_\lambda) + g_L(z_\lambda) = \tilde{f}(z_\lambda) + (\operatorname{cl} g_L)(z_\lambda)$$
$$\leq (1-\lambda)\tilde{f}(z) + \lambda \tilde{f}(z_0) + (1-\lambda)(\operatorname{cl} g_L)(z) + \lambda(\operatorname{cl} g_L)(z_0).$$

As $\tilde{f}(z_0) < +\infty$ and $(\operatorname{cl} g_L)(z_0) = g(z_0) < +\infty$, by taking limits over $\lambda \downarrow 0$, we get

$$\operatorname{cl}(f+g)(z) \leq \liminf_{\lambda \downarrow 0} \operatorname{cl}(f+g)(z_\lambda) \leq \tilde{f}(z) + (\operatorname{cl} g_L)(z).$$

Since this last inequality also holds when $z \notin \operatorname{cl}(L \cap \operatorname{dom} f \cap \operatorname{dom} g)$, we deduce that

$$\operatorname{cl}(f+g)(z) \leq \tilde{f}(z) + (\operatorname{cl} g_L)(z) \text{ for all } z \in X. \tag{7.27}$$

Consequently, by taking the infimum over $L \in \mathcal{F}$, we get $\operatorname{cl}(f+g) \leq \tilde{f} + \inf_{L \in \mathcal{F}}(\operatorname{cl} g_L)$, and this implies the desired inequality "$\leq$" in (7.25). ∎

Theorem 7.3.2 *Let $f_t, g : X \to \mathbb{R}_\infty$, $t \in T$, be proper convex functions, $f := \sup_{t \in T} f_t$, and suppose that (7.24) holds. Then, for every $x \in X$, we have*

$$\partial(f+g)(x) = \bigcap_{\varepsilon>0,\, L \in \mathcal{F}(x)} \overline{\operatorname{co}} \left\{ \bigcup_{t \in T_\varepsilon(x)} \partial_\varepsilon f_t(x) + \partial(g + \mathrm{I}_{L \cap \operatorname{dom} f})(x) \right\}. \tag{7.28}$$

Proof. The proof of the inclusion "$\supset$" follows easily (as in Exercise 113). Thus, we only need to prove "$\subset$" in the nontrivial case when $\partial(f+g)(x) \neq \emptyset$; hence by Exercise 62,

$$\operatorname{cl}(f+g)(x) = (f+g)(x) \text{ and } \partial(f+g)(x) = \partial(\operatorname{cl}(f+g))(x). \tag{7.29}$$

The idea of the proof is to look for an appropriate family of lsc convex functions giving rise to a tight approximation of the subdifferential of $f+g$. To this aim, we fix $L \in \mathcal{F}(x)$ and consider the functions $\tilde{f}$ and g_L defined previously in (7.26); i.e., $\tilde{f} = \sup_{t \in T}(\operatorname{cl} f_t)$, $g_L := g + \mathrm{I}_{L \cap \operatorname{dom} f}$. Then, by Proposition 7.3.1,

$$\operatorname{cl}(f+g) \leq \tilde{f} + \operatorname{cl} g_L. \tag{7.30}$$

Moreover, since $\tilde{f}(x) + (\operatorname{cl} g_L)(x) \leq (f+g)(x) = \operatorname{cl}(f+g)(x)$ by (7.29), the inequality above ensures that

$$\operatorname{cl}(f+g)(x) = (f+g)(x) = \tilde{f}(x) + (\operatorname{cl} g_L)(x), \tag{7.31}$$

which in turn yields, due to the relations $\tilde{f} \leq f$ and $(\operatorname{cl} g_L)(x) \leq g(x)$,

$$f(x) = \tilde{f}(x) \text{ and } (\operatorname{cl} g_L)(x) = g_L(x) = g(x) \in \mathbb{R}. \tag{7.32}$$

In particular, g_L is lsc at x and $\operatorname{cl} g_L \in \Gamma_0(X)$, and Proposition 4.1.10 guarantees that $\partial_\varepsilon g_L(x) = \partial_\varepsilon(\operatorname{cl} g_L)(x) \neq \emptyset$ for all $\varepsilon > 0$. Now, we take

7.3. SUPREMUM-SUM SUBDIFFERENTIAL ...

$x^* \in \partial(f+g)(x) = \partial(\operatorname{cl}(f+g))(x)$, so that $x^* \in \partial(\tilde{f} + (\operatorname{cl} g_L))(x)$ due to (7.30) and (7.31). Let us define the lsc convex functions

$$h_t := (\operatorname{cl} f_t) + \operatorname{cl} g_L, \ t \in T, \text{ and } h := \sup_{t \in T} h_t.$$

Observe that $x \in \operatorname{dom} h \subset L$, and so $\operatorname{ri}(\operatorname{cone}(\operatorname{dom} h - x)) \neq \emptyset$. Therefore, since $h \leq f + g + I_L$ and $L \cap \operatorname{dom} f \cap \operatorname{dom} g \subset \operatorname{dom} h$, Proposition 5.3.1 gives rise to

$$x^* \in \partial h(x) = \bigcap_{\varepsilon > 0} \overline{\operatorname{co}} \left\{ \bigcup_{t \in \tilde{T}_\varepsilon(x)} \partial_\varepsilon h_t(x) + N_{\operatorname{dom} h}(x) \right\}$$

$$\subset \bigcap_{\varepsilon > 0} \overline{\operatorname{co}} \left\{ \bigcup_{t \in \tilde{T}_\varepsilon(x)} \partial_\varepsilon h_t(x) + N_{L \cap \operatorname{dom} f \cap \operatorname{dom} g}(x) \right\}, \quad (7.33)$$

where, thanks to (7.32),

$$\tilde{T}_\varepsilon(x) := \{t \in T : (\operatorname{cl} f_t)(x) + (\operatorname{cl} g_L)(x) \geq h(x) - \varepsilon\} \subset T_\varepsilon(x). \quad (7.34)$$

Moreover, for each $t \in \tilde{T}_\varepsilon(x)$, one has

$$f_t(x) \geq (\operatorname{cl} f_t)(x) \geq f(x) - \varepsilon \geq f_t(x) - \varepsilon, \quad (7.35)$$

implying that $\partial_\varepsilon(\operatorname{cl} f_t)(x) \subset \partial_{2\varepsilon} f_t(x)$. Consequently, thanks to the following inclusions coming from (4.45),

$$\partial_\varepsilon h_t(x) \subset \operatorname{cl}\left(\partial_\varepsilon(\operatorname{cl} f_t)(x) + \partial_\varepsilon(\operatorname{cl} g_L)(x)\right) \subset \operatorname{cl}\left(\partial_{2\varepsilon} f_t(x) + \partial_{2\varepsilon}(\operatorname{cl} g_L)(x)\right),$$

and by (7.32) and (7.33), we obtain

$$x^* \in \bigcap_{\varepsilon > 0} \overline{\operatorname{co}} \left\{ \bigcup_{t \in \tilde{T}_\varepsilon(x)} \operatorname{cl}\left(\partial_{2\varepsilon} f_t(x) + \partial_{2\varepsilon}(\operatorname{cl} g_L)(x)\right) + N_{L \cap \operatorname{dom} f \cap \operatorname{dom} g}(x) \right\}$$

$$\subset \bigcap_{\varepsilon > 0} \overline{\operatorname{co}} \left\{ \bigcup_{t \in \tilde{T}_\varepsilon(x)} \partial_{2\varepsilon} f_t(x) + \partial_{2\varepsilon}(\operatorname{cl} g_L)(x) + N_{L \cap \operatorname{dom} f \cap \operatorname{dom} g}(x) \right\}$$

$$= \bigcap_{\varepsilon > 0} \overline{\operatorname{co}} \left\{ \bigcup_{t \in \tilde{T}_\varepsilon(x)} \partial_\varepsilon f_t(x) + \partial_\varepsilon(\operatorname{cl} g_L)(x) \right\}, \quad (7.36)$$

where in the last equality we took into account that

$$\operatorname{dom}(\operatorname{cl} g_L) \subset \operatorname{cl}(\operatorname{dom} g_L) = \operatorname{cl}(L \cap \operatorname{dom} f \cap \operatorname{dom} g).$$

Finally, we introduce the closed convex sets

$$A_\varepsilon := \overline{\mathrm{co}}\left\{\bigcup_{t\in \tilde{T}_\varepsilon(x)} \partial_\varepsilon f_t(x)\right\}, \varepsilon > 0.$$

For every $z \in \mathrm{ri}(\mathrm{dom}(\mathrm{cl}\, g_L)) = \mathrm{ri}(\mathrm{dom}\, g_L) = \mathrm{ri}(L \cap \mathrm{dom}\, f \cap \mathrm{dom}\, g)$ and $\varepsilon > 0$, by (7.35) we have

$$\sigma_{A_\varepsilon}(z - x) \leq \sup_{t\in \tilde{T}_\varepsilon(x)} (f_t(z) - f_t(x) + \varepsilon) \leq f(z) - f(x) + 2\varepsilon < +\infty,$$

showing that $(\mathrm{ri}(\mathrm{dom}(\mathrm{cl}\, g_L)) - x) \cap \mathrm{dom}\, \sigma_{A_\varepsilon} \neq \emptyset$. Consequently, since $\mathrm{cl}\, g_L \in \Gamma_0(X)$ and $(\mathrm{cl}\, g_L)|_{\mathrm{aff}(\mathrm{dom}\, g_L)}$ is continuous on $\mathrm{ri}(\mathrm{dom}(\mathrm{cl}\, g_L))$, by Exercise 112 and (7.32), we get

$$x^* \in \bigcap_{\varepsilon > 0} \mathrm{cl}\left(A_\varepsilon + \partial_\varepsilon(\mathrm{cl}\, g_L)(x)\right) = \bigcap_{\varepsilon > 0} \mathrm{cl}\left(A_\varepsilon + \partial(\mathrm{cl}\, g_L)(x)\right)$$

$$= \bigcap_{\varepsilon > 0} \overline{\mathrm{co}}\left\{\bigcup_{t\in \tilde{T}_\varepsilon(x)} \partial_\varepsilon f_t(x) + \partial g_L(x)\right\}.$$

Therefore, since $\tilde{T}_\varepsilon(x) \subset T_\varepsilon(x)$ by (7.34), we deduce that

$$x^* \in \bigcap_{\varepsilon > 0} \overline{\mathrm{co}}\left\{\bigcup_{t\in T_\varepsilon(x)} \partial_\varepsilon f_t(x) + \partial g_L(x)\right\},$$

which leads us to the desired inclusion. ∎

The following corollary is a particular instance of Theorem 7.3.2, with g being the indicator of a convex set.

Corollary 7.3.3 *Let $f_t : X \to \mathbb{R}_\infty$, $t \in T$, be proper convex functions and consider $f := \sup_{t\in T} f_t$. Let $D \subset X$ be a non-empty convex set such that*

$$\mathrm{cl}\,(f + \mathrm{I}_D)(x) = \sup_{t\in T}(\mathrm{cl}\, f_t)(x) \text{ for all } x \in \mathrm{dom}\, f \cap D.$$

Then, for every $x \in X$, we have

$$\partial(f + \mathrm{I}_D)(x) = \bigcap_{\varepsilon > 0,\, L\in\mathcal{F}(x)} \overline{\mathrm{co}}\left\{\bigcup_{t\in T_\varepsilon(x)} \partial_\varepsilon f_t(x) + \mathrm{N}_{L\cap D\cap \mathrm{dom}\, f}(x)\right\}.$$

7.3. SUPREMUM-SUM SUBDIFFERENTIAL ...

The intersection over the L's in Theorem 7.3.2 can obviously be omitted in the finite-dimensional setting; in fact, we have the following result which is more general.

Corollary 7.3.4 *Let $f_t, g : X \to \mathbb{R}_\infty$, $t \in T$, be proper convex functions, $f := \sup_{t \in T} f_t$, and suppose that (7.24) holds. Additionally, we assume that $\mathrm{ri}(\mathrm{dom}\, f \cap \mathrm{dom}\, g) \neq \emptyset$ and $g_{|\mathrm{aff}(\mathrm{dom}\, f \cap \mathrm{dom}\, g)}$ is continuous on $\mathrm{ri}(\mathrm{dom}\, f \cap \mathrm{dom}\, g)$. Then, for every $x \in X$, we have*

$$\partial(f+g)(x) = \bigcap_{\varepsilon > 0} \overline{\mathrm{co}} \left\{ \bigcup_{t \in T_\varepsilon(x)} \partial_\varepsilon f_t(x) + \partial(g + \mathrm{I}_{\mathrm{dom}\, f})(x) \right\}. \quad (7.37)$$

Proof. If $x \notin \mathrm{dom}\, f \cap \mathrm{dom}\, g = \mathrm{dom}\, f \cap \mathrm{dom}(g + \mathrm{I}_{\mathrm{dom}\, f})$, then $\partial(f+g)(x) = \partial(g + \mathrm{I}_{\mathrm{dom}\, f})(x) = \emptyset$, and (7.37) holds trivially; so, we take $x \in \mathrm{dom}\, f \cap \mathrm{dom}\, g$. Given $U \in \mathcal{N}_{X^*}$, we choose $L \in \mathcal{F}(x)$ such that $L^\perp \subset U$, and take $L_1 \in \mathcal{F}(x)$ satisfying $L \subset L_1$ and $L_1 \cap \mathrm{ri}(\mathrm{dom}\, f \cap \mathrm{dom}\, g) \neq \emptyset$. Then, by Theorem 7.3.2,

$$\partial(f+g)(x) \subset \bigcap_{\varepsilon > 0} \overline{\mathrm{co}} \left(\bigcup_{t \in T_\varepsilon(x)} \partial_\varepsilon f_t(x) + \partial(g + \mathrm{I}_{L_1 \cap \mathrm{dom}\, f})(x) \right), \quad (7.38)$$

and we provide next a simplified expression for $\partial(g + \mathrm{I}_{L_1 \cap \mathrm{dom}\, f})(x)$. To this aim, we introduce the functions

$$\varphi := \mathrm{I}_{L_1} \text{ and } \psi := g + \mathrm{I}_{\mathrm{dom}\, f},$$

and check that they satisfy the conditions of Theorem 7.2.2(iii). Observe that $\varphi_{|L_1}$ is continuous on L_1, and the equality $\psi_{|\mathrm{aff}(\mathrm{dom}\, \psi)} = g_{|\mathrm{aff}(\mathrm{dom}\, f \cap \mathrm{dom}\, g)}$ holds on $\mathrm{ri}(\mathrm{dom}\, f \cap \mathrm{dom}\, g) \neq \emptyset$. So, $\psi_{|\mathrm{aff}(\mathrm{dom}\, \psi)}$ is continuous on $\mathrm{ri}(\mathrm{dom}\, \psi)$ by assumption. Then, Theorem 7.2.2(iii) yields

$$\partial(g + \mathrm{I}_{L_1 \cap \mathrm{dom}\, f})(x) = \partial(\varphi + \psi)(x) = \mathrm{cl}(\partial\varphi(x) + \partial\psi(x)) = \mathrm{cl}(L_1^\perp + \partial(g + \mathrm{I}_{\mathrm{dom}\, f})(x)).$$

Plugging this equality into (7.38) yields

$$\partial(f+g)(x) \subset \bigcap_{\varepsilon > 0} \overline{\mathrm{co}} \left(\bigcup_{t \in T_\varepsilon(x)} \partial_\varepsilon f_t(x) + \mathrm{cl}\left(\partial(g + \mathrm{I}_{\mathrm{dom}\, f})(x) + L_1^\perp \right) \right)$$

$$= \bigcap_{\varepsilon > 0} \overline{\mathrm{co}} \left(\bigcup_{t \in T_\varepsilon(x)} \partial_\varepsilon f_t(x) + \partial(g + \mathrm{I}_{\mathrm{dom}\, f})(x) + L_1^\perp \right)$$

$$\subset \bigcap_{\varepsilon>0} \overline{\mathrm{co}} \left(\bigcup_{t \in T_\varepsilon(x)} \partial_\varepsilon f_t(x) + \partial(g + \mathrm{I}_{\mathrm{dom}\, f})(x) + U \right).$$

Finally, since U was arbitrarily chosen, we get

$$\partial(f+g)(x) \subset \bigcap_{\varepsilon>0} \overline{\mathrm{co}} \left(\bigcup_{t \in T_\varepsilon(x)} \partial_\varepsilon f_t(x) + \partial(g + \mathrm{I}_{\mathrm{dom}\, f})(x) \right),$$

and the inclusion "$\subset$" of (7.37) follows. This finishes the proof since the opposite inclusion is straightforward from Theorem 7.3.2, in virtue of the relation $\partial(g + \mathrm{I}_{\mathrm{dom}\, f})(x) \subset \partial(g + \mathrm{I}_{L \cap \mathrm{dom}\, f})(x)$. ∎

Another way to avoid finite-dimensional linear subspaces, as was achieved in Corollary 7.3.4, is to use the ε-subdifferential of the function $g + \mathrm{I}_{\mathrm{dom}\, f}$.

Corollary 7.3.5 *Let $f_t, g : X \to \mathbb{R}_\infty$, $t \in T$, be proper convex functions, $f := \sup_{t \in T} f_t$, and suppose that (7.24) holds. If $g + \mathrm{I}_{\mathrm{dom}\, f} \in \Gamma_0(X)$, then for every $x \in X$ we have*

$$\partial(f+g)(x) = \bigcap_{\varepsilon>0} \overline{\mathrm{co}} \left\{ \bigcup_{t \in T_\varepsilon(x)} \partial_\varepsilon f_t(x) + \partial_\varepsilon(g + \mathrm{I}_{\mathrm{dom}\, f})(x) \right\}. \quad (7.39)$$

Proof. Fix $x \in \mathrm{dom}\, f \cap \mathrm{dom}\, g$, $\varepsilon > 0$ and $L \in \mathcal{F}(x)$. By (4.45), one has

$$\partial(g + \mathrm{I}_{L \cap \mathrm{dom}\, f})(x) = \partial(g + \mathrm{I}_{\mathrm{dom}\, f} + \mathrm{I}_L)(x) \subset \mathrm{cl}\left(\partial_\varepsilon(g + \mathrm{I}_{\mathrm{dom}\, f})(x) + L^\perp\right),$$

and Theorem 7.3.2 gives rise to

$$\partial(f+g)(x) \subset \overline{\mathrm{co}} \left\{ \bigcup_{t \in T_\varepsilon(x)} \partial_\varepsilon f_t(x) + \partial(g + \mathrm{I}_{L \cap \mathrm{dom}\, f})(x) \right\}$$

$$\subset \overline{\mathrm{co}} \left\{ \bigcup_{t \in T_\varepsilon(x)} \partial_\varepsilon f_t(x) + \mathrm{cl}\left(\partial_\varepsilon(g + \mathrm{I}_{\mathrm{dom}\, f})(x) + L^\perp \right) \right\}$$

$$= \overline{\mathrm{co}} \left\{ \bigcup_{t \in T_\varepsilon(x)} \partial_\varepsilon f_t(x) + \partial_\varepsilon(g + \mathrm{I}_{\mathrm{dom}\, f})(x) + L^\perp \right\}.$$

Thus, using Exercise 10(iv) and taking the intersection over $L \in \mathcal{F}(x)$ and $\varepsilon > 0$, we get the nontrivial inclusion in (7.39). ∎

7.4 Exercises

Exercise 108 *Give a proof of (7.2) based on formula (4.45).*

Exercise 109 *Consider two functions $f : Y \to \overline{\mathbb{R}}$ and $g : X \to \overline{\mathbb{R}}$ (non-necessarily convex), where Y is another lcs, and a continuous linear mapping $A : X \to Y$ such that $\overline{\mathrm{co}}(g + f \circ A) = (\overline{\mathrm{co}}g) + (\overline{\mathrm{co}}f) \circ A$. Prove that, for every $x \in X$,*

$$\partial(g + f \circ A)(x) = \bigcap_{\varepsilon > 0} \mathrm{cl}\left(\partial_\varepsilon g(x) + A^* \partial_\varepsilon f(Ax)\right),$$

where A^* is the adjoint mapping of A.

Exercise 110 *Let $f, g : X \to \mathbb{R}_\infty$ be such that $\mathrm{cl}\, f$ and $\mathrm{cl}\, g$ are proper. If f and $\mathrm{cl}\, g$ are convex, and f is finite and continuous at some point in $\mathrm{dom}(\mathrm{cl}\, g)$, prove that $\mathrm{cl}(f + g) = (\mathrm{cl}\, f) + (\mathrm{cl}\, g)$ and $\partial(f + g) = \partial f + \partial g$.*

Exercise 111 *Prove Corollary 7.2.6.*

Exercise 112 *Let $(A_\varepsilon)_{\varepsilon > 0}$ be a non-decreasing family of non-empty closed convex sets of X^*; that is, if $\varepsilon_1 \leq \varepsilon_2$, then $A_{\varepsilon_1} \subset A_{\varepsilon_2}$. Given a function $g \in \Gamma_0(X)$, we assume that $(\mathrm{ri}(\mathrm{dom}\, g) - x) \cap \mathrm{dom}\, \sigma_{A_\varepsilon} \neq \emptyset$, for every small $\varepsilon > 0$, and $g_{|\mathrm{aff}(\mathrm{dom}\, g)}$ is continuous on $\mathrm{ri}(\mathrm{dom}\, g)$. Prove that*

$$\bigcap_{\varepsilon > 0} \mathrm{cl}\left(A_\varepsilon + \partial_\varepsilon g(x)\right) = \bigcap_{\varepsilon > 0} \mathrm{cl}\left(A_\varepsilon + \partial g(x)\right).$$

Exercise 113 *Given lsc convex functions $f_t, f, g : X \to \overline{\mathbb{R}}, t \in T$, such that $f := \sup_{t \in T} f_t$, prove that, for every $x \in X$,*

$$\partial(f + g)(x) = \bigcap_{\varepsilon > 0,\, L \in \mathcal{F}(x)} \overline{\mathrm{co}}\left\{\bigcup_{t \in T_\varepsilon(x)} \partial_\varepsilon f_t(x) + \partial_\varepsilon g(x) + \mathrm{N}_{L \cap \mathrm{dom}\, f \cap \mathrm{dom}\, g}(x)\right\}. \quad (7.40)$$

Exercise 114 *Let $f_t : X \to \mathbb{R}_\infty$, $t \in T$, be proper convex functions and consider $f := \sup_{t \in T} f_t$ such that*

$$(\mathrm{cl}\, f)(x) = \sup_{t \in T}(\mathrm{cl}\, f_t)(x) \text{ for all } x \in \mathrm{dom}\, f.$$

(i) Prove that formula (5.26) holds; that is, for every $x \in X$,

$$\partial f(x) = \bigcap_{\varepsilon > 0,\, L \in \mathcal{F}(x)} \overline{\mathrm{co}}\left\{\bigcup_{t \in T_\varepsilon(x)} \partial_\varepsilon f_t(x) + \mathrm{N}_{L \cap \mathrm{dom}\, f}(x)\right\}. \quad (7.41)$$

(ii) If $x \in X$ is such that either $\mathrm{ri}(\mathrm{cone}(\mathrm{dom}\, f - x)) \neq \emptyset$ or cone $(\mathrm{dom}\, f - x)$ is closed, prove that (5.65) holds; that is,

$$\partial f(x) = \bigcap_{\varepsilon > 0} \overline{\mathrm{co}} \left\{ \bigcup_{t \in T_\varepsilon(x)} \partial_\varepsilon f_t(x) + \mathrm{N}_{\mathrm{dom}\, f}(x) \right\}.$$

(iii) If $x \in X$ and $\mathrm{dom}\, f$ is closed, prove that

$$\partial f(x) = \bigcap_{\varepsilon > 0} \overline{\mathrm{co}} \left\{ \bigcup_{t \in T_\varepsilon(x)} \partial_\varepsilon f_t(x) + \mathrm{N}^\varepsilon_{\mathrm{dom}\, f}(x) \right\}.$$

Exercise 115 *Let $f_t : X \to \mathbb{R}_\infty$, $t \in T$, be proper convex functions and consider $f := \sup_{t \in T} f_t$ and a convex set $D \subset X$. Prove that $\mathrm{cl}\,(f + \mathrm{I}_D) = \sup_{t \in T}(\mathrm{cl}\, f_t)$ holds on D if and only if it holds on the larger set $\cup_{L \in \mathcal{F}} \mathrm{cl}(L \cap D)$. (Remember that $\mathcal{F}$ is the family of finite-dimensional subspaces of X.)*

7.5 Bibliographical notes

Main references for this chapter are [49] and [50]. Theorem 7.1.1 is a slight extension of the Hiriart-Urruty–Phelps formula [111] (see (4.45)), which relaxes the lower semicontinuity assumption. It was established in [103, Theorem 13]. Corollary 7.1.3 is the classical chain rule by Moreau and Rockafellar for the sum and composition with a continuous linear mapping (see, e.g., [161]). Theorem 7.2.1, yielding an asymmetric chain rule, was established in [49, Theorem 12], and the symmetric version given in Theorem 7.2.2 is Theorem 15 of the same paper. Indeed, Theorem 7.2.2 is an infinite-dimensional extension of [174, Theorem 23.8]. In particular, Theorem 7.2.2(iii) makes use of an assumption that can be regarded as a counterpart of the Attouch-Brzis condition ([7]) for general locally convex spaces. Theorem 7.2.5, given in [49, Corollary 23], provides an asymmetric version of the results in [19, Theorem 4.2] (compare, also, with [19, Corollary 4.3] where the so-called quasi-relative interior is involved). Corollary 7.2.7, given in [49, Corollary 24], is in the spirit of the sequential calculus rules provided in [123], [165], and [188]. Theorem 7.3.2, establishing a mixed supremum-sum rule, is given in [49, Theorem 4]. Corollary 7.3.5, particularly its second statement, is related to [113, Theorem 5.1] and applies when f is a polyhedral function (see [49, Lemma 8]). A characterization of

7.5. BIBLIOGRAPHICAL NOTES

this closedness property can also be found in [152]. Exercise 110 provides an extension of Corollary 7.1.3 to non-convex functions whose closures are in $\Gamma_0(X)$. In fact, functions with convex closures are frequently used in variational analysis even though they are non-convex (see, e.g., [80]). Exercise 115 is related to [134, Theorem 3.1].

Chapter 8
Miscellaneous

This last chapter addresses several issues related to the previous chapters. The first part is mainly aimed at deriving optimality conditions for a convex optimization problem, posed in an lcs, with an arbitrary number of constraints. The approach taken is to replace the set of constraints with a unique constraint via the supremum function. Subsequently, we appeal to the properties of the subdifferential of the supremum function that has been exhaustively studied in the previous chapters. With this goal, we extend to infinite convex systems two constraint qualifications that are crucial in linear semi-infinite programming. The first, called the Farkas–Minkowski property, is global in nature, while the other is a local property, called locally Farkas–Minkowski. We obtain two types of Karush–Kuhn–Tucker (KKT, in brief) optimality conditions: asymptotic and non-asymptotic.

In section 8.3, we analyze the relationship between the optimal solutions of a given optimization problem and those of its convex regularization. This will be performed by exact or approximate solutions of the original problem. Furthermore, in the same section, we give different formulas for the subdifferential of the conjugate function, which are extensions of the Fenchel formula (4.18). In section 8.4, we develop an integration theory for the exact and approximate subdifferentials of non-convex functions. This section builds on chapters 5 and 6 to provide extensions of the integration results of section 4.4 that were limited to convex functions. Section 8.5 establishes some variational characterizations of convexity of functions and sets, while last section 8.6 is devoted to the so-called Chebychev sets.

Also, in this chapter, X is an lcs with a dual X^* that is endowed with a compatible topology (unless otherwise stated). The associated bilinear form is represented by $\langle \cdot, \cdot \rangle$.

8.1 Convex systems and Farkas-type qualifications

In this section, we deal with the *convex optimization problem*

$$\text{(P) Min } g(x) \\ \text{s.t. } f_t(x) \leq 0, \ t \in T, \\ x \in C, \tag{8.1}$$

where T is an arbitrary (possibly infinite) index set, C is a non-empty closed convex subset of X, and $\{g; f_t, t \in T\} \subset \Gamma_0(X)$. We consider the *constraint system*

$$\mathcal{S} := \{f_t(x) \leq 0, \ t \in T; \ x \in C\}, \tag{8.2}$$

and denote by F the corresponding set of solutions, also called *feasible set*. Observe that F is a closed convex set in X. When $F \neq \emptyset$, we say that $\mathcal{S}$ is a *consistent* system. The constraint $x \in C$ is referred to as the abstract constraint, whereas $f_t(x) \leq 0$, $t \in T$, are the explicit constraints. We assume that $F \cap \text{dom}\, g \neq \emptyset$. We say that (P) is *solvable* when it has optimal solutions. An important particular case is that when the explicit constraints are affine and there is no abstract constraint, i.e.,

$$\mathcal{S} := \{\langle a_t^*, x \rangle \leq b_t, \ t \in T\}, \tag{8.3}$$

with $a_t^* \in X^*$, $t \in T$. If T is infinite, the *objective function* g is linear, and $X = \mathbb{R}^n$, we have the so-called *linear semi-infinite optimization problem* (*LSIP*, in brief).

It is obvious that (P) can be written equivalently as follows:

$$\text{(P) Min } g(x) \\ \text{s.t. } f(x) \leq 0, \tag{8.4}$$

8.1. CONVEX SYSTEMS AND FARKAS-TYPE ...

where
$$f := \sup\{f_t,\ t \in T;\ I_C\}, \tag{8.5}$$

and, consequently, $F = \{x \in X : f(x) \leq 0\}$.

The aim of the next section is to establish KKT optimality conditions for (P). For this purpose, we extend to infinite convex systems two constraint qualifications, which play a crucial role in linear semi-infinite programming. The first, called the Farkas–Minkowski property, is of global nature, whereas the second, called locally Farkas–Minkowski, is a local property.

Let us recall that the space of generalized finite sequences, $\mathbb{R}^{(T)}$, was defined in chapter 3 (see (2.3)) as the (topological) dual of $\mathbb{R}^T$, and that $\mathbb{R}_+^{(T)}$ is its nonnegative cone. If $\lambda \in \mathbb{R}_+^{(T)}$ and we are dealing with proper functions (as in this chapter, where g, f_t, $t \in T$, belong to $\Gamma_0(X)$), we define

$$\left(\sum_{t \in T} \lambda_t f_t\right)(x) := \sum_{t \in \text{supp}\,\lambda} \lambda_t f_t(x) + I_{\bigcap_{t \in T} \text{dom}\, f_t}(x) \tag{8.6}$$

with the convention $\sum_\emptyset = 0$. Analogously, if $\{Y_t,\ t \in T\}$ is a family of subsets of some linear space Y, whose zero is also denoted by θ, for $\lambda \in \mathbb{R}_+^{(T)}$, we have

$$\sum_{t \in T} \lambda_t Y_t := \sum_{t \in \text{supp}\,\lambda} \lambda_t Y_t,$$

with the convention $\sum_\emptyset = \{\theta\}$. Particularly, cone co $A = \{\sum_{t \in T} \lambda_t A : \lambda \in \mathbb{R}_+^{(T)}\}$ for every non-empty set $A \subset Y$.

Definition 8.1.1 *The characteristic cone of* $S = \{f_t(x) \leq 0,\ t \in T;\ x \in C\}$ *is the convex cone*

$$K := \text{cone co}\left\{\bigcup_{t \in T} \text{epi}\, f_t^* \cup \text{epi}\, \sigma_C\right\}. \tag{8.7}$$

Taking into account that epi σ_C is a convex cone, we can write

$$K = \text{cone co}\left\{\bigcup_{t \in T} \text{epi}\, f_t^*\right\} + \text{epi}\, \sigma_C.$$

For the linear system (8.3), epi $f_t^* = (a_t^*, b_t) + \mathbb{R}_+(\theta, 1)$, $t \in T$, and epi $\sigma_C \equiv$ epi $\sigma_X = \mathbb{R}_+(\theta, 1)$. Hence

$$K = \operatorname{cone}\operatorname{co}\{(a_t^*, b_t),\ t \in T;\ (\theta, 1)\}. \tag{8.8}$$

Next, in Theorem 8.1.4, we will prove the so-called *generalized Farkas lemma* in a simple way by applying the following lemma.

Lemma 8.1.2 *If* $F \neq \emptyset$, *then* $\operatorname{epi} \sigma_F = \operatorname{cl} K$.

Proof. Since $F = \{x \in X : f(x) \leq 0\}$, with $f := \sup\{f_t,\ t \in T;\ \mathrm{I}_C\}$ $(\in \Gamma_0(X))$, Exercise 32(b) applies and yields $\operatorname{epi} \sigma_F = \operatorname{cl}(\operatorname{cone}(\operatorname{epi} f^*))$. Now, since we are assuming $F \neq \emptyset$, by Proposition 3.2.6,

$$f^* = \overline{\operatorname{co}}\left(\inf\{f_t^*,\ t \in T;\ \mathrm{I}_C^*\}\right) = \overline{\operatorname{co}}\left(\inf\{f_t^*,\ t \in T;\ \sigma_C\}\right),$$

entailing $\operatorname{epi} f^* = \overline{\operatorname{co}}\{\cup_{t \in T} \operatorname{epi} f_t^* \cup \operatorname{epi} \sigma_C\}$. Therefore,

$$\operatorname{epi} \sigma_F = \overline{\operatorname{cone}}\left(\overline{\operatorname{co}}\left\{\bigcup_{t \in T} \operatorname{epi} f_t^* \cup \operatorname{epi} \sigma_C\right\}\right) = \operatorname{cl} K.$$

∎

By applying Exercise 32(a), we may also characterize the feasibility of $\mathcal{S}$; actually,

$$F \neq \emptyset \iff (\theta, -1) \notin \operatorname{cl} K. \tag{8.9}$$

Definition 8.1.3 *Given the functions* $h, \ell : X \to \overline{\mathbb{R}}$, *we say that the inequality* $\ell \leq h$ *is consequence of the consistent system* $\mathcal{S}$ *with feasible set* F, *if and only if* $\ell(x) \leq h(x)$, *for every* $x \in F$.

The following result can be regarded as a generalized Farkas lemma. It turns out to be the key stone for deriving different characterizations of consequent relations in Theorem 8.1.5.

Theorem 8.1.4 *Let* $h, \ell \in \Gamma_0(X)$, *consider the consistent system* $\mathcal{S}$ *with feasible set* F, *and suppose that* $F \cap \operatorname{dom} h \neq \emptyset$. *Then* $\ell \leq h$ *is consequence of* $\mathcal{S}$ *if and only if*

$$\operatorname{epi} \ell^* \subset \operatorname{cl}(\operatorname{epi} h^* + K). \tag{8.10}$$

Proof. The inequality $\ell \leq h$ is the consequence of the consistent system $\mathcal{S}$ if and only if $\ell \leq h + \mathrm{I}_F$. This happens if and only if $(h + \mathrm{I}_F)^* \leq \ell^*$, equivalently if and only if, taking into account (4.44) and Lemma 8.1.2,

$$\operatorname{epi} \ell^* \subset \operatorname{epi}(h + \mathrm{I}_F)^* = \operatorname{cl}(\operatorname{epi} h^* + \operatorname{epi} \sigma_F)$$
$$= \operatorname{cl}(\operatorname{epi} h^* + \operatorname{cl} K) = \operatorname{cl}(\operatorname{epi} h^* + K).$$

∎

8.1. CONVEX SYSTEMS AND FARKAS-TYPE ...

The following theorem provides several Farkas-type results.

Theorem 8.1.5 *Suppose that S in (8.2) is consistent. Then the following statements hold:*

(i) *If $(a^*, \alpha) \in X^* \times \mathbb{R}$, the inequality $\langle a^*, \cdot \rangle \leq \alpha$ is consequence of S if and only if*

$$(a^*, \alpha) \in \operatorname{cl} K. \tag{8.11}$$

(ii) *If $\ell \in \Gamma_0(X)$ and $\alpha \in \mathbb{R}$, the inequality $\ell \leq \alpha$ is a consequence of S if and only if $(\theta, \alpha) + \operatorname{epi} \ell^* \subset \operatorname{cl} K$.*

(iii) *If $h \in \Gamma_0(X), \gamma \in \mathbb{R}$, and $F \cap \operatorname{dom} h \neq \emptyset$, the inequality $h \geq \gamma$ is a consequence of S if and only if*

$$(\theta, -\gamma) \in \operatorname{cl}(\operatorname{epi} h^* + K). \tag{8.12}$$

(iv) *If $h \in \Gamma_0(X)$, $\gamma \in \mathbb{R}$, $F \cap \operatorname{dom} h \neq \emptyset$, and $\operatorname{epi} h^* + K$ is w^*-closed, then $h \geq \gamma$ is a consequence of S if and only if there exists $\lambda \in \mathbb{R}_+^{(T)}$ such that*

$$h(x) + \sum_{t \in T} \lambda_t f_t(x) \geq \gamma \text{ for all } x \in C. \tag{8.13}$$

(v) *If $h \in \Gamma_0(X)$ is finite and continuous somewhere in F, $\gamma \in \mathbb{R}$, and K is w^*-closed, then $h \geq \gamma$ is a consequence of S if and only if (8.13) holds for some $\lambda \in \mathbb{R}_+^{(T)}$.*

Proof. (i) It is a straightforward consequence of Theorem 8.1.4 with $\ell := \langle a^*, \cdot \rangle - \alpha$ and $h \equiv 0$. If the inequality $\langle a^*, \cdot \rangle \leq \alpha$ is consequence of S, since $\operatorname{epi} h^* = \mathbb{R}_+(\theta, 1)$ we get

$$(a^*, \alpha) \in \operatorname{epi} \ell^* \subset \operatorname{cl}(\operatorname{epi} h^* + K) = \operatorname{cl}(\mathbb{R}_+(\theta, 1) + K) = \operatorname{cl} K.$$

Conversely, if (8.11) holds,

$$\operatorname{epi} \ell^* = (a^*, \alpha) + \mathbb{R}_+(\theta, 1) \subset (\operatorname{cl} K) + \mathbb{R}_+(\theta, 1) = \operatorname{cl} K = \operatorname{cl}(\operatorname{epi} h^* + K),$$

and Theorem 8.1.4 yields the desired conclusion.

(ii) Using Theorem 8.1.4, $\ell \leq \alpha$ is consequence of S if and only if

$$\operatorname{epi} \ell^* \subset \operatorname{cl}((\theta, -\alpha) + \mathbb{R}_+(\theta, 1) + K)$$
$$= \operatorname{cl}((\theta, -\alpha) + K) = (\theta, -\alpha) + \operatorname{cl} K.$$

(iii) Now take $\ell \equiv \gamma$. Then, by Theorem 8.1.4, $h \geq \gamma$ is consequence of $\mathcal{S}$ if and only if

$$(\theta, -\gamma) \in (\theta, -\gamma) + \mathbb{R}_+(\theta, 1) = \operatorname{epi} \ell^* \subset \operatorname{cl}(\operatorname{epi} h^* + K).$$

(iv) Suppose that $\operatorname{epi} h^* + K$ is w^*-closed. If $h \geq \gamma$ is consequence of $\mathcal{S}$, by (iii) there will exist $a^* \in \operatorname{dom} h^*$, $\rho \geq 0$, $x_t^* \in \operatorname{dom} f_t^*$, $\alpha_t \geq 0$, $t \in T$, $z^* \in \operatorname{dom} \sigma_C$, $\beta \geq 0$ and $\lambda \in \mathbb{R}_+^{(T)}$ such that

$$(\theta, -\gamma) = (a^*, h^*(a^*) + \rho) + \sum_{t \in \operatorname{supp} \lambda} \lambda_t(x_t^*, f_t^*(x_t^*) + \alpha_t) + (z^*, \sigma_C(z^*) + \beta).$$

Equivalently,

$$\begin{cases} \theta = a^* + \sum_{t \in \operatorname{supp} \lambda} \lambda_t x_t^* + z^*, \\ \gamma = -h^*(a^*) - \sum_{t \in \operatorname{supp} \lambda} \lambda_t(f_t^*(x_t^*) + \alpha_t) - \sigma_C(z^*) - \rho - \beta. \end{cases} \quad (8.14)$$

Since, for all $x \in X$, $h^*(a^*) \geq \langle a^*, x \rangle - h(x)$, $f_t^*(x_t^*) \geq \langle x_t^*, x \rangle - f_t(x)$ for all $t \in T$, and $\sigma_C(z^*) \geq \langle z^*, x \rangle$ for all $x \in C$, it follows from the two equalities in (8.14) that, for all $x \in C$,

$$\gamma = \langle a^*, x \rangle - h^*(a^*) + \sum_{t \in \operatorname{supp} \lambda} \lambda_t(\langle x_t^*, x \rangle - f_t^*(x_t^*))$$
$$+ (\langle z^*, x \rangle - \sigma_C(z^*)) - \sum_{t \in \operatorname{supp} \lambda} \lambda_t \alpha_t - \rho - \beta$$
$$\leq h(x) + \sum_{t \in \operatorname{supp} \lambda} \lambda_t f_t(x) - \sum_{t \in \operatorname{supp} \lambda} \lambda_t \alpha_t - \rho - \beta$$
$$\leq h(x) + \sum_{t \in \operatorname{supp} \lambda} \lambda_t f_t(x),$$

which is (8.13). Finally, if (8.13) is satisfied and $x \in F \; (\subset C)$, then $\gamma \leq h(x) + \sum_{t \in \operatorname{supp} \lambda} \lambda_t f_t(x) \leq h(x)$, and we are done as $h \geq \gamma$ is consequence of $\mathcal{S}$.

(v) Due to Exercise 119, the continuity assumption ensures that $\operatorname{epi} h^* + K = \operatorname{epi} h^* + \operatorname{cl} K$ is w^*-closed. Thus, it suffices to apply (iv). ∎

The following property is crucial in getting (exact) KKT conditions for problem (P). In fact, it constitutes the first constraint qualification of system $\mathcal{S}$ in problem (P).

Definition 8.1.6 *We say that the consistent system $\mathcal{S}$ in (8.2) is Farkas–Minkowski (FM, in brief) if K is w^*-closed.*

8.1. CONVEX SYSTEMS AND FARKAS-TYPE ...

If $\hat{T}$ is a finite subset of T, we say that

$$\hat{\mathcal{S}} := \{f_t(x) \leq 0,\ t \in \hat{T};\ x \in C\} \tag{8.15}$$

is a *finite subsystem* of $\mathcal{S}$, and its feasible set is correspondingly denoted by $\hat{F}$. The finite subsystems always include the abstract constraint $x \in C$, and so $\hat{F} \subset C$.

Theorem 8.1.7 *If $\mathcal{S}$ is FM, then every continuous affine consequence of $\mathcal{S}$ is also consequence of a finite subsystem of it. The converse statement holds if $\mathcal{S}$ is linear.*

Proof. Since $\mathcal{S}$ is FM, if $\langle a^*, x \rangle \leq \alpha$ is consequence of $\mathcal{S}$, then $(a^*, \alpha) \in \operatorname{cl} K = K$, by Theorem 8.1.5($i$). This implies the existence of a (possibly empty) finite subset $\hat{T} \subset T$, $\{x_t^*,\ t \in \hat{T};\ z^*\} \subset X^*$, and $\left\{ \lambda_t,\ t \in \hat{T};\ \alpha_t,\ t \in \hat{T};\ \beta \right\} \subset \mathbb{R}_+$ such that

$$(a^*, \alpha) = \sum_{t \in \operatorname{supp} \lambda} \lambda_t \left(x_t^*, f_t^*(x_t^*) + \alpha_t\right) + (z^*, \sigma_C(z^*) + \beta) \in \hat{K},$$

where $\hat{K}$ is the characteristic cone of the subsystem $\hat{\mathcal{S}}$ in (8.15); i.e.,

$$\hat{K} = \operatorname{cone\ co} \left\{ \bigcup_{t \in \hat{T}} \operatorname{epi} f_t^* \cup \operatorname{epi} \sigma_C \right\}.$$

Since $(a^*, \alpha) \in \hat{K} \subset \operatorname{cl} \hat{K}$, the inequality $\langle a^*, \cdot \rangle \leq \alpha$ is consequence of $\hat{\mathcal{S}}$, again by Theorem 8.1.5(i). Now, we consider a linear system; i.e., $C = X$ and $f_t(x) = \langle a_t^*, x \rangle - b_t$, with $a_t^* \in X^*$ and $b_t \in \mathbb{R}$, $t \in T$. Take any $(a^*, \alpha) \in \operatorname{cl} K$ and let us prove that $(a^*, \alpha) \in K$. Theorem 8.1.5(i) establishes that $\langle a^*, \cdot \rangle \leq \alpha$ is consequence of $\mathcal{S}$. By assumption, there exists a finite set $\hat{T} \subset T$ such that $\langle a^*, \cdot \rangle \leq \alpha$ is consequence of $\hat{\mathcal{S}}$, so that $(a^*, \alpha) \in \operatorname{cl} \hat{K}$, where $\hat{K}$ is the characteristic cone of $\hat{\mathcal{S}}$ as defined above. Since this cone is a polyhedral set, it is w^*-closed (Exercise 2) and $(a^*, \alpha) \in \hat{K} \subset K$. Thus, K is w^*-closed as we have proved that $\operatorname{cl} K \subset K$. ∎

The following example shows that the converse statement of Theorem 8.1.7 is not true for a very simple convex system.

Example 8.1.8 Let $C = X = \mathbb{R}$, $T = \{1\}$ and $\mathcal{S} = \{f_1(x) := (1/2)x^2 \leq 0\}$. Since $f_1^*(u) = (1/2)u^2$, the characteristic cone $K = (\mathbb{R} \times]0, +\infty[) \cup \{0_2\}$ *is not closed. Thus, $\mathcal{S}$ is a finite non-FM convex system.*

The following result is used later.

Proposition 8.1.9 *Let S be an FM system and $(a^*, \alpha) \in X^* \times \mathbb{R}$. Then the inequality $\langle a^*, \cdot \rangle \leq \alpha$ is consequence of S if and only if there exists $\lambda \in \mathbb{R}_+^{(T)}$ such that*

$$\langle a^*, x \rangle - \alpha \leq \sum_{t \in \mathrm{supp}\,\lambda} \lambda_t f_t(x) \text{ for all } x \in C. \tag{8.16}$$

Proof. It is a mere application of Theorem 8.1.5(iv) with $h \equiv \langle -a^*, \cdot \rangle$ and $\gamma = -\alpha$, and observing that

$$\mathrm{epi}\, h^* + K = (-a^*, 0) + (\{\theta\} \times \mathbb{R}_+) + K = (-a^*, 0) + K;$$

that is, $\mathrm{epi}\, h^* + K$ is a w^*-closed set because S is FM. ∎

Now, we introduce another constraint qualification. Given $x \in X$, consider the set of indices

$$A(x) := \{t \in T : f_t(x) = 0\}.$$

If $x \in F$, $A(x)$ corresponds to the so-called *active constraints at* x, and it is easily verified (see Exercise 116) that

$$\mathrm{N}_C(x) + \mathrm{cone\, co}\left(\bigcup_{t \in A(x)} \partial f_t(x)\right) \subseteq \mathrm{N}_F(x). \tag{8.17}$$

Definition 8.1.10 *We say that the consistent system S in (8.2) is locally Farkas–Minkowski (LFM, in short) at $x \in F$ if*

$$\mathrm{N}_C(x) + \mathrm{cone\, co}\left(\bigcup_{t \in A(x)} \partial f_t(x)\right) = \mathrm{N}_F(x). \tag{8.18}$$

And S is said to be LFM *if it is LFM at every feasible point.*

Thanks to (8.17), S is LFM at $x \in F$ if and only if

$$\mathrm{N}_F(x) \subset \mathrm{N}_C(x) + \mathrm{cone\, co}\left(\bigcup_{t \in A(x)} \partial f_t(x)\right).$$

The LFM property is closely related to the following condition, involving the set of indices

$$T(x) := \{t \in T : f_t(x) = \tilde{f}(x)\},\ x \in C,$$

8.1. CONVEX SYSTEMS AND FARKAS-TYPE ...

where
$$\tilde{f} := \sup_{t \in T} f_t. \tag{8.19}$$

The following theorem is a LFM counterpart to Theorem 8.1.7.

Theorem 8.1.11 *If $\mathcal{S}$ is LFM at $x \in F$, and for certain $a^* \in X^*$ we have*
$$\langle a^*, z \rangle \leq \langle a^*, x \rangle \text{ for all } z \in F, \tag{8.20}$$

then $\langle a^*, \cdot \rangle \leq \langle a^*, x \rangle$ is also consequence of a finite subsystem of $\mathcal{S}$. The converse statement holds provided that $\mathcal{S}$ is linear.

Proof. We only consider the non-trivial case $a^* \neq \theta$. Then (8.20) is equivalent to $a^* \in N_F(x) \setminus \{\theta\}$, and (8.18) entails the existence of

$$y^* \in N_C(x) \text{ and } x^* \in \text{cone co} \left(\bigcup_{t \in A(x)} \partial f_t(x) \right),$$

such that $a^* = y^* + x^*$.

If $x^* = \theta$, then $a^* = y^* \in N_C(x)$ and $\langle a^*, \cdot \rangle \leq \langle a^*, x \rangle$ is consequence of any possible subsystems of $\mathcal{S}$, whose solution set is always included in C.

If $x^* \neq \theta$, the convexity of the subdifferential set entails the existence of $\lambda \in \mathbb{R}_+^{(T)}$ and $u_t^* \in \partial f_t(x)$, $t \in \text{supp}\,\lambda \subset A(x)$, such that $x^* = \sum_{t \in \text{supp}\,\lambda} \lambda_t u_t^*$. Let

$$\widehat{\mathcal{S}} := \{f_t(z) \leq 0,\ t \in \text{supp}\,\lambda;\ z \in C\},$$

and let $z \in \hat{F}$, where $\hat{F}$ is the solution set of $\widehat{\mathcal{S}}$. We have, for every $t \in \text{supp}\,\lambda \subset A(x)$,

$$0 \geq f_t(z) \geq f_t(x) + \langle u_t^*, z - x \rangle = \langle u_t^*, z - x \rangle,$$

and so

$$0 \geq \sum_{t \in \text{supp}\,\lambda} \lambda_t f_t(z) \geq \sum_{t \in \text{supp}\,\lambda} \lambda_t \langle u_t^*, z - x \rangle = \langle x^*, z - x \rangle$$
$$= \langle a^* - y^*, z - x \rangle = \langle a^*, z - x \rangle + \langle -y^*, z - x \rangle \geq \langle a^*, z - x \rangle,$$

where the last inequality comes from $y^* \in N_C(x)$. Thus, we have proved that $\langle a^*, z \rangle \leq \langle a^*, x \rangle$ for every $z \in \hat{F}$, and we finished the proof of the first statement.

Now, let $C = X$ and $f_t(z) = \langle a_t^*, z \rangle - b_t$, with $a_t^* \in X^*$ and $b_t \in \mathbb{R}$, $t \in T$. Let $a^* \in \mathrm{N}_F(x) \setminus \{\theta\}$; i.e., $\langle a^*, z - x \rangle \leq 0$ for all $z \in F$. By assumption, there exists a finite set $\hat{T} \subset T$ such that $\langle a^*, z \rangle \leq \langle a^*, x \rangle$ for all $z \in \hat{F}$, where

$$\hat{F} := \{z \in X : \langle a_t^*, z \rangle \leq b_t \text{ for all } t \in \hat{T}\}.$$

Then, by Theorem 8.1.5(i),

$$(a^*, \langle a^*, x \rangle) \in \mathrm{cl}\,\hat{K} = \hat{K} = \mathrm{cone\,co}\left\{(a_t^*, b_t),\ t \in \hat{T};\ (\theta, 1)\right\};$$

i.e., there exist $\lambda_t \geq 0$, $t \in \hat{T}$, not all of them equal to zero, and $\mu \geq 0$ such that $(a^*, \langle a^*, x \rangle) = \sum_{t \in \hat{T}} \lambda_t (a_t^*, b_t) + \mu(\theta, 1)$ so that, making the scalar product of $(a^*, \langle a^*, x \rangle)$ and $(x, -1)$,

$$0 = \sum_{t \in \hat{T}} \lambda_t (\langle a_t^*, x \rangle - b_t) - \mu = \sum_{t \in \mathrm{supp}\,\lambda} \lambda_t (\langle a_t^*, x \rangle - b_t) - \mu.$$

Since $x \in F$, it must be $\mu = 0$ and $\mathrm{supp}\,\lambda \subset A(x)$, entailing

$$a^* \in \mathrm{cone\,co}\{a_t^*,\ t \in A(x)\} = \mathrm{cone\,co}\left(\bigcup_{t \in A(x)} \partial f_t(x)\right),$$

and (8.18) holds. ∎

The converse statement in the last theorem does not hold in general for convex systems without any additional assumption, as the same Example 8.1.8 shows. An example of an infinite convex system illustrating this fact is given in Exercise 118. The following example shows that the class of LFM systems is strictly larger than the FM class.

Example 8.1.12 *Consider the system*

$$\mathcal{S} := \{-x_1 - t^2 x_2 \leq -2t,\ t > 0\}.$$

Observe that every consequent linear inequality supporting F is a consequence of a finite subsystem of $\mathcal{S}$, namely the subsystem formed by the same inequality. However, the inequality $x_2 \geq 0$ is a consequence of $\mathcal{S}$, but for every finite linear subsystem $\widehat{\mathcal{S}}$, it happens that $\hat{F} \cap \{(x_1, x_2) \in \mathbb{R}^2 : x_2 < 0\} \neq \emptyset$, and so $x_2 \geq 0$ is not a consequence of $\widehat{\mathcal{S}}$. Thus, by Theorem 8.1.7, $\mathcal{S}$ is not FM.

8.2. OPTIMALITY AND DUALITY IN ... 317

The LFM property is related to the so-called basic constraint qualification:

Definition 8.1.13 *We say that the basic constraint qualification (BCQ, in short) is fulfilled at $x \in F$ if*

$$N_F(x) \subseteq N_C(x) + \text{cone co}\left(\bigcup_{t \in T(x)} \partial f_t(x)\right). \tag{8.21}$$

Exercise 117 establishes the equivalence between LFM and BCQ under the continuity of $\tilde{f}$ at the reference point x when $x \in \text{int}\,C$.

8.2 Optimality and duality in (semi)infinite convex optimization

Our aim in this second section is to present KKT optimality conditions for problem (P) introduced previously. We do it by exploiting the advantages of representation (8.4), and appealing to the properties of the supremum function f and the characterizations of its subdifferential provided in the previous chapters. The following result, which is known as the *Pshenichnyi–Rockafellar theorem*, constitutes the simplest generalization of the Fermat rule for problem (P) in the absence of explicit constraints.

Theorem 8.2.1 *Suppose that either $(\text{dom}\,g) \cap (\text{int}\,C) \neq \emptyset$ or there exists $x_0 \in (\text{dom}\,g) \cap C$ where g is continuous. Then, under the absence of explicit constraints, $\overline{x} \in C$ is an optimal solution of* (P) *if and only if*

$$\partial g(\overline{x}) \cap (-N_C(\overline{x})) \neq \emptyset.$$

Proof. We apply Proposition 4.1.20 to the convex functions g and I_C. We obtain that $\overline{x} \in C$ is an optimal solution of (P) if and only if $\theta \in \partial(g + I_C)(\overline{x}) = \partial g(\overline{x}) + N_C(\overline{x})$, if and and only if $\partial g(\overline{x}) \cap (-N_C(\overline{x})) \neq \emptyset$. ∎

The following result gives *approximate KKT conditions*, avoiding the qualifications of Theorem 8.2.1.

Theorem 8.2.2 *A feasible solution of* (P) *is optimal if and only if, for all $\varepsilon > 0$,*

$$\theta \in \mathrm{cl}\left(\bigcup_{\mu>0,\;\lambda\in\Delta(T)} \partial_\varepsilon g(\bar{x}) + \mu\partial_{\left(\frac{\varepsilon}{\mu} + \sum_{t\in\mathrm{supp}\,\lambda}\lambda_t f_t(x)\right)}\left(\sum_{t\in\mathrm{supp}\,\lambda}\lambda_t f_t\right)(\bar{x}) + \mathrm{N}^\varepsilon_C(\bar{x})\right),$$

where $\Delta(T)$ is defined in (2.46).

Proof. Observe that $\bar{x} \in C$ is an optimal solution of (P) if and only if it is optimal for the unconstrained optimization problem

$$\inf_{x\in X}\left\{g(x) + \mathrm{I}_{[\sup_{t\in T} f_t \leq 0]}(x) + \mathrm{I}_C(x)\right\};$$

hence, if and only if $\theta \in \partial(g + \mathrm{I}_{[\sup_{t\in T} f_t \leq 0]}(x) + \mathrm{I}_C)(\bar{x})$. Therefore, the conclusion follows by combining Proposition 4.1.20 and Example 5.1.13. ∎

The following theorem provides *KKT-type optimality conditions* for problem (P). These conditions are in contrast to those of Theorem 8.2.5 below, which are of a fuzzy type.

Theorem 8.2.3 *Provided that g is finite and continuous somewhere in F, under LFM at $\bar{x} \in F \cap \mathrm{dom}\, g$, the point $\bar{x}$ is a (global) minimum of (P) if and only if there exists $\lambda \in \mathbb{R}_+^{(T)}$ such that*

$$\theta \in \partial g(\bar{x}) + \sum_{t\in\mathrm{supp}\,\lambda}\lambda_t \partial f_t(\bar{x}) + \mathrm{N}_C(\bar{x}) \text{ and } \lambda_t f_t(\bar{x}) = 0 \text{ for all } t \in T.$$
(8.22)

Proof. Notice that $\bar{x} \in F \cap \mathrm{dom}\, g$ is a minimum of (P) if and only if $\theta \in \partial(g + \mathrm{I}_F)(\bar{x})$, if and only if $\theta \in \partial g(\bar{x}) + \partial \mathrm{I}_F(\bar{x}) = \partial g(\bar{x}) + \mathrm{N}_F(\bar{x})$ due to Proposition 4.1.20. In other words, if and only if there exists $a^* \in \partial g(\bar{x})$ such that $\langle -a^*, \cdot\rangle \leq \langle -a^*, \bar{x}\rangle$ is a consequence of $\mathcal{S}$. Therefore, if $\bar{x}$ is a minimum of (P), then Theorem 8.1.11 entails the existence of some finite set $\hat{T} \subset T$ such that

$$-(a^*, \langle a^*, \bar{x}\rangle) \in \mathrm{cone}\,\mathrm{co}\left\{\bigcup_{t\in\hat{T}}\mathrm{epi}\,f_t^*\right\} + \mathrm{epi}\,\sigma_C;$$

that is, there exist $\lambda \in \mathbb{R}_+^{(T)}$, $x_t^* \in \mathrm{dom}\, f_t^*$, $\alpha_t \geq 0$, $t \in \mathrm{supp}\,\lambda$ (a possibly empty subset of $\hat{T}$), $z^* \in \mathrm{dom}\,\sigma_C$, and $\beta \geq 0$, satisfying

$$-(a^*, \langle a^*, \bar{x}\rangle) = \sum_{t\in T}\lambda_t\left(x_t^*, f_t^*(x_t^*) + \alpha_t\right) + \left(z^*, \sigma_C(z^*) + \beta\right).$$

From this last equation, if $\eta := \sum_{t\in T}\lambda_t\alpha_t + \beta$, we get

8.2. OPTIMALITY AND DUALITY IN ...

$$0 = \sum_{t \in T} \lambda_t (\langle x_t^*, \overline{x} \rangle - f_t^* (x_t^*)) + (\langle z^*, \overline{x} \rangle - \sigma_C (z^*)) - \eta$$
$$\leq \sum_{t \in T} \lambda_t f_t(\overline{x}) - \eta \leq 0,$$

but this implies, from the one side, $\operatorname{supp} \lambda \subset A(\overline{x})$, $\beta = 0$ and $\alpha_t = 0$ for all $t \in \operatorname{supp} \lambda$, and from the other side

$$\langle x_t^*, \overline{x} \rangle - f_t^* (x_t^*) = f_t(\overline{x}) \text{ for all } t \in \operatorname{supp} \lambda, \quad (8.23)$$

and

$$\langle z^*, \overline{x} \rangle - \sigma_C (z^*) = \mathrm{I}_C(\overline{x}). \quad (8.24)$$

Therefore,

$$x_t^* \in \partial f_t(\overline{x}) \text{ for all } t \in \operatorname{supp} \lambda, \text{ and } z^* \in \partial \mathrm{I}_C(\overline{x}) = \mathrm{N}_C(\overline{x}),$$

and

$$-a^* = \sum_{t \in T} \lambda_t x_t^* + z^* \in \sum_{t \in \operatorname{supp} \lambda} \lambda_t \partial f_t(\overline{x}) + \mathrm{N}_C(\overline{x})$$

leads us to (8.22). Conversely, we show that (8.22) implies that $\overline{x}$ is a minimum of (P). Indeed, (8.22) gives rise to some $\lambda \in \mathbb{R}_+^{(T)}$ and $a^* \in X^*$ such that $-a^* \in \mathrm{N}_C(\overline{x})$ and (using the convention $0 f_t = \mathrm{I}_{\operatorname{dom} f_t}$)

$$a^* \in \partial g(\overline{x}) + \sum_{t \in \operatorname{supp} \lambda} \lambda_t \partial f_t(\overline{x}) \subset \partial \left(g + \sum_{t \in T} \lambda_t f_t \right)(\overline{x}),$$

so that

$$g(x) + \sum_{t \in T} \lambda_t f_t(x) \geq g(\overline{x}) + \sum_{t \in T} \lambda_t f_t(\overline{x}) + \langle a^*, x - \overline{x} \rangle \text{ for all } x \in X.$$
(8.25)

Since $\lambda_t f_t(\overline{x}) = 0$ for all $t \in \operatorname{supp} \lambda$, and $-a^* \in \mathrm{N}_C(\overline{x})$, (8.25) implies

$$g(x) + \sum_{t \in T} \lambda_t f_t(x) - g(\overline{x}) \geq \langle a^*, x - \overline{x} \rangle \geq 0 \text{ for all } x \in C.$$

In particular, for $x \in F$, we get $g(x) \geq g(x) + \sum_{t \in T} \lambda_t f_t(x) \geq g(\overline{x})$, which proves that $\overline{x}$ is a minimum of (P). ∎

The following example illustrates Theorem 8.2.3.

Example 8.2.4 *Consider problem* (P) *in* $\mathbb{R}^2$ *where*

$$\text{(P) Min } x_2$$
$$\text{s.t. } tx_1 - x_2 \leq (1/2)t^2, \ t > 0,$$

whose optimal set is $\{x \in \mathbb{R}^2 : x_1 \leq 0, \ x_2 = 0\}$. *Then*

$$\mathcal{S} := \{tx_1 - x_2 \leq (1/2)t^2, \ t > 0\}$$

is neither FM nor LFM at every point in the optimal set. Indeed, the inequality $x_2 \geq 0$ *defines a supporting half-space to* F *at every optimal solution* $\bar{x}$ *of* (P), *but it is not consequence of any finite subsystem of* $\mathcal{S}$. *Hence,* $\mathcal{S}$ *is not LFM and, a fortiori, is not FM (by Exercise 122). Additionally, the KKT optimality conditions at* $\bar{x} = 0_2$ *are not satisfied as* $A(0_2) = \emptyset$ *and*

$$0_2 \notin \partial g(0_2) + N_{\mathbb{R}^2}(0_2) = \{(0,1)\}.$$

Observe that if we consider the enlarged inequality system

$$\mathcal{S}' := \left\{ tx_1 - x_2 \leq (1/2)t^2, \ t \in [0, \infty[\right\},$$

then $\mathcal{S}'$ *turns out to be LFM as every supporting half-space to* F' *(the feasible set of* $\mathcal{S}'$*) is consequence of a subsystem composed by a unique inequality constraint (the same inequality). Moreover, every affine consequence of* $\mathcal{S}'$ *is of the form* $-x_2 \leq -tx_1 - \alpha$ *with* $\alpha \leq -t^2/2$, $t \geq 0$, *and therefore, it is also consequence of the finite subsystem*

$$\widehat{\mathcal{S}'} := \left\{ -x_2 \leq -tx_1 + (1/2)t^2 \right\}.$$

Consequently, Theorem 8.2.3 implies that, for every optimal solution $\bar{x} \in \{x \in \mathbb{R}^2 : x_1 \leq 0, \ x_2 = 0\}$, *we have that* $A(\bar{x}) = \{0\}$, *and the KKT optimality conditions are fulfilled as* $0_2 \in (0,1) + \mathbb{R}_+(0,-1)$.

Next, we establish *fuzzy KKT optimality conditions* for problem (P) under, again, the LFM property.

Theorem 8.2.5 *Let us assume that* $\mathcal{S}$ *is LFM and* $(\operatorname{ri} F) \cap \operatorname{dom} g \neq \emptyset$. *Then* $\bar{x} \in F$ *is a minimum of* (P) *if and only if, for each* $\varepsilon > 0$ *and* $U \in \mathcal{N}_{X^*}$, *there exists* $\lambda = \lambda(\varepsilon, U) \in \mathbb{R}_+^{(T)}$ *such that* $\operatorname{supp} \lambda \subset A(\bar{x})$ *and*

$$\theta \in \partial_\varepsilon g(\bar{x}) + \sum_{t \in \operatorname{supp} \lambda} \lambda_t \partial f_t(\bar{x}) + N_C(\bar{x}) + U. \tag{8.26}$$

8.2. OPTIMALITY AND DUALITY IN ...

Proof. Suppose that $\bar{x} \in F \cap \operatorname{dom} g$ is a minimum of (P). Since $(\operatorname{ri} F) \cap \operatorname{dom} g \neq \emptyset$, Theorem 7.2.1(ii) yields

$$\partial(g + \mathrm{I}_F)(\bar{x}) = \bigcap_{\varepsilon > 0} \operatorname{cl}(\partial_\varepsilon g(\bar{x}) + \mathrm{N}_F(\bar{x})).$$

Then,

$$\bar{x} \text{ is optimal for (P)} \Leftrightarrow \theta \in \bigcap_{\varepsilon > 0} \operatorname{cl}(\partial_\varepsilon g(\bar{x}) + \mathrm{N}_F(\bar{x})).$$

So, $\theta \in \partial_\varepsilon g(\bar{x}) + \mathrm{N}_F(\bar{x}) + U$ for every given $\varepsilon > 0$ and $U \in \mathcal{N}_{X^*}$. Thus, by the LFM property, we have that

$$0 \in \partial_\varepsilon g(\bar{x}) + \operatorname{cone co} \left(\bigcup_{t \in A(\bar{x})} \partial f_t(\bar{x}) \right) + \mathrm{N}_C(\bar{x}) + U,$$

and we are done with the necessity statement. Conversely, we fix $x \in F$ ($\subset C$). Given $\varepsilon > 0$, we choose $U \in \mathcal{N}_{X^*}$ such that $|\langle u^*, x - \bar{x} \rangle| \leq \varepsilon$ for all $u^* \in U$. If (8.26) holds, then there exists $u_\varepsilon^* \in U$ such that

$$u_\varepsilon^* \in \partial_\varepsilon g(\bar{x}) + \sum_{t \in \operatorname{supp} \lambda} \lambda_t \partial f_t(\bar{x}) + \mathrm{N}_C(\bar{x}) \subset \partial_\varepsilon \left(g + \sum_{t \in \operatorname{supp} \lambda} \lambda_t f_t + \mathrm{I}_C \right)(\bar{x}),$$

and we deduce

$$g(x) + \sum_{t \in \operatorname{supp} \lambda} \lambda_t f_t(x) \geq g(\bar{x}) + \sum_{t \in \operatorname{supp} \lambda} \lambda_t f_t(\bar{x}) + \langle u_\varepsilon^*, x - \bar{x} \rangle - \varepsilon.$$
(8.27)

Hence, since $\operatorname{supp} \lambda_\varepsilon \subset A(\bar{x})$,

$$g(x) \geq g(x) + \sum_{t \in \operatorname{supp} \lambda} \lambda_t f_t(x) \geq g(\bar{x}) + \langle u_\varepsilon^*, x - \bar{x} \rangle - \varepsilon \geq g(\bar{x}) - 2\varepsilon,$$

the desired conclusion follows by taking limits for $\varepsilon \to 0$. ∎

We proceed by giving alternative fuzzy-type optimality conditions for problem (P), using the following *strong Slater qualification condition*:

$$\tilde{f}(x_0) = \sup_{t \in T} f_t(x_0) < 0, \text{ for some } x_0 \in C; \quad (8.28)$$

the point x_0 is called a *strong Slater point*. For the sake of simplicity, the set C is supposed to be the whole space X.

Theorem 8.2.6 *Assume that X is a reflexive Banach space. Given problem* (P), *suppose that $C = X$, T is compact and the mappings $t \mapsto f_t(z)$, $z \in X$, are usc, and assume that* (8.28) *holds. Let $\bar{x} \in F$ be a minimum of* (P) *such that $\inf_{t \in T} f_t(\bar{x}) > -\infty$ and $A(\bar{x}) \neq \emptyset$. Then, for every $\varepsilon > 0$, there exists $(\lambda_0, \lambda) = (\lambda_0(\varepsilon), \lambda(\varepsilon)) \in \mathbb{R}_+ \times \mathbb{R}_+^{(T)}$ such that $\lambda_0 > 0$, $\lambda_0 + \sum_{t \in A(\bar{x})} \lambda_t = 1$, $\sum_{t \in T \setminus A(\bar{x})} \lambda_t \leq \varepsilon$ and*

$$\theta \in \lambda_0 \partial_\varepsilon g(\bar{x}) + \sum_{t \in A(\bar{x})} \lambda_t \partial_\varepsilon f_t(\bar{x}) + \sum_{t \in T \setminus A(\bar{x})} \lambda_t \partial_\varepsilon f_t(\bar{x}) + \varepsilon B_{X^*}. \quad (8.29)$$

Proof. It is well-known that $\bar{x}$ is a minimum of the convex supremum function $\varphi : X \to \mathbb{R}_\infty$, defined as

$$\varphi(x) := \sup\{g(x) - g(\bar{x}); \ f_t(x), \ t \in T\}; \quad (8.30)$$

that is, $\theta \in \partial \varphi(\bar{x})$, and Theorem 6.4.3 yields

$$\theta \in \partial \varphi(\bar{x})$$

$$= \bigcap_{\varepsilon > 0} \overline{\text{co}} \left(\left(\partial_\varepsilon g(\bar{x}) \cup \left(\bigcup_{t \in A(\bar{x})} \partial_\varepsilon f_t(\bar{x}) \right) \right) + \left(\bigcup_{t \in T \setminus A(\bar{x})} \{0, \varepsilon\} \partial_\varepsilon f_t(\bar{x}) \right) \right)$$

$$= \bigcap_{\varepsilon > 0} \text{cl}^{\|\cdot\|_*} \left(\text{co} \left(\partial_\varepsilon g(\bar{x}) \cup \left(\bigcup_{t \in A(\bar{x})} \partial_\varepsilon f_t(\bar{x}) \right) \right) + \text{co} \left(\bigcup_{t \in T \setminus A(\bar{x})} [0, \varepsilon] \partial_\varepsilon f_t(\bar{x}) \right) \right), \quad (8.31)$$

where $\|\cdot\|_*$ is the dual norm. Then, given $\varepsilon > 0$, we obtain

$$\theta \in \text{co} \left(\partial_\varepsilon g(\bar{x}) \cup \left(\bigcup_{t \in A(\bar{x})} \partial_\varepsilon f_t(\bar{x}) \right) \right) + \text{co} \left(\bigcup_{t \in T \setminus A(\bar{x})} [0, \varepsilon] \partial_\varepsilon f_t(\bar{x}) \right) + \varepsilon B_{X^*},$$

and there exists an associated $(\mu_{\varepsilon, 0}, \mu_\varepsilon) \in \mathbb{R}_+ \times \mathbb{R}_+^{(T)}$ such that $\mu_{\varepsilon, 0} + \sum_{t \in A(\bar{x})} \mu_{\varepsilon, t} = 1$, $\sum_{t \in T \setminus A(\bar{x})} \mu_{\varepsilon, t} = 1$ and

$$\theta \in \mu_{\varepsilon, 0} \partial_\varepsilon g(\bar{x}) + \sum_{t \in A(\bar{x})} \mu_{\varepsilon, t} \partial_\varepsilon f_t(\bar{x}) + \sum_{t \in T \setminus A(\bar{x})} [0, \varepsilon \mu_{\varepsilon, t}] \partial_\varepsilon f_t(\bar{x}) + \varepsilon B_{X^*}. \quad (8.32)$$

Let us show that $\mu_{\varepsilon, 0} > 0$ for all $\varepsilon > 0$ small enough. Arguing by contradiction, we suppose the existence of some $\varepsilon_k \downarrow 0$ such that $\mu_{\varepsilon_k, 0} = 0$ for all $k \geq 1$. Consequently, taking into account Proposition 4.1.10, (8.32) becomes, for all $k \geq 1$,

8.2. OPTIMALITY AND DUALITY IN ...

$$\theta \in \sum_{t\in A(\bar{x})} \mu_{\varepsilon_k,t}\partial_{\varepsilon_k} f_t(\bar{x}) + \sum_{t\in T\setminus A(\bar{x})} [0, \varepsilon_k\mu_{\varepsilon_k,t}]\partial_{\varepsilon_k} f_t(\bar{x}) + \varepsilon_k B_{X^*}$$
$$\subset \operatorname{co}\left(\left(\bigcup_{t\in A(\bar{x})} \partial_{\varepsilon_k} f_t(\bar{x})\right) + \left(\bigcup_{t\in T\setminus A(\bar{x})} [0,\varepsilon_k]\partial_{\varepsilon_k} f_t(\bar{x})\right)\right) + \varepsilon_k B_{X^*};$$

that is,

$$\theta \in \bigcap_{k\geq 1} \left(\operatorname{co}\left(\left(\bigcup_{t\in A(\bar{x})} \partial_{\varepsilon_k} f_t(\bar{x})\right) + \left(\bigcup_{t\in T\setminus A(\bar{x})} [0,\varepsilon_k]\partial_{\varepsilon_k} f_t(\bar{x})\right)\right) + \varepsilon_k B_{X^*}\right).$$

Next, given any $\varepsilon > 0$, we can find some $k_0 \geq 1$ such that $\varepsilon_{k_0} < \varepsilon$ and the last relation yields

$$\theta \in \operatorname{co}\left(\left(\bigcup_{t\in A(\bar{x})} \partial_{\varepsilon_{k_0}} f_t(\bar{x})\right) + \left(\bigcup_{t\in T\setminus A(\bar{x})} [0,\varepsilon_{k_0}]\partial_{\varepsilon_k} f_t(\bar{x})\right)\right) + \varepsilon_{k_0} B_{X^*}$$
$$\subset \operatorname{co}\left(\left(\bigcup_{t\in A(\bar{x})} \partial_{\varepsilon} f_t(\bar{x})\right) + \left(\bigcup_{t\in T\setminus A(\bar{x})} [0,\varepsilon]\partial_{\varepsilon} f_t(\bar{x})\right)\right) + \varepsilon B_{X^*},$$

and we conclude that

$$\theta \in \bigcap_{\varepsilon>0} \left(\operatorname{co}\left(\left(\bigcup_{t\in A(\bar{x})} \partial_{\varepsilon} f_t(\bar{x})\right) + \left(\bigcup_{t\in T\setminus A(\bar{x})} [0,\varepsilon]\partial_{\varepsilon} f_t(\bar{x})\right)\right) + \varepsilon B_{X^*}\right)$$
$$= \bigcap_{\varepsilon>0} \overline{\operatorname{co}}\left(\left(\bigcup_{t\in A(\bar{x})} \partial_{\varepsilon} f_t(\bar{x})\right) + \left(\bigcup_{t\in T\setminus A(\bar{x})} [0,\varepsilon]\partial_{\varepsilon} f_t(\bar{x})\right)\right)$$
$$= \bigcap_{\varepsilon>0} \overline{\operatorname{co}}\left(\left(\bigcup_{t\in A(\bar{x})} \partial_{\varepsilon} f_t(\bar{x})\right) + \left(\bigcup_{t\in T\setminus A(\bar{x})} \{0,\varepsilon\}\partial_{\varepsilon} f_t(\bar{x})\right)\right).$$

Hence, applying once again Theorem 6.4.3, we obtain $\theta \in \partial \tilde{f}(\bar{x})$, where $\tilde{f} = \sup_{t\in T} f_t$, contradicting the strong Slater condition. Finally, the conclusion follows by taking $\lambda_0(\varepsilon) := \mu_{\varepsilon,0}$, $\lambda_t(\varepsilon) := \mu_{\varepsilon,t}$ for $t \in A(\bar{x})$, and $\lambda_t(\varepsilon) := \gamma_{\varepsilon,t}\mu_{\varepsilon,t}$ with $\gamma_{\varepsilon,t} \in [0,\varepsilon]$, for $t \in T \setminus A(\bar{x})$. ∎

We proceed by deriving other optimality conditions for the *semi-infinite convex optimization problem* (P); i.e., now $X = \mathbb{R}^n$ and $C \subset \mathbb{R}^n$ is convex, not necessarily closed, T is a Hausdorff topological space, and the functions g, f_t, $t \in T$, are proper convex, possibly not lsc. Let us denote

$$D := \operatorname{dom} g \cap \operatorname{dom} \tilde{f}, \tag{8.33}$$

where we have denoted $\tilde{f} = \sup_{t \in T} f_t$ (see (8.19)). The following theorems provide *Fritz-John optimality conditions* for problem (P).

Theorem 8.2.7 *Given the semi-infinite problem* (P) *above, assume that* T *is compact and the mappings* $t \mapsto f_t(z)$, $z \in X$, *are usc. If* $\bar{x}$ *is optimal for* (P) *such that* $A(\bar{x}) \neq \emptyset$, *then*

$$0_n \in \mathrm{co}\left\{\partial(g + \mathrm{I}_{D \cap C})(\bar{x}) \cup \bigcup_{t \in A(\bar{x})} \partial(f_t + \mathrm{I}_{D \cap C})(\bar{x})\right\}. \tag{8.34}$$

Moreover, we have

$$0_n \in \mathrm{co}\left\{\partial g(\bar{x}) \cup \bigcup_{t \in A(\bar{x})} \partial f_t(\bar{x})\right\} + \mathrm{N}_{D \cap C}(\bar{x}), \tag{8.35}$$

provided that, for all $t \in A(\bar{x})$,

$$\mathrm{ri}(\mathrm{dom}\, g) \cap \mathrm{ri}(D \cap C) \neq \emptyset, \quad \mathrm{ri}(\mathrm{dom}\, f_t) \cap \mathrm{ri}(D \cap C) \neq \emptyset. \tag{8.36}$$

Proof. We know that if $\bar{x}$ is optimal for (P), then $\bar{x}$ is an unconstrained minimum of the supremum function $\varphi : \mathbb{R}^n \to \mathbb{R}_\infty$, defined as

$$\varphi(x) := \sup\{g(x) - g(\bar{x}),\ \mathrm{I}_C(x) - 2\varepsilon_0,\ f_t(x), t \in T\},$$

where $\varepsilon_0 > 0$ is fixed. Consequently, according to Remark 14, we obtain

$$0_n \in \partial \varphi(\bar{x}) = \mathrm{co}\left\{\partial(g + \mathrm{I}_{D \cap C})(\bar{x}) \cup \bigcup_{t \in A(\bar{x})} \partial(f_t + \mathrm{I}_{D \cap C})(\bar{x})\right\};$$

that is, (8.34) holds. Finally, (8.35) follows from (8.34) by applying Proposition 4.1.26. ∎

In the following result, we use the *strong conical hull intersection property (strong CHIP)* at $x \in C \cap D$ of the family $\{C, \mathrm{dom}\, g, \mathrm{dom}\, f_t, t \in T\}$, stating that

$$\mathrm{N}_{C \cap \mathrm{dom}\, g \cap (\cap_{t \in T} \mathrm{dom}\, f_t)}(x) = \mathrm{N}_C(x) + \mathrm{N}_{\mathrm{dom}\, g}(x) + \sum_{t \in T} \mathrm{N}_{\mathrm{dom}\, f_t}(x), \tag{8.37}$$

where

$$\sum_{t \in T} \mathrm{N}_{\mathrm{dom}\, f_t}(x) := \left\{\sum_{t \in J} a_t :\ a_t \in \mathrm{N}_{\mathrm{dom}\, f_t}(x),\ J \text{ being a finite subset of } T\right\}. \tag{8.38}$$

8.2. OPTIMALITY AND DUALITY IN ...

Observe, for instance, that (8.37) holds in the finite-dimensional setting when T is finite and

$$(\operatorname{ri} C) \cap \operatorname{ri}(\operatorname{dom} g) \cap \left(\bigcap_{t \in T} \operatorname{ri}(\operatorname{dom} f_t) \right) \neq \emptyset, \qquad (8.39)$$

as a consequence of Proposition 4.1.26 when applied to the corresponding indicator functions.

Theorem 8.2.8 *Assume that T is compact, the mappings $t \mapsto f_t(z)$, $z \in X$, are usc, and that (8.36) holds. Let $\bar{x}$ be optimal for (P) such that $A(\bar{x}) \neq \emptyset$. If (8.37) holds at $\bar{x}$, then*

$$0_n \in \operatorname{co}\left\{ \partial g(\bar{x}) \cup \bigcup_{t \in A(\bar{x})} \partial f_t(\bar{x}) \right\} + \operatorname{N}_C(\bar{x}) + \operatorname{N}_{\operatorname{dom} g}(\bar{x}) + \sum_{t \in T} \operatorname{N}_{\operatorname{dom} f_t}(\bar{x}).$$

Proof. The compactness and the upper semicontinuity assumptions imply that $\operatorname{dom} f = \cap_{t \in T} \operatorname{dom} f_t$ (Exercise 9), entailing

$$D \cap C = \operatorname{dom} g \cap C \cap \left(\bigcap_{t \in T} \operatorname{dom} f_t \right).$$

Then, using (8.35), we get

$$0_n \in \operatorname{co}\left\{ \partial g(\bar{x}) \cup \left(\bigcup_{t \in A(\bar{x})} \partial f_t(\bar{x}) \right) \right\} + \operatorname{N}_{D \cap C}(\bar{x}),$$

and the result comes by applying (8.37). ■

Next, we give additional KKT optimality conditions for problem (P) under the strong Slater qualification.

Theorem 8.2.9 *If in Theorem 8.2.8 we assume, additionally, the existence of a strong Slater point in $\operatorname{dom} g$, then the optimality of $\bar{x}$ implies the existence of a (possibly empty) finite set $\widehat{A}(\bar{x}) \subset A(\bar{x})$ and scalars $\lambda_t > 0$ for $t \in \widehat{A}(\bar{x})$, satisfying*

$$0_n \in \partial g(\bar{x}) + \sum_{t \in \widehat{A}(\bar{x})} \lambda_t \partial f_t(\bar{x}) + \operatorname{N}_C(\bar{x}) + \sum_{t \in T} \operatorname{N}_{\operatorname{dom} f_t}(\bar{x}). \qquad (8.40)$$

Proof. According to Theorem 8.2.8, we have

$$0_n \in \operatorname{co}\left\{\partial g(\bar{x}) \cup \left(\bigcup_{t \in A(\bar{x})} \partial f_t(\bar{x})\right)\right\} + N_C(\bar{x}) + N_{\operatorname{dom} g}(\bar{x}) + \sum_{t \in T} N_{\operatorname{dom} f_t}(\bar{x}). \tag{8.41}$$

If $\partial g(\bar{x})$ does not intervene within the convex combination of 0_n coming from (8.41), then we get

$$0_n \in \operatorname{co}\left\{\bigcup_{t \in A(\bar{x})} \partial f_t(\bar{x})\right\} + N_C(\bar{x}) + N_{\operatorname{dom} g}(\bar{x}) + \sum_{t \in T} N_{\operatorname{dom} f_t}(\bar{x})$$

$$\subset \operatorname{co}\left\{\bigcup_{t \in A(\bar{x})} \partial f_t(\bar{x})\right\} + N_C(\bar{x}) + N_{\operatorname{dom} g}(\bar{x}) + N_{\operatorname{dom} \tilde{f}}(\bar{x}). \tag{8.42}$$

Therefore,

$$0_n \in \partial \tilde{f}(\bar{x}) + N_C(\bar{x}) + N_{\operatorname{dom} g}(\bar{x}) + N_{\operatorname{dom} \tilde{f}}(\bar{x}) \subset \partial(\tilde{f} + I_{C \cap \operatorname{dom} g})(\bar{x}),$$

and $\bar{x}$ is a minimum of $\tilde{f} + I_{C \cap \operatorname{dom} g}$, constituting a contradiction with the existence of a strong Slater point $x_0 \in C \cap \operatorname{dom} g$,

$$0 = \tilde{f}(\bar{x}) = (\tilde{f} + I_{C \cap \operatorname{dom} g})(\bar{x}) \le (\tilde{f} + I_{C \cap \operatorname{dom} g})(x_0) = \tilde{f}(x_0) < 0.$$

Otherwise, if $\partial g(\bar{x})$ intervenes in (8.41), then there would exist scalars $\alpha > 0$ and $\beta \in \mathbb{R}_+^{(T)}$ such that $\operatorname{supp} \beta \subset A(\bar{x})$, $\alpha + \sum_{t \in A(\bar{x})} \beta_t = 1$ and

$$0_n \in \alpha \partial g(\bar{x}) + \sum_{t \in \operatorname{supp} \beta} \beta_t \partial f_t(\bar{x}) + N_C(\bar{x}) + N_{\operatorname{dom} g}(\bar{x}) + \sum_{t \in T} N_{\operatorname{dom} f_t}(\bar{x})$$

$$= \alpha \partial g(\bar{x}) + \sum_{t \in \operatorname{supp} \beta} \beta_t \partial f_t(\bar{x}) + N_C(\bar{x}) + \sum_{t \in T} N_{\operatorname{dom} f_t}(\bar{x}),$$

because $\alpha \partial g(\bar{x}) + N_{\operatorname{dom} g}(\bar{x}) = \alpha \partial g(\bar{x})$ and $\partial g(\bar{x}) \ne \emptyset$. Hence, dividing by α,

$$0_n \in \partial g(\bar{x}) + \sum_{t \in \operatorname{supp} \beta} \alpha^{-1} \beta_t \partial f_t(\bar{x}) + N_C(\bar{x}) + \sum_{t \in T} N_{\operatorname{dom} f_t}(\bar{x}),$$

and we are done. ∎

We illustrate Theorem 8.2.9 by means of the following example.

Example 8.2.10 *In* (P) *suppose* $n = 1, C = \mathbb{R}, T = \{1\}, g(x) = x$, *and* $f_1(x) = -\sqrt{x}$ *if* $x \ge 0$ *and* $+\infty$ *if not. Then* $\bar{x} = 0$ *is the unique optimal point of* (P) *and we have that* $\operatorname{ri}(\operatorname{dom} g) \cap \operatorname{ri}(\operatorname{dom} f_1) = \mathbb{R}_+^*$. *Hence,* (8.36) *and* (8.37) *hold, due to Proposition 4.1.26. Then, since*

8.2. OPTIMALITY AND DUALITY IN ... 327

$A(\bar{x}) = \{1\}$, $N_C(0) = \{0\}$ and the strong Slater condition holds, we can apply Theorem 8.2.9 with $\widehat{A}(0) = \emptyset$:

$$0 \in \partial g(0) + N_{\mathbb{R}_+}(0) = 1 + \,]-\infty, 0\,].$$

It turns out that we cannot get rid of the term $N_{\mathrm{dom}\, f_1}(0) = N_{\mathbb{R}_+}(0)$.

Corollary 8.2.11 *Let $\bar{x}$ be a feasible point of the semi-infinite problem* (P) *such that $C = \mathrm{dom}\, g = \mathrm{dom}\, f_t = \mathbb{R}^n$ for all $t \in T$. Assume that T is compact, the mappings $t \mapsto f_t(z)$, $z \in X$, are usc, and the strong Slater condition (8.28) holds. Then $\bar{x}$ is optimal for* (P) *if and only if there exists $\lambda \in \mathbb{R}_+^{(T)}$ such that*

$$0 \in \partial g(\bar{x}) + \sum_{t \in \mathrm{supp}\, \lambda} \lambda_t \partial f_t(\bar{x}) \text{ and } \lambda_t f_t(\bar{x}) = 0 \text{ for all } t \in T. \quad (8.43)$$

Proof. Conditions (8.36) and (8.37) are fulfilled in the current setting. Assume that $\bar{x}$ is optimal for (P). If $A(\bar{x}) \neq \emptyset$, then (8.43) comes from Theorem 8.2.9. Otherwise, if $A(\bar{x}) = \emptyset$, then by the continuity of $\tilde{f} = \max_{t \in T} f_t$ ($\mathrm{dom}\,\tilde{f} = \cap_{t \in T} \mathrm{dom}\, f_t = \mathbb{R}^n$), $\bar{x}$ is an interior point of the feasible set of (P), and so it satisfies $0_n \in \partial g(\bar{x})$. The proof of the converse statement is the same as the one in Theorem 8.2.3. ■

In the last part of this section, we specify the basic perturbational duality scheme developed in section 4.2 to convex infinite programming. We introduce a family of perturbed problems associated to (P) together with the corresponding Lagrange dual problem (D). Under the assumption that the constraint system S is FM, we will get strong duality between (P) and (D).

Let us consider the perturbed primal problem

$$(\mathrm{P}_y) \text{ Min } g(x)$$
$$\text{s.t. } f_t(x) \leq y_t, \ t \in T, \ x \in C,$$

where $y := (y_t) \in \mathbb{R}^T$. Recall that T is an arbitrary index set, C is a non-empty closed convex set, and $\{g; f_t, t \in T\} \subset \Gamma_0(X)$. The feasible set of (P_y), represented by F_y, can be empty for some $y \neq 0_T$, where 0_T represents both the zero of $\mathbb{R}^T$ and the zero of $\mathbb{R}^{(T)}$.

If we represent by $v(y)$ the optimal value of (P_y), then in particular $v(0_T) = v(\mathrm{P})$, and the perturbed optimal value function $v : \mathbb{R}^T \to \overline{\mathbb{R}}$ is convex, and possibly non-proper. The function $F \in \Gamma_0(X \times \mathbb{R}^T)$ defined by

$$F(x, y) := g(x) + I_{F_y}(x)$$

allows us to write

$$(\mathrm{P}_y) \, \underset{x \in X}{\mathrm{Min}} \, F(x,y), \text{ and } (\mathrm{P}) \equiv (\mathrm{P}_{0_T}) \, \underset{x \in X}{\mathrm{Min}} \, F(x, 0_T).$$

For each $(x^*, y^*) \in X^* \times (\mathbb{R}^T)^* = X^* \times \mathbb{R}^{(T)}$, we have

$$\begin{aligned}
F^*(x^*, y^*) &= \sup_{x \in X, \, y \in \mathbb{R}^T} \{\langle x^*, x \rangle + \langle y^*, y \rangle - F(x,y)\} \\
&= \sup_{x \in X} \sup_{y \in \mathbb{R}^T} \left\{ \langle x^*, x \rangle + \sum_{t \in T} y_t^* y_t - \mathrm{I}_{F_y}(x) - g(x) \right\} \\
&= \begin{cases} \sup_{x \in C} \left\{ \langle x^*, x \rangle + \sum_{t \in T} y_t^* f_t(x) - g(x) \right\}, & \text{if } y_t^* \le 0, \, \forall t \in \mathrm{supp}\, y^*, \\ +\infty, & \text{otherwise.} \end{cases}
\end{aligned}$$

In particular, if $\lambda := -y^*$,

$$F^*(\theta, \lambda) = \begin{cases} -\inf_{x \in C} \left\{ g(x) + \sum_{t \in T} \lambda_t f_t(x) \right\}, & \text{if } \lambda \in \mathbb{R}_+^{(T)}, \\ +\infty, & \text{otherwise.} \end{cases}$$

Additionally, we have that

$$\begin{aligned}
v^*(\lambda) &= \sup_{y \in \mathbb{R}^T} \left\{ \langle \lambda, y \rangle - \inf_{x \in X} F(x,y) \right\} \\
&= \sup_{x \in X, \, y \in \mathbb{R}^T} \{\langle \lambda, y \rangle - F(x,y)\} = F^*(\theta, \lambda).
\end{aligned} \quad (8.44)$$

The perturbed dual problem of (P) is defined as

$$(\mathrm{D}_{x^*}) \, \underset{\lambda \in \mathbb{R}_+^{(T)}}{\mathrm{Min}} \, F^*(x^*, \lambda), \text{ and } (\mathrm{D}) \equiv (\mathrm{D}_\theta) \, \underset{\lambda \in \mathbb{R}_+^{(T)}}{\mathrm{Min}} \, F^*(0, \lambda).$$

Since the Lagrangian for (P) is the function $L : X \times \mathbb{R}^{(T)} \to \mathbb{R}_\infty$ defined by

$$L(x, \lambda) := \begin{cases} g(x) + \sum_{t \in T} \lambda_t f_t(x), & \text{if } x \in C \text{ and } \lambda \in \mathbb{R}_+^{(T)}, \\ +\infty, & \text{otherwise,} \end{cases}$$

it turns out that

$$(\mathrm{D}) - \underset{\lambda \in \mathbb{R}_+^{(T)}}{\mathrm{Max}} \, \inf_{x \in C} L(x, \lambda).$$

8.2. OPTIMALITY AND DUALITY IN ...

Then

$$v(\mathrm{P})=v(0_T) \geq v^{**}(0_T) = \sup_{\lambda \in \mathbb{R}_+^{(T)}} (-v^*(\lambda)) = \sup_{\lambda \in \mathbb{R}_+^{(T)}} (-F^*(0, \lambda)) = -v(\mathrm{D}),$$

and the weak duality holds between (P) and (D), $v(\mathrm{P}) + v(\mathrm{D}) \geq 0$. The next result shows that the strong duality between (P) and (D) also holds under the standard assumptions used in this section.

Theorem 8.2.12 *Let us assume that problem* (P) *satisfies* $v(0_T) \in \mathbb{R}$. *Suppose that* $\mathcal{S}$ *is FM and* g *is finite and continuous at some feasible point of* (P). *Then* (D) *is solvable and* $v(\mathrm{P}) + v(\mathrm{D}) = 0$; *i.e., strong duality holds.*

Proof. Because g is finite and continuous at some feasible point of (P) and $\mathcal{S}$ is FM, Proposition 4.1.20 and Lemma 8.1.2 give rise to

$$\mathrm{epi}(g + \mathrm{I}_F)^* = \mathrm{epi}\, g^* + \mathrm{epi}\, \mathrm{I}_F^* = \mathrm{epi}\, g^* + \mathrm{cl}\, K = \mathrm{epi}\, g^* + K,$$

and $\mathrm{epi}\, g^* + K$ is w^*-closed. Moreover, $v(0_T) \in \mathbb{R}$ entails the implication

$$f_t(x) \leq 0,\ t \in T,\ x \in C \;\Rightarrow\; g(x) \geq v(0_T),$$

and Theorem 8.1.5(*iv*) yields some $\bar\lambda \in \mathbb{R}_+^{(T)}$ such that $g(x) + \sum_{t \in T} \bar\lambda_t f_t(x) \geq v(0_T)$ for all $x \in C$; that is,

$$-v(\mathrm{D}) \geq \inf_{x \in C} L(x, \bar\lambda) \geq v(0_T) = v(\mathrm{P}).$$

This, together with the weak duality, gives rise to $v(\mathrm{P}) = -v(\mathrm{D})$, and to the conclusion that $\bar\lambda$ is an optimal solution of (D). ∎

Let us consider, instead of the Lagrangian function L introduced above, the function $L_s : X \times \mathbb{R} \to \mathbb{R}_\infty$ defined by

$$L_s(x, \lambda) := \begin{cases} g(x) + \lambda \tilde f(x), & \text{if } x \in C \text{ and } \lambda \geq 0, \\ +\infty, & \text{otherwise}, \end{cases}$$

where $\tilde f$ is defined in (8.19); i.e., $\tilde f := \sup_{t \in T} f_t$. Correspondingly, we consider the following dual problem of (P),

$$(\mathrm{D}_s) - \mathrm{Max}_{\lambda \geq 0} \inf_{x \in C} L_s(x, \lambda);$$

in other words, (D_s) has the format of (D) when the primal problem (P) is written in the equivalent form

(P) Min $g(x)$
s.t. $\tilde{f}(x) \leq 0$, $x \in C$.

We can verify that $v(\mathrm{D}) \geq v(\mathrm{D}_s)$. Then, arguing as above but considering now this last representation, we consider the perturbation function $F_s \in \Gamma_0(X \times \mathbb{R})$ defined by

$$F_s(x, y) := g(x) + \mathrm{I}_{[\tilde{f} \leq y]}(x),$$

whose conjugate satisfies

$$F_s^*(0, \lambda) = \begin{cases} -\inf_{x \in C}\{g(x) + \lambda \tilde{f}(x)\}, & \text{if } \lambda \geq 0, \\ +\infty, & \text{otherwise.} \end{cases}$$

Therefore, similarly as above, we get

$$v(\mathrm{P}) = v(0) \geq v^{**}(0) = \sup_{\lambda \geq 0}(-v^*(\lambda)) = \sup_{\lambda \geq 0}(-F_s^*(0, \lambda)) = -v(\mathrm{D}_s),$$

and the weak duality holds between (P) and (D_s); that is, $v(\mathrm{P}) + v(\mathrm{D}_s) \geq 0$. The following result easily comes from Theorem 4.2.2.

Corollary 8.2.13 *Assume that problem* (P) *satisfies the strong Slater condition (8.28) at some point in* $\mathrm{dom}\, g$. *Then strong duality holds for the pair* (P) *and* (D_s).

8.3 Convexification processes in optimization

Given a proper function $f : X \to \mathbb{R}_\infty$, we consider the optimization problem

$$(\mathrm{P}_f)\ \text{Min}\ f(x) \qquad \text{s.t.}\ x \in X, \tag{8.45}$$

together with associated *convex relaxed problem*

$$(\mathrm{P}_g)\ \text{Min}\ g(x) \qquad \text{s.t.}\ x \in X,$$

where $g : X \to \overline{\mathbb{R}}$ is a proper convex function that somehow approximates f, for instance, the convex, closed or closed convex hulls of f.

8.3. CONVEXIFICATION PROCESSES IN ...

More precisely, we are interested in comparing the optimal sets of both problems (P_f) and its relaxation (P_g). To this aim, if h is one of these functions, f and g, we denote by $v(P_h)$ and $\varepsilon\text{-sol}(P_h)$ the optimal value and the set of ε-optimal solutions of (P_h), $\varepsilon \geq 0$, respectively; that is,

$$\varepsilon\text{-sol}(P_h) := (\partial_\varepsilon h)^{-1}(\theta) = \{x \in X : h(x) \leq v(P_h) + \varepsilon\}.$$

In addition, we write $\text{sol}(P_h) := 0\text{-sol}(P_h)$ to denote the set of optimal solutions of (P_h). Due to the properness assumption, we have $\varepsilon\text{-sol}(P_h) = \emptyset$ whenever $v(P_h) = -\infty$, while $\varepsilon\text{-sol}(P_h) \neq \emptyset$ when $\varepsilon > 0$ and $v(P_h) > -\infty$.

The following simple lemma shows that the convexification process does not alter the optimal value of the original problem (P_f).

Lemma 8.3.1 *For every function $f : X \to \mathbb{R}_\infty$, the values $v(P_f)$ and $v(P_g)$ coincide for every function $g : X \to \mathbb{R}_\infty$ satisfying $\overline{\text{co}} f \leq g \leq f$.*

Proof. Take $g : X \to \mathbb{R}_\infty$ such that $\overline{\text{co}} f \leq g \leq f$. Then the inequality $v(P_f) \geq v(P_g)$ is a consequence of the relation $g \leq f$. In particular, if $v(P_f) = -\infty$, then we get $v(P_f) = v(P_g) = -\infty$. Otherwise, if $v(P_f) > -\infty$, then $f(x) \geq v(P_f)$ for all $x \in X$, and we deduce that $g(x) \geq (\overline{\text{co}} f)(x) \geq v(P_f)$ for all $x \in X$; that is, $v(P_g) \geq v(P_f)$. ∎

It is also clear that every solution of (P_f) is a solution of (P_g), for any function $g : X \to \mathbb{R}_\infty$ satisfying $\overline{\text{co}} f \leq g \leq f$. A general purpose would be to express the optimal set of (P_g) in terms of the approximate and/or exact optimal solutions of (P_f), but here we will focus on ($P_{\overline{\text{co}} f}$). As before, we use $\mathcal{F}_{X^*}$ to represent the family of finite-dimensional linear subspaces of X^*.

Theorem 8.3.2 *For every function $f : X \to \mathbb{R}_\infty$ with a proper conjugate, we have*

$$\text{sol}(P_{\overline{\text{co}} f}) = \bigcap_{\varepsilon > 0,\ L \in \mathcal{F}_{X^*}} \overline{\text{co}}\left(\varepsilon\text{-sol}(P_f) + N_{L \cap \text{dom} f^*}(\theta)\right). \tag{8.46}$$

Moreover, the following statements hold true:

(i) *If $\text{ri}(\text{cone}(\text{dom } f^*)) \neq \emptyset$ or if $\text{cone}(\text{dom } f^*)$ is closed, then*

$$\text{sol}(P_{\overline{\text{co}} f}) = \bigcap_{\varepsilon > 0} \overline{\text{co}}\left(\varepsilon\text{-sol}(P_f) + N_{\text{dom} f^*}(\theta)\right),$$

and, when additionally $\text{cone}(\text{dom } f^*) = X^*$,

$$\text{sol}(P_{\overline{\text{co}} f}) = \bigcap_{\varepsilon > 0} \overline{\text{co}}\left(\varepsilon\text{-sol}(P_f)\right).$$

(ii) *If* $\operatorname{int}(\operatorname{cone}(\operatorname{dom} f^*)) \neq \emptyset$, *then*

$$\operatorname{sol}(P_{\overline{co}f}) = N_{\operatorname{dom} f^*}(\theta) + \bigcap_{\varepsilon>0} \overline{co}\,(\varepsilon\text{-}\operatorname{sol}(P_f)).$$

Proof. Since f^* is assumed proper, the function f admits an affine minorant and, according to Theorem 3.2.2, we have that $f^{**} = \overline{co}f$. So, (4.18) together with (4.8) yields

$$\partial f^*(\theta) = (\partial f^{**})^{-1}(\theta) = (\partial(\overline{co}f))^{-1}(\theta) = \operatorname{sol}(P_{\overline{co}f}). \qquad (8.47)$$

Now the inclusion "$\supset$" in (8.46) follows easily, because of (4.8), (4.17), and (4.9) we have

$$\varepsilon\text{-}\operatorname{sol}(P_f) + N_{\operatorname{dom} f^*}(\theta) = (\partial_\varepsilon f)^{-1}(\theta) + N_{\operatorname{dom} f^*}(\theta)$$
$$\subset \partial_\varepsilon f^*(\theta) + N_{\operatorname{dom} f^*}(\theta) = \partial_\varepsilon f^*(\theta).$$

Hence, using (4.15) and (8.47),

$$\bigcap_{\varepsilon>0} \overline{co}\,(\varepsilon\text{-}\operatorname{sol}(P_f) + N_{\operatorname{dom} f^*}(\theta)) \subset \bigcap_{\varepsilon>0} \partial_\varepsilon f^*(\theta) = \partial f^*(\theta) = \operatorname{sol}(P_{\overline{co}f})$$

and, particularly, the inclusion becomes an equality when $\operatorname{sol}(P_{\overline{co}f}) = \emptyset$. To see that such an equality also holds when $\operatorname{sol}(P_{\overline{co}f}) \neq \emptyset$; hence, $f^*(\theta) \in \mathbb{R}$, we apply Example 5.3.8 to the supremum function $f^* := \sup_{x \in X}(\langle \cdot, x \rangle - f(x))$:

$$\operatorname{sol}(P_{\overline{co}f}) = \partial f^*(\theta) = \bigcap_{\varepsilon>0,\ L \in \mathcal{F}_{X^*}} \overline{co}\,(E_\varepsilon + N_{L \cap \operatorname{dom} f^*}(\theta)), \qquad (8.48)$$

where

$$E_\varepsilon := \{x \in \operatorname{dom} f : -f(x) \geq f^*(\theta) - \varepsilon\}. \qquad (8.49)$$

In other words, thanks (4.18) and (4.8), we have $E_\varepsilon = (\partial_\varepsilon f)^{-1}(\theta) = \varepsilon\text{-}\operatorname{sol}(P_f)$, and from (8.48), we obtain (8.46). The other assertions in (i) and (ii) follow similarly by using, instead of Example 5.3.8, Proposition 5.3.1 and Corollary 5.3.2, respectively. ∎

Additional relationships between $\operatorname{sol}(P_{\overline{co}f})$ and $\operatorname{sol}(P_f)$ are given in the following theorem. Remember that the continuity assumption on the conjugate function f^* assumed below is with respect to a compatible topology given in X^*.

8.3. CONVEXIFICATION PROCESSES IN ...

Theorem 8.3.3 *Given a weakly lsc function $f : X \to \mathbb{R}_\infty$, we assume that f^* is finite and continuous somewhere. Then we have that*

$$\text{sol}(\text{P}_{\overline{\text{co}}f}) = \text{N}_{\text{dom } f^*}(\theta) + \overline{\text{co}}\left(\text{sol}(\text{P}_f)\right),$$

and, when $X = \mathbb{R}^n$,

$$\text{sol}(\text{P}_{\overline{\text{co}}f}) = \text{N}_{\text{dom } f^*}(\theta) + \text{co}\left(\text{sol}(\text{P}_f)\right).$$

Proof. We consider the compatible dual pair $((X^*, \mathfrak{T}_{X^*}), (X, \sigma(X, X^*)))$, where $\mathfrak{T}_{X^*}$ is a compatible topology given in X^* for which f^* is continuous at some point. We apply Corollary 6.3.6 in our dual pair to the family of continuous convex functions $f_x : X^* \to \mathbb{R}$, $x \in X$, defined by

$$f_x(\cdot) := \langle \cdot, x \rangle - f(x),$$

whose supremum is f^*. Furthermore, taking into account (8.47) and the fact that f^* is proper, we show as in the proof of Theorem 8.3.2 that $\text{sol}(\text{P}_{\overline{\text{co}}f}) = \partial f^*(\theta)$. Therefore, since f^* is assumed to be continuous somewhere, Corollary 6.3.6 applies and yields

$$\text{sol}(\text{P}_{\overline{\text{co}}f}) = \partial f^*(\theta) = \text{N}_{\text{dom } f^*}(\theta) + \overline{\text{co}}\left\{\bigcap_{\varepsilon > 0} (\text{cl } E_\varepsilon)\right\}, \quad (8.50)$$

where E_ε is defined as in (8.49); that is, $E_\varepsilon = \{x \in X : -f(x) \geq f^*(\theta) - \varepsilon\}$, and $\text{cl } E_\varepsilon$ is the closure of E_ε with respect to the weak topology in X. The same Corollary 6.3.6 shows that (8.50) holds with co instead of $\overline{\text{co}}$ when $X = \mathbb{R}^n$. Moreover, since f is assumed to be weakly lsc, the set E_ε is weakly closed, and so

$$\bigcap_{\varepsilon > 0} \text{cl}\,(E_\varepsilon) = \bigcap_{\varepsilon > 0} E_\varepsilon = \{x \in X : f^*(\theta) + f(x) = 0\} = \text{sol}(\text{P}_f).$$

The conclusion comes from (8.50). ∎

The continuity assumption of the conjugate function in Theorems 8.3.2 and 8.3.3 is taken with respect to any topology in X^* that is compatible with the duality pair (X^*, X). Of course, the Mackey topology provides the less restrictive continuity condition, while the choice of the w^*-continuity would be very restrictive as it forces, for instance in the Hilbert setting, the original function f to have a finite-dimensional effective domain. However, it is possible to relax this assumption of continuity in the case of normed spaces, allowing to consider the continuity with respect to the norm topology.

Given a function $f : X \to \overline{\mathbb{R}}$, where X is assumed to be Banach, we consider the function $\mathrm{cl}^{w^{**}} \hat{f} : X^{**} \to \overline{\mathbb{R}}$ introduced in section 4.3, where $\hat{f}$ is the extension of f to X^{**} (see (4.103)). Then, by Theorem 4.3.3, we know that

$$(\mathrm{cl}^{w^{**}} \hat{f})(z) = \liminf_{x \to^{w^{**}} z,\ x \in X} f(x), \qquad (8.51)$$

where "$\to^{w^{**}}$" refers to the convergence with respect to the topology $\sigma(X^{**}, X^*)$ in X^{**}. The issue here is that the norm topology in X^* is not necessarily compatible for the duality pair (X, X^*), entailing that the conjugate function of f^* should be defined on the whole space X^{**}:

$$(f^*)^*(z) = \sup\{\langle z, x^*\rangle - f^*(x^*),\ x^* \in X^*\},\ z \in X^{**}.$$

Abusing the notation, we write $(f^*)^* = f^{**}$ and use $\mathrm{N}_{\mathrm{dom}\, f^*}(\theta)$ to denote the cone of vectors in X^{**} that are normal to $\mathrm{dom}\, f^*$ at x^*. Accordingly, instead of $(\mathrm{P}_{\overline{\mathrm{co}}f})$ used above, we consider the problems $(\mathrm{P}_{f^{**}})$ and $(\mathrm{P}_{\mathrm{cl}^{w^{**}} \hat{f}})$ posed in X^{**}. Then, in the following theorem, we establish different relationships among the optimal solutions of problems (P_f), $(\mathrm{P}_{\mathrm{cl}^{w^{**}} \hat{f}})$ and $(\mathrm{P}_{f^{**}})$. These results involve the normal cone $\mathrm{N}_{\mathrm{dom}\, f^*}(\theta)\ (\subset X^{**})$ and the w^{**}-closed convex hull $\overline{\mathrm{co}}^{w^{**}}$.

The need for such an analysis that requires the passage to the bidual space could be motivated by certain regularization methods in applied topics such as problems issued from calculus of variations. In such a case, the associated bidual consists of a good framework to search for generalized solutions. After that, the problem arises of how to relate these new solutions to those of the initial problem. The following results establish an abstract general framework to address such an approach.

Theorem 8.3.4 *Assume that X is a Banach space. Given a proper function $f : X \to \mathbb{R}_\infty$, we suppose that f^* is finite and norm-continuous somewhere. Then the following statements hold true, provided that X^* is endowed with its dual norm topology:*
(i)
$$\mathrm{sol}(\mathrm{P}_{f^{**}}) = \mathrm{N}_{\mathrm{dom}\, f^*}(\theta) + \bigcap_{\varepsilon > 0} \overline{\mathrm{co}}^{w^{**}} (\varepsilon\text{-}\mathrm{sol}(\mathrm{P}_f)).$$

8.3. CONVEXIFICATION PROCESSES IN ...

(ii) $$\mathrm{sol}(\mathrm{P}_{f^{**}}) = \mathrm{N}_{\mathrm{dom}\, f^*}(\theta) + \overline{\mathrm{co}}^{w^{**}}\left(\mathrm{sol}(\mathrm{P}_{\mathrm{cl}^{w^{**}}\hat{f}})\right). \qquad (8.52)$$

(iii) If f^* is norm-continuous at θ, then

$$\mathrm{sol}(\mathrm{P}_{f^{**}}) = \bigcap_{\varepsilon>0} \overline{\mathrm{co}}^{w^{**}}(\varepsilon\text{-}\mathrm{sol}(\mathrm{P}_f)) \qquad (8.53)$$

$$= \overline{\mathrm{co}}^{w^{**}}\left(\mathrm{sol}(\mathrm{P}_{\mathrm{cl}^{w^{**}}\hat{f}})\right), \qquad (8.54)$$

and, consequently,

$$\mathrm{sol}(\mathrm{P}_{\overline{\mathrm{co}}\, f}) = \bigcap_{\varepsilon>0} \overline{\mathrm{co}}\,(\varepsilon\text{-}\mathrm{sol}(\mathrm{P}_f)) \qquad (8.55)$$

$$= X \cap \left(\overline{\mathrm{co}}^{w^{**}}\left(\mathrm{sol}(\mathrm{P}_{\mathrm{cl}^{w^{**}}\hat{f}})\right)\right). \qquad (8.56)$$

Proof. Given the extension $\hat{f}$ of f to X^{**}, as defined in (4.103), and thanks to the identification of X as a linear subspace of X^{**}, the definition of $\hat{f}$ implies that

$$\varepsilon\text{-}\mathrm{sol}(\mathrm{P}_f) = \varepsilon\text{-}\mathrm{sol}(\mathrm{P}_{\hat{f}}) \text{ for all } \varepsilon \geq 0, \qquad (8.57)$$

and

$$(\hat{f})^* = f^* \text{ and } \mathrm{dom}(\hat{f})^* = \mathrm{dom}\, f^*. \qquad (8.58)$$

In addition, the continuity assumption on f^* implies that $\hat{f}$ is minorized by a w^{**}-continuous affine mapping. So, we get

$$\|\cdot\|_* - \mathrm{int}(\mathrm{cone}(\mathrm{dom}(\hat{f})^*)) = \|\cdot\|_* - \mathrm{int}(\mathrm{cone}(\mathrm{dom}\, f^*)) \neq \emptyset$$

and Theorem 8.3.2(ii), applied to the function $\hat{f}$ in the dual pair $((X^{**}, w^{**}), (X^*, \|\cdot\|_*))$, together with (4.105) entails

$$\mathrm{sol}(\mathrm{P}_{f^{**}}) = \mathrm{sol}(\mathrm{P}_{\overline{\mathrm{co}}^{w^{**}}(\hat{f})}) = \mathrm{N}_{\mathrm{dom}(\hat{f})^*}(\theta) + \bigcap_{\varepsilon>0} \overline{\mathrm{co}}^{w^{**}}\left(\varepsilon\text{-}\mathrm{sol}(\mathrm{P}_{\hat{f}})\right).$$

Then, taking into account (8.57) and (8.58), we get

$$\mathrm{sol}(\mathrm{P}_{f^{**}}) = \mathrm{sol}(\mathrm{P}_{\overline{\mathrm{co}}^{w^{**}}(\hat{f})}) = \mathrm{N}_{\mathrm{dom}\, f^*}(\theta) + \bigcap_{\varepsilon>0} \overline{\mathrm{co}}^{w^{**}}(\varepsilon\text{-}\mathrm{sol}(\mathrm{P}_f)) \quad (8.59)$$

and the formula in (i) follows. If f^* is additionally *norm*-continuous at θ, then

$$\mathrm{N}_{\mathrm{dom}(\hat{f})^*}(\theta) = \mathrm{N}_{\mathrm{dom}\, f^*}(\theta) = \{\theta\}$$

and (8.59) reduces to

$$\text{sol}(P_{f^{**}}) = \text{sol}(P_{\overline{co}^{w^{**}}(\hat{f})}) = \bigcap_{\varepsilon>0} \overline{co}^{w^{**}}(\varepsilon\text{-}\text{sol}(P_f)), \qquad (8.60)$$

showing that (8.53) holds. At the same time, due to the relation $(f^{**})_{|X} = \overline{co} f$ coming from (4.106), relation (8.60) leads us to

$$\text{sol}(P_{\overline{co} f}) = X \cap \text{sol}(P_{\overline{co}^{w^{**}}(\hat{f})}) = X \cap \left(\bigcap_{\varepsilon>0} \overline{co}^{w^{**}}(\varepsilon\text{-}\text{sol}(P_f))\right)$$

$$= \bigcap_{\varepsilon>0} \left(X \cap \overline{co}^{w^{**}}(\varepsilon\text{-}\text{sol}(P_f))\right) = \bigcap_{\varepsilon>0} \overline{co}(\varepsilon\text{-}\text{sol}(P_f)),$$

and we are done with (8.55).

We now turn to the proof of formulas (8.52), (8.54), and (8.56). To show (8.52), we proceed as in the proof of Theorem 8.3.3 but, similarly as in the paragraph above, we apply Corollary 6.3.6 in the dual pair $((X^*, \|\cdot\|_*), (X^{**}, w^{**}))$ to the family of *norm*-continuous convex functions $f_x(\cdot) = \langle \cdot, x \rangle - f(x)$, $x \in X$. Then, taking into account (4.105), we get

$$\text{sol}(P_{f^{**}}) = \text{sol}(P_{\overline{co}^{w^{**}}(\hat{f})}) = \partial(\hat{f})^*(\theta) = N_{\text{dom}(\hat{f})^*}(\theta) + \overline{co}^{w^{**}}\left\{\bigcap_{\varepsilon>0} \text{cl}^{w^{**}}(E_\varepsilon)\right\}, \qquad (8.61)$$

where

$$E_\varepsilon := \left\{z \in X^{**} : -\hat{f}(z) \geq (\hat{f})^*(\theta) - \varepsilon\right\} = \{z \in X : -f(z) \geq f^*(\theta) - \varepsilon\}.$$

Observe that $\text{cl}^{w^{**}}(E_\varepsilon) \subset \left\{z \in X^{**} : -(\text{cl}^{w^{**}} \hat{f})(z) \geq f^*(\theta) - \varepsilon\right\}$ and, using (8.58) and (3.7),

$$f^*(\theta) = (\hat{f})^*(\theta) = (\text{cl}^{w^{**}} \hat{f})^*(\theta) = -\inf_{z \in X^{**}}(\text{cl}^{w^{**}} \hat{f})(z),$$

so that

$$\bigcap_{\varepsilon>0} \text{cl}^{w^{**}}(E_\varepsilon) \subset \bigcap_{\varepsilon>0} \left\{z \in X^{**} : -(\text{cl}^{w^{**}} \hat{f})(z) \geq f^*(\theta) - \varepsilon\right\}$$

$$= \left\{z \in X^{**} : -(\text{cl}^{w^{**}} \hat{f})(z) = f^*(\theta)\right\} = \text{sol}(P_{\text{cl}^{w^{**}} \hat{f}}).$$

Therefore, using again (8.58), (8.61) implies that

8.3. CONVEXIFICATION PROCESSES IN ...

$$\text{sol}(P_{f^{**}}) \subset N_{\text{dom}\,f^*}(\theta) + \overline{\text{co}}^{w^{**}}(\text{sol}(P_{\text{cl}^{w^{**}}\hat{f}})).$$

The converse of this last inclusion also holds because $\text{sol}(P_{\text{cl}^{w^{**}}\hat{f}}) \subset \text{sol}(P_{\overline{\text{co}}^{w^{**}}(\hat{f})}) = \partial(\hat{f})^*(\theta)$ and, due to (4.9),

$$N_{\text{dom}\,f^*}(\theta) + \overline{\text{co}}^{w^{**}}(\text{sol}(P_{\text{cl}^{w^{**}}\hat{f}})) \subset N_{\text{dom}(\hat{f})^*}(\theta) + \partial(\hat{f})^*(\theta)$$
$$= \partial(\hat{f})^*(\theta) = \text{sol}(P_{\overline{\text{co}}^{w^{**}}(\hat{f})});$$

that is, (8.52) holds.

Finally, formula (8.54) is easily derived from (8.52), while (8.56) follows from (8.54) due to the fact seen above, $\text{sol}(P_{\overline{\text{co}}f}) = X \cap \text{sol}(P_{\overline{\text{co}}^{w^{**}}(\hat{f})}) = X \cap \text{sol}(P_{f^{**}})$. ∎

The key in the proof of the previous results, giving different expressions for $\text{sol}(P_{\overline{\text{co}}f})$ and $\text{sol}(P_{f^{**}})$, is the characterization of the subdifferential of the conjugate at θ. More generally, the following corollary gives the subdifferential of the conjugate at each point in terms of the operators $\partial_\varepsilon f$, $\varepsilon \geq 0$, in the setting of compatible dual pairs (X, X^*). The formulas below can also be considered as non-convex extensions of the Fenchel relation given in (4.18). For the sake of simplicity, we denote $\mathcal{F}(x^*)$ the family of finite-dimensional linear subspaces of X^* containing $x^* \in X^*$.

Corollary 8.3.5 *Given a function* $f : X \to \mathbb{R}_\infty$ *with a proper conjugate, for all* $x^* \in X^*$, *we have*

$$\partial f^*(x^*) = \bigcap_{\varepsilon > 0,\ L \in \mathcal{F}(x^*)} \overline{\text{co}}\left((\partial_\varepsilon f)^{-1}(x^*) + N_{L \cap \text{dom}\,f^*}(x^*)\right). \quad (8.62)$$

Moreover, the following assertions are true:

(i) If $\text{ri}(\text{cone}(\text{dom}\,f^* - x^*)) \neq \emptyset$ *or if* $\text{cone}(\text{dom}\,f^* - x^*)$ *is closed, then*

$$\partial f^*(x^*) = \bigcap_{\varepsilon > 0} \overline{\text{co}}\left((\partial_\varepsilon f)^{-1}(x^*) + N_{\text{dom}\,f^*}(x^*)\right). \quad (8.63)$$

(ii) If f *is weakly lsc and* f^* *is continuous somewhere, then*

$$\partial f^*(x^*) = N_{\text{dom}\,f^*}(x^*) + \overline{\text{co}}\left((\partial f)^{-1}(x^*)\right). \quad (8.64)$$
$$= N_{\text{dom}\,f^*}(x^*) + \text{co}\left((\partial f)^{-1}(x^*)\right) \ (\text{when } X = \mathbb{R}^n). \quad (8.65)$$

Proof. Fix $x^* \in X^*$ and observe that, due to (4.18),

$$\partial f^*(x^*) = \text{sol}(\text{P}_{f^{**}-x^*}) = \text{sol}(\text{P}_{(f-x^*)^{**}}),$$

where $(\text{P}_{f^{**}-x^*})$ is the optimization problem whose objective function is $f^{**}(\cdot) - \langle x^*, \cdot \rangle$, which coincides with the biconjugate of the function $g(\cdot) := f(\cdot) - \langle x^*, \cdot \rangle$. Then, applying (8.46) to the function g, we obtain that

$$\partial f^*(x^*) = \partial g^*(\theta) = \text{sol}(\text{P}_{\overline{\text{co}}g}) = \bigcap_{\varepsilon>0,\ L\in\mathcal{F}_{X^*}} \overline{\text{co}}\left(\varepsilon\text{-sol}(\text{P}_g) + \text{N}_{L\cap\text{dom }g^*}(\theta)\right),$$

where $\mathcal{F}_{X^*} = \mathcal{F}(\theta)$. Thus, since $\text{dom }g^* = (\text{dom }f^*) - x^*$ and

$$\varepsilon\text{-sol}(\text{P}_g) = \{x \in X : -g(x) \geq g^*(\theta) - \varepsilon\}$$
$$= \{x \in X : -f(x) + \langle x^*, x\rangle \geq f^*(x^*) - \varepsilon\} = (\partial_\varepsilon f)^{-1}(x^*),$$

we get

$$\partial f^*(x^*) = \bigcap_{\varepsilon>0,\ L\in\mathcal{F}_{X^*}} \overline{\text{co}}\left((\partial_\varepsilon f)^{-1}(x^*) + \text{N}_{L\cap(\text{dom }f^*-x^*)}(\theta)\right)$$
$$= \bigcap_{\varepsilon>0,\ L\in\mathcal{F}(x^*)} \overline{\text{co}}\left((\partial_\varepsilon f)^{-1}(x^*) + \text{N}_{L\cap\text{dom }f^*}(x^*)\right),$$

proving (8.62).

The proof of the remaining formulas follows the same pattern as above using the corresponding statements in Theorems 8.3.2, 8.3.3, and 8.3.4. ∎

The following corollary characterizes $\partial f^*(x^*)$ as a subset of X^{**} in the framework of Banach spaces.

Corollary 8.3.6 *Let X be a Banach space with X^* being endowed with the dual norm topology. Given a function $f : X \to \mathbb{R}_\infty$ such that f^* is norm-continuous somewhere, for every $x^* \in X^*$, we have*

$$\partial f^*(x^*) = \text{N}_{\text{dom }f^*}(x^*) + \overline{\text{co}}^{w^{**}}\left((\partial(\text{cl}^{w^{**}}\hat{f}))^{-1}(x^*)\right), \qquad (8.66)$$

and, when X is reflexive and f is weakly lsc,

$$\partial f^*(x^*) = \text{N}_{\text{dom }f^*}(x^*) + \overline{\text{co}}\left((\partial f)^{-1}(x^*)\right). \qquad (8.67)$$

Proof. Formula (8.66) follows as in the proof of Corollary 8.3.5, applying (8.52). Formula (8.67) is easily derived from (8.66). ∎

8.3. CONVEXIFICATION PROCESSES IN ...

The previous formulas giving the subdifferential of the conjugate become simpler when the latter is Fréchet-differentiable. To certain extent, the formulas below extend to non-convex functions the well-known relation $\partial f^* = (\partial f)^{-1}$ satisfied by functions in $\Gamma_0(X)$ (see (4.18)).

Theorem 8.3.7 *Let X be a Banach space with X^* being endowed with the dual norm topology. Given a function $f : X \to \mathbb{R}_\infty$ and $x^* \in X^*$, the following assertions hold:*

(i) If f^* is Fréchet-differentiable at x^* and f is lsc at $(f^*)'(x^*)$ ($\in X$), then

$$\{(f^*)'(x^*)\} = \partial f^*(x^*) = (\partial f)^{-1}(x^*).$$

(ii) If f^* is Gâteaux-differentiable at x^*, with Gâteaux-derivative $(f^*)'_G(x^*) \in X$, and f is weakly lsc at $(f^*)'_G(x^*)$, then

$$\{(f^*)'_G(x^*)\} = \partial f^*(x^*) = (\partial f)^{-1}(x^*).$$

Proof. (i) The Fréchet differentiability of f^* at x^*, with derivative $\bar{x} := (f^*)'(x^*)$ ($\bar{x} \in X$, due to Proposition 4.3.10), implies its continuity at x^* and (8.66) simplifies to

$$\{\bar{x}\} = \partial f^*(x^*) = \overline{\text{co}}^{w^{**}} \left\{(\partial(\text{cl}^{w^{**}} \hat{f}))^{-1}(x^*)\right\}$$
$$= X \cap (\partial(\text{cl}^{w^{**}} \hat{f}))^{-1}(x^*) = (\partial(\text{cl}^w f))^{-1}(x^*),$$

where the last equality comes from Lemma 4.3.2. Hence, $x^* \in \partial(\text{cl}^w f)(\bar{x})$ and Proposition 4.1.6(v) gives us

$$(\text{cl}^w f)(\bar{x}) + f^*(x^*) = (\text{cl}^w f)(\bar{x}) + (\text{cl}^w f)^*(x^*) = \langle \bar{x}, x^* \rangle. \qquad (8.68)$$

Remembering the definition of $\text{cl}^w f$ in (2.34), for each $n \geq 1$, we find $x_n \in X$ such that $|\langle \bar{x} - x_n, x^* \rangle| \leq 1/n$ and

$$f(x_n) \leq (\text{cl}^w f)(\bar{x}) + 1/n = \langle \bar{x}, x^* \rangle - f^*(x^*) + 1/n \leq \langle x_n, x^* \rangle - f^*(x^*) + 2/n,$$

entailing that

$$f^*(x^*) \leq \liminf_n \left(\langle x_n, x^* \rangle - f(x_n) + 2/n\right) = \liminf_n \left(\langle x_n, x^* \rangle - f(x_n)\right)$$
$$\leq \limsup_n \left(\langle x_n, x^* \rangle - f(x_n)\right) \leq f^*(x^*);$$

that is, $\lim_n (\langle x_n, x^* \rangle - f(x_n)) = f^*(x^*)$. Thus, by Proposition 4.3.10, the Fréchet differentiability of f^* at x^* implies that $x_n \to \bar{x}$ and so, using the lower semicontinuity of f at $\bar{x}$, the last equality yields

$$f^*(x^*) = \lim_n (\langle x_n, x^* \rangle - f(x_n)) \leq \langle \bar{x}, x^* \rangle - f(\bar{x}) \leq f^*(x^*).$$

Therefore, $\bar{x} \in (\partial f)^{-1}(x^*) \subset \partial f^*(x^*) = \{\bar{x}\}$ and the conclusion of (i) follows.

(ii) The Gâteaux-differentiability assumption on the w^*-lsc (hence, $\|\cdot\|_*$-lsc) convex function f^* implies that $x^* \in \|\cdot\|_*\text{-int}(\text{dom } f^*)$, so that f^* is norm-continuous at x^* by Corollary 2.2.8. Thus, denoting $\bar{z} := (f^*)'_G(x^*)$, (8.66) reduces to

$$\{\bar{z}\} = \overline{\text{co}}^{w^{**}}\left((\partial(\text{cl}^{w^{**}} \hat{f}))^{-1}(x^*)\right) = (\partial(\text{cl}^{w^{**}} \hat{f}))^{-1}(x^*).$$

Since $\bar{z} \in X$ by the current assumption, Lemma 4.3.2 entails

$$\{\bar{z}\} = X \cap (\partial(\text{cl}^{w^{**}} \hat{f}))^{-1}(x^*) = (\partial(\text{cl}^w f))^{-1}(x^*),$$

and the desired conclusion comes, as in the paragraph above, from the weak lower semicontinuity of f at $\bar{z}$. ∎

The following corollary gives a slight extension of the Stegall variational principle (see Theorem 4.3.13), under the *norm*-continuity of the conjugate function; see Exercise 126 for a sufficient condition for such a continuity property. Formula (8.69) below shows that the (Fenchel) relation $\partial f^* = (\partial f)^{-1}$ holds in a dense subset of dom f^* when working in Banach spaces enjoying the RNP.

Corollary 8.3.8 *Let X be a Banach space with the RNP, and let the function $f : X \to \mathbb{R}_\infty$ be lsc and such that f^* is finite and norm-continuous somewhere. Then there exists some $\mathcal{G}_\delta$-set $D \subset X^*$, dense in dom f^*, and satisfying*

$$\{(f^*)'(x^*)\} = \partial f^*(x^*) = (\partial f)^{-1}(x^*) \text{ for all } x^* \in D. \quad (8.69)$$

Consequently, the functions $f - x^*$, $x^* \in D$, attain a strong minimum on X.

Proof. By the RNP property, X^* is w^*-Asplund, and so the w^*-lsc convex function f^* is Fréchet-differentiable on a $\mathcal{G}_\delta$-set D, which is dense in dom f^*. Hence, using Theorem 8.3.7(i), for each $x^* \in D$, we have

$$\{(f^*)'(x^*)\} = \partial f^*(x^*) = (\partial f)^{-1}(x^*),$$

8.3. CONVEXIFICATION PROCESSES IN ...

particularly showing that $(f^*)'(x^*)$ is a minimum of $f - x^*$ for all $x^* \in D$. Moreover, for every fixed $x^* \in D$ and every sequence $(x_n)_n \subset X$ such that $f(x_n) - \langle x^*, x_n \rangle \to \inf_X(f - x^*) = -f^*(x^*)$, by Proposition 4.3.10 the Fréchet differentiability of f^* at x^* implies that $x_n \to (f^*)'(x^*)$. This proves that the element $(f^*)'(x^*)$ is, indeed, a strong minimum of $f - x^*$. ∎

The following corollary shows, under the continuity of the conjugate function and the RNP, that the function f and its closed convex hull $\overline{co}f$ have the same optimal solutions up to tilt perturbations.

Corollary 8.3.9 *Assume that X is Banach with the RNP, and endow X^* with the dual norm topology. Let $f : X \to \mathbb{R}_\infty$ be an lsc function such that $\inf_X f \in \mathbb{R}$, and suppose that f^* is finite and norm-continuous somewhere. Then, for every $\varepsilon > 0$, there exists $x^* \in \varepsilon B_{X^*}$ such that*

$$\mathrm{sol}(\mathrm{P}_{(\overline{co}f)-x^*}) = \mathrm{sol}(\mathrm{P}_{f-x^*}).$$

Proof. On the one hand, the current assumption ensures that $\overline{co}f \in \Gamma_0(X)$, so that $(\partial(\overline{co}f))^{-1} = \partial(\overline{co}f)^* = \partial f^*$ by (4.18). On the other hand, by Corollary 8.3.8, there exists some $\mathcal{G}_\delta$-set $D \subset X^*$, which is dense in $\mathrm{dom}\, f^*$ and satisfies $\partial f^*(x^*) = (\partial f)^{-1}(x^*)$ for all $x^* \in D$; that is, $(\partial(\overline{co}f))^{-1}(x^*) = (\partial f)^{-1}(x^*)$ for all $x^* \in D$. Therefore, as $(\partial(\overline{co}f))^{-1}(x^*) = \mathrm{sol}(\mathrm{P}_{(\overline{co}f)-x^*})$ and $(\partial f)^{-1}(x^*) = \mathrm{sol}(\mathrm{P}_{f-x^*})$, we obtain $\mathrm{sol}(\mathrm{P}_{(\overline{co}f)-x^*}) = \mathrm{sol}(\mathrm{P}_{f-x^*})$ for all $x^* \in D$. Finally, since $\theta \in \mathrm{dom}\, f^*$ thanks to the assumption $\inf_X f \in \mathbb{R}$, for every $\varepsilon > 0$ there exists $x^* \in D \cap (\varepsilon B_{X^*})$, and we are done. ∎

The following result extends Corollary 4.3.9 to non-necessarily convex functions.

Corollary 8.3.10 *Let X be a Banach space with the RNP, and let $f : X \to \overline{\mathbb{R}}$ be an lsc function such that f^* is norm-continuous somewhere. Then we have*

$$\mathrm{cl}^{\|\cdot\|}(\mathrm{dom}\, f^*) = \mathrm{cl}^{\|\cdot\|}(\mathrm{Im}(\partial f)).$$

Moreover, if X is reflexive and f is weakly lsc, then

$$(\partial f^*)^{-1}(X) = \mathrm{Im}(\partial f).$$

Proof. It is clear that $\mathrm{Im}\,\partial f \subset \mathrm{dom}\, f^*$, and so $\mathrm{cl}^{\|\cdot\|}(\mathrm{Im}\,\partial f) \subset \mathrm{cl}^{\|\cdot\|}(\mathrm{dom}\, f^*)$. Conversely, take $x^* \in \mathrm{dom}\, f^* \subset \mathrm{cl}^{\|\cdot\|}(\mathrm{dom}\, f^*) = \mathrm{cl}^{\|\cdot\|}(D)$, where $D \subset X^*$ is as in Corollary 8.3.8; that is, a dense set in $\mathrm{dom}\, f^*$ such that the equality $\{(f^*)'(z^*)\} = (\partial f)^{-1}(z^*)$ holds for all $z^* \in D$. Then there exists a sequence $(x_n^*) \subset D$ such that $x_n^* \to x^*$ in norm. Since $x_n := (f^*)'(x_n^*) = (\partial f)^{-1}(x_n^*)$, we infer that $x_n^* \in \partial f(x_n) \subset \operatorname{Im} \partial f$, and so $x^* \in \operatorname{cl}^{\|\cdot\|}(\operatorname{Im} \partial f)$. This yields the first part of the corollary.

When X is reflexive and f is weakly lsc, we have $x^* \in \operatorname{dom} \partial f^*$ if and only if $\partial f^*(x^*) \neq \emptyset$; thus by (8.67) and taking into account (2.6), if and only if $(\partial f)^{-1}(x^*) \neq \emptyset$, if and only if $x^* \in \operatorname{Im} \partial f$. ∎

8.4 Non-convex integration

In this section, we apply the material in chapters 5 and 6 to extend the integration results of section 4.4 that were limited to convex functions. We will need the following lemma.

Lemma 8.4.1 *Assume that X is a Banach space, and let $f, g \in \Gamma_0(X)$. Given a non-empty open set $V \subset (\partial f)^{-1}(X^*)$, we suppose the existence of a dense subset D of V such that $\partial f(x) \subset \partial g(x)$ for all $x \in D$. Then*
$$\partial f(x) \subset \partial g(x) \text{ for all } x \in V.$$

Proof. First, since $V \subset (\partial f)^{-1}(X^*) \subset \operatorname{dom} f$, $V \subset V \cap (\partial f)^{-1}(X^*) \subset (\partial g)^{-1}(X^*) \subset \operatorname{dom} g$ and V is open, Corollary 2.2.8 guarantees that f and g are continuous on V. We fix $x \in V$. Proceeding by contradiction, we assume that $x^* \in \partial f(x) \setminus \partial g(x)$ exists. Then, using the separation theorem in (X^*, w^*), we find $z_0 \in X \setminus \{\theta\}$ and $\alpha_0 \in \mathbb{R}$ such that

$$\langle x^*, z_0 \rangle < \alpha_0 < \langle y^*, z_0 \rangle \text{ for all } y^* \in \partial g(x); \tag{8.70}$$

that is,
$$\partial g(x) \subset W := \{y^* \in X^* : \langle y^*, z_0 \rangle > \alpha_0\}.$$

Furthermore, since W is (X^*, w^*)-open, Proposition 4.1.7 ((i) and (iii)) establishes the existence of $U \in \mathcal{N}_X$ and $\varepsilon > 0$ such that $B(x, 2\varepsilon) \subset V$ and

$$\partial g(y) \subset U^\circ \cap W \text{ for all } y \in B(x, 2\varepsilon). \tag{8.71}$$

We choose $\delta > 0$ small enough such that $y := x - \delta z_0 \in B(x, \varepsilon) (\subset V)$. Then, by the density of D in V, we also choose $(y_n)_n \subset D \cap B(x, 2\varepsilon)$ such that $y_n \to y$; therefore, the current assumption and Proposition 4.1.22 imply the existence of a sequence $(y_n^*)_n \subset X^*$ such that $y_n^* \in \partial f(y_n) \subset \partial g(y_n)$ for all $n \geq 1$. Then, due to Theorem 2.1.9, (8.71)

8.4. NON-CONVEX INTEGRATION

gives rise to a subnet $(y_i^*)_i$ of $(y_n^*)_n$ and $y^* \in X^*$ such that

$$y_i^* \to^{w^*} y^* \text{ and } y_i^* \in \partial f(y_i) \subset \partial g(y_i) \subset U^\circ \text{ for all } i.$$

Therefore, since $y_i \to y$, Proposition 4.1.6(ix) implies that $y^* \in \partial f(y) \cap \partial g(y)$ ($\subset \partial f(y) \cap W$, by (8.71)). Thus, (8.70) leads us to

$$\langle x^* - y^*, x - y \rangle = \delta \langle x^* - y^*, z_0 \rangle < 0,$$

which is a contradiction with $\langle x^* - y^*, x - y \rangle \geq 0$, coming from the monotonicity of ∂f. The proof is over. ∎

Next, we introduce the concept of epi-pointed functions.

Definition 8.4.2 *A function $f : X \to \mathbb{R}_\infty$ is called $\mathfrak{T}$-epi-pointed, for a locally convex topology $\mathfrak{T}$ in X^*, if f^* is $\mathfrak{T}$-continuous at some point of its effective domain.*

The first result of this section, given in Theorem 8.4.3 below, provides an integration result that is valid in Banach spaces enjoying the RNP. The β-epi-pointedness somewhat replaces the convexity of the involved functions in section 4.4 (see, for instance, Proposition 4.4.8).

Theorem 8.4.3 *Let X be a Banach space with the RNP and consider functions $f, g : X \to \overline{\mathbb{R}}$ such that f is lsc and $\|\cdot\|_*$-epi-pointed. If*

$$\partial f(x) \subset \partial g(x) \text{ for all } x \in X,$$

then there exists some $c \in \mathbb{R}$ such that

$$f^{**} = g^{**} \square \sigma_{\text{dom } f^*} + c, \tag{8.72}$$

with an exact inf-convolution. Equivalently,

$$f^*(x^*) = g^*(x^*) - c \text{ for all } x^* \in \text{cl}(\text{dom } f^*). \tag{8.73}$$

Proof. First, from the relations $\partial g \subset \partial(\overline{\text{co}}g)$ (Exercise 62) and $g^{**} = (\overline{\text{co}}g)^{**}$, we may assume without loss of generality that g is convex and lsc. Furthermore, since $\text{Im}(\partial f) \subset \text{dom } f^* \subset \text{cl}^{w^*}(\text{dom } f^*) = \partial \sigma_{\text{dom } f^*}(\theta)$ by (4.13), we have that (Exercise 49)

$$\partial f(x) \subset \partial g(x) \cap \partial \sigma_{\text{dom } f^*}(\theta) \subset \partial h(x), \tag{8.74}$$

where we denote $h := g \square \sigma_{\text{dom } f^*}$. Then we will assume without loss of generality that $\theta \in \|\cdot\|_*\text{-int}(\text{dom } f^*)$. By the RNP (see page 35), the dual space X^* is a w^*-Asplund space. Since the w^*-lsc convex function f^* is (*norm-*) continuous on $\|\cdot\|_*\text{-int}(\text{dom } f^*)$, it is Fréchet-differentiable on a (*norm-*) dense subset D of the last set. Thus, using the identification of X as a subspace of X^{**}, Proposition 4.1.22 and Theorem 8.3.7(*i*), together with (8.74), entail for all $x^* \in D$

$$\emptyset \neq \partial f^*(x^*) = (\partial f)^{-1}(x^*) \subset \partial g(x) \cap \partial \sigma_{\text{dom } f^*}(\theta) \subset (\partial h)^{-1}(x^*) \subset \partial h^*(x^*); \tag{8.75}$$

hence, in particular, $f^*, g^* \in \Gamma_0(X^*)$. Moreover, since D is (*norm-*) dense in $\|\cdot\|_*\text{-int}(\text{dom } f^*)$, applying Lemma 8.4.1 in the Banach space $(X^*, \|\cdot\|_*)$, we get

$$\partial f^*(x^*) \subset \partial h^*(x^*) \text{ for all } x^* \in \|\cdot\|_*\text{-int}(\text{dom } f^*).$$

Therefore, Corollary 4.4.4 gives rise to some $c \in \mathbb{R}$ such that $f^* \equiv h^* + c$ on $\text{cl}^{\|\cdot\|}(\text{dom } f^*)$, which in turn implies that

$$f^* - c = h^* + \mathrm{I}_{\text{cl}^{\|\cdot\|}(\text{dom } f^*)} = \left(g^* + \mathrm{I}_{\text{cl}^{\|\cdot\|}(\text{dom } f^*)}\right) + \mathrm{I}_{\text{cl}^{\|\cdot\|}(\text{dom } f^*)} = h^*.$$

Thus, taking the conjugate in each side, and since $(\|\cdot\|_*\text{-int}(\text{dom } f^*)) \cap \text{dom } g^* \neq \emptyset$ (from (8.75)), Proposition 4.1.20(*i*) entails

$$f^{**} + c = h^{**} = \left(g^* + \mathrm{I}_{\text{cl}^{\|\cdot\|}(\text{dom } f^*)}\right)^* = g^{**} \square \sigma_{\text{dom } f^*},$$

where the inf-convolution is exact. Therefore, (8.72) follows. Finally, the claimed equivalence is also a consequence of Proposition 4.1.20. ∎

We give a series of simple examples to justify the conditions used in Theorem 8.4.3. The first example shows the necessity of the lower semicontinuity assumption of f.

Example 8.4.4 *We consider the functions* $f, g : \mathbb{R} \to \mathbb{R}$ *given by*

$$f(x) := \begin{cases} |x| - 1, & \text{if } |x| > 1, \\ 1, & \text{if } -1 \leq x \leq 1, \end{cases} \quad \text{and } g(x) := |x|,$$

so that $f^* = |\cdot| + \mathrm{I}_{[-1,1]}$; *hence,* $\text{dom } f^* = [-1, 1]$ *and*

$$f^{**}(x) := \begin{cases} |x| - 1, & \text{if } |x| > 1, \\ 0, & \text{if } -1 \leq x \leq 1. \end{cases}$$

Moreover, we have

8.4. NON-CONVEX INTEGRATION

$$\partial f(x) := \begin{cases} \emptyset, & \text{if } -1 \leq x \leq 1, \\ 1, & \text{if } x > 1, \\ -1, & \text{if } x < 1, \end{cases} \text{ and } \partial g(x) := \begin{cases} 1, & \text{if } x > 0, \\ -1, & \text{if } x < 0, \\ [-1,1], & \text{if } x = 0, \end{cases}$$

showing that $\partial f(x) \subset \partial g(x)$ for all $x \in \mathbb{R}$. In other words, all the conditions in Theorem 8.4.3 hold except the lower semicontinuity of f. It is clear that f^{**} and $g^{**} = g$ do not coincide up to an additive constant. Formula (8.72) is also not satisfied, since $\sigma_{\text{dom } f^*} = |\cdot|$ and $g^{**}\square\sigma_{\text{dom } f^*} = |\cdot|$ (Exercise 17).

The following example shows that the term $\sigma_{\text{dom } f^*}$ cannot be ignored within the conclusion of Theorem 8.4.3.

Example 8.4.5 We consider the functions $f, g : \mathbb{R} \to \mathbb{R}_\infty$ given by

$$f(x) := \begin{cases} e^{-x}, & \text{if } x \geq 1, \\ x, & \text{if } 0 \leq x < 1, \\ +\infty, & \text{if } x < 0, \end{cases} \text{ and } g(x) := \begin{cases} x, & \text{if } x \geq 0, \\ +\infty, & \text{if } x < 0. \end{cases}$$

Then $\text{dom } f^* = \mathbb{R}_-$ and we have

$$\partial f(x) := \begin{cases} \emptyset, & \text{if } x \neq 0, \\ \{0\}, & \text{if } x = 0, \end{cases} \text{ and } \partial g(x) := \begin{cases} 1, & \text{if } x > 0, \\ [0,1], & \text{if } x = 0, \\ \emptyset, & \text{if } x < 0. \end{cases}$$

Here, we have that $\partial f(x) \subset \partial g(x)$ for all $x \in \mathbb{R}$, but obviously $f^{**} = I_{\mathbb{R}_+}$ and $g^{**}\ (=g)$ do not coincide up to an additive constant. But, for all $x \in \mathbb{R}$, we have

$$\left(g^{**}\square\sigma_{\text{dom } f^*}\right)(x) = \inf_{x_1}\left(g(x_1) + I_{\mathbb{R}_+}(x - x_1)\right) = \inf_{0 \leq x_1 \leq x} x_1 = f^{**}(x).$$

The following corollary gives conditions under which the term $\sigma_{\text{dom } f^*}$ is dropped out from (8.72).

Corollary 8.4.6 Let X be a Banach space with the RNP and let $f, g : X \to \overline{\mathbb{R}}$ be functions such that f is lsc and $\|\cdot\|_*$-epi-pointed. If

$$\partial f(x) \subset \partial g(x) \text{ for every } x \in X, \tag{8.76}$$

then f^{**} and g^{**} coincide up to an additive constant, provided that one of the following conditions holds:
 (i) $g \leq f$,
 (ii) $\|\cdot\|_*$-$\text{int}(\text{dom } g^*) \subset \text{cl}^{\|\cdot\|_*}(\text{dom } f^*)$.

Proof. First, according to Theorem 8.4.3, there exists a constant $c \in \mathbb{R}$ such that $f^{**} = g^{**} \square \sigma_{\mathrm{dom}\, f^*} + c$. We also proved there that $f^*, g^* \in \Gamma_0(X^*)$. Then, applying Theorem 3.2.2 to f^* and g^* in the pair $((X^*, \|\cdot\|_*), (X^{**}, w^{**}))$, we obtain

$$f^* = (f^*)^{**} = (g^*)^{**} + \mathrm{I}_{\mathrm{cl}^{\|\cdot\|_*}(\mathrm{dom}\, f^*)} - c = g^* + \mathrm{I}_{\mathrm{cl}^{\|\cdot\|_*}(\mathrm{dom}\, f^*)} - c. \tag{8.77}$$

Now, under (i), we have $f^* \leq g^*$, and so $\mathrm{dom}\, g^* \subset \mathrm{dom}\, f^*$. Thus, by (8.77), $f^* = g^* + \mathrm{I}_{\mathrm{cl}^{\|\cdot\|_*}(\mathrm{dom}\, f^*)} - c = g^* - c$, and the conclusion follows by taking the conjugate.

Under (ii), we have

$$\|\cdot\|_* - \mathrm{int}(\mathrm{dom}\, g^*) \subset \mathrm{cl}^{\|\cdot\|_*}(\mathrm{dom}\, f^*) \subset \mathrm{cl}^{\|\cdot\|_*}(\mathrm{dom}\, g^*),$$

due to (8.77), and we deduce that $\mathrm{cl}^{\|\cdot\|_*}(\mathrm{dom}\, g^*) = \mathrm{cl}^{\|\cdot\|_*}(\mathrm{dom}\, f^*)$. Therefore, the conclusion follows as above, using again (8.77). ∎

Remark 20 *As one can expect, the term $\sigma_{\mathrm{dom}\, f^*}$ is also dropped out from (8.72) in the convex setting. In fact, using Exercise 62, (8.76) reads $\partial f(x) \subset \partial g(x) \subset \partial(\mathrm{cl}\, g)(x)$, for every $x \in X$, and Proposition 4.4.8, applied to the functions $f \in \Gamma_0(X)$ and the lsc function $\mathrm{cl}\, g$, establishes the equality (up to some additive constant) of f and $\mathrm{cl}\, g$ in any Banach space not necessarily with RNP. In particular, this implies that $f^{**} = (\mathrm{cl}\, g)^{**} + c = g^{**} + c$.*

The following theorem gives a general integration criterion in the lcs X, when the latter is not necessarily Banach, and without using the epi-pointedness condition. As in (8.73), the conclusion now relies on the equality between the conjugates f^* and g^* (up to some constant) but only on the set $\cup_{L \in \mathcal{F}_{X^*}} \mathrm{cl}(L \cap \mathrm{dom}\, f^*)$ rather than the set $\mathrm{cl}(\mathrm{dom}\, f^*)$. Remember here that $\mathcal{F}_{X^*}$ denotes the family of all finite-dimensional linear subspaces of X^*. As Exercise 5(iii) shows, the inclusion $\cup_{L \in \mathcal{F}_{X^*}} \mathrm{cl}(L \cap \mathrm{dom}\, f^*) \subset \mathrm{cl}(\mathrm{dom}\, f^*)$ may be strict even in simple cases such as ℓ_2.

Theorem 8.4.7 *Consider two functions $f, g : X \to \mathbb{R}_\infty$ such that f^* is proper, and assume the existence of some $\delta > 0$ such that*

$$\partial_\varepsilon f(x) \subset \partial_\varepsilon g(x) \text{ for all } x \in X \text{ and } \varepsilon \in\,]0, \delta[. \tag{8.78}$$

Then there exists some c satisfying

$$f^*(x^*) = g^*(x^*) + c \text{ for all } x^* \in \bigcup_{L \in \mathcal{F}_{X^*}} \mathrm{cl}(L \cap \mathrm{dom}\, f^*). \tag{8.79}$$

8.4. NON-CONVEX INTEGRATION

Proof. Let us first verify that

$$\operatorname{dom} f^* \subset \operatorname{dom} g^*. \tag{8.80}$$

Indeed, if $x^* \in \operatorname{dom} f^*$, then, for every $\varepsilon \in \,]0, \delta[$, there exists some $x \in X$ such that $x^* \in \partial_\varepsilon f(x) \subset \partial_\varepsilon g(x)$, and $x^* \in \operatorname{dom} g^*$. The rest of the proof is divided into two steps.

1st step. We prove (8.79) in the case where the set $\operatorname{dom} f^*$ is finite-dimensional. Hence, given $x^* \in \operatorname{dom} f^*$, we have that

$$\operatorname{ri}(\operatorname{cone}(\operatorname{dom} f^* - x^*)) \neq \emptyset$$

and (8.63) in Corollary 8.3.5 together with the current assumption yields

$$\partial f^*(x^*) = \bigcap_{0<\varepsilon<\delta} \overline{\operatorname{co}}\left((\partial_\varepsilon f)^{-1}(x^*) + \operatorname{N}_{\operatorname{dom} f^*}(x^*)\right)$$
$$\subset \bigcap_{0<\varepsilon<\delta} \overline{\operatorname{co}}\left((\partial_\varepsilon g)^{-1}(x^*) + \operatorname{N}_{\operatorname{dom} f^*}(x^*)\right).$$

But $(\partial_\varepsilon g)^{-1}(x^*) \subset \partial_\varepsilon g^*(x^*)$ by (4.17), so we get

$$\partial f^*(x^*) \subset \bigcap_{0<\varepsilon<\delta} \partial_\varepsilon (g^* + \operatorname{I}_{\operatorname{cl}(\operatorname{dom} f^*)})(x^*) = \partial (g^* + \operatorname{I}_{\operatorname{cl}(\operatorname{dom} f^*)})(x^*).$$

Therefore, there exists some c such that (Exercise 61)

$$f^* = g^* + \operatorname{I}_{\operatorname{cl}(\operatorname{dom} f^*)} + \operatorname{I}_{\operatorname{aff}(\operatorname{dom} f^*)} + c = g^* + \operatorname{I}_{\operatorname{cl}(\operatorname{dom} f^*)} + c.$$

Consequently, f^* and $g^* + c$ coincide on the set $\operatorname{cl}(\operatorname{dom} f^*)$, which in the current case is equal to $\cup_{L \in \mathcal{F}_{X^*}} \operatorname{cl}(L \cap \operatorname{dom} f^*)$ (Exercise 5(*i*)).

2nd step. Fix $x_0^* \in \operatorname{dom} f^*$ ($\subset \operatorname{dom} g^*$, by (8.80)), $L \in \mathcal{F}(x_0^*)$ and consider the functions

$$f_1 := f \square \sigma_L, \quad g_1 := g \square \sigma_L,$$

so that $f_1^* = f^* + \operatorname{I}_L$ and the function f_1^* is proper and has a finite-dimensional (effective) domain. Moreover, due to (4.38) in Proposition 4.1.13, for every $x \in X$ and $\varepsilon \in (0, \delta/2)$, we have

$$\partial_\varepsilon f_1(x) = \bigcap_{0<\alpha<\delta/2} \bigcup_{\substack{x_1+x_2=x \\ \varepsilon_1,\varepsilon_2>0,\ \varepsilon_1+\varepsilon_2=\varepsilon+\alpha}} \partial_{\varepsilon_1} f(x_1) \cap \partial_{\varepsilon_2} \sigma_L(x_2)$$

$$\subset \bigcap_{0<\alpha<\delta/2} \bigcup_{\substack{x_1+x_2=x \\ \varepsilon_1,\varepsilon_2>0,\ \varepsilon_1+\varepsilon_2=\varepsilon+\alpha}} \partial_{\varepsilon_1} g(x_1) \cap \partial_{\varepsilon_2} \sigma_L(x_2) = \partial_\varepsilon g_1(x).$$

Thus, by the first step, there exists some c_L such that

$$f^* + I_L = f_1^* = g_1^* + I_{\mathrm{cl}(L \cap \mathrm{dom}\, f^*)} + c_L$$
$$= (g^* + I_L) + I_{\mathrm{cl}(L \cap \mathrm{dom}\, f^*)} + c_L = g^* + I_{\mathrm{cl}(L \cap \mathrm{dom}\, f^*)} + c_L.$$

In particular, the evaluation of this last relation at x_0^* gives us $c_L = f^*(x_0) - g^*(x_0) =: c$. Therefore, since $L \in \mathcal{F}(x_0^*)$ was arbitrarily chosen, we infer that

$$f^* = \inf_{L \in \mathcal{F}(x_0^*)} (f^* + I_L) = \inf_{L \in \mathcal{F}(x_0^*)} \left(g^* + I_{\mathrm{cl}(L \cap \mathrm{dom}\, f^*)}\right) + c \qquad (8.81)$$
$$= \inf_{L \in \mathcal{F}_{X^*}} \left(g^* + I_{\mathrm{cl}(L \cap \mathrm{dom}\, f^*)}\right) + c = g^* + I_{\cup_{L \in \mathcal{F}_{X^*}} \mathrm{cl}(L \cap \mathrm{dom}\, f^*)} + c,$$

and (8.79) is proved. ∎

Remark 21 *In terms of primal objects, Theorem 8.4.7 yields the following relation between the closed convex hulls of the functions f and g (Exercise 128):*

$$\overline{\mathrm{co}}\, f = \sup_{L \in \mathcal{F}_{X^*}} \mathrm{cl}\left((\overline{\mathrm{co}}\, g) \square \sigma_{L \cap \mathrm{dom}\, f^*}\right) + c. \qquad (8.82)$$

The following corollary simplifies Theorem 8.4.7.

Corollary 8.4.8 *With the assumptions of Theorem 8.4.7, the following statements hold for some scalar c:*
(i) If $\mathrm{dom}\, f^*$ is closed or $\mathrm{ri}(\mathrm{dom}\, f^*) \neq \emptyset$, then

$$\overline{\mathrm{co}}\, f = \mathrm{cl}\left((\overline{\mathrm{co}}\, g) \square \sigma_{\mathrm{dom}\, f^*}\right) + c.$$

(ii) If g^* is continuous somewhere in $\mathrm{dom}\, f^*$, then

$$\overline{\mathrm{co}}\, f = (\overline{\mathrm{co}}\, g) \square \sigma_{\mathrm{dom}\, f^*} + c.$$

(iii) If $g \leq f$ or $\mathrm{dom}\, g^* \subset \mathrm{dom}\, f^*$, then $\overline{\mathrm{co}}\, f = \overline{\mathrm{co}}\, g + c$.

Proof. (i) In the current case, $A := \cup_{L \in \mathcal{F}_{X^*}} \mathrm{cl}(L \cap \mathrm{dom}\, f^*) = \mathrm{cl}(\mathrm{dom}\, f^*)$, by Exercise 5(i). So, the conclusion follows as in Exercise 128.

(ii) Since $\mathrm{dom}\, f^* \subset A$, (8.79) implies $f^* = g^* + I_{\mathrm{dom}\, f^*} - c$. Therefore, due to (4.56) and Theorem 3.2.2, we get $\overline{\mathrm{co}}\, f = (\overline{\mathrm{co}}\, g) \square \sigma_{\mathrm{dom}\, f^*} + c$.

(iii) We have $\mathrm{dom}\, g^* \subset \mathrm{dom}\, f^* \subset A$. Then $f^* = g^* + I_A - c = g^* - c$, by (8.79), and the conclusion follows once again by Theorem 3.2.2. ∎

8.4. NON-CONVEX INTEGRATION

The equality between $\overline{\text{co}} f$ and $\overline{\text{co}} g$ below follows provided that the inclusion in (8.78) is reinforced for all $\varepsilon > 0$.

Corollary 8.4.9 *Given two proper functions $f, g : X \to \mathbb{R}_\infty$ with f^* being proper, we assume that*

$$\partial_\varepsilon f(x) \subset \partial_\varepsilon g(x) \text{ for all } x \in X \text{ and all } \varepsilon > 0.$$

Then $\overline{\text{co}} f$ and $\overline{\text{co}} g$ are equal up to an additive constant.

Proof. By Exercise 75, for every $x^* \in X^*$ and $\varepsilon \geq 0$, we have

$$\partial_\varepsilon f^*(x^*) = \bigcap_{\delta > \varepsilon} \text{cl} \left\{ \bigcup_{\substack{\lambda \in \Delta_k \\ k \geq 1}} \sum_{1 \leq i \leq k} \lambda_i (\partial_{\varepsilon_i} f)^{-1}(x^*) : \varepsilon_i \geq 0, \sum_{1 \leq i \leq k} \lambda_i \varepsilon_i \leq \delta \right\}$$

$$\subset \bigcap_{\delta > \varepsilon} \text{cl} \left\{ \bigcup_{\substack{\lambda \in \Delta_k \\ k \geq 1}} \sum_{1 \leq i \leq k} \lambda_i (\partial_{\varepsilon_i} g)^{-1}(x^*) : \varepsilon_i \geq 0, \sum_{1 \leq i \leq k} \lambda_i \varepsilon_i \leq \delta \right\} = \partial_\varepsilon g^*(x^*).$$

Thus, since f^{**} is also proper (by Proposition 3.1.4, as f^* is assumed proper), by applying Theorem 8.4.7 to the functions $f^*, g^* \in \Gamma_0(X^*)$ in the compatible dual pair $((X^*, w^*), (X, \mathfrak{T}_X))$ we find some c such that

$$f^{**} = g^{**} + \inf_{L \in \mathcal{F}_X} I_{\text{cl}(L \cap \text{dom } f^{**})} + c \geq g^{**} + c.$$

Then $f^* \leq g^* - c$, using Theorem 3.2.2, and we deduce that $\text{dom } g^* \subset \text{dom } f^*$. The conclusion follows then by Corollary 8.4.8(iii). ■

Let us illustrate the above integration criteria by means of a simple example.

Example 8.4.10 *Consider the functions $f, g : \mathbb{R} \to \mathbb{R}_\infty$ defined by $g(x) := I_{\{0\}}$ and*

$$f(x) := \begin{cases} |x| + 1, & \text{if } x \neq 0 \\ 0, & \text{if } x = 0. \end{cases}$$

Then $f^* = I_{[-1,1]}$, $g^* \equiv 0$, $(\overline{\text{co}} f)(x) = |x|$, and $(\overline{\text{co}} g) \square \sigma_{\text{dom } f^*}(x) = |x|$. The ε-subdifferential of f and g are given by

$$\partial_\varepsilon f(x) = \begin{cases} \emptyset, & \text{if } \varepsilon \in [0, 1[\text{ and } x \neq 0, \\ [-1, 1], & \text{if } \varepsilon \in [0, 1[\text{ and } x = 0, \\ \left[1 - \frac{\varepsilon - 1}{x}, 1\right], & \text{if } \varepsilon \geq 1 \text{ and } x > 0, \\ \left[-1, 1 + \frac{1 - \varepsilon}{x}\right], & \text{if } \varepsilon \geq 1 \text{ and } x < 0, \end{cases}$$

and $\partial_\varepsilon g(x) = \emptyset$, if $x \neq 0$, and $\partial_\varepsilon g(x) = \mathbb{R}$, if $x = 0$. Observe that the hypothesis of Theorem 8.4.7 holds but not the one of Corollary 8.4.9.

As applications of the previous integration results, one can obtain constructive characterizations of the closed convex envelope $\overline{\mathrm{co}} f$ by means of ε-subdifferentials or subdifferentials of f. See Exercises 129 and 130, which provide extensions to non-convex functions of Corollary 4.4.9 and its consequences in section 4.4.

8.5 Variational characterization of convexity

We give in this section different criteria ensuring the convexity of functions defined on the lcs X. We will need the following lemma in order to compare the subdifferential of an epi-pointed function and that of its biconjugate.

Lemma 8.5.1 *Given a τ-epi-pointed weakly lsc function $f : X \to \mathbb{R}_\infty$, we denote*

$$M_f := \{x^* \in X^* : \mathrm{sol}(\mathrm{P}_{f-x^*}) \text{ is convex}\},$$

where $\mathrm{sol}(\mathrm{P}_{f-x^*})$ is the optimal set of problem (P_{f-x^*}) (see (8.45)). Then, for every $x \in X$, we have

$$\partial f^{**}(x) \cap (\tau\text{-}\mathrm{int}(\mathrm{dom}\, f^*)) \cap M_f \subset \partial f(x). \tag{8.83}$$

Consequently, for every non-empty w^*-compact convex set C contained in τ-$\mathrm{int}\,(\mathrm{dom}\, f^*) \cap M_f$, we have

$$\sigma_C \square f^{**} = \sigma_C \square f.$$

Proof. Take $x \in X$ and $x^* \in \partial f^{**}(x) \cap (\tau\text{-}\mathrm{int}(\mathrm{dom}\, f^*)) \cap M_f$. Since the set $(\partial f)^{-1}(x^*) = \mathrm{sol}(\mathrm{P}_{f-x^*})$ is convex and (weakly) closed, applying Corollary 8.3.5(ii) in the dual pair $((X, \tau(X, X^*)), (X^*, \tau(X^*, X)))$, we get

$$\partial f^*(x^*) = \mathrm{N}_{\mathrm{dom}\, f^*}(x^*) + \overline{\mathrm{co}}\left((\partial f)^{-1}(x^*)\right) = (\partial f)^{-1}(x^*).$$

Consequently, using Theorem 3.2.2, since $f^* \in \Gamma_0(X^*)$ we obtain

$$x \in \partial (f^{**})^*(x^*) = \partial f^*(x^*) = (\partial f)^{-1}(x^*)$$

showing that $x^* \in \partial f(x)$.

8.5. VARIATIONAL CHARACTERIZATION OF ...

To show the second statement, we choose a non-empty w^*-compact convex set $C \subset (\tau\text{-}\operatorname{int}(\operatorname{dom} f^*)) \cap M_f$. Since f^* is τ-continuous somewhere, it is τ-continuous in C, and taking into account Proposition 3.1.4, we infer from Proposition 4.1.20 that $(I_C + f^*)^* = \sigma_C \square f^{**} \in \Gamma_0(X)$ and there are $x_1, x_2 \in X$ such that $x = x_1 + x_2$ and

$$(\sigma_C \square f^{**})(x) = \sigma_C(x_1) + f^{**}(x_2).$$

Observe that $\sigma_C \leq \sigma_{C \cup (-C)} =: \rho_C$ and this latter is a $\tau(X, X^*)$-continuous seminorm in X, so Proposition 2.2.6 entails that σ_C, and *a fortiori* $\sigma_C \square f^{**}$, are also $\tau(X, X^*)$-continuous. Therefore, $\partial(\sigma_C \square f^{**})(x) \neq \emptyset$, due to Proposition 4.1.22, and (see Exercise 49)

$$\emptyset \neq \partial(\sigma_C \square f^{**})(x) = \partial \sigma_C(x_1) \cap \partial f^{**}(x_2). \qquad (8.84)$$

Consequently, using (5.2), we have

$$\partial \sigma_C(x_1) \subset C \subset (\tau - \operatorname{int}(\operatorname{dom} f^*)) \cap M_f$$

and (8.83) together with (8.84) implies that

$$\emptyset \neq \partial(\sigma_C \square f^{**})(x) \subset \partial \sigma_C(x_1) \cap \partial f(x_2) \subset \partial(\sigma_C \square f)(x); \qquad (8.85)$$

that is,

$$\emptyset \neq \partial(\sigma_C \square f^{**})(x) \subset \partial(\sigma_C \square f)(x) \text{ for all } x \in X.$$

In particular, $\partial(\sigma_C \square f)(x) \neq \emptyset$ for all $x \in X$, and $\sigma_C \square f \in \Gamma_0(X)$. Therefore, applying Theorem 4.4.3, we find γ such that $\sigma_C \square f^{**} = \sigma_C \square f + \gamma$. More precisely, taking the conjugates and using Theorem 3.2.2, we get $I_C + f^* = I_C + (f^{**})^* = I_C + f^* - \gamma$, and we deduce that $\gamma = 0$ because $\emptyset \neq C \subset \tau\text{-}\operatorname{int}(\operatorname{dom} f^*)$; that is, $\sigma_C \square f^{**} = \sigma_C \square f$. ∎

Theorem 8.5.2 *Let $f : X \to \mathbb{R}_\infty$ be a τ-epi-pointed weakly lsc function, and let D be a convex dense subset of $\operatorname{dom} f^*$ such that $\operatorname{sol}(P_{f-x^*})$ is convex for every $x^* \in D$. Then we have that*

$$f^{**} = \sigma_{\operatorname{dom} f^*} \square f.$$

Consequently, f is convex provided that $\operatorname{dom} f^* = X^*$.

Proof. We may assume that $D \subset \tau\text{-int}(\operatorname{dom} f^*)$; otherwise, instead of D, we should work with the set $D \cap \tau\text{-int}(\operatorname{dom} f^*)$ which is also a dense subset in $\operatorname{dom} f^*$. Pick $x_0^* \in \tau\text{-int}(\operatorname{dom} f^*)$ and define the family

$$\mathfrak{C} := \{\operatorname{co} F : F \text{ is a finite subset of } D,\ x_0^* \in F\},$$

endowed with the order given by ascending inclusions. Observe that each $C \in \mathfrak{C}$ is convex τ-compact and satisfies $C \subset D \subset \tau\text{-int}(\operatorname{dom} f^*) \cap M_f$, and Lemma 8.5.1 entails $\sigma_C \square f = \sigma_C \square f^{**}$. Moreover, since f^* is τ-continuous at $x_0^* \in C = \operatorname{dom} I_C$, by Propositions 4.1.20 and 3.2.5, we have $\sigma_C \square f = \sigma_C \square f^{**} = (I_C + f^*)^* \in \Gamma_0(X)$ and $\sup_{C \in \mathfrak{C}} (\sigma_C \square f) \in \Gamma_0(X)$. Thus, for every $x \in X$, we obtain

$$\sup_{C \in \mathfrak{C}} (\sigma_C \square f)(x) = \sup_{C \in \mathfrak{C}} (\sigma_C \square f^{**})(x) \leq (\sigma_{\operatorname{dom} f^*} \square f^{**})(x) \leq (\sigma_{\operatorname{dom} f^*} \square f)(x),$$
(8.86)

with an equality when $\alpha := \sup_{C \in \mathfrak{C}} (\sigma_C \square f)(x) = +\infty$. Thus, we suppose that $\alpha < +\infty$. Moreover, the function $f^{**} - x_0^*$ is weakly inf-compact, thanks to Proposition 3.1.3 applied in the dual pair $((X^*, \tau), (X, \mathfrak{T}_X))$. Also, for each $C \in \mathfrak{C}$, every sequence $(y_k)_k \subset X$ such that $\sigma_C(x - y_k) + f(y_k) \to (\sigma_C \square f)(x)$ $(\in \mathbb{R})$ satisfies, for all k sufficiently large,

$$\langle x_0^*, x - y_k \rangle + f(y_k) \leq \sigma_C(x - y_k) + f(y_k) \leq (\sigma_C \square f)(x) + 1 \leq \alpha + 1;$$

that is,
$$f(y_k) - \langle x_0^*, y_k \rangle \leq \alpha + 1 - \langle x_0^*, x \rangle =: \gamma \in \mathbb{R}.$$

So, we can assume without loss of generality that $(y_k)_k \subset [f - x_0^* \leq \gamma] \subset [f^{**} - x_0^* \leq \gamma] =: A$, with A being weakly compact. Therefore,

$$\sup_{C \in \mathfrak{C}} (\sigma_C \square f)(x) = \sup_{C \in \mathfrak{C}} \min_{y \in A} (\sigma_C(x - y) + f(y)),$$

and Proposition 3.4.1 implies that

$$\sup_{C \in \mathfrak{C}} (\sigma_C \square f)(x) = \min_{y \in A} \sup_{C \in \mathfrak{C}} (\sigma_C(x - y) + f(y)) = \min_{y \in A} (\sigma_D(x - y) + f(y))$$
$$= \min_{y \in A} (\sigma_{\operatorname{dom} f^*}(x - y) + f(y)) \geq (\sigma_{\operatorname{dom} f^*} \square f)(x).$$

Consequently, combining with (8.86),

$$\sup_{C \in \mathfrak{C}} (\sigma_C \square f) = \sigma_{\operatorname{dom} f^*} \square f = \sigma_{\operatorname{dom} f^*} \square f^{**},$$

8.5. VARIATIONAL CHARACTERIZATION OF ... 353

and the fact that $\sigma_C \square f \in \Gamma_0(X)$ implies that all these functions are in $\Gamma_0(X)$. Thus, the proof is done because Theorem 3.2.2 applies again to conclude that $(\sigma_{\operatorname{dom} f^*} \square f^{**})^* = I_{\operatorname{cl}(\operatorname{dom} f^*)} + f^* = f^*$ and, so, $f^{**} = (\sigma_{\operatorname{dom} f^*} \square f^{**})^{**} = \sigma_{\operatorname{dom} f^*} \square f^{**}$; that is, $\sigma_{\operatorname{dom} f^*} \square f = \sigma_{\operatorname{dom} f^*} \square f^{**} = f$. ∎

The following example illustrates the necessity of considering the support function of dom f^* in Theorem 8.5.2.

Example 8.5.3 *Let $f : \mathbb{R} \to \mathbb{R}$ be the (non-convex) lsc function defined by*

$$f(x) = \begin{cases} |x|, & \text{if } x \in [-1,1], \\ |x| + e^{-|x|}, & \text{if } x \in \mathbb{R}\setminus[-1,1]. \end{cases}$$

We have $f^* = I_{[-1,1]}$; hence f is epi-pointed, and

$$\operatorname{sol}(\mathrm{P}_{f-\alpha}) = \begin{cases} \{0\}, & \text{if } \alpha \in \,]-1,1[, \\ [0,1], & \text{if } \alpha = 1, \\ [-1,0], & \text{if } \alpha = -1, \\ \emptyset, & \text{if } \alpha \notin [-1,1]. \end{cases}$$

Therefore, Theorem 8.5.2 applies and yields $f^{**} = |\cdot| = \sigma_{\operatorname{dom} f^*} \square f$. However, the equality $f^{**} = f$ obviously fails as f is not convex.

The following example shows the necessity of assuming the convexity of the set D in Theorem 8.5.2.

Example 8.5.4 *Assume that X is a reflexive Banach space, and consider a function $h \in \Gamma_0(X)$ such that $\operatorname{dom} h^* = X^*$; hence h^* is (norm-) continuous on X by Corollary 2.2.8, and so h is τ-epi-pointed, as τ coincides with the dual norm topology. Then we choose a non-convex weakly lsc and nonnegative function g such that $f := h + g$ is not convex. Then f is weakly lsc and τ-epi-pointed because $h^* \geq f^*$, by Proposition 2.2.6. Furthermore, we have $\operatorname{dom} f^* = X^*$, and so f^* is also τ-continuous on X^*. Consequently, f^* is Fréchet-differentiable in a (G_δ)-dense subset $D \subset X^*$, and Theorem 8.3.7(i) entails for all $x^* \in D$*

$$\operatorname{sol}(\mathrm{P}_{f-x^*}) = (\partial f)^{-1}(x^*) = \{(f^*)'(x^*)\};$$

that is, $\operatorname{sol}(\mathrm{P}_{f-x^*})$ is convex for all $x^* \in D$. At the same time, the equality $f^{**} = \sigma_{\operatorname{dom} f^*} \square f \,(= f)$ obviously fails because f is not convex.

8.6 Chebychev sets and convexity

In his section, we study the convexity of Chebychev sets when X is a Hilbert space with an inner product $\langle \cdot, \cdot \rangle$.

Definition 8.6.1 *A non-empty set $A \subset X$ is said to be a* Chebychev set *if the projection set $\pi_A(x)$ is a singleton for every $x \in X$.*

We also introduce the concept of *weak projection*.

Definition 8.6.2 *Let $A \subset X$ be a non-empty set. Given $x \in X$, the set $\tilde{\pi}_A(x) \subset X$ given by*

$$\tilde{\pi}_A(x) := \{w\text{-}\lim_k x_k : x_k \in A, \ \|x_k - x\| \to d_A(x)\},$$

is called weak projection set *of x on the set A.*

It is clear that $\tilde{\pi}_A(x)$ is weakly closed, and that every projection is a weak projection; that is,

$$\pi_A(x) \subset \tilde{\pi}_A(x) \text{ for all } x \in X, \tag{8.87}$$

with a possibly strict inclusion in infinite-dimensional spaces. In particular, if A is weakly closed, then every weak projection $y = w\text{-}\lim_k x_k$, such that $(x_k)_k \subset A$ and $\|x_k - x\| \to d_A(x)$, satisfies $y \in A$ and

$$d_A(x) = \lim_k \|x_k - x\| \geq \|y - x\| \geq d_A(x),$$

due to the weak lower semicontinuity of the norm function. Therefore, $d_A(x) = \|y - x\|$ and y is a projection of x on A. Consequently, we have that

$$\tilde{\pi}_A(x) = \pi_A(x). \tag{8.88}$$

A similar relation holds for the so-called approximately compact sets.

Definition 8.6.3 *A non-empty set $A \subset X$ is said to be* approximately compact *if, for every $x \in X$ and $(x_k)_k \subset A$ satisfying $\|x_k - x\| \to d_A(x)$, the sequence $(x_k)_k$ has a (norm-) convergent subsequence.*

Proposition 8.6.4 *If $A \subset X$ is approximately compact, then $\tilde{\pi}_A(x) = \pi_{\operatorname{cl} A}(x)$ for every $x \in X$.*

Proof. Take $y \in \tilde{\pi}_A(x)$; that is, $y = w\text{-}\lim_k x_k$ for a sequence $(x_k)_k \subset A$ such that $\|x_k - x\| \to d_A(x)$. Then there exists a *norm*-convergent subsequence $(x_{k_m})_m$ such that $y = \lim_m x_{k_m} \in \operatorname{cl} A$ and

8.6. CHEBYCHEV SETS AND CONVEXITY

$$d_A(x) = \lim_m \|x_{k_m} - x\| = \|y - x\| \geq d_{\mathrm{cl}\, A}(x) = d_A(x).$$

Thus, the Kadec–Klee property entails that (x_{k_m}) (norm-) converges to y, and we conclude that every subsequence of $(x_k)_k$ has a subsequence which norm-converges to the same limit y. Consequently, the whole sequence $(x_k)_k$ norm-converges to y, and we deduce that $\tilde{\pi}_A(x) \subset \pi_{\mathrm{cl}\,A}(x)$. Conversely, if $y \in \pi_{\mathrm{cl}\,A}(x)$ ($\subset \mathrm{cl}\,A$), then $\|x - y\| = d_{\mathrm{cl}\,A}(x)$, and every sequence $(x_k)_k \subset A$ which norm-converges (hence, weak converges) to y satisfies $\|x_k - x\| \to \|y - x\| = d_{\mathrm{cl}\,A}(x) = d_A(x)$; that is, $y \in \tilde{\pi}_A(x)$, and the inclusion $\pi_{\mathrm{cl}\,A}(x) \subset \tilde{\pi}_A(x)$ follows. ∎

The above concepts are related to the function $f_A : X \to \mathbb{R}_\infty$ defined by

$$f_A(x) := \mathrm{I}_A(x) + (1/2)\|x\|^2. \tag{8.89}$$

The following lemma expresses the conjugate and the subdifferential of f_A. Other properties are gathered in Exercise 25. The function in (i) below is the so-called *Asplund function*.

Lemma 8.6.5 *Let* $A \subset X$ *be a non-empty set. Then* $\mathrm{dom}\, f_A^* = X$, f^* *is (norm-) continuous and, for all* $x \in X$,
  (i) $(f_A)^*(x) = \frac{1}{2}(\|x\|^2 - d_A^2(x))$.
  (ii) $\partial(\mathrm{cl}^w f_A)(x) = (\tilde{\pi}_A)^{-1}(x)$.
  (iii) $\partial(f_A)^*(x) = \overline{\mathrm{co}}(\tilde{\pi}_A(x))$.
  (iv) $\partial(f_A)^*(x) = \mathrm{co}(\tilde{\pi}_A(x))$ when $X = \mathbb{R}^n$.

Proof. Relation (i) is easily verified and implies that $\mathrm{dom}(f_A)^* = X$, so that f^* is norm-continuous by Proposition 2.2.6. Moreover, $x \in (\partial(\mathrm{cl}^w f_A))^{-1}(y)$ if and only if

$$(\mathrm{cl}^w f_A)(x) + (f_A)^*(y) = (\mathrm{cl}^w f_A)(x) + (\mathrm{cl}^w f_A)^*(y) = \langle y, x \rangle \, ;$$

that is, by assertion (i), if and only if

$$(\mathrm{cl}^w f_A)(x) + (1/2)\|y\|^2 - (1/2)d_A^2(y) = \langle y, x \rangle \, .$$

Therefore, $x \in (\partial(\mathrm{cl}^w f_A))^{-1}(y)$ if and only if there exists a sequence $(x_k)_k \subset A$ that weakly converges to x and such that

$$\lim_k (1/2)\|x_k - y\|^2 = \lim_k \left((1/2)\|x_k\|^2 - \langle y, x_k \rangle\right) + (1/2)\|y\|^2 = (1/2)d_A^2(y).$$

In other words, $x \in (\partial(\mathrm{cl}^w f_A))^{-1}(y)$ if and only if $x \in \tilde{\pi}_A(y)$. This proves assertion (ii), while assertion (iii) comes by combining Corollary 8.3.5(ii) and assertion (ii). Finally, assertion (iv) comes from (8.65). ∎

Corollary 8.6.6 *Let $A \subset X$ be a non-empty set, and suppose that A is either weakly closed or approximately compact. Then, for every $x \in X$,*

$$\partial(f_A)^*(x) = \overline{\mathrm{co}}(\pi_{\mathrm{cl}\,A}(x)) \tag{8.90}$$

and, when $X = \mathbb{R}^n$,

$$\partial(f_A)^*(x) = \mathrm{co}(\pi_{\mathrm{cl}\,A}(x)). \tag{8.91}$$

Proof. Relation (8.90) follows by combining Lemma 8.6.5(iii), (8.88) (when A is weakly closed) and Proposition 8.6.4 (when A is approximately compact). Relation (8.91) follows similarly but using assertion (iv) in Lemma 8.6.5 instead of (iii). ∎

We now proceed by characterizing the ($norm, norm$)-continuity of the projection mapping.

Proposition 8.6.7 *The following properties are equivalent, for every non-empty closed set $A \subset X$,*
  (i) π_A is ($norm, norm$)-continuous.
  (ii) d_A^2 is Fréchet-differentiable on X.

Proof. According to Lemma 8.6.5(i), d_A^2 is Fréchet-differentiable on X if and only if $(f_A)^*$ is so. Moreover, using Lemma 8.6.5(iii), for all $x \in X$, we have

$$\pi_A(x) \in \overline{\mathrm{co}}\{\tilde{\pi}_A(x)\} = \partial(f_A)^*(x);$$

that is, $\pi_A(\cdot)$ is a selection of $\partial(f_A)^*$. Then the equivalence of (i) and (ii) comes from Proposition 4.1.8. ∎

The following theorem establishes the convexity of the weak closure of sets having convex weak projection sets.

Theorem 8.6.8 *Assume that X is Hilbert. Let $A \subset X$ be a non-empty set, and consider the following assertions:*
  (i) $\tilde{\pi}_A(x)$ is convex, for every $x \in X$.
  (ii) $\tilde{\pi}_A(x)$ is convex, for every x in a convex dense subset of X.
  (iii) The function $\mathrm{cl}^w f_A$ is convex.
  (iv) The set $\mathrm{cl}^w A$ is convex.
  Then (i) $\Longleftrightarrow$ (ii) $\Longleftrightarrow$ (iii) $\Longrightarrow$ (iv).

8.6. CHEBYCHEV SETS AND CONVEXITY

Proof. $(i) \Rightarrow (ii)$ This is obvious.

$(ii) \Rightarrow (iii)$ Denote $g := \text{cl}^w f_A$ and let D be as in (ii). Take $x \in D$. Thanks to Lemma 8.6.5(ii) we have that $\partial g = (\tilde{\pi}_A)^{-1}$ and, so, $\tilde{\pi}_A(x) = (\partial g)^{-1}(x)$. Thus, $\tilde{\pi}_A(x)$ is weakly closed due to the weak lower semicontinuity of g. At the same time, Lemma 8.6.5(iii) ensures that $\partial (f_A)^*(x) = \overline{\text{co}}(\tilde{\pi}_A(x))$ and, therefore,

$$\partial (f_A)^*(x) = \text{cl}(\tilde{\pi}_A(x)) = \tilde{\pi}_A(x) = (\partial g)^{-1}(x);$$

that is, $\text{sol}(\text{P}_{g-x}) = (\partial g)^{-1}(x)$ is convex for every $x \in D$. Furthermore, using Lemma 8.6.5, we have $\text{dom}\, g^* = \text{dom}(f_A)^* = X$ and, in particular, g is continuous in X thanks to Corollary 2.2.8; that is, g is *norm*-epi-pointed. Therefore, the convexity of the weakly lsc function $g = \text{cl}^w f_A$ follows from Theorem 8.5.2.

$(iii) \Rightarrow (i)$ Assume that the function $\text{cl}^w f_A$ is convex. Then Lemma 8.6.5(ii) implies that the set $\tilde{\pi}_A(x) = (\partial (\text{cl}^w f_A))^{-1}(x)$ is convex for every $x \in X$.

$(iii) \Rightarrow (iv)$ See Exercise 25(iv). ∎

The convexity of $\text{cl}^w A$ is the maximum that can be obtained from Theorem 8.6.8. Here is an example.

Example 8.6.9 *Let A be the unit sphere in ℓ_2. Then we verify that*

$$\tilde{\pi}_A(x) := \begin{cases} \pi_A(x), & \text{if } x \neq \theta, \\ B_{\ell_2}, & \text{if } x = \theta, \end{cases}$$

so that $\tilde{\pi}_A(x)$ is convex for all $x \in \ell^2$. However, the set A is obviously not convex, but $\text{cl}^w A = B_{\ell_2}$ is convex.

Now we give the characterization of the convexity of a set A by means of the single-valuedness of $\tilde{\pi}_A$, as well as other criteria related to the continuity of π_A or to the differentiability of d_A^2.

Theorem 8.6.10 *Given a closed set $A \subset X$, the following conditions are equivalent:*
  (i) A is convex.
  (ii) $\tilde{\pi}_A$ is single valued.
  (iii) d_A^2 is Gâteaux differentiable.
  (iv) d_A^2 is Fréchet differentiable.
  (v) π_A has a (norm, weak)-continuous selection.
  (vi) π_A has a (norm, norm)-continuous selection.

Proof. $(iv) \Rightarrow (iii)$ and $(vi) \Rightarrow (v)$: They are evident.

$(vi) \Rightarrow (iv)$: Notice that π_A is a selection of $\partial(f_A)^*$, combining (8.87) and Lemma 8.6.5(iii). Then, thanks to Proposition 4.1.8(ii), assertion (vi) implies the Fréchet differentiability of $(f_A)^*$ on X, which in turn implies the Fréchet differentiability of d_A^2, due to Lemma 8.6.5(i).

$(v) \Rightarrow (iii)$: This is proved similarly as $(vi) \Rightarrow (iv)$, but applying assertion (i) in Proposition 4.1.8 instead of assertion (ii) there.

$(iii) \Rightarrow (ii)$ By Lemma 8.6.5(i), assertion (iii) implies the Gâteaux-differentiability of the function $(f_A)^*$. So, according to Lemma 8.6.5(iii), for all $x \in X$ we have that $\partial(f_A)^*(x) = \overline{\text{co}}(\tilde{\pi}_A(x))$ and $\tilde{\pi}_A(x)$ is a singleton.

$(ii) \Rightarrow (i)$ Fix $x \in X$. Due to the relation $\emptyset \neq \pi_A(x) \subset \tilde{\pi}_A(x)$ coming from (8.87), by assertion (ii) we have that $\pi_A(x) = \tilde{\pi}_A(x)$. So, Lemma 8.6.5$(iii)$ and Exercise 25(v) imply that

$$\partial(f_A)^*(x) = \overline{\text{co}}(\tilde{\pi}_A(x)) = \pi_A(x) = (\partial f_A)^{-1}(x),$$

which in turn yields, using (4.18),

$$\partial(f_A)^{**}(x) = (\partial(f_A)^*)^{-1}(x) \subset \partial f_A(x) \text{ for all } x \in X.$$

Therefore, since f_A is lsc (because A is closed) and $(f_A)^{**} \in \Gamma_0(X)$, Proposition 4.4.8 implies the equality of f_A and $(f_A)^{**}$ up to some constant. This obviously entails that f_A is convex and (i) follows.

$(i) \Rightarrow (vi)$: It comes from the 1-Lipschitz continuity of π_A when A is a (closed) convex set. ∎

By combining the equivalence $(i) \Leftrightarrow (ii)$ in Theorem 8.6.10 and (8.88) together with Proposition 8.6.4, we get the following characterization of Chebychev sets.

Corollary 8.6.11 *Let $A \subset X$ be a non-empty closed set which is either weakly closed or approximately compact. Then A is convex if and only if it is a Chebychev set.*

8.7 Exercises

Exercise 116 *Given problem* (P) *in (8.1), if x is a feasible point of this problem (i.e., $x \in F$), and $A(x)$ is the set of active indices at x, prove that*

8.7. EXERCISES

$$N_C(x) + \text{cone co}\left(\bigcup_{t \in A(x)} \partial f_t(x)\right) \subseteq N_F(x).$$

Exercise 117 *Suppose that, in problem* (P), $\tilde{f} := \sup_{t \in T} f_t$ *is continuous at* $x \in F$ *and* $x \in \operatorname{int} C$. *Prove that the conditions LFM and BCQ at* x *are equivalent.*

Exercise 118 *Let* $X = C = \mathbb{R}$, *and*

$$S = \left\{f_t(x) := \max\{0, x^{2t+1}\} \leq 0, \ t \in \mathbb{N}\right\}.$$

Prove that S *is not LFM despite that the condition in Theorem 8.1.11 is satisfied.*

Exercise 119 *Assume that the feasible set* F *of problem* (P) *in (8.1) is non-empty. Prove that* $\operatorname{epi} g^* + \operatorname{cl} K$ *is* w^*-*closed if any of the following conditions is satisfied:*
 (a) *The set* $\operatorname{epi} g^* + K$ *is* w^*-*closed.*
 (b) *The function* g *is linear.*
 (c) g *is continuous at some point of* F.
Consequently, prove that if g *is continuous at some point of* F *and* S *is FM, then* $\operatorname{epi} g^* + K$ *is* w^*-*closed.*

Exercise 120 *Given the convex problem* (P) *such that* $C = \operatorname{dom} g = \operatorname{dom} f_t = \mathbb{R}^n$, *for all* $t \in T$, *consider the linear system*

$$S_L := \{\langle x^*, x \rangle \leq \langle x^*, y \rangle - f_t(y), \ (t, y) \in T \times \mathbb{R}^n, \ x^* \in \partial f_t(y)\}.$$

Prove the following assertions:
 (a) *The set of solutions of* S_L *coincides with the feasible set of* (P).
 (b) *The constraint system* S *of* (P) *is LFM if and only if* S_L *is LFM.*

Exercise 121 *Prove Theorem 8.2.3 when the consistent system* $S := \{f_t(x) \leq 0, \ t \in T, \ x \in C\}$ *is LFM at* $\bar{x} \in F \cap \operatorname{dom} g$, *by applying Theorem 8.2.1.*

Exercise 122 *Prove that if the consistent system* $S := \{f_t(x) \leq 0, \ t \in T, \ x \in C\}$ *in (8.2) is FM, then it is also LFM at every feasible point* $\bar{x}$.

Exercise 123 *Suppose that* $v(0) \in \mathbb{R}$, S *is FM, and* g *is continuous at some point of* F. *Prove that* $\bar{x} \in F$ *is minimum of* (P) *if and only if there exists* $\bar{\lambda} \in \mathbb{R}_+^{(T)}$ *such that* $(\bar{x}, \bar{\lambda})$ *is a saddle point of the Lagrangian function* L; *that is,*

$$L(\bar{x}, \lambda) \leq L(\bar{x}, \bar{\lambda}) \leq L(x, \bar{\lambda}), \quad \forall \lambda \in \mathbb{R}_+^{(T)} \text{ and } \forall x \in C. \quad (8.92)$$

In such a case, $\bar{\lambda}$ is a maximizer of $(\mathcal{D})$.

Exercise 124 *Prove that for any function $f: X \to \overline{\mathbb{R}}$ such that $\operatorname{dom} f^* \neq \emptyset$ one has*

$$\operatorname{argmin} f^{**} = \bigcap_{\varepsilon > 0, \, x^* \in \operatorname{dom} f^*} \overline{\operatorname{co}} \left(\varepsilon\text{-}\operatorname{argmin} f + \{x^*\}^- \right). \quad (8.93)$$

Exercise 125 *Let X be a Banach space with the RNP, and let $f: X \to \mathbb{R}_\infty$ be an lsc function such that, for some $a > 0$ and $b \in \mathbb{R}$,*

$$f(x) \geq a \|x\| + b \text{ for all } x \in X. \quad (8.94)$$

Prove that there exists some $\mathcal{G}_\delta$-set $D \subset aB_{X^*}$, which is dense in aB_{X^*}, and such that the functions $f - x^*$, $x^* \in D$, attain a strong minimum on X.

Exercise 126 *Given a normed space X and a function $f: X \to \overline{\mathbb{R}}$, prove that the following conditions are equivalent:*
  (i) *f is β-epi-pointed; that is, f^* is $\|\cdot\|_*$-continuous somewhere.*
  (ii) *There are $\alpha > 0$, $\mu \in \mathbb{R}$ and $x_0^* \in X^*$ such that*

$$f(x) \geq \alpha \|x\| + \langle x_0^*, x \rangle + \mu \text{ for all } x \in X. \quad (8.95)$$

Exercise 127 *Assume that X is Banach with the RNP. Let $C \subset X$ be a closed bounded convex set, and let $f: C \to \mathbb{R}$ be an lsc function such that $\inf_C f \in \mathbb{R}$. Prove that, for every $\varepsilon > 0$, there exists $x^* \in \varepsilon B_{X^*}$ such that*

$$\operatorname{argmin}((\overline{\operatorname{co}} f) - x^*) = \operatorname{argmin}(f - x^*).$$

Exercise 128 *Consider two functions $f, g: X \to \mathbb{R}_\infty$ such that f^* is proper, and assume the existence of some $\delta > 0$ such that $\partial_\varepsilon f(x) \subset \partial_\varepsilon g(x)$ for all $x \in X$ and $\varepsilon \in \,]0, \delta[$. Prove the existence of some c such that*

$$\overline{\operatorname{co}} f = \sup_{L \in \mathcal{F}_{X^*}} \operatorname{cl}((\overline{\operatorname{co}} g) \square \sigma_{L \cap \operatorname{dom} f^*}) + c.$$

Exercise 129 *Given a function $f: X \to \mathbb{R}_\infty$, $x_0 \in (\partial f)^{-1}(X^*)$ and $\delta > 0$, prove that*

$$(\overline{\operatorname{co}} f)(x) = f(x_0) + \sup \left\{ \sum_{i=0}^{n-1} \langle x_i^*, x_{i+1} - x_i \rangle + \langle x_n^*, x - x_n \rangle - \sum_{i=0}^{n} \varepsilon_i \right\},$$

where the supremum is taken over $n \geq 0$, $\varepsilon_i \in (0, \delta)$, $x_i \in X$ and $x_i^* \in \partial_{\varepsilon_i} f(x_i)$, $i = 0, \ldots, n$ (with the convention that $\sum_{i=0}^{-1} = \sum_{i=1}^{0} = 0$).

Exercise 130 Assume that X is a Banach space with the RNP. Let $f : X \to \mathbb{R}_\infty$ be lsc and $\|\cdot\|_*$-epi-pointed, and let $x_0 \in (\partial f)^{-1}(X^*)$. Prove that

$$(\overline{\text{co}} f)(x) = f(x_0) + \sup \left\{ \sum_{i=0}^{n-1} \langle x_i^*, x_{i+1} - x_i \rangle + \langle x_n^*, x - x_n \rangle \right\}, \quad (8.96)$$

where the supremum is taken over $n \geq 0$, $x_i \in X$, $x_i^* \in \partial f(x_i)$, $i = 0, \ldots, n$.

Exercise 131 Let $f_1, f_2 : X \to \mathbb{R}_\infty$ be two functions such that $\text{dom}(f_1 + f_2)^* \neq \emptyset$, and consider the following assertions:
(i) There exists some $\delta > 0$ such that, for all $\varepsilon \in \,]0, \delta\,]$ and $x \in X$,

$$\partial_\varepsilon (f_1 + f_2)(x) = \bigcap_{\delta > 0} \text{cl} \left(\bigcup_{\substack{\varepsilon_1 + \varepsilon_2 \leq \varepsilon + \delta \\ \varepsilon_1, \varepsilon_2 \geq 0}} \partial_{\varepsilon_1} f_1(x) + \partial_{\varepsilon_2} f_2(x) \right). \quad (8.97)$$

(ii) For all $x \in X$,

$$\partial (f_1 + f_2)(x) = \bigcap_{\varepsilon > 0} \text{cl} \left(\partial_\varepsilon f_1(x) + \partial_\varepsilon f_2(x) \right). \quad (8.98)$$

(iii) $\overline{\text{co}}(f_1 + f_2) = (\overline{\text{co}} f_1) + (\overline{\text{co}} f_2)$.

Prove that $(i) \Rightarrow (iii) \Rightarrow (ii)$. If, in addition, X is Banach with the RNP, f_1, f_2 are lsc, and $\text{int}(\text{dom}(f_1 + f_2)^*) \neq \emptyset$, then prove that $(ii) \iff (iii)$.

8.8 Bibliographical notes

The material in section 8.1 has been mainly extracted from [49], [71], [72], and [116, Lemma 2.1]. Since the literature on the Farkas lemma (Theorem 8.1.5) is vast (see, e.g., the survey in [117]), we mention here some works giving Farkas-type results for the kind of systems considered in the section: [27], [92], [118], and [132] for semi-infinite systems; [71], [116], [121], and [133] for infinite systems; and [74], [90], [119], and [120] for cone convex systems. The condition of the closedness of the characteristic cone K in (8.7) was introduced in [37] as a very general assumption for the duality theorem in LSIP, where it plays a crucial role (see, e.g., [92]). The FM property for convex systems was first studied in [119] with X being Banach and all the functions being finite-valued, under the name of closed cone constraint qualification. It is known that the FM property is strictly weaker than several known interior-type regularity conditions, and gives rise to the so-called non-asymptotic optimality conditions for (P) (Theorem 8.2.3). The characterization of feasibility given in (8.9) was first established for linear systems in [84, Theorem 1], and the convex version is provided in [71, Theorem 3.1]. Proposition 8.1.9 is Theorem 4.4(iii') in [71]. The LFM property, related to BCQ (Definition 8.1.13), appeared in [108, p. 307] relative to ordinary convex programming problems with equality and inequality constraints. It was extended in [171] to the setting of linear semi-infinite systems, and intensively studied in [92, Chapter 5]. The consequences of its extension to convex semi-infinite systems were analyzed in [82]. For a deep analysis of BCQ and related conditions, see, also, [132] and [133]. Conditions BCQ and LFM are practically equivalent, as it is pointed out in Exercise 117 (which is Proposition 2 in [72]). For other Farkas-type qualification conditions, see [27]. An extensive comparative analysis of constraint qualifications for (P) is given in [133] and [136]. Other contributions to constraint qualifications and optimality conditions in nonlinear semi-infinite and infinite programming are [158] and [159] (see, also, [155, chapter 8]). In [159] the authors dealt with the Banach space setting in contrast to the Asplund space one in [157].

The different statements in Theorem 8.1.5 can be found in [71, Proposition 3.4], extending [116, Theorem 2.1], [118, Theorem 4.1], and [27, Theorem 5.6], where the last two deal with finite-valued convex functions in the Euclidean space. Theorem 8.1.5(iii) applies to the problem of set containment, which is key in solving large scale knowledge-based data classification problems (e.g., [147]). Theorem 8.1.5(iv) is related to [74, Theorem 2.2], which deals with conical constraints in Banach spaces. In the presence of a set constraint C, and assuming the continuity of the involved functions, Theorem 8.1.5(iv) is given in [98], under a different closedness condition which is related to the FM property ([74, p. 93]).

The notion of strongly CHIP, used in Theorem 8.2.8, was introduced in [70], and extended to infinite families of convex sets in [135] and [136]. Previously, in [92, Chapter 7], KKT conditions for convex semi-infinite optimization were derived for finite-valued functions (see, also, [140] for differentiable convex functions). The original version of

8.8. BIBLIOGRAPHICAL NOTES

Corollary 8.2.11 can be found in [92, Theorems 7.8 and 7.9], where it is proved that, under the corresponding assumptions, the constraint system $\mathcal{S}$ is LFM. Apart from this, many KKT conditions exist in the literature which are obtained via different approaches: approximate subdifferentials of the data functions ([49], [55], [111]), exact subdifferentials at close points ([188]), asymptotic KKT conditions ([137]) for linear semi-infinite programming, Farkas–Minkowski-type closedness criteria ([71]) in convex semi-infinite optimization, strong CHIP-like qualifications for convex optimization with non-necessarily convex $\mathcal{C}^1$-constraints ([38]) (see, also, [78] for locally Lipschitz constraints), among others. We also refer to [205], and references therein, for KKT conditions in the framework of sub-smooth semi-infinite optimization, and to [85] for analyzing the relationship among KKT rules and Lagrangian dualities. We finally refer to [92], [94], [138], etc., and references therein, for theory, algorithms and applications of semi-infinite optimization.

The relaxation arguments used in section 8.3 are very useful in practice, namely, in calculus of variations, in mathematical programming problems, as well as in many other theoretical and numerical purposes (see, e.g., [80] and [108]). This fundamental topic has been considered by many researchers in recent years. The approach in [14] is based on the subdifferential analysis of the closed convex hull. The different formulas in Theorem 8.3.2 were given in [141], based on the subdifferential of the Fenchel conjugate, which itself relies on the subdifferential of the supremum function. Theorem 8.3.4, given in [46], adapts Theorem 8.3.2 to the setting of Banach spaces. Theorem 8.3.3 is given in [45]. A variant of Theorem 8.3.3 involves the asymptotic theory from [68] and [69], and was first established in [14, Theorem 4.6]. Theorem 8.3.7 provides criteria for the Gâteaux and Fréchet differentiability of the conjugate function as in [5], [22], and [127]. A characterization of the w^*-continuity of the conjugate function can be found in [161, Corollary 8.g] (remember that the continuity of the conjugate function in Theorems 8.3.2 and 8.3.3 is required with respect to a compatible topology). The concept of strong minima (or maxima) in Definition 4.3.12 is also called Tychonoff well-posedness (e.g., [92]). Theorem 8.3.7(ii) can be found, for instance, in [3, page 52]. Theorem 8.4.3 is given in [44] (see, also, [143] and [144]). Theorem 8.4.7 is provided in [43] (see, also, [62]). The convex counterpart to Corollary 8.4.9 is given in [148] for locally convex spaces, and in [193] for normed spaces. Theorem 8.5.2 is given in [59] (extending [179, Theorem 10] to the setting of locally convex spaces). The definition of Asplund function in Lemma 8.6.5(i) can be found [4]. The notion of approximately compact sets was introduced in

[194]. It amounts to saying that the associated problem $\inf_{y\in A}\|x-y\|$ is well-posed at x (see [106]). It is still unknown whether the weak projection in Theorem 8.6.10(ii) can be replaced with the standard projection. Exercise 118 is similar to [82, Example 2.1]. Exercise 120 is [92, Theorem 7.10]. Exercise 125 is stated in [22, Corollary 2.3]. Exercise 129 is [43, Theorem 4.1]. Exercise 130 can be found in [44, Corollary 15]. A related result relying on the construction of the closed convex hull was given in [13]. The implication (iii) $\Rightarrow$ (ii) in Exercise 131 is known (see, e.g., [43]).

Chapter 9

Exercises - Solutions

9.1 Exercises of chapter 2

Exercise 1: (i) First, observe that $p_C(\theta) = \inf\{\lambda \geq 0 : \theta \in \lambda C\} = 0$. To check that p_C is positively homogeneous on $\operatorname{dom} p_C$, we fix $x \in \operatorname{dom} p_C$ and $\alpha \geq 0$. If $\alpha = 0$, then $p_C(0x) = p_C(\theta) = 0 = 0 p_C(x)$. If $\alpha > 0$, then

$$p_C(\alpha x) = \inf\{\lambda \geq 0 : \alpha x \in \lambda C\} = \alpha \inf\{\lambda \geq 0 : x \in \lambda C\} = \alpha p_C(x).$$

(ii) To prove the inclusion "$\subset$" in (2.63) take $(x, \alpha) \in \operatorname{epi}_s p_C$, so that $p_C(x) < \alpha$. If $p_C(x) > 0$, then there exists $\lambda > 0$ such that $x \in \lambda C$ and $\alpha > \lambda$. So,

$$(x, \alpha) \subset (\lambda C) \times \ ]\lambda, +\infty[- \lambda \left(C \times \ ]1, +\infty[\right)$$
$$\subset \mathbb{R}_+^* \left(C \times \ ]1, +\infty[\right) \subset \mathbb{R}_+^* \left((C \cup \{\theta\}) \times \ ]1, +\infty[\right).$$

If $p_C(x) = 0$ and $x \in \lambda_k C$ for some sequence $\lambda_k \downarrow 0$, then, for large enough k such that $\lambda_k < \alpha$, we obtain

$$(x, \alpha) \in (\lambda_k C) \times \ ]\lambda_k, +\infty[= \lambda_k \left(C \times \ ]1, +\infty[\right)$$
$$\subset \mathbb{R}_+^* \left(C \times \ ]1, +\infty[\right) \subset \mathbb{R}_+^* \left((C \cup \{\theta\}) \times \ ]1, +\infty[\right).$$

If $p_C(x) = 0$ and there is no sequence $\lambda_k \downarrow 0$ such that $x \in \lambda_k C$, then $x \in 0C$, entailing that $x = \theta$. Thus,

$$(x, \alpha) = (\theta, \alpha) \in \{\theta\} \times \mathbb{R}_+^* \subset \mathbb{R}_+^* ((C \cup \{\theta\}) \times \,]1, +\infty[).$$

In order to prove the opposite inclusion in (2.63), we take $\mu > 0$ and $(x, \alpha) \in \mu((C \cup \{\theta\}) \times \,]1, +\infty[)$; that is, $(x, \alpha) = \mu(a, \beta)$ for some $a \in C \cup \{\theta\}$ and $\beta > 1$. If $a = \theta$, then $x = \theta$ and $p_C(x) = 0 < \alpha$, and we get $(x, \alpha) \in \text{epi}_s\, p_C$. Otherwise, if $a \in C$, then, by the positive homogeneity of p_C on $\text{dom}\, p_C$, and the fact that $p_C(a) \leq 1$ and $\mu < \alpha$, we conclude that $p_C(x) = p_C(\mu a) = \mu p_C(a) \leq \mu < \alpha$, entailing that $(x, \alpha) \in \text{epi}_s\, p_C$.

(iii) If C is convex, then we can show that $\text{epi}_s\, p_C$ is convex by observing that $\text{co}(\text{epi}_s\, p_C) \subset \mathbb{R}_+^* (C \times \,]1, +\infty[) \subset \text{epi}_s\, p_C$; that is, $\text{co}(\text{epi}_s\, p_C) = \text{epi}_s\, p_C$.

Exercise 2: Given $x \in C$ let us denote $I_x := \{i \in \{1, .., m\} : \langle a_i^*, x \rangle = b_i\}$. If $I_x = \emptyset$, then $x \in \text{int}\, C$, and so $\mathbb{R}_+(C - x) = X$. Otherwise, i.e., if $I_x \neq \emptyset$, we prove that

$$\mathbb{R}_+(C - x) = \{u \in X : \langle a_i^*, u \rangle \leq 0, \text{ for } i \in I_x\}.$$

The inclusion "$\subset$" is clear. To verify the other inclusion, we take $u(\neq \theta)$ in the set of the right-hand side and choose an $\alpha > 0$ such that $\langle a_i^*, \alpha u + x \rangle \leq b_i$ for all $i \in \{1, .., m\} \setminus J_x$. Then, since $\langle a_i^*, \alpha u + x \rangle = \langle a_i^*, \alpha u \rangle + b_i \leq b_i$ for all $i \in J_x$, we get $\alpha u + x \in C$, and so $u \in \alpha^{-1}(C - x) \subset \mathbb{R}_+(C - x)$.

Exercise 3: (i) Suppose that $\text{int}\, A \neq \emptyset$. Since we only need to prove that $A^i \subset \text{int}\, A$, we pick $a \in A^i$. Proceeding by contradiction, we suppose that $a \notin \text{int}\, A$. Then, by the separation theorem, we find a proper closed half-space $S \subset X$ such that $X = \mathbb{R}_+(A - a) \subset S$, yielding a contradiction.

(ii) Suppose that X is finite-dimensional and take $a \in A^i$. Then $\mathbb{R}_+(A - a) = X$, and so $\text{aff}\, A = X$ and we infer that $\text{int}\, A = \text{ri}\, A \neq \emptyset$. Then we apply assertion (i).

(iii) We denote $A := \cup_{m \geq 1} A_m$ for closed convex sets $A_m \subset X$, $m \geq 1$. Due to the inclusion $\text{int}\, A \subset A^i$ the equality $\text{int}\, A = A^i$ holds when $A^i = \emptyset$. Otherwise, we take $a \in A^i$. Since $A - a$ is absorbing and $\theta \in A - a$, we have that $X = \cup_{n=1}^\infty n(A - a) = \cup_{n,m=1}^\infty n(A_m - a)$. Then, by the Baire theorem, there exist $n_0, m_0 \geq 1$ such that the set $\text{int}(n_0(A_{m_0} - x))$ is non-empty, entailing $\emptyset \neq \text{int}\, A_{m_0} \subset \text{int}\, A$. Hence, $A^i = \text{int}\, A$, by assertion (i).

9.1 EXERCISES OF CHAPTER 2

Finally, given a proper lsc convex function $f : X \to \mathbb{R}_\infty$ defined on the Banach space X, we show that $\operatorname{int}(\operatorname{dom} f) = (\operatorname{dom} f)^i$. Indeed, it suffices to apply assertion (iii) above to the set $\operatorname{dom} f = \cup_{n \geq 1}[f \leq n]$, since $[f \leq n]$ is convex and closed.

Exercise 4: Observe that each A_k is convex but not necessarily closed. Also, the set A is convex. Assume that $x \in A^i$ ($\subset \ell_2$), and so there exists some $k_0 \geq 1$ such that $x \in A_{k_0}$; that is, $x = \sum_{i=1}^{k_0} \alpha_i e_i$, $\alpha_{k_0} > 0$. Let $t > 0$ such that $-e_{k_0+1} \in t(A - x)$. Then $-(1/t)e_{k_0+1} + x = -(1/t)e_{k_0+1} + \sum_{1 \leq k_0} \alpha_i e_i \in A$, and a contradiction occurs. Therefore, $A^i = \operatorname{int} A = \emptyset$.

Exercise 5: (i) Let $C \subset X$ be a non-empty convex set. If C is closed, then the conclusion is trivial. In the second case, pick $x_0 \in \operatorname{ri}(C)$ and denote $\tilde{C} := \cup_{L \in \mathcal{F}_X} \operatorname{cl}(L \cap C)$. Given $x \in \operatorname{cl}(C)$, if $x = x_0$, then obviously $x \in \tilde{C}$. Otherwise, $x \neq x_0$, and (2.15) implies that $]x, x_0] \subset L \cap C$, where $L := \operatorname{span}\{x, x_0\}$. Thus, $x \in \operatorname{cl}(L \cap C) \subset \tilde{C}$.

(ii) If C is the set A of Exercise 4, then $\operatorname{cl}(C) = \ell_2$ and $\cup_{L \in \mathcal{F}_X} \operatorname{cl}(L \cap C) = C$ because, for each $L \in \mathcal{F}_X$, there exists some k_0 such that $L \subset A_{k_0}$.

(iii) Let $(e_n)_n$ be the canonical basis of ℓ_2, denote $C := \operatorname{span}\{e_n, n \geq 1\}$ and consider the function $g : \ell_2 \to \mathbb{R}_\infty$ defined by

$$g(x) := \begin{cases} \sum_{k \geq 1}(|x_k|+1)k, & \text{if } x := \sum_{k \geq 1} x_k e_k \in C, \\ +\infty, & \text{if not.} \end{cases}$$

The function g is easily shown to be convex, entailing the convexity of its closed hull $f := \operatorname{cl} g$. Next, we verify that $\operatorname{dom} f = C$. It is clear that $C \subset \operatorname{dom} f$. Conversely, take $x \in \operatorname{dom} f \setminus C$ and let $(x^m)_m \subset C$, $x^m := \sum_{k \geq 1} x_k^m e_k$, be a sequence converging to x such that $f(x) = \lim_m g(x^m)$. So, we can find $m_j \to +\infty$ together with $k_j \geq m_j$ such that $\left|x_{k_j}^{m_j}\right| > 0$. Therefore,

$$f(x) = \lim_j g(x^{m_j}) \geq \limsup_j \left(\left|x_{k_j}^{m_j}\right| + 1\right) k_j \geq \limsup_j m_j = +\infty,$$

and we deduce that $x \notin \operatorname{dom} f$. Finally, we verify that

$$\bigcup_{L \in \mathcal{F}_{X^*}} \operatorname{cl}(L \cap \operatorname{dom} f) = \bigcup_{L \in \mathcal{F}_{X^*}} \operatorname{cl}(L \cap C) = C \subsetneq \ell_2 = \operatorname{cl}(C) = \operatorname{cl}(\operatorname{dom} f).$$

Exercise 6: Denote $f := \inf_{\alpha>0}(\mathrm{I}_{\alpha A^\circ} + \alpha)$. If $x \notin \mathrm{dom}\,\sigma_A$, then by (2.49), we have $x \notin \mathbb{R}_+ A^\circ$, and so $f(x) = +\infty = \sigma_A(x)$. If $x \in \mathrm{dom}\,\sigma_A$ and $(0 = \langle \theta, x \rangle \leq) \sigma_A(x) < \lambda$ for a given $\lambda \in \mathbb{R}$, then $\lambda^{-1}x \in A^\circ$, and so $f(x) \leq \mathrm{I}_{\lambda A^\circ}(x) + \lambda = \lambda$. This entails $f(x) \leq \sigma_A(x)$ when $\lambda \downarrow \sigma_A(x)$. To show the opposite inequality, we suppose that $f(x) < \lambda$ for a given $\lambda \in \mathbb{R}$. Then there exists some $\alpha_0 > 0$ such that $x \in \alpha_0 A^\circ$ and $\alpha_0 < \lambda$. Hence, $\sigma_A(x) \leq \alpha_0 < \lambda$ and we deduce that $\sigma_A(x) \leq f(x)$ as $\lambda \downarrow f(x)$.

Exercise 7: It is clear that $h := \inf_{i,p\in\mathcal{P}} \sigma_{A_{i,p}}$ is positively homogeneous. To prove its subadditivity, we fix $x, y \in X$ and $\alpha, \beta \in \mathbb{R}$ such that $h(x) < \alpha$ and $h(y) < \beta$. Then there are i_1, i_2 and $p_1, p_2 \in \mathcal{P}$ such that $\sigma_{A_{i_1,p_1}}(x) < \alpha$ and $\sigma_{A_{i_2,p_2}}(y) < \beta$. Take i_0 such that $i_1 \preccurlyeq i_0$ and $i_2 \preccurlyeq i_0$ and choose $p_0 \in \mathcal{P}$ such that $p_0 \geq \max\{p_1, p_2\}$ ($\mathcal{P}$ is saturated). Then $A_{i_0,p_0} \subset A_{i_1,p_1} \cap A_{i_2,p_2}$ and we deduce

$$h(x+y) \leq \sigma_{A_{i_0,p_0}}(x+y) \leq \sigma_{A_{i_0,p_0}}(x) + \sigma_{A_{i_0,p_0}}(y)$$
$$\leq \sigma_{A_{i_1,p_1}}(x) + \sigma_{A_{i_2,p_2}}(y) < \alpha + \beta.$$

Thus, as $\alpha \to h(x)$ and $\beta \to h(y)$, we obtain $h(x+y) \leq h(x) + h(y)$. We are done since this last inequality also holds if $h(x) = +\infty$ or $h(y) = +\infty$.

Exercise 8: Assume that $f(\theta) = 0$ and $f(\lambda x) = \lambda f(x)$ for all $x \in \mathrm{dom}\,f$ and $\lambda > 0$. Given $(x, \alpha) \in \mathrm{epi}\,f$ (which is non-empty by the properness of f) and $\lambda > 0$; hence $x \in \mathrm{dom}\,f$, we have that $f(\lambda x) = \lambda f(x) \leq \lambda \alpha$; that is, $\lambda(x, \alpha) = (\lambda x, \lambda \alpha) \in \mathrm{epi}\,f$. Thus, since $(\theta, 0) \in \mathrm{epi}\,f$, we deduce that $\mathbb{R}_+^*(\mathrm{epi}\,f \cup \{(\theta, 0)\}) \subset \mathrm{epi}\,f$, and $\mathrm{epi}\,f$ is a cone.

Conversely, assume that $\mathrm{epi}\,f$ is a cone. So, $(\theta, 0) \in \mathrm{epi}\,f$ and $f(\theta) \in \mathbb{R}$ because f is proper. Moreover, given $x \in \mathrm{dom}\,f$ and $\lambda > 0$, we have $(\lambda x, \lambda f(x)) = \lambda(x, f(x)) \in \mathbb{R}_+^* \mathrm{epi}\,f \subset \mathrm{epi}\,f$, entailing that $f(\lambda x) \leq \lambda f(x)$, and hence $\lambda x \in \mathrm{dom}\,f$. Hence, $f(\lambda x) \leq \lambda f(x) = \lambda f(\lambda^{-1}\lambda x) \leq \lambda \lambda^{-1} f(\lambda x) = f(\lambda x)$ and we deduce that $f(\lambda x) = \lambda f(x)$.

Finally, if $\theta \in \mathrm{dom}\,f$ and $f(\lambda x) = \lambda f(x)$ for all $x \in \mathrm{dom}\,f$ and $\lambda > 0$, then $f(\theta) = f(\lambda \theta) = \lambda f(\theta)$ for all $\lambda > 0$, and $f(\theta) = 0$ follows when $\lambda \downarrow 0$.

Exercise 9: It is clear that $\mathrm{dom}\,f \subset \bigcap_{t\in T} \mathrm{dom}\,f_t$. If $x \in \bigcap_{t\in T} \mathrm{dom}\,f_t$, then the current assumption implies that $f(x) = \max_{t\in T} f_t(x) \in \mathbb{R}$, giving rise to the converse inclusion.

To prove the second statement, we take $z \in \bigcap_{t\in T} \mathbb{R}_+(\mathrm{dom}\,f_t - x)$. If $z = \theta$, then we are obviously done. Otherwise, for each $t \in T$, there exist $\alpha_t, m_t > 0$ and $z_t \in \mathrm{dom}\,f_t$ such that $z = \alpha_t(z_t - x)$ and $f_t(z_t) < m_t$. By arguing as above, the upper semicontinuity assumption yields

9.1 EXERCISES OF CHAPTER 2

some neighborhood V_t of t such that $f_s(z_t) < m_t$ for all $s \in V_t$. Consider a finite covering $V_{t_1}, \ldots, V_{t_k}$ of T, so that $z_{t_i} = \alpha_{t_i}^{-1} z + x \in \operatorname{dom} f_s$ for all $s \in V_{t_i}$. Next, for $\bar{\alpha} := \max\{\alpha_{t_i}, \, i = 1, \ldots, k\}$ we obtain that $\bar{\alpha}^{-1} z \in \bar{\alpha}^{-1} \alpha_{t_i}(\cap_{s \in V_{t_i}} (\operatorname{dom} f_s - x)) \subset \cap_{s \in V_{t_i}} (\operatorname{dom} f_s - x)$, $1 \leq i \leq k$, where the last inclusion comes from the convexity of the set $\cap_{s \in V_{t_i}} (\operatorname{dom} f_s - x)$ and the fact that $\theta \in \cap_{s \in V_{t_i}} (\operatorname{dom} f_s - x)$. Hence,

$$\bar{\alpha}^{-1} z \in \cap_{i=1,\ldots,k} \cap_{s \in V_{t_i}} (\operatorname{dom} f_s - x)$$
$$= \cap_{s \in V_{t_i}, \, i=1,\ldots,k} (\operatorname{dom} f_s - x) = \cap_{t \in T} (\operatorname{dom} f_t - x),$$

and the first statement of the lemma leads us to $z \in \mathbb{R}_+ (\cap_{t \in T} (\operatorname{dom} f_t - x)) = \mathbb{R}_+ (\operatorname{dom} f - x)$.

Exercise 10: (i) Due to the fact that the family of sets

$$\{x^* \in X^* : \ |\langle x_i, x^* \rangle| \leq \delta, \ i \in 1, \ldots, k\}, \quad x_i \in X, \ k \in \mathbb{N}, \ \delta > 0,$$

is a base of θ-neighborhoods of the w^*-topology, and observing that

$$\{x_i \in X, \ i \in 1, \ldots, k\}^\perp \subset \{x^* \in X^* \mid |\langle x_i, x^* \rangle| \leq \delta, \ i \in 1, \ldots, k\},$$

(2.11) allows us to write $\operatorname{cl}(A) \subset \cap_{L \in \mathcal{F}} \operatorname{cl}(A + L^\perp) \subset \cap_{U \in \mathcal{N}_{X^*}} \operatorname{cl}(A + U) = \operatorname{cl}(A)$. If $x \in X$, then we have that $\cap_{L \in \mathcal{F}} \operatorname{cl}(A + L^\perp) \subset \cap_{L \in \mathcal{F}(x)} \operatorname{cl}(A + L^\perp)$. Conversely, we have that $\cap_{L \in \mathcal{F}} \operatorname{cl}(A + L^\perp) = \cap_{L \in \mathcal{F}} \operatorname{cl}(A + (L + \mathbb{R}x)^\perp) \subset \cap_{L \in \mathcal{F}} \operatorname{cl}(A + L^\perp)$.

(ii) Let A be closed and let B be convex and compact with $\theta \in B$. Then $A \subset \cap_{\varepsilon > 0} (A + \varepsilon B)$. Conversely, take $x \in \cap_{\varepsilon > 0} (A + \varepsilon B)$. Hence, for each $0 < \varepsilon < 1$, we write $x = a_\varepsilon + \varepsilon b_\varepsilon$ for some $a_\varepsilon \in A$ and $b_\varepsilon \in B$. Since $\varepsilon b_\varepsilon \in \varepsilon B \subset B$, we may assume without loss of generality that $b_\varepsilon \to b \in B$ and $\varepsilon b_\varepsilon \to c \in B$ as $\varepsilon \downarrow 0$. Moreover, for every $x^* \in X^*$, we have that $\lim_{\varepsilon \downarrow 0} \langle \varepsilon b_\varepsilon, x^* \rangle = 0$; that is, $\varepsilon b_\varepsilon \to^w \theta$ and so $c = \theta$. Hence, $x = \lim_{\varepsilon \downarrow 0} (a_\varepsilon + \varepsilon b_\varepsilon) = \lim_{\varepsilon \downarrow 0} a_\varepsilon \in \operatorname{cl} A = A$.

In relation to the second statement in this part, we write

$$\bigcap_{\varepsilon > 0} (A_\varepsilon + \varepsilon B) = \bigcap_{\delta > 0} \bigcap_{\varepsilon > 0} (A_\delta + \varepsilon B) = \bigcap_{\delta > 0} A_\delta.$$

(iii) For every $L_0 \in \mathcal{F}$, we have

$$\bigcap_{L \in \mathcal{F}} \operatorname{cl}\left(C_L + L^\perp\right) \subset \bigcap_{L \in \mathcal{F}, \, L \supset L_0} \operatorname{cl}\left(C_L + L_0^\perp\right)$$
$$\subset \bigcap_{L \in \mathcal{F}} \operatorname{cl}\left(C_{\operatorname{span}\{L \cup L_0\}} + L_0^\perp\right) \subset \bigcap_{L \in \mathcal{F}} \operatorname{cl}\left(C_L + L_0^\perp\right),$$

as $C_{\text{span}\{L\cup L_0\}} \subset C_L$. Thus, as L_0 is arbitrary in $\mathcal{F}$, by (2.67)

$$\bigcap_{L\in\mathcal{F}} \text{cl}\left(C_L + L^\perp\right) \subset \bigcap_{L_0\in\mathcal{F}}\bigcap_{L\in\mathcal{F}} \text{cl}\left(C_L + L_0^\perp\right) = \bigcap_{L\in\mathcal{F}}\bigcap_{L_0\in\mathcal{F}} \text{cl}\left(C_L + L_0^\perp\right) = \bigcap_{L\in\mathcal{F}} \text{cl}\left(C_L\right),$$

and we conclude that $\bigcap_{L\in\mathcal{F}} \text{cl}\left(C_L + L^\perp\right) = \bigcap_{L\in\mathcal{F}} \text{cl}\left(C_L\right)$. Moreover, as $\{C_L, L \in \mathcal{F}\}$ is non-increasing, for each $x \in X$, we have

$$\bigcap_{L\in\mathcal{F}} \text{cl}\left(C_L + L^\perp\right) \subset \bigcap_{L\in\mathcal{F}} \text{cl}\left(C_{\text{span}\{L,x\}} + L^\perp\right) \subset \bigcap_{L\in\mathcal{F}} \text{cl}\left(C_L + L^\perp\right).$$

Finally, the desired relation comes by observing that $\bigcap_{L\in\mathcal{F}} \text{cl}\left(C_L\right) = \bigcap_{L\in\mathcal{F}(x)} \text{cl}\left(C_L\right)$.

(iv) The first statement comes from assertions (i) and (iii). To verify the last statement, take $x \in \bigcap_{\varepsilon>0} \overline{\text{co}}(A_\varepsilon + \varepsilon B_{X^*})$. Applying (2.10) in X^* and taking into account that B_{X^*} is w^*-compact, we obtain that $x \in \bigcap_{\varepsilon>0} \text{cl}(\text{co}(A_\varepsilon) + \varepsilon B_{X^*}) = \bigcap_{\varepsilon>0} (\overline{\text{co}}(A_\varepsilon) + \varepsilon B_{X^*})$. Since $(\overline{\text{co}}(A_\varepsilon))_{\varepsilon>0} \subset X^*$ is also non-decreasing with respect to ε, assertion (ii) applies in (X^*, w^*) and entails $x \in \bigcap_{\varepsilon>0} \overline{\text{co}}(A_\varepsilon)$, yielding the inclusion "$\subset$".

Exercise 11: The conclusion follows from the equality $\Lambda(\text{co}\,A) = \text{co}(\Lambda A)$ together with its consequence $\Lambda(\overline{\text{co}}A) = \text{cl}(\Lambda(\text{co}\,A)) = \overline{\text{co}}(\Lambda A)$, which are true under the assumption $0 \notin \Lambda$.

Exercise 12: The inclusion "$\supset$" is obvious since $1 \in \Lambda_\varepsilon$ entails $A_\varepsilon \subset \Lambda_\varepsilon A_\varepsilon$. Let us prove the inclusion "$\subset$". If $a \in \bigcap_{\varepsilon>0} \Lambda_\varepsilon A_\varepsilon$, we have $a = \lambda_\varepsilon a_\varepsilon$ with $\lambda_\varepsilon \in \Lambda_\varepsilon$ and $a_\varepsilon \in A_\varepsilon$ for all $\varepsilon > 0$. Since $\lim_{\varepsilon\downarrow 0} \lambda_\varepsilon = 1$, we can assume that $\lambda_\varepsilon > 0$. For each $\varepsilon > 0$, the net $(a_\delta)_{0<\delta<\varepsilon}$ is contained in A_ε, and the closedness of A_ε implies that $\lim_{\delta\downarrow 0} a_\delta = \lim_{\delta\downarrow 0} (\lambda_\delta)^{-1} a = a \in A_\varepsilon$.

Exercise 13: Using (2.17), (2.69) implies that, for all $t \in T$,

$$\text{ri}(C_i) \cap \text{ri}(C) = \text{ri}(C_i) \cap \text{ri}\left(\bigcap_{1\leq i\leq m} C_i\right)$$

$$= \text{ri}(C_i) \cap \left(\bigcap_{1\leq i\leq m} \text{ri}(C_i)\right) = \bigcap_{1\leq i\leq m} \text{ri}(C_i) \neq \emptyset,$$

and (2.70) follows. Conversely, if condition (2.70) holds, we choose $x_i \in \text{ri}(C_i) \cap C$ and denote $\bar{x} := \sum_{i=1}^m \frac{1}{m} x_i$. Then for each $i_0 \in \{1, ..., m\}$, we have $\sum_{1\leq i\leq m,\ i\neq i_0} \frac{1}{m-1} x_i \in C \subset C_{i_0}$ and so, by (2.15), $\bar{x} = \frac{1}{m} x_{i_0} +$

9.1 EXERCISES OF CHAPTER 2

$\left(\frac{m-1}{m}\right) \sum_{1\leq i\leq m,\ i\neq i_0} \frac{1}{m-1} x_i \in \mathrm{ri}(C_{i_0})$. In other words, $\bar{x} \in \bigcap_{1\leq i\leq m} \mathrm{ri}(C_i)$ and (2.69) holds.

Exercise 14: We only need to prove that $\inf_{\mathrm{ri}\,A} f \leq f(y)$ for any $y \in \mathrm{cl}\,A$ such that $\inf_{\mathrm{ri}\,A} f > -\infty$ and $f(y) < +\infty$. Taking $z \in (\mathrm{ri}\,A) \cap \mathrm{dom}\,f$, (2.15) would yield $z_\lambda := \lambda z + (1-\lambda)y \in \mathrm{ri}\,A$ for all $\lambda \in {]0,1]}$. So, $-\infty < \inf_{\mathrm{ri}\,A} f \leq f(z_\lambda) \leq \lambda f(z) + (1-\lambda)f(y) < +\infty$ for all $\lambda \in {]0,1]}$. Accordingly, $f(z) \in \mathbb{R}$, $f(y) \in \mathbb{R}$ and we are done by taking $\lambda \downarrow 0$.

Exercise 15: Since $f(x) \geq (\overline{\mathrm{co}}\,f)(x)$ for all $x \in X$, the liminf in (2.71) is nonnegative (possibly $+\infty$). Otherwise, there would exist $c > 0$ and $A > 0$ such that

$$\inf_{\|y\|\geq \|x\|} \frac{f(y) - (\overline{\mathrm{co}}\,f)(y)}{\|y\|} \geq c \text{ for all } x \in X \text{ such that } \|x\| > A.$$

Thus, $f(x) - (\overline{\mathrm{co}}\,f)(x) \geq c\|x\|$ when $\|x\| > A$, while $f(x) - (\overline{\mathrm{co}}\,f)(x) \geq 0$ otherwise. In short, $f \geq (\overline{\mathrm{co}}\,f)(\cdot) + c(\|\cdot\| - A)$. Since the function on the right-hand side is convex, we have $\overline{\mathrm{co}}\,f \geq (\overline{\mathrm{co}}\,f)(\cdot) + c(\|\cdot\| - A)$. But this is false for $\|x\| > A$, yielding a contradiction. Now observe that, certainly, the gap between $(\overline{\mathrm{co}}\,f)(x)$ and $f(x)$ when $\|x\| \to \infty$ may be larger and larger; this is the case of $f: x \in \mathbb{R} \mapsto f(x) := \sqrt{|x|}$.

Exercise 16: Fix $t_1, t_2, t_3 \in \mathrm{dom}\,f$ such that $t_1 < t_2 < t_3$. Then, taking $\lambda := \frac{t_2-t_1}{t_3-t_1}$, we have $\lambda \in {]0,1[}$ and $(1-\lambda)t_1 + \lambda t_3 = t_2$. Thus, $f(t_2) \leq (1-\lambda)f(t_1) + \lambda f(t_3)$ by the convexity of f, and we obtain $f(t_2) - f(t_1) \leq \lambda(f(t_3) - f(t_1))$ and $f(t_2) - f(t_3) \leq (1-\lambda)(f(t_1) - f(t_3))$. The desired inequalities follows by rearranging these terms.

Exercise 17: Apply (2.57) and (2.58).

Exercise 18: Applying (2.17) to $A = \mathrm{dom}\,f$ and $B = \mathrm{dom}\,g$, we get (2.73). Moreover, since $f_{|\mathrm{aff}(A)}$ is continuous on $\mathrm{ri}(A)$, and $\mathrm{aff}(A \cap B) \subset \mathrm{aff}(A)$, it follows that $f_{|\mathrm{aff}(A\cap B)}$ is continuous on $\mathrm{ri}(A) \cap \mathrm{aff}(A \cap B)$. Similarly, $g_{|\mathrm{aff}(A\cap B)}$ is continuous on $\mathrm{ri}(B) \cap \mathrm{aff}(A \cap B)$ so that $(f+g)_{|\mathrm{aff}(A\cap B)} = f_{|\mathrm{aff}(A\cap B)} + g_{|\mathrm{aff}(A\cap B)}$ is continuous on $\mathrm{ri}(A) \cap \mathrm{ri}(B) \cap \mathrm{aff}(A \cap B) = \mathrm{ri}(A \cap B) \cap \mathrm{aff}(A \cap B) = \mathrm{ri}(A \cap B)$.

Exercise 19: $(i) \Rightarrow (ii)$ If $x_0 \in \mathrm{dom}\,f \cap \mathrm{ri}(\mathrm{dom}\,g)$ and $g_{|\mathrm{aff}(\mathrm{dom}\,g)}$ is continuous on $\mathrm{ri}(\mathrm{dom}\,g)$, then, obviously, $(x_0, g(x_0)+1) \in \mathrm{ri}(\mathrm{epi}\,g)$. Now, for any $x \in \mathrm{dom}\,g$, one has $(x_0, g(x_0)+1) - (x, g(x)) \in \mathrm{ri}(\mathrm{epi}\,g - (x, g(x)))$ and $(x_0 - x, g(x_0) - g(x) + 1) \in (\mathrm{dom}\,f - x) \times \mathbb{R}$.

$(iii) \Rightarrow (i)$ Take

$$(z, \mu) \in (\mathrm{ri}(\mathrm{epi}\,g - (x_0, g(x_0)))) \cap ((\mathrm{dom}\,f - x_0) \times \mathbb{R})$$

for $x_0 \in \text{dom } g$. Since $(z, \mu) + (x_0, g(x_0)) \in \text{ri}(\text{epi } g)$, there exist $V \in \mathcal{N}_X$ and $\varepsilon > 0$ such that

$$(V \cap \{(\text{aff dom } g) - (z + x_0)\}) \times]g(x_0) + \mu - \varepsilon, g(x_0) + \mu + \varepsilon[\subset \text{epi } g.$$

Hence, $g_{|\text{aff}(\text{dom } g)}$ is continuous at $z + x_0 \in \text{ri}(\text{dom } g)$, and then on $\text{ri}(\text{dom } g)$, thanks to Proposition 2.2.6. In addition, since $(z, \mu) \in (\text{dom } f - x_0) \times \mathbb{R}$, we get $z + x_0 \in \text{dom } f$, and so $z + x_0 \in \text{ri}(\text{dom } g) \cap \text{dom } f$. We are done since $(ii) \Rightarrow (iii)$ obviously.

Exercise 20: The own definition of the lsc hull of a function yields $\text{cl}(f + g) \geq \text{cl } f + \text{cl } g$. To prove the opposite inequality, we take $x \in \text{dom}(\text{cl } f) \cap \text{dom}(\text{cl } g) \subset \text{cl}(\text{dom } f) \cap \text{cl}(\text{dom } g)$.

(i) We pick $x_0 \in \text{ri}(\text{dom } f) \cap \text{ri}(\text{dom } g)$ and denote $x_\lambda := \lambda x + (1 - \lambda)x_0$, $\lambda \in [0, 1[$. Then, since $x_\lambda \in \text{ri}(\text{dom } f) \cap \text{ri}(\text{dom } g)$, by the continuity assumption, we infer that $(\text{cl } f)(x_\lambda) = f(x_\lambda)$ and $(\text{cl } g)(x_\lambda) = g(x_\lambda)$. Consequently, by the convexity of f and g, for every $\lambda \in [0, 1[$

$$f(x_\lambda) + g(x_\lambda) = (\text{cl } f)(x_\lambda) + (\text{cl } g)(x_\lambda)$$
$$\leq \lambda((\text{cl } f)(x) + (\text{cl } g)(x)) + (1 - \lambda)((\text{cl } f)(x_0) + (\text{cl } g)(x_0)).$$

Thus, since $x_0 \in \text{dom } f \cap \text{dom } g \subset \text{dom}(\text{cl } f) \cap \text{dom}(\text{cl } g)$ and $\text{cl } f, \text{cl } g$ are necessarily proper, by taking limits for $\lambda \uparrow 1$, we get the desired inequality

$$\text{cl}(f+g)(x) \leq \liminf_{\lambda \uparrow 1}(f+g)(x_\lambda)$$
$$\leq \liminf_{\lambda \uparrow 1} \lambda((\text{cl } f)(x) + (\text{cl } g)(x)) + (1-\lambda)((\text{cl } f)(x_0) + (\text{cl } g)(x_0))$$
$$= (\text{cl } f)(x) + (\text{cl } g)(x).$$

(ii) We choose $x_0 \in \text{dom } f \cap \text{ri}(\text{dom } g)$ and denote x_λ as above so that $x_\lambda \in \text{dom}(\text{cl } f) \cap \text{ri}(\text{dom } g)$ and $(\text{cl } g)(x_\lambda) = g(x_\lambda)$. If $\text{cl } f$ is not proper, then $(\text{cl } f)(x_\lambda) = -\infty$ for all $\lambda \in [0, 1[$ and we get $\text{cl}(f + g)(x) = \text{cl}((\text{cl } f) + g)(x) \leq \liminf_{\lambda \uparrow 1}\{(\text{cl } f)(x_\lambda) + g(x_\lambda)\} = -\infty$, as required. Moreover, when $\text{cl } f$ is proper, we shall argue as in the proof of (i).

9.2 Exercises of chapter 3

Exercise 21: It suffices to prove the equality $\sigma_{\mathrm{dom}\, f} = \sigma_{\mathrm{dom}\, f^{**}}$ when f is convex and lsc. To show that it is sufficient to take $f^*(\theta) = 0$, we pick $x_0^* \in \mathrm{dom}\, f^*$ so that $f^*(x_0^*) \in \mathbb{R}$, as f^* is supposed to be proper. We consider the function $g := f - x_0^* + f^*(x_0^*)$. Then $g^* = f^*(\cdot + x_0^*) - f^*(x_0^*)$ and $g^{**} = f^{**} - x_0^* + f^*(x_0^*)$, and we obtain $g^*(\theta) = 0$, $\mathrm{dom}\, f = \mathrm{dom}\, g$, and $\mathrm{dom}\, f^{**} = \mathrm{dom}\, g^{**}$. Thus, providing that $\sigma_{\mathrm{dom}\, g^{**}} = \sigma_{\mathrm{dom}\, g}$, we deduce that $\sigma_{\mathrm{dom}\, f^{**}} = \sigma_{\mathrm{dom}\, g^{**}} = \sigma_{\mathrm{dom}\, g} = \sigma_{\mathrm{dom}\, f}$.

Exercise 22: It is clear that $\max\{\sigma_{A_{k+1}}, \ldots, \sigma_{A_m}\} = \sigma_{A_{k+1} \cup \ldots \cup A_m}$ and, knowing that $\sigma_A + \sigma_B = \sigma_{A+B}$ for every pair of sets $A, B \subset X$,

$$\sigma_{A_1} + \ldots + \sigma_{A_k} + \max\{\sigma_{A_{k+1}}, \ldots, \sigma_{A_m}\} = \sigma_{A_1 + \ldots + A_k} + \sigma_{A_{k+1} \cup \ldots \cup A_m}$$
$$= \sigma_{A_1 + \ldots + A_k + (A_{k+1} \cup \ldots \cup A_m)}.$$

Next, observing that

$$\mathrm{dom}\, \sigma_{A_1 + \ldots + A_m} = \mathrm{dom}(\sigma_{A_1} + \ldots + \sigma_{A_m})$$
$$= \mathrm{dom}(\max\{\sigma_{A_i}, i = 1, \ldots, m\}) = \mathrm{dom}(\sigma_{A_1 \cup \ldots \cup A_m}),$$

by (3.52), we get

$$([\overline{\mathrm{co}}(A_1 \cup \ldots \cup A_m)]_\infty)^- = \mathrm{cl}(\mathrm{dom}\, \sigma_{A_1 \cup \ldots \cup A_m})$$
$$= \mathrm{cl}(\mathrm{dom}\, \sigma_{A_1 + \ldots + A_m}) = ([\overline{\mathrm{co}}(A_1 + \ldots + A_m)]_\infty)^-.$$

Thus, the first equality in (3.72) follows from (3.51). The second equality in (3.72) comes from the first one and the following observation

$$[\overline{\mathrm{co}}(A_1 \cup \ldots \cup A_m)]_\infty = [\overline{\mathrm{co}}(A_1 \cup \ldots \cup A_k \cup (A_{k+1} \cup \ldots \cup A_m))]_\infty$$
$$= [\overline{\mathrm{co}}(A_1 + \ldots + A_k + (A_{k+1} \cup \ldots \cup A_m))]_\infty.$$

Exercise 23: Denoting $T := T_1 \cup T_2$ and $A := \overline{\mathrm{co}}\left(\bigcup_{t \in T} A_t\right)$, the functions $\varphi_1 := \sup_{t \in T_1} \sigma_{A_t}$ and $\varphi_2 := \sup_{t \in T_2} \sigma_{\rho A_t}$ satisfy $\max\{\varphi_1, \varphi_2\} = \sigma_{(\bigcup_{t \in T_1} A_t) \cup (\bigcup_{t \in T_2} \rho A_t)}$ and

$$\begin{aligned}\varphi_1 + \varphi_2 &= \sup_{t_1 \in T_1, t_2 \in T_2} (\sigma_{A_{t_1}} + \sigma_{\rho A_{t_2}}) \\ &= \sup_{t_1 \in T_1, t_2 \in T_2} \sigma_{A_{t_1} + \rho A_{t_2}} = \sigma_{\bigcup_{t_1 \in T_1, t_2 \in T_2}(A_{t_1} + \rho A_{t_2})}.\end{aligned} \quad (9.1)$$

Moreover, since $\text{dom}(\varphi_1 + \varphi_2) = \text{dom}(\max\{\varphi_1, \varphi_2\}) = \text{dom}(\max\{\varphi_1, \rho^{-1}\varphi_2\})$, (3.52) yields

$$\left(\left[\overline{\text{co}}\left(\bigcup_{t_1 \in T_1,\, t_2 \in T_2}(A_{t_1} + \rho A_{t_2})\right)\right]_\infty\right)^- = \left(\left[\overline{\text{co}}\left(\left(\bigcup_{t \in T_1} A_t\right) \cup \left(\bigcup_{t \in T_2} \rho A_t\right)\right)\right]_\infty\right)^-$$

$$= \left(\left[\overline{\text{co}}\left(\left(\bigcup_{t \in T_1} A_t\right) \cup \left(\bigcup_{t \in T_2} A_t\right)\right)\right]_\infty\right)^-,$$

and we are done; thanks to (3.51).

Exercise 24: Fix $x \in \text{dom}\,\sigma_A$ and $\varepsilon \geq 0$. The first equality comes from (3.52). Take $x^* \in \text{N}^\varepsilon_{([\overline{\text{co}}A]_\infty)^-}(x)$, so that $\langle x^*, \alpha y - x \rangle \leq \varepsilon$ for all $y \in ([\overline{\text{co}}A]_\infty)^-$ and $\alpha > 0$. Then, dividing by α and next making $\alpha \uparrow +\infty$, by (3.51), we obtain $x^* \in ([\overline{\text{co}}A]_\infty)^{--} = [\overline{\text{co}}A]_\infty$. Moreover, since $\langle x^*, (\alpha - 1)x \rangle \leq \varepsilon$ and $x \in \text{dom}\,\sigma_A \subset ([\overline{\text{co}}A]_\infty)^-$ (by (3.52)), it follows that $\langle x^*, (\alpha - 1)x \rangle \leq \varepsilon$ for all $\alpha \geq 0$. Hence, taking $\alpha = 0$, we get $-\varepsilon \leq \langle x^*, x \rangle$, while $\langle x^*, x \rangle \leq 0$ when $\alpha \to +\infty$. By summarizing, we have proved that $x^* \in [\overline{\text{co}}A]_\infty \cap \{x^* \in X^* : -\varepsilon \leq \langle x^*, x \rangle \leq 0\}$. Conversely, if $x^* \in [\overline{\text{co}}A]_\infty \cap \{x^* : -\varepsilon \leq \langle x^*, x \rangle \leq 0\}$, then $\langle x^*, y - x \rangle \leq \langle x^*, -x \rangle \leq \varepsilon$ for all $y \in ([\overline{\text{co}}A]_\infty)^-$, and so $x^* \in \text{N}^\varepsilon_{([\overline{\text{co}}A]_\infty)^-}(x)$.

Exercise 25: (i) We have $\text{epi}\,f_A = (A \times \mathbb{R}) \times \text{epi}(\frac{1}{2}\|\cdot\|^2)$, and so f_A is convex if and only if A is convex.

(ii) If A is closed, then obviously f_A is lsc. Conversely, if f is lsc, then $f_A = \text{cl}\,f_A = \text{cl}(\text{I}_A + \frac{1}{2}\|\cdot\|^2)$ and the continuity of the norm ensures that $\text{I}_A + \frac{1}{2}\|\cdot\|^2 = \text{cl}(\text{I}_A) + \frac{1}{2}\|\cdot\|^2 = \text{I}_{\text{cl}\,A} + \frac{1}{2}\|\cdot\|^2$; that is, A is closed.

(iii) From the relation $\text{cl}(\text{I}_A + \frac{1}{2}\|\cdot\|^2) = \text{I}_{\text{cl}\,A} + \frac{1}{2}\|\cdot\|^2$, it follows that $\text{cl}\,f_A$ is convex if and only if $\text{cl}\,A$ is convex.

(iv) Assume that $\text{cl}^w f_A$ is convex. Given $\lambda \in\,]0, 1[$ and $x_1, x_2 \in A$, the relation $\text{cl}^w f \geq \text{I}_{\text{cl}^w A} + \frac{1}{2}\|\cdot\|^2$ yields

$$\begin{aligned}\text{I}_{\text{cl}^w A}(\lambda x_1 + (1-\lambda)x_2) &\leq (\text{cl}^w f)(\lambda x_1 + (1-\lambda)x_2) \\ &\leq \lambda(\text{cl}^w f)(x_1) + (1-\lambda)(\text{cl}^w f)(x_2) \\ &\leq \lambda f(x_1) + (1-\lambda)f(x_2) \\ &\leq \frac{\lambda}{2}\|x_1\|^2 + \frac{(1-\lambda)}{2}\|x_2\|^2,\end{aligned}$$

so that $\lambda x_1 + (1-\lambda)x_2 \in \text{cl}^w A$. More generally, if $x_1, x_2 \in \text{cl}^w A$ and the nets $(x_{1,i})_i$, $(x_{2,j})_j \subset A$ weakly converge to x_1 and x_2, respectively, then the last arguments show that $\lambda x_{1,i} + (1-\lambda)x_{2,j} \in \text{cl}^w A$ for all i, j. Hence, by taking the limits on i and j, we obtain $\lambda x_1 + (1-\lambda)x_2 \in \text{cl}^w A$, proving the convexity of $\text{cl}^w A$.

9.2 EXERCISES OF CHAPTER 3

(v) We have $x \in (\partial f_A)^{-1}(y)$ if and only if $f_A(x) + (f_A)^*(y) = \langle y, x \rangle$; that is, by assertion (i), if and only if $f_A(x) + (1/2)\|y\|^2 - (1/2)d_A^2(y) = \langle y, x \rangle$. Therefore, $x \in (\partial f_A)^{-1}(y)$ if and only if $x \in A$ and

$$(1/2)\|x-y\|^2 = (1/2)\|x\|^2 - \langle y, x \rangle + (1/2)\|y\|^2 = (1/2)d_A^2(y).$$

In other words, $x \in (\partial f_A)^{-1}(y)$ if and only if $x \in \pi_A(y)$.

Exercise 26: First, taking into account that U is absorbing and U° is w^*-compact, for each $U \in \mathcal{N}_X$, we have $\mathrm{dom}(f \square \sigma_{U^\circ}) = \mathrm{dom}\, f + \mathrm{dom}\, \sigma_{U^\circ} = \mathrm{dom}\, f + X$ and $(f \square \sigma_{U^\circ})(x+y) \le f(x) + \sigma_{U^\circ}(y) \le f(x) + 1 < +\infty$ for all $x \in \mathrm{dom}\, f$ and $y \in U$; that is, $f \square \sigma_{U^\circ}$ is continuous and $g := \sup_{U \in \mathcal{N}_X} f \square \sigma_{U^\circ}$ is lsc on X. Let $\langle x_0^*, \cdot \rangle + \alpha_0$ be such that $f + \langle x_0^*, \cdot \rangle + \alpha_0 \ge 0$. Given $x \in \mathrm{dom}\, g$ and $m \ge 1$, we choose $n \ge 1$ such that

$$\frac{n}{2(n-1)} \le 1 \text{ and } \max\{g(x), -m\} + \frac{1}{n} + \langle x_0^*, x \rangle + \alpha_0 \le \frac{n}{2}. \quad (9.2)$$

Next, given $U \in \mathcal{N}_X^0 := \{V \in \mathcal{N}_X : \sigma_V(-x_0^*) \le 1\}$, we choose $y_{n,U} \in X$ such that

$$\langle x_0^*, y_{n,U} - x \rangle - \alpha_0 + \sigma_{(\frac{1}{n}U)^\circ}(y_{n,U}) \le f(x - y_{n,U}) + \sigma_{(\frac{1}{n}U)^\circ}(y_{n,U})$$
$$\le \max\{f \square \sigma_{(\frac{1}{n}U)^\circ}(x), -m\} + \frac{1}{n}. \quad (9.3)$$

Thus, since $\langle -x_0^*, y_{n,U} \rangle \le \sigma_U(-x_0^*)\sigma_{U^\circ}(y_{n,U}) \le \sigma_{U^\circ}(y_{n,U})$ due to (3.54), (9.2) entails

$$\sigma_{(\frac{1}{n}U)^\circ}(y_{n,U}) \le \max\{f \square \sigma_{(\frac{1}{n}U)^\circ}(x), -m\} + \frac{1}{n} +$$
$$+ \langle x_0^*, x \rangle + \alpha_0 + \sigma_{U^\circ}(y_{n,U}) \le \frac{n}{2} + \sigma_{U^\circ}(y_{n,U}),$$

and we deduce that $\sigma_{U^\circ}(y_{n,U}) = \frac{1}{n}\sigma_{(\frac{1}{n}U)^\circ}(y_{n,U}) \le \frac{1}{2} + \frac{1}{n}\sigma_{U^\circ}(y_{n,U})$, and so $\sigma_{U^\circ}(y_{n,U}) \le \frac{n}{2(n-1)} \le 1$; that is, $y_{n,U} \in U$. Next, since the set $\mathbb{N} \times \mathcal{N}_X^0$ can be made a directed set through the partial order given by $(n_1, U_1) \preccurlyeq (n_2, U_2)$ if and only if $n_1 \le n_2$ and $U_2 \subset U_1$, we infer that the net $(y_{n,U})_{n,U}$ converges to θ. Consequently, using the second inequality in (9.3), we get

$$\max\{g(x), -m\} \geq \max\left\{\sup_{n\geq 1,\ U\in\mathcal{N}_X} (f\Box\sigma_{(\frac{1}{n}U)^\circ})(x), -m\right\}$$

$$\geq \sup_{n\geq 1,\ U\in\mathcal{N}_X^0} \max\left\{(f\Box\sigma_{(\frac{1}{n}U)^\circ})(x), -m\right\}$$

$$\geq \liminf_{n\to\infty,\ U\in\mathcal{N}_X^0} \left(f(x-y_{n,U}) + \sigma_{(\frac{1}{n}U)^\circ}(y_{n,U}) - \frac{1}{n}\right)$$

$$\geq \liminf_{n\to\infty} f(x-y_{n,U}) \geq (\operatorname{cl} f)(x).$$

Thus, letting $m \to +\infty$ we obtain that $g(x) \geq (\operatorname{cl} f)(x)$. Since this inequality obviously holds when $g(x) = +\infty$, we deduce that $\operatorname{cl} f \leq g \leq f$. Finally, by taking the closure in each side of these inequalities, and remembering that g is lsc, we conclude that $\operatorname{cl} f = \operatorname{cl} g = g$.

Exercise 27: Because $\emptyset \neq \operatorname{dom}(\operatorname{cl} f) \supset \operatorname{dom} f$ and $\operatorname{cl} f$ is not proper, $\operatorname{cl} f$ takes the value $-\infty$, and so $(\alpha f)^* = (\alpha(\operatorname{cl} f))^* \equiv +\infty$ for all $\alpha > 0$. Thus, $\min_{\alpha\geq 0}(\alpha f)^* = (0f)^* = (I_{\operatorname{dom} f})^* = \sigma_{\operatorname{dom} f}$. Moreover, since $(\operatorname{cl} f)(x) = -\infty$ for all $z \in \operatorname{cl}(\operatorname{dom} f)\ (\neq \emptyset)$, we have that $[\operatorname{cl} f \leq 0] = \operatorname{cl}(\operatorname{dom} f)$. In addition, we must have that $\inf_X f < 0$; otherwise, we would have the contradiction $-\infty = \inf_X(\operatorname{cl} f) = \inf_X f \geq 0$. Therefore, $\sigma_{[f\leq 0]} = \sigma_{\operatorname{cl}[f\leq 0]} = \sigma_{[\operatorname{cl} f\leq 0]} = \sigma_{\operatorname{cl}(\operatorname{dom} f)} = \sigma_{\operatorname{dom} f}$, by Lemma 3.3.3 and the first statement follows. At the same time, since $\operatorname{epi} f^* = \emptyset$, we deduce that $\operatorname{epi} \sigma_{[f\leq 0]} = \operatorname{epi} \sigma_{\operatorname{dom} f} = \left(\mathbb{R}_+^* \operatorname{epi} f^*\right) \cup \operatorname{epi} \sigma_{\operatorname{dom} f}$, giving rise to the second statement.

Exercise 28: The current assumption implies that $\sigma_A + \sigma_B = \sigma_{A+B} \leq \sigma_{A+C} = \sigma_A + \sigma_C$. Since A is bounded, σ_A is finite-valued and we deduce $\sigma_B \leq \sigma_C$. Then, Theorem 3.2.2 entails $I_{\overline{\operatorname{co}}C} = (\sigma_C)^* \leq (\sigma_B)^* = (\sigma_{\overline{\operatorname{co}}B})^* = I_{\overline{\operatorname{co}}B}$, and we get $B \subset \overline{\operatorname{co}}B \subset \overline{\operatorname{co}}C$.

Exercise 29: First, we prove that $[\overline{\operatorname{co}}(A_1 + A_2)]_\infty \subset \mathbb{R}_+\{e_1\}$. Let $u := (u_i)_{i\geq 1} \in [\overline{\operatorname{co}}(A_1 + A_2)]_\infty = [\overline{\operatorname{co}}(A_1 \cup A_2)]_\infty$, by Exercise 22, so that $-e_1 + \alpha u \in \overline{\operatorname{co}}(A_1 \cup A_2)$ for all $\alpha > 0$. Hence, for each $\alpha > 0$,

$$u = \alpha^{-1}e_1 + \lim_k \left(\sum_{j\in J_{\alpha,k}} \alpha^{-1}\lambda_{\alpha,k,j}(i_j e_1 + 2i_j e_{i_j}) - \sum_{l\in L_{\alpha,k}} \alpha^{-1}\lambda_{\alpha,k,l}i_l e_{i_l}\right),$$

where $J_{\alpha,k}$ and $L_{\alpha,k}$ are finite subsets of $\mathbb{N}$, such that one of them can be empty, and $\lambda_{\alpha,k,j}, \lambda_{\alpha,k,l}$ are positive real numbers such that $\sum_{j\in J_{\alpha,k}} \lambda_{\alpha,k,j} + \sum_{l\in L_{\alpha,k}} \lambda_{\alpha,k,l} = 1$.

Let us check that $u_i = 0$ for all $i \geq 2$, for instance, that $u_2 = 0$. We distinguish two cases: a) First, if $2 \notin J_{\alpha,k} \cup L_{\alpha,k}$ for some $\alpha > 0$, then for infinitely many k, we get $u_2 = 0$. b) If for each $\alpha > 0$, we have $2 \in J_{\alpha,k} \cup L_{\alpha,k}$ eventually on k, then $u_2 = 0 + \lim_k$

9.2 EXERCISES OF CHAPTER 3

$\left(\sum_{j\in J_{\alpha,k},\ i_j=2} 4\alpha^{-1}\lambda_{\alpha,k,j} + \sum_{l\in L_{\alpha,k},\ i_l=2} 2\alpha^{-1}\lambda_{\alpha,k,l}\right)$ and we deduce that $0 \le u_2 \le 6\alpha^{-1}$; that is, $u_2 = 0$ (by taking the limit as $\alpha \to \infty$). Next, we prove that $\mathbb{R}_+\{e_1\} \subset [\overline{\text{co}}(A_1 + A_2)]_\infty$ or, equivalently, $e_1 \in [\overline{\text{co}}(A_1 + A_2)]_\infty$. Suppose for the contrary that $e_1 \notin [\overline{\text{co}}(A_1 + A_2)]_\infty$. Since, $[\overline{\text{co}}(A_1 + A_2)]_\infty = (\operatorname{dom}\sigma_{A_1+A_2})^-$, by (3.77), we find some $z \in \operatorname{dom}\sigma_{A_1+A_2}$ ($\subset \ell_\infty$) such that $z_1 := \langle z, e_1 \rangle > 0$. Observe that $\operatorname{dom}\sigma_{A_1+A_2} = \operatorname{dom}(\max(\sigma_{A_1}, \sigma_{A_2}))$, and so $z \in \operatorname{dom}(\max(\sigma_{A_1}, \sigma_{A_2}))$; that is, $\sigma_{A_1}(z), \sigma_{A_2}(z) \in \mathbb{R}$. More specifically, we have that $\sigma_{A_1}(z) = \sup_{i\ge 1}\{iz_1 + 2iz_i\}$ and $\sigma_{A_2}(z) = \sup_{i\ge 1}\{-iz_i\}$. Hence,

$$\sigma_{A_1}(z) = \sup_{i\ge 1}\{iz_1 + 2iz_i\} \ge \sup_{i\ge 1}\{iz_1\} - \sup_{i\ge 1}\{-2iz_i\}$$
$$= \sup_{i\ge 1}\{iz_1\} - 2\sigma_{A_2}(z) = +\infty,$$

and this constitutes a contradiction since $\sigma_{A_1}(z) < +\infty$.

Exercise 30: Since $\operatorname{dom}\sigma_A$ is a cone containing z, we write $[\overline{\text{co}}A]_\infty \cap \{z\}^\perp = (\operatorname{cl}(\operatorname{dom}\sigma_A))^\circ \cap (\mathbb{R}z)^\circ = (\operatorname{dom}\sigma_A)^\circ \cap (\mathbb{R}z)^\circ$, and so

$$[\overline{\text{co}}A]_\infty \cap \{z\}^\perp = ((\operatorname{dom}\sigma_A) + \mathbb{R}z)^\circ = ((\operatorname{dom}\sigma_A) - \mathbb{R}_+z)^\circ = N_{\operatorname{dom}\sigma_A}(z).$$

Exercise 31: Since f is proper, we have that $(\inf_{t\in T} f_t^*)^* = \sup_{t\in T} f_t^{**} = \sup_{t\in T} f_t = f$; hence $f^* = (\inf_{t\in T} f_t^*)^{**} = \overline{\text{co}}(\inf_{t\in T} f_t^*)$ and
$\operatorname{epi} f^* = \overline{\text{co}}(\bigcup_{t\in T} \operatorname{epi} f_t^*)$. Thus, since $[\operatorname{epi} f^*]_\infty = \operatorname{epi}(\sigma_{\operatorname{dom} f})$,

$$N_{\operatorname{dom} f}(x) = \{x^* \in X^* : (x^*, \langle x^*, x \rangle) \in \operatorname{epi}(\sigma_{\operatorname{dom} f})\}$$
$$= \left\{ x^* \in X^* : (x^*, \langle x^*, x \rangle) \in \left[\overline{\text{co}}\left(\bigcup_{t\in T} \operatorname{epi} f_t^*\right)\right]_\infty \right\};$$

that is, (3.79) holds. Now we denote by $E_1(x)$ and $E_2(x)$ the sets appearing in the right-hand sides of (3.78) and (3.79), respectively. To prove the inclusion $E_2(x) \subset E_1(x)$, observe that $[\overline{\text{co}}(\bigcup_{t\in T} \operatorname{gph} f_t^*)]_\infty \cap (-[\mathbb{R}_+(\theta, 1)]_\infty) = \{(\theta, 0)\}$ and

$$\overline{\text{co}}\left(\bigcup_{t\in T}\operatorname{gph} f_t^*\right) \subset \operatorname{cl}\left[\overline{\text{co}}\left(\bigcup_{t\in T}\operatorname{gph} f_t^*\right) + \mathbb{R}_+(\theta, 1)\right] = \overline{\text{co}}\left(\bigcup_{t\in T}\operatorname{epi} f_t^*\right) = \operatorname{epi} f^*.$$

Thus, since f^* is proper the set $\overline{\text{co}}(\bigcup_{t\in T} \operatorname{gph} f_t^*) + \mathbb{R}_+(\theta, 1)$ is closed. Whence,

$$\left[\overline{\mathrm{co}}\left(\bigcup_{t\in T}\mathrm{epi}\,f_t^*\right)\right]_\infty = \left[\overline{\mathrm{co}}\left(\bigcup_{t\in T}\mathrm{gph}\,f_t^*\right)+\mathbb{R}_+(\theta,1)\right]_\infty = \left[\overline{\mathrm{co}}\left(\bigcup_{t\in T}\mathrm{gph}\,f_t^*\right)\right]_\infty + \mathbb{R}_+(\theta,1).$$

Take $x^* \in E_2(x)$, so that $(x^*, \langle x^*, x\rangle) = (u^*, \eta + \lambda)$ for some $(u^*, \eta) \in [\overline{\mathrm{co}}\,(\bigcup_{t\in T}\mathrm{gph}\,f_t^*)]_\infty$ and $\lambda \geq 0$. Then, since

$$\mathrm{dom}\,f \times \{-1\} \subset \mathrm{dom}\,(\sigma_{\mathrm{epi}\,f^*}) \subset ([\mathrm{epi}\,f^*]_\infty)^-,$$

we obtain

$$\mathrm{dom}\,f \times \{-1\} \subset ([\overline{\mathrm{co}}\,(\bigcup_{t\in T}\mathrm{epi}\,f_t^*)]_\infty)^- \subset ([\overline{\mathrm{co}}\,(\bigcup_{t\in T}\mathrm{gph}\,f_t^*)]_\infty)^-,$$

which yields $\langle (u^*, \eta), (x, -1)\rangle \leq 0$. In addition, we have $x^* = u^*$ and so

$$\lambda = \langle (x^*, \eta), (x, -1)\rangle = \langle (u^*, \eta), (x, -1)\rangle \leq 0;$$

that is, $\lambda = 0$, and so $(x^*, \langle x^*, x\rangle) = (u^*, \eta) \in [\overline{\mathrm{co}}\,(\bigcup_{t\in T}\mathrm{gph}\,f_t^*)]_\infty$; that is, $x^* \in E_1(x)$.

Exercise 32: (a) If $F \neq \emptyset$, then $\mathrm{I}_F(x) \geq f(x)$ for all $x \in X$, and consequently, $\sigma_F = \mathrm{I}_F^* \leq f^*$, entailing $\mathrm{epi}\,f^* \subset \mathrm{epi}\,\sigma_F$. Then, as $\mathrm{epi}\,\sigma_F$ is a closed convex cone, $\mathrm{cl}(\mathrm{cone}(\mathrm{epi}\,f^*)) \subset \mathrm{epi}\,\sigma_F$. Now, since $(\theta, -1) \notin \mathrm{epi}\,\sigma_F$, we get $(\theta, -1) \notin \mathrm{cl}(\mathrm{cone}(\mathrm{epi}\,f^*))$. Conversely, if $(\theta, -1) \notin \mathrm{cl}(\mathrm{cone}(\mathrm{epi}\,f^*))$, by the separation theorem there will exist $(x, \lambda) \in X \times \mathbb{R}$ such that $\langle (\theta, -1), (x, \lambda)\rangle = -\lambda < 0$ and

$$\langle (x^*, \alpha), (x, \lambda)\rangle = \langle x^*, x\rangle + \alpha\lambda \geq 0 \text{ for all } (x^*, \alpha) \in \mathrm{cl}(\mathrm{cone}(\mathrm{epi}\,f^*)). \tag{9.4}$$

Now, if $\overline{x} := (1/\lambda)x$, (9.4) yields $\langle x^*, \overline{x}\rangle + \alpha \geq 0$ for all $(x^*, \alpha) \in \mathrm{cl}(\mathrm{cone}(\mathrm{epi}\,f^*))$. Then, if $x^* \in \mathrm{dom}\,f^*$ (remember that $f^* \in \Gamma_0(X^*)$), we have $\langle x^*, \overline{x}\rangle + f^*(x^*) \geq 0$; equivalently, $\langle x^*, -\overline{x}\rangle - f^*(x^*) \leq 0$. Hence, $f(-\overline{x}) = f^{**}(-\overline{x}) = \sup_{x^* \in \mathrm{dom}\,f^*}\{\langle x^*, -\overline{x}\rangle - f^*(x^*)\} \leq 0$ and $-\overline{x} \in F$.

(b) In (a), we proved the inclusion $\mathrm{cl}(\mathrm{cone}\,\mathrm{epi}\,f^*) \subset \mathrm{epi}\,\sigma_F$. To establish the converse inclusion, suppose that $(x^*, \alpha) \notin \mathrm{cl}(\mathrm{cone}\,\mathrm{epi}\,f^*)$. Also, from (a), we have that $(\theta, -1) \notin \mathrm{cl}(\mathrm{cone}(\mathrm{epi}\,f^*))$. We consider the segment $B := \mathrm{co}\{(x^*, \alpha), (\theta, -1)\}$ so that $B \cap \mathrm{cl}(\mathrm{cone}(\mathrm{epi}\,f^*)) = \emptyset$; otherwise, $B \cap \mathrm{cl}(\mathrm{cone}\,\mathrm{epi}\,f^*) \neq \emptyset$ would yield $\delta_0(x^*, \alpha) + (1 - \delta_0)(\theta, -1) \in \mathrm{cl}(\mathrm{cone}(\mathrm{epi}\,f^*))$ for some $\delta_0 \in]0, 1[$. Thus, since $(\theta, 1) \in \mathrm{cl}(\mathrm{cone}\,\mathrm{epi}\,f^*)$, we deduce that

$$\delta_0(x^*, \alpha) = \delta_0(x^*, \alpha) + (1 - \delta_0)(\theta, -1) + (1 - \delta_0)(\theta, 1) \in \mathrm{cl}(\mathrm{cone}(\mathrm{epi}\,f^*))$$

9.2 EXERCISES OF CHAPTER 3

and we get the contradiction $(x^*, \alpha) \in \text{cl}(\text{cone}(\text{epi } f^*))$ as this set is a cone. Next, taking into account that B is compact, we apply the separation theorem to conclude the existence of $(x, \rho) \in X \times \mathbb{R}$ satisfying $\langle (z^*, \gamma), (x, \rho) \rangle \geq 0$, for all $(z^*, \gamma) \in \text{cl}(\text{cone}(\text{epi } f^*))$, and

$$\begin{aligned}\langle \delta(x^*, \alpha) + (1-\delta)(\theta, -1), (x, \rho) \rangle &= \delta \langle x^*, x \rangle \\ +(\delta\alpha - (1-\delta))\rho &< 0 \text{ for all } \delta \in [0,1],\end{aligned} \quad (9.5)$$

In particular, $\rho > 0$ and $\langle x^*, x \rangle + \alpha\rho < 0$; equivalently, $\langle x^*, (1/\rho)(-x) \rangle > \alpha$. Additionally, it follows that $\langle (z^*, f^*(z^*)), ((1/\rho)(-x), -1) \rangle \leq 0$ for all $z^* \in \text{dom } f^*$; i.e., $\langle (z^*, (1/\rho)(-x)) \rangle - f^*(z^*) \leq 0$ for all $z^* \in \text{dom } f^*$, entailing $(1/\rho)(-x) \in F$ and

$$f((1/\rho)(-x)) = f^{**}((1/\rho)(-x)) = \sup_{z^* \in \text{dom } f^*} \{\langle z^*, (1/\rho)(-x) \rangle - f^*(z^*)\} \leq 0.$$

Finally, $\sigma_F(x^*) \geq \langle x^*, (1/\rho)(-x) \rangle > \alpha$ and $(x^*, \alpha) \notin \text{epi } \sigma_F$, and the inclusion $\text{epi } \sigma_F \subset \text{cl}(\text{cone}(\text{epi } f^*))$ is proved.

Exercise 33: Since $f^+ = \max\{f, 0\} = \sup_{\lambda \in]0,1]} \lambda f$ and $\lambda f \in \Gamma_0(X)$ for all $0 < \lambda \leq 1$, Proposition 3.2.6 yields $(f^+)^* = \overline{\text{co}}(\inf_{\lambda \in]0,1]}(\lambda f)^*)$.

Exercise 34: Let $g : X^* \to \overline{\mathbb{R}}$ denote the function defined by

$$g(x^*) := \inf\left\{\mathrm{I}_{\left\{\sum_{1 \leq i \leq k} \gamma_i a_i\right\}}(x^*) + \sum_{1 \leq i \leq k} \gamma_i b_i : \gamma \in \Delta_k\right\},$$

which is easily shown to be in $\Gamma_0(X^*)$, due to the compactness of Δ_k. Also, the last infimum is attained, and so

$$g(x^*) = \inf\left\{\sum_{1 \leq i \leq k} \gamma_i b_i : \sum_{1 \leq i \leq k} \gamma_i a_i = x^*, \ \gamma \in \Delta_k\right\}$$

$$= \min\left\{\sum_{1 \leq i \leq k} \gamma_i b_i : \sum_{1 \leq i \leq k} \gamma_i a_i = x^*, \ \gamma \in \Delta_k\right\}.$$

Then, by (3.10) and (2.45),

$$g^*(x) = \sup\left\{\left\langle \sum_{1 \leq i \leq k} \gamma_i a_i, x \right\rangle - \sum_{1 \leq i \leq k} \gamma_i b_i : \gamma \in \Delta_k\right\} = f(x),$$

and Theorem 3.2.2 leads us to $f^*(x^*) = g^{**}(x^*) = g(x^*)$.

Exercise 35: By the current assumption on A_0,

$$\sup_{x \in A} f(x, y) = +\infty = \sup_{x \in A_0} f(x, y) = +\infty$$

for all $y \in B$, and the upper semicontinuity and concavity of the functions $f(\cdot, y)$, $y \in B$, yield elements $x(y) \in A$, $y \in B$, such that $f(x(y), y) = +\infty$. Hence, the usc concave functions $f(\cdot, y)$, $y \in B$, are non-proper (equivalently, the convex function $-f(\cdot, y)$ is non-proper) and, consequently, they only take infinite values (either $+\infty$ or $-\infty$). But, since we are also assuming that $A_0 \neq \emptyset$, we have $f(x, y) > -\infty$ for all $x \in A_0$ and $y \in B$, and so $f(x, y) = +\infty$ for all $x \in A_0$ and $y \in B$. Therefore, using again the upper semicontinuity and concavity assumptions on f,

$$\max_{x \in A} \inf_{y \in B} f(x, y) = \sup_{x \in A} \inf_{y \in B} f(x, y)$$
$$\geq \sup_{x \in A_0} \inf_{y \in B} f(x, y) = +\infty = \inf_{y \in B} \sup_{x \in A_0} f(x, y),$$

and the proof is finished.

9.3 Exercises of chapter 4

Exercise 36: If $m = \inf_X f \notin \mathbb{R}$, we have $\varepsilon - \operatorname{argmin} f = \emptyset$, and so $\sigma_{\varepsilon-\operatorname{argmin} f} = -\infty$, which entails $\operatorname{dom}(\sigma_{\varepsilon-\operatorname{argmin} f}) = X^*$, and the inclusion $\operatorname{cone}(\operatorname{dom} f^*) \subset \operatorname{dom}(\sigma_{\varepsilon-\operatorname{argmin} f})$ is trivially satisfied. Assume, then, that $m \in \mathbb{R}$. For any $x^* \in \operatorname{dom} f^*$, if we take $x \in \varepsilon - \operatorname{argmin} f$, we have

$$\langle x^*, x \rangle \leq f(x) + f^*(x^*) \leq m + \varepsilon + f^*(x^*).$$

Therefore, $\sigma_{\varepsilon-\operatorname{argmin} f}(x^*) \leq m + \varepsilon + f^*(x^*) < +\infty$. Hence,

$$\operatorname{dom} f^* \subset \operatorname{dom}(\sigma_{\varepsilon-\operatorname{argmin} f})$$

and we are done since the set on the right-hand side is a cone.

Exercise 37: By (3.10), we have that

$$(f'_\varepsilon(x; \cdot))^* = \sup_{s > 0} \left(\frac{f(x + s \cdot) - f(x) + \varepsilon}{s} \right)^*$$
$$= \sup_{s > 0} \left((s^{-1} f(x + s \cdot))^* + \frac{f(x) - \varepsilon}{s} \right).$$

9.3 EXERCISES OF CHAPTER 4

At the same time, we have

$$\left(\frac{f(x+s\cdot)}{s}\right)^*(x^*) = s^{-1}\sup_{z\in X}\{\langle x^*, x+sz\rangle - f(x+sz)\} - s^{-1}\langle x^*, x\rangle$$
$$= s^{-1}(f^*(x^*) - \langle x, x^*\rangle),$$

and we are done.

Exercise 38: Take $u^* \in N_{\text{dom }f}(x)$ and fix $x_0^* \in \partial_\varepsilon f(x)$. Then, for every $\lambda > 0$, we have that $\langle x_0^* + \lambda u^*, y - x\rangle \leq \langle x_0^*, y - x\rangle \leq f(y) - f(x) + \varepsilon$ for all $y \in \text{dom } f$; that is, $x_0^* + \lambda u^* \in \partial_\varepsilon f(x)$ and $u^* \in [\partial_\varepsilon f(x)]_\infty$. Conversely, if $u^* \in [\partial_\varepsilon f(x)]_\infty$, then $\langle x_0^* + \lambda u^*, y - x\rangle \leq f(y) - f(x) + \varepsilon$ for all $y \in \text{dom } f$, $\lambda \geq 0$ and, by dividing by λ and making $\lambda \to +\infty$, we deduce that $u^* \in N_{\text{dom }f}(x)$.

Exercise 39: First, assume that $X = \mathbb{R}^n$. If f is proper, then $\text{ri}(\text{dom } f) \neq \emptyset$ and $\partial f(x) \neq \emptyset$ for all $x \in \text{ri}(\text{dom } f)$. Hence, if $u \in \partial f(x) = \partial(\text{cl } f)(x)$, $x \in \text{ri}(\text{dom } f)$, then $\text{cl } f$ is minorized by the (continuous) affine mapping $f(x) + \langle u, \cdot - x\rangle$, and so it is proper. Conversely, if $\text{cl } f$ is proper, then from the inequality $f \geq \text{cl } f$, we deduce that f is also proper. Now, consider the space $c_0 := \{(x_n)_{n\geq 1} : x_n \in \mathbb{R}, x_n \to 0\}$, which is a subspace of ℓ_∞ such that $(c_0)^* = \ell_1$, and define the proper convex function $f : c_0 \to \overline{\mathbb{R}}$ as

$$f(x) := \begin{cases} \sum_n x_n, & x \in c_{00}, \\ +\infty, & \text{otherwise,} \end{cases}$$

where $c_{00} := \{(x_n)_{n\geq 1} : x_n \in \mathbb{R}, x_n = 0 \text{ except for finitely many } n\text{'s}\}$. If $\text{cl } f$ were proper, so that $\text{cl } f \in \Gamma_0(c_0)$, then there would exist $x^* \in \ell_1$ and $\alpha \in \mathbb{R}$ such that $\sum_n x_n^* x_n + \alpha \leq (\text{cl } f)(x)$ for all $x \in c_0$ (by Theorem 3.2.2). Because f is homogeneous, it can easily be seen that $\text{cl } f$ is positively homogeneous and we write $\sum_n x_n^*(\gamma x_n) + \alpha \leq (\text{cl } f)(\gamma x)$ for all $x \in c_0$ and all $\gamma > 0$ so that $\sum_n x_n^* x_n \leq (\text{cl } f)(x) \leq \sum_n x_n$ for all $x \in c_{00}$, by dividing over γ and after making $\gamma \uparrow \infty$. In particular, replacing x by e_n and $-e_n$, $n \geq 1$, in the last inequality, we obtain that $x_n^* = 1$ for all $n \geq 1$, and we get the contradiction $x^* \notin \ell^1$.

Exercise 40: (i) We may assume that $x = \theta \in f^{-1}(\mathbb{R}) \cap g^{-1}(\mathbb{R})$. Let $U \in \mathcal{N}_X$ such that $f(z) = g(z)$ for all $z \in U$. Given $x^* \in \partial f(\theta)$, we get $\langle x^*, z\rangle \leq f(z) - f(\theta) = g(z) - g(\theta)$ for all $z \in U$. Now, given $z \in X \setminus U$, there exists some $\lambda \in]0,1]$ such that $\lambda z \in U$, and the last inequality yields

$$\langle x^*, \lambda z\rangle \leq g(\lambda z) - g(\theta) \leq \lambda g(z) + (1-\lambda)g(\theta) - g(\theta) = \lambda g(z) - \lambda g(x).$$

Hence, dividing by λ, we get $\langle x^*, z\rangle \leq g(z) - g(\theta)$ and $x^* \in \partial g(\theta)$. This shows that $\partial f(\theta) \subset \partial g(\theta)$, and by symmetry, we have the other inclusion.

(ii) By the lower semicontinuity of f, there exists some $U \in \mathcal{N}_X$ such that f and g coincide on $x + U$, and the first assertion applies.

Exercise 41: (i) To simplify the proof, we take $m = 2$. Using (3.52), (4.28), (4.1), and taking into account that $\mathbb{R}_+(B \cap C) = (\mathbb{R}_+ B) \cap (\mathbb{R}_+ C)$, when B and C are convex sets containing θ, we get

$$([\operatorname{cl}(\partial_{\varepsilon_1} f_1(x) + \partial_{\varepsilon_2} f_2(x) + A^*(\partial_\varepsilon g(Ax)))]_\infty)^-$$
$$= \operatorname{cl}(\operatorname{dom}(\sigma_{\partial_{\varepsilon_1} f_1(x)} + \sigma_{\partial_{\varepsilon_2} f_2(x)} + \sigma_{A^*(\partial_\varepsilon g(Ax))}))$$
$$= \operatorname{cl}(\operatorname{dom}(\sigma_{\partial_{\varepsilon_1} f_1(x)} + \sigma_{\partial_{\varepsilon_2} f_2(x)} + \sigma_{\partial_\varepsilon g(Ax)} \circ A))$$
$$= \operatorname{cl}\left((\operatorname{dom}((f_1)'_{\varepsilon_1}(x,\cdot)) \cap (\operatorname{dom}((f_2)'_{\varepsilon_2}(x,\cdot)) \cap A^{-1}(\operatorname{dom} g'_\varepsilon(Ax,\cdot))\right)$$
$$= \operatorname{cl}\left(\mathbb{R}_+(\operatorname{dom} f_1 - x) \cap \mathbb{R}_+(\operatorname{dom} f_2 - x) \cap \mathbb{R}_+ A^{-1}(\operatorname{dom} g - Ax)\right)$$
$$= \operatorname{cl}\left(\mathbb{R}_+(\operatorname{dom} f_1 - x) \cap \mathbb{R}_+(\operatorname{dom} f_2 - x) \cap \mathbb{R}_+(\operatorname{dom}(g \circ A) - x)\right)$$
$$= \operatorname{cl}\left(\mathbb{R}_+(\operatorname{dom}(f_1 + f_2 + g \circ A) - x)\right).$$

Whence the conclusion follows using (3.51).

(ii) It is enough to apply (3.72).

Exercise 42: For every $\varepsilon > 0$ and $k \geq 1$, we have

$$\partial_\varepsilon f(0) = \{\alpha \in \mathbb{R} : \alpha x \leq x^2 + \varepsilon \text{ for all } x \in \mathbb{R}\} = [-2\sqrt{\varepsilon}, 2\sqrt{\varepsilon}],$$

$$\mathrm{N}^\varepsilon_{[-1/k,1/k]}(0) = \{\alpha \in \mathbb{R} : \alpha x \leq \varepsilon \text{ for all } x \in [-1/k, 1/k]\} = [-k\varepsilon, k\varepsilon].$$

Then, using (4.58), we have

$$\partial_\varepsilon(f + \mathrm{I}_{[-1/k,1/k]})(0) = \bigcup_{\substack{\varepsilon_1+\varepsilon_2=\varepsilon \\ \varepsilon_1,\varepsilon_2 \geq 0}} (\partial_{\varepsilon_1} f(0) + \mathrm{N}^{\varepsilon_2}_{[-1/k,1/k]}(0))$$
$$= \bigcup_{\substack{\varepsilon_1+\varepsilon_2=\varepsilon \\ \varepsilon_1,\varepsilon_2 \geq 0}} ([-2\sqrt{\varepsilon_1}, 2\sqrt{\varepsilon_1}] + [-k\varepsilon_2, k\varepsilon_2])$$
$$= \bigcup_{0 \leq \varepsilon_1 \leq \varepsilon} [-2\sqrt{\varepsilon_1} - k(\varepsilon - \varepsilon_1), 2\sqrt{\varepsilon_1} + k(\varepsilon - \varepsilon_1)].$$

9.3 EXERCISES OF CHAPTER 4

In particular, taking $\varepsilon_1 = \frac{\varepsilon}{2}$, the assumption $k > \frac{2(2-\sqrt{2})}{\sqrt{\varepsilon}}$ yields $\alpha_k := 2\sqrt{\varepsilon_1} + k(\varepsilon - \varepsilon_1) = \sqrt{2\varepsilon} + \frac{k\varepsilon}{2} > 2\sqrt{\varepsilon}$, so that $\alpha_k \in \partial_\varepsilon(f + I_{[-1/k,1/k]})(0) \setminus \partial_\varepsilon f(0)$.

Exercise 43: (i) Obviously, $z^* \in [\partial_\varepsilon f(x)]_\infty$ if and only if $x^* + \lambda z^* \in \partial_\varepsilon f(x)$ for all $\lambda \geq 0$ and $x^* \in \partial_\varepsilon f(x)$; in other words, $f(y) \geq f(x) + \langle x^* + \lambda z^*, y - x\rangle - \varepsilon \;\;\forall y \in \text{dom } f, \;\; \lambda \geq 0$ and $x^* \in \partial_\varepsilon f(x)$. With y and x^* fixed, dividing by λ and taking limits for $\lambda \uparrow \infty$, we conclude that $z^* \in N_{\text{dom } f}(x)$. Conversely, for $z^* \in N_{\text{dom } f}(x)$ and $x^* \in \partial_\varepsilon f(x)$, we can write, $f(y) \geq f(x) + \langle x^*, y - x\rangle - \varepsilon \geq f(x) + \langle x^* + \lambda z^*, y - x\rangle - \varepsilon$ for any $y \in \text{dom } f$ and $\lambda \geq 0$. In other words, $x^* + \lambda z^* \in \partial_\varepsilon f(x)$ for all $\lambda \geq 0$, and $z^* \in [\partial_\varepsilon f(x)]_\infty$.

(ii) Left to the reader.

Exercise 44: We prove first that (4.134) holds. It is clear that $\text{epi } f - (x, f(x)) \subset \text{epi } f'(x; \cdot)$, and since $f'(x; \cdot)$ is sublinear, $\mathbb{R}_+(\text{epi } f - (x, f(x))) \subset \text{epi } f'(x; \cdot)$. Conversely, for $(u, \lambda) \in \text{epi } f'(x; \cdot)$ and $\delta > 0$, there will exist $t > 0$ such that $t^{-1}(f(x + tu) - f(x)) < \lambda + \delta$; that is, $(u, \lambda + \delta) \in t^{-1}(\text{epi } f - (x, f(x))) \subset \mathbb{R}_+(\text{epi } f - (x, f(x)))$ and $(u, \lambda) \in \text{cl}\,\mathbb{R}_+(\text{epi } f - (x, f(x))) = \mathbb{R}_+(\text{epi } f - (x, f(x)))$. Thus, $\text{epi } f'(x; \cdot) = \mathbb{R}_+(\text{epi } f - (x, f(x))$ so that $f'(x; \cdot)$ is lsc. Consequently, $\partial f(x) \neq \emptyset$ and (4.10) implies that $f'(x; \cdot) = \sigma_{\partial f(x)}$. Therefore, $(\text{cl } f)(x) = f(x) \in \mathbb{R}$ and $\text{cl } f$ is proper. Moreover, since

$$\mathbb{R}_+(\text{epi}(\text{cl } f) - (x, f(x))) = \mathbb{R}_+(\text{cl}(\text{epi } f) - (x, f(x))) = \mathbb{R}_+(\text{epi } f - (x, f(x)),$$

we infer that $f'(x; \cdot) = (\text{cl } f)'(x; \cdot) = \sigma_{\partial(\text{cl } f)(x)} = \sigma_{\partial f(x)}$.

Exercise 45: Consider a net $(\alpha_i, (x_i, \lambda_i))_i \subset \mathbb{R}_+ \times \text{epi}(f + g)$ such that $\alpha_i((x_i, \lambda_i) - (x, f(x) + g(x))))$ converges to some $(u, \mu) \in X \times \mathbb{R}$; we shall prove that $(u, \mu) \in \mathbb{R}_+(\text{epi}(f + g) - (x, f(x) + g(x)))$. If $(u, \mu) = (\theta, 0)$, we are done, and therefore, we shall examine only the cases $u \neq \theta$ or $u = \theta, \mu \neq 0$. Obviously, the net $\alpha_i(x_i - x)$ converges to u and $\alpha_i(\lambda_i - f(x) - g(x))$ converges to μ. If $y^* \in \partial f(x)$ and $z^* \in \partial g(x)$ (due to Exercise 44, $\partial f(x)$ and $\partial g(x)$ are both non-empty), by taking into account that $f(x_i) + g(x_i) \leq \lambda_i$, we can easily prove that

$$\langle y^*, \alpha_i(x_i - x)\rangle \leq \alpha_i(f(x_i) - f(x)) \leq \alpha_i(\lambda_i - f(x) - g(x)) - \langle z^*, \alpha_i(x_i - x)\rangle,$$

and so, for every $\varepsilon > 0$, the net $\alpha_i(f(x_i) - f(x))$ is eventually contained in the interval $[\langle y^*, u\rangle - \varepsilon, \mu - \langle z^*, u\rangle + \varepsilon]$. Similarly, the net

$\alpha_i(g(x_i) - g(x))$ is eventually contained in the interval $[\langle z^*, u\rangle - \varepsilon, \mu - \langle y^*, u\rangle + \varepsilon]$. Therefore, we may suppose that the (eventually bounded) nets $(\alpha_i(f(x_i) - f(x)))_i$ and $(\alpha_i(g(x_i) - g(x)))_i$ converge to some $\mu_1, \mu_2 \in \mathbb{R}$, respectively. Moreover, we have $\alpha_i(f(x_i) - f(x)) + \alpha_i(g(x_i) - g(x)) \le \alpha_i(\lambda_i - f(x) - g(x))$ for every i, entailing $\mu_1 + \mu_2 \le \mu$. In other words, $\alpha_i((x_i, f(x_i)) - (x, f(x))) \to (u, \mu_1)$ and $\alpha_i((x_i, g(x_i)) - (x, g(x))) \to (u, \mu_2)$. Thus, by the current assumption, there are $\gamma_1, \gamma_2 \ge 0$ such that $(u, \mu_1) \in \gamma_1(\operatorname{epi} f - (x, f(x)))$ and $(u, \mu_2) \in \gamma_2(\operatorname{epi} g - (x, g(x)))$. If the scalars γ_1 and γ_2 are positive, so that $(u, \mu_1) \ne (\theta, 0)$ and $(u, \mu_2) \ne (\theta, 0)$, by the convexity of the sets $\operatorname{epi} f - (x, f(x))$ and $\operatorname{epi} g - (x, g(x))$ and the fact that they contain $(\theta, 0)$, we may assume that $\gamma_1 = \gamma_2 \ge 1$. Then, denoting $\gamma := \gamma_1 = \gamma_2$, we conclude that $(x, f(x)) + \gamma^{-1}(u, \mu_1) \in \operatorname{epi} f$ and $(x, g(x)) + \gamma^{-1}(u, \mu_2) \in \operatorname{epi} g$. This relation is also valid if $(u, \mu_1) = (\theta, 0)$ or $(u, \mu_2) = (\theta, 0)$. Hence,

$$f(x + \gamma^{-1}u) + g(x + \gamma^{-1}u) \le f(x) + \gamma^{-1}\mu_1 + g(x) + \gamma^{-1}\mu_2$$
$$\le f(x) + g(x) + \gamma^{-1}\mu.$$

Therefore, $(x + \gamma^{-1}u, f(x) + g(x) + \gamma^{-1}\mu) \in \operatorname{epi}(f + g)$, and so $(u, \mu) \in \gamma(\operatorname{epi}(f + g) - (x, f(x) + g(x)))$.

Exercise 46: We have that $x^* \in \partial(f \circ A)(x)$ if and only if $\langle x^*, z - x\rangle \le f(Az) - f(Ax)$ for every $z \in X$; hence, if and only if $\langle x^*, z - x\rangle \le f(y) + I_{\operatorname{gph} A}(z, y) - f(Ax)$ for every $(z, y) \in X \times Y$, if and only if $\langle (x^*, \theta), (z - x, y - Ax)\rangle \le g(z, y) + h(z, y) - g(x, Ax) - h(x, Ax)$ for every $(z, y) \in X \times Y$. In other words, $x^* \in \partial(f \circ A)(x)$ if and only if $(x^*, \theta) \in \partial(g + h)(x, Ax)$.

Exercise 47: If $x^* \in \partial_\varepsilon f(x)$ with $\varepsilon \ge 0$, we have $s^{-1}(f(x + su) - f(x) + \varepsilon) \ge \langle x^*, u\rangle$ for all $u \in X$ and $s > 0$, and so

$$f'(x; u) - f'(x; \theta) = f'(x; u) = \inf_{s>0} \frac{f(x + su) - f(x) + \varepsilon}{s} \ge \langle x^*, u\rangle,$$

entailing $x^* \in \partial(f'_\varepsilon(x; \cdot))(\theta)$. Conversely, if $x^* \in \partial(f'_\varepsilon(x; \cdot))(\theta)$, we have $f(x + u) - f(x) + \varepsilon \ge f'_\varepsilon(x; u) \ge f'_\varepsilon(x; \theta) + \langle x^*, u\rangle = \langle x^*, u\rangle$ for all $u \in X$ and $x^* \in \partial_\varepsilon f(x)$. Moreover, $u \in \operatorname{dom} f'_\varepsilon(x; \cdot)$ if and only if $\inf_{s>0} s^{-1}(f(x + su) - f(x) + \varepsilon) < +\infty$, if and only if there exists $s_0 > 0$ such that $f(x + s_0 u) \in \mathbb{R}$; i.e., if and only if $u \in (1/s_0)(\operatorname{dom} f - x) \subset \mathbb{R}_+(\operatorname{dom} f - x)$.

9.3 EXERCISES OF CHAPTER 4

Exercise 48: We start the proof by assuming that $m = 1$, where the current hypothesis yields the existence of some $x_0 \in \text{dom } f_0$ such that $f_1(x_0) < 0$. Let us define the function $F : X \times \mathbb{R} \to \mathbb{R}_\infty$ by $F(x,y) := f_0(x) + I_{[f_1 \leq y_1]}(x)$. We are going to apply Corollary 4.2.5 to F and the dual pair $(X \times \mathbb{R}, X^* \times \mathbb{R})$ given via the linear pairing $\langle (x^*, y^*), (x,y) \rangle = \langle x^*, x \rangle + y^* y$, where $y^* y$ is the inner product in $\mathbb{R}$. It is clear that $\inf_{x \in X} F(x, 0) = \inf_{[f_1 \leq 0]} f_0$ and $F \in \Gamma_0(X \times \mathbb{R})$. Moreover, we have

$$F^*(\theta, y^*) = \sup\{y^* y - F(x,y) : x \in X, \ y \in \mathbb{R}\}$$
$$= \sup\{y^* y - f_0(x) : x \in X, \ f_1(x) \leq y\},$$

so that $F^*(\theta, y^*) = +\infty$ if $y^* > 0$ and otherwise, when $y^* \leq 0$,

$$F^*(\theta, y^*) = \sup_{x \in X}\{y^* f_1(x) - f_0(x)\} = -\inf_{x \in X}\{f_0(x) - y^* f_1(x)\}.$$

Therefore, because $\inf_{x \in X} F(x, 0) \leq F(x_0, 0) = f_0(x_0) < +\infty$ and $F(x_0, \cdot)$ is continuous at $y_0 = 0 \in \mathbb{R}$, as a consequence of Proposition 2.2.6, Corollary 4.2.5 entails

$$\inf_{[f_1 \leq 0]} f_0 = -\min_{y^* \in \mathbb{R}} F^*(\theta, y^*)$$
$$= -\min_{y^* \leq 0}\left(-\inf_{x \in X}\{f_0(x) - y^* f_1(x)\}\right) = \max_{\lambda \geq 0} \inf_{x \in X}\{f_0(x) + \lambda f_1(x)\}.$$

Now, to deal with the general case of m constraints, we write $\inf_{[\max_{1 \leq i \leq m} f_i \leq 0]} f_0 = \inf_{[f_m \leq 0]} f_0(x) + I_{[f_1 \leq 0]}(x) + \ldots + I_{[f_{m-1} \leq 0]}(x)$, and the first part of the proof yields some $\lambda_m \geq 0$ such that

$$\inf_{[f_1 \leq 0]} f_0 = \inf_{x \in X}\{f_0(x) + I_{[f_1 \leq 0]}(x) + \ldots + I_{[f_{m-1} \leq 0]}(x) + \lambda_m f_m(x)\}.$$

Thus, the conclusion follows by successively applying this argument.

Exercise 49: To verify the inclusion (4.135), in the non-trivial case, take $x^* \in \partial_{\varepsilon_1} f(x) \cap \partial_{\varepsilon_2} g(y)$. For every $z, z_1, z_2 \in X$ such that $z_1 + z_2 = z$, we have $\langle x^*, z_1 - x \rangle \leq f(z_1) - f(x) + \varepsilon_1$ and $\langle x^*, z_2 - y \rangle \leq g(z_2) - g(y) + \varepsilon_2$ and, by summing up, we get

$$\langle x^*, z - (x+y) \rangle \leq f(z_1) + g(z_2) - f(x) - g(y) + \varepsilon_1 + \varepsilon_2 \quad (9.6)$$
$$\leq f(z_1) + g(z_2) - (f \square g)(x+y) + \varepsilon_1 + \varepsilon_2; \quad (9.7)$$

in particular, taking the infimum over z_1 and z_2 in (9.6), we get

$$(f\Box g)(x+y) \geq f(x) + g(y) - \varepsilon_1 - \varepsilon_2.$$

Moreover, after taking the infimum over z_1 and z_2 in (9.7),

$$\langle x^*, z - (x+y)\rangle \leq \inf_{z_1+z_2=z}(f(z_1) + g(z_2)) - (f\Box g)(x+y) + \varepsilon_1 + \varepsilon_2$$
$$= (f\Box g)(z) - (f\Box g)(x+y) + \varepsilon_1 + \varepsilon_2,$$

that is, $x^* \in \partial_{\varepsilon_1+\varepsilon_2}(f\Box g)(x+y)$.

Suppose now that $(f\Box g)(x+y) \geq f(x) + g(y) - \varepsilon$ and take $x^* \in \partial_\delta(f\Box g)(x+y)$. Then, for all $z \in X$,

$$(f\Box g)(z) \geq (f\Box g)(x+y) + \langle x^*, z - (x+y)\rangle - \delta$$
$$\geq f(x) + g(y) + \langle x^*, z - (x+y)\rangle - \delta - \varepsilon.$$

Then

$$f(z-y) + g(y) \geq (f\Box g)(z) \geq f(x) + g(y) + \langle x^*, z - (x+y)\rangle - \delta - \varepsilon, \tag{9.8}$$

in other words, since $y \in \text{dom}\, g$, we deduce that $f(z-y) \geq f(x) + \langle x^*, (z-y) - x\rangle - \delta - \varepsilon$ for all $z \in X$, implying that $x^* \in \partial_{\delta+\varepsilon}f(x)$. Analogously, we show that $x^* \in \partial_{\delta+\varepsilon}g(y)$.

Finally, if $x^* \in \partial(f\Box g)(x+y)$ and $(x_n, y_n)_n$ satisfies $x_n + y_n = x + y$ and $f(x_n) + g(y_n) \leq (f\Box g)(x+y) + \frac{1}{n}$, we write, for all $z \in X$,

$$(f\Box g)(z) \geq (f\Box g)(x+y) + \langle x^*, z - (x+y)\rangle$$
$$\geq f(x_n) + g(y_n) + \langle x^*, z - (x+y)\rangle - \frac{1}{n}.$$

Thus, if $y^* \in \partial f(x) \cap \partial g(y)$, we have

$$(f\Box g)(z) \geq f(x) + g(y) + \langle y^*, (x_n + y_n) - (x+y)\rangle + \langle x^*, z - (x+y)\rangle - \frac{1}{n}$$
$$= f(x) + g(y) + \langle x^*, z - (x+y)\rangle - \frac{1}{n}.$$

Taking limits for $n \to \infty$, one gets (9.8), and the proof is finished as above.

Exercise 50: Since f is proper, its extension $\hat{f}$ is also proper. Then, by Proposition 4.1.14, we have

9.3 EXERCISES OF CHAPTER 4

$$\partial(\mathrm{cl}^{w^{**}} \hat{f})(z) = \bigcap_{\substack{\varepsilon>0 \\ U \in \mathcal{N}_{X^{**}}}} \bigcup_{y \in z+U} \partial_\varepsilon \hat{f}(y),$$

and the definition of $\hat{f}$ together with the identification of X as a subspace of X^{**} yields

$$\partial(\mathrm{cl}^{w^{**}} \hat{f})(z) = \bigcap_{\substack{\varepsilon>0 \\ U \in \mathcal{N}_{X^{**}}}} \bigcup_{y \in z+U,\, y \in X} \partial_\varepsilon f(y).$$

Consequently, again by Proposition 4.1.14,

$$\left(\partial(\mathrm{cl}^{w^{**}} \hat{f})\right)^{-1}(x^*) = \bigcap_{\varepsilon>0} \mathrm{cl}^{w^{**}}((\partial_\varepsilon \hat{f})^{-1}(x^*)) = \bigcap_{\varepsilon>0} \mathrm{cl}^{w^{**}}((\partial_\varepsilon f)^{-1}(x^*)).$$

Exercise 51: Given convex functions $f : X \to \overline{\mathbb{R}}$, $g : Y \to \overline{\mathbb{R}}$, and a continuous linear mapping $A : X \to Y$, we suppose that g is finite and continuous at some point of the form Ax_0 with $x_0 \in \mathrm{dom}\, f$. We define the convex function $F : X \times Y \to \mathbb{R}_\infty$ by $F(x, y) := f(x) + g(Ax + y)$, so that $F(x_0, \cdot)$ is finite and continuous at θ. Therefore, by Proposition 4.1.24, the conjugate of the function $\varphi(x) := F(x, \theta)$ is computed as $\varphi^* = \min_{y^* \in Y^*} F^*(\cdot, y^*)$, where F^* is expressed as, for all $x^* \in X^*$ and $y^* \in Y^*$,

$$F^*(x^*, y^*) = \sup_{x \in X,\, y \in Y} \{\langle x^*, x\rangle + \langle y^*, y - Ax\rangle - f(x) + g(y)\}$$
$$= f^*(x^* - A^* y^*) + g^*(y^*).$$

Thus, taking into account (4.44), we infer that $\varphi^*(x^*) = \min_{y^* \in Y^*} \{f^*(x^* - A^* y^*) + g^*(y^*)\}$ for all $x^* \in X^*$. Moreover, using the same arguments (taking, for example, $f \equiv 0$), we show that the conjugate of the function $(g \circ A)$ satisfies $(g \circ A)^*(x^*) = \min_{y^* \in Y^*,\, A^* y^* = x^*} g^*(y^*) = (A^* g^*)(x^*)$ for all $x^* \in X^*$, proving (4.56).

Exercise 52: Formula (4.136) follows easily from (4.137), because

$$\partial(f \circ A)(x) = \bigcap_{\delta>0} \partial_\delta(f \circ A)(x) = \bigcap_{\delta>0} \mathrm{cl}\left(A_0^* \partial_\delta f(Ax)\right),$$

and so, we only need to show (4.137). With this aim, we introduce the continuous linear mapping $\hat{A} : X \times \mathbb{R} \to Y$ defined by $\hat{A}(x, \lambda) := A_0 x + \lambda b$, and the proper lsc convex function $h : X \times \mathbb{R} \to \mathbb{R}_\infty$ defined by $h := f \circ \hat{A}$. Hence, for every given $x \in X$ and $\varepsilon > 0$, by Proposition 4.1.16, we have $\partial_\varepsilon h(x,1) = \mathrm{cl}\left(\hat{A}^* \partial_\varepsilon f(A_0 x + b)\right)$, where $\hat{A}^* : Y^* \to X^* \times \mathbb{R}$ is the adjoint of $\hat{A}$, satisfying for every $(x, \lambda) \in X \times \mathbb{R}$

$$\left\langle (x,\lambda), (\hat{A})^* y^* \right\rangle = \langle A_0 x + \lambda b, y^* \rangle$$
$$= \langle x, A_0^* y^* \rangle + \lambda \langle b, y^* \rangle = \langle (x,\lambda), (A_0^* y^*, \langle b, y^* \rangle) \rangle,$$

that is, $(\hat{A})^* y^* = (A_0^* y^*, \langle b, y^* \rangle)$. In other words,

$$\partial_\varepsilon h(x,1) = \mathrm{cl}\left\{ (A_0^* y^*, \langle b, y^* \rangle) : y^* \in \partial_\varepsilon f(A_0 x + b) \right\}.$$

At the same time, taking into account (4.28), for every $u \in X$

$$\sigma_{\partial_\varepsilon h(x,1)}(u,0) = h'_\varepsilon((x,1);(u,0)) = \inf_{s>0} \frac{h((x,1) + s(u,0)) - h(x,1) + \varepsilon}{s}$$
$$= \inf_{s>0} \frac{f(A_0(x+su)+b) - f(A_0 x + b) + \varepsilon}{s} = \sigma_{\partial_\varepsilon (f \circ A)(x)}(u),$$

and we get

$$\sigma_{\partial_\varepsilon (f \circ A)(x)}(u) = \sigma_{\{(A_0^* y^*, \langle b, y^* \rangle) : y^* \in \partial_\varepsilon f(A_0 x + b)\}}(u,0) = \sigma_{A_0^* \partial_\varepsilon f(A_0 x + b)}(u).$$

The last relation leads us, due to Corollary 3.2.9, to $\partial_\varepsilon (f \circ A)(x) = \mathrm{cl}\left(A_0^* \partial_\varepsilon f(A_0 x + b) \right)$.

Exercise 53: Define the functions $f := I_L$ and $g := I_M \circ A$. Then $f \in \Gamma_0(X)$, $g \in \Gamma_0(Y)$, and $\theta \in \mathrm{dom}\, f \cap A^{-1}(\mathrm{dom}\, g)$, and we verify that $x^* \in (L \cap A^{-1}(M))^-$ if and only if $\langle x^*, x \rangle \le 0$ for all $x \in L \cap A^{-1}(M)$, if and only if $\langle x^*, x \rangle \le I_L(x) + I_M \circ A(x)$ for all $x \in X$. In other words, using formula (4.45) in Proposition 4.1.16, we obtain

$$(L \cap A^{-1}(M))^- = \partial(I_L + I_M \circ A)(\theta)$$
$$= \bigcap_{\varepsilon > 0} \mathrm{cl}\left(\bigcup_{\substack{\varepsilon_1 + \varepsilon_2 = \varepsilon \\ \varepsilon_1, \varepsilon_2 \ge 0}} (\partial_{\varepsilon_1} f(\theta) + A^* \partial_{\varepsilon_2} g(A\theta)) \right) = \mathrm{cl}\left(L^- + A^*(M^-) \right).$$

Exercise 54: Denote $f := \sigma_A$. Since $\mathrm{cl}(\mathrm{dom}\, f) = [(\overline{\mathrm{co}} A)_\infty]^-$ by (3.52), for each $x \in \mathrm{dom}\, f$, we have

$$N_{\mathrm{dom}\, f}(x) = N_{\mathrm{cl}(\mathrm{dom}\, f)}(x)$$
$$= (\mathrm{cl}(\mathbb{R}_+ (\mathrm{cl}(\mathrm{dom}\, f) - x)))^- = \left([(\overline{\mathrm{co}} A)_\infty]^- - x \right)^-.$$

9.3 EXERCISES OF CHAPTER 4

But we have $([(\overline{\text{co}}A)_\infty]^- - x)^- = ([(\overline{\text{co}}A)_\infty]^- - \mathbb{R}_+ x)^-$, and so Corollary 3.3.6 together with Exercise 53 ensures that $([(\overline{\text{co}}A)_\infty]^- - \mathbb{R}_+ x)^- = (\overline{\text{co}}A)_\infty \cap \{-x\}^-$; that is, $N_{\text{dom} f}(x) = (\overline{\text{co}}A)_\infty \cap \{-x\}^-$. Moreover, any $y \in (\overline{\text{co}}A)_\infty \cap \{-x\}^-$ satisfies $y \in (\overline{\text{co}}A)_\infty = ([(\overline{\text{co}}A)_\infty]^-)^- = (\text{dom } f)^-$; that is, $0 \le \langle y, x \rangle \le 0$ as $x \in \text{dom } f$, and we conclude that $y \in (\overline{\text{co}}A)_\infty \cap \{x\}^\perp$.

Exercise 55: (i) We have

$$x^* \in \partial_\varepsilon f(x) \Rightarrow \langle x^*, y - x \rangle \le f(y) - f(x) + \varepsilon \text{ for all } y \in X$$
$$\Rightarrow \left\langle x^*_{|L}, y - x \right\rangle = \langle x^*, y - x \rangle \le f_{|L}(y) - f_{|L}(x) + \varepsilon \text{ for all } y \in L$$
$$\Rightarrow x^*_{|L} \in \partial_\varepsilon f_{|L}(x).$$

Suppose now that $\text{dom } f \subset L$. If $z^* \in \partial_\varepsilon f_{|L}(x)$, then $\langle z^*, y - x \rangle \le f_{|L}(y) - f_{|L}(x) + \varepsilon = f(y) - f(x) + \varepsilon$ for all $y \in L$. Consequently, for every extension x^* of z^* to X, the inclusion $\text{dom } f \subset L$ yields $\langle x^*, y - x \rangle \le f(y) - f(x) + \varepsilon$ for all $y \in X$, and so $x^* \in \partial_\varepsilon f(x)$. Relation (4.139) follows from (4.138) as $\text{dom}(f + I_L) \subset L$. To prove (4.140), take $z^* \in \partial_\varepsilon (f + I_L)(x)$. Then $z^*_{|L} \in \partial(f + I_L)_{|L}(x)$, by (4.139), and we obtain $z^* = z^* + \theta \in \{x^* + L^\perp : x^*_{|L} \in \partial_\varepsilon (f + I_L)_{|L}(x)\}$.
Conversely, if $z^* \in x^* + L^\perp$ with $x^*_{|L} \in \partial_\varepsilon (f + I_L)_{|L}(x)$, then for all $z \in L$, we get

$$\langle x^*, z - x \rangle = \langle x^*_{|L}, z - x \rangle \le (f + I_L)_{|L}(z) - (f + I_L)_{|L}(x) + \varepsilon$$
$$= (f + I_L)(z) - (f + I_L)(x) + \varepsilon;$$

that is, $z^* \in \partial_\varepsilon (f + I_L)(x) + L^\perp \subset \partial_\varepsilon (f + I_L)(x)$.

(ii) By (4.140), we have

$$\partial(f + g + I_L)(x) = \{x^* + L^\perp : x^*_{|L} \in \partial(f_{|L} + g_{|L})(x)\}$$
$$= \{x^* + L^\perp : x^*_{|L} \in \partial f_{|L}(x) + \partial g_{|L}(x)\}$$
$$= \{x^* + L^\perp : x^*_{|L} = \tilde{z}^*_1 + \tilde{z}^*_2, \ \tilde{z}^*_1 \in \partial f_{|L}(x), \ \tilde{z}^*_2 \in \partial g_{|L}(x)\}.$$

Then, using (4.139), for $\tilde{z}^*_1$ and $\tilde{z}^*_2$ there are $z^*_1 \in \partial(f + I_L)(x)$ and $z^*_2 \in \partial(g + I_L)(x)$ such that $\tilde{z}^*_1 = z^*_{1|L}$ and $\tilde{z}^*_2 = z^*_{2|L}$. Hence,

$$\partial(f + g + I_L)(x) = \{x^* + L^\perp : x^*_{|L} = z^*_{1|L} + z^*_{2|L}, \ z^*_1 \in \partial(f + I_L)(x), \ z^*_2 \in \partial(g + I_L)(x)\}$$
$$= \{x^* + L^\perp : x^* \in \partial(f + I_L)(x) + \partial(g + I_L)(x) + L^\perp\}$$
$$= \partial(f + I_L)(x) + \partial(g + I_L)(x) + L^\perp.$$

(iii) Given $x \in L \cap M$, we pick $x^* \in \partial(f + g + I_{L\cap M})(x)$. Since $(f + I_L + g)_{|M}$ is also the restriction of the function $f + g + I_{L\cap M}$ to M, by part (i) and (4.141), we obtain

$$x^*_{|M} \in \partial(f + I_L + g)_{|M}(x) = \partial(f + I_L)_{|M}(x) + \partial g_{|M}(x).$$

Let $u^* \in \partial(f + I_L)_{|M}(x)$ and $v^* \in \partial g_{|M}(x)$ such that $x^*_{|M} = u^* + v^*$. Then, since $(f + I_L)_{|M}$ is the restriction of $f + I_{L\cap M}$ to M and $\text{dom}(f + I_{L\cap M}) \subset M$, by (i) we find $\hat{z}^* \in \partial(f + I_{L\cap M})(x)$ such that $\hat{z}^*_{|M} = u^*$. Similarly, we find $y^* \in \partial(g + I_M)(x)$ such that $y^*_{|M} = v^*$. More precisely, arguing similarly, since $\hat{z}^*_{|L} \in \partial(f + I_M)_{|L}(x) = \partial(f + I_L)_{|L}(x) + \partial(I_M)_{|L}(x)$, by (4.142), we find $\tilde{z}^* \in \partial(f + I_L)_{|L}(x)$ and $w^* \in \partial(I_M)_{|L}(x)$ satisfying $\hat{z}^*_{|L} = \tilde{z}^* + w^*$, together with $z^* \in \partial(f + I_L)(x)$ such that $z^*_{|L} = \tilde{z}^*$. In particular, we have

$$\langle x^*, z \rangle = \langle u^* + v^*, z \rangle = \langle \hat{z}^* + y^*, z \rangle$$
$$= \langle \tilde{z}^* + y^*, z \rangle = \langle z^* + y^*, z \rangle \text{ for all } z \in L \cap M,$$

which, using the relation $(L \cap M)^\perp = \text{cl}(L^\perp + M^\perp)$ (see Exercise 53), entails $x^* \in \partial(f + I_L)(x) + \partial(g + I_M)(x) + \text{cl}(L^\perp + M^\perp)$. This finishes the proof of (iii) as the opposite inclusion is straightforward.

The statement in (iv) follows by applying statement (i) to the indicator function of A.

Exercise 56: For the sake of simplicity, as in the proof of (4.58) in Proposition 4.1.26, we prove (4.56) when $\varepsilon = 0$, $m = n$, A is the identity mapping, and $\theta \in \text{dom} f \cap \text{dom} g$. Take $x^* \in \mathbb{R}^n$. Then, due to Proposition 4.1.16, we have $(f + g)^*(x^*) = \text{cl}(f^* \square g^*)(x^*)$, and so by Proposition 3.1.4, we may assume that $(f + g)^*(x^*) \in \mathbb{R}$; otherwise, $+\infty = (f + g)^*(x^*) = \text{cl}(f^* \square g^*)(x^*)$ and (4.56) obviously holds.

Arguing as in the proof of (4.58) in Proposition 4.1.26, we denote $E := \text{span}(\text{dom} g)$, and consider the restrictions $\tilde{f}$ and $\tilde{g}$ of the functions $f + I_E$ and g to the subspace E, respectively. By the current assumption, we can show that $\tilde{g}$ is (finite and) continuous somewhere in $\text{dom} \tilde{f}$, and Proposition 4.1.20 together with the Banach extension theorem implies the existence of some $y^*, z^* \in \mathbb{R}^n$ such that $x^* \in y^* + z^* + E^\perp$ and

$$(f + g)^*(x^*) = (\tilde{f} + \tilde{g})^*(x^*_{|E}) = \left(\tilde{f}^* \square \tilde{g}^*\right)(x^*_{|E}) = (f + I_E)^*(y^*) + g^*(z^*). \quad (9.9)$$

Similarly, we denote $F := \text{span}(\text{dom } f)$ and consider the restrictions $\hat{f}$ and $\hat{I}_{E\cap F}$ of the functions f and $I_{F\cap E}$ to F, respectively. Then we can show that $\hat{f}$ is continuous at $x_0 \in \text{dom}\,\hat{I}_{E\cap F}$, and therefore, as above, we find some $u^*, v^* \in \mathbb{R}^n$ such that $y^* \in u^* + v^* + F^\perp$ and

$$(f + I_E)^*(y^*) = (\hat{f} + \hat{I}_{E\cap F})^*(y^*_{|F}) = \left(\hat{f}^*\square\hat{I}^*_{E\cap F}\right)(y^*_{|F})$$
$$= f^*(u^*) + I^*_{E\cap F}(v^*) = f^*(u^*) + I_{E^\perp + F^\perp}(v^*), \quad (9.10)$$

where the last equality comes from the relation $I^*_{E\cap F} = \sigma_{E\cap F} = I_{(E\cap F)^\perp}$, and the identity $(E\cap F)^\perp = \text{cl}(E^\perp + F^\perp) = E^\perp + F^\perp$ (see Exercise 53). Therefore, combining (9.9) and (9.10), we get $(f+g)^*(x^*) = f^*(u^*) + I_{E^\perp + F^\perp}(v^*) + g^*(z^*)$. But we have $(f+g)^*(x^*) \in \mathbb{R}$, and so we must have $I_{E^\perp + F^\perp}(v^*) = 0$, which implies that $v^* \in E^\perp + F^\perp$. So, $x^* \in y^* + z^* + E^\perp \subset u^* + v^* + F^\perp + z^* + E^\perp = u^* + z^* + E^\perp + F^\perp$. Let $x_1^* \in u^* + F^\perp$ and $x_2^* \in z^* + E^\perp$ such that $x_1^* + x_2^* = x^*$. Then, by the definition of the conjugate, we easily check that $f^*(u^*) = f^*(x_1^*)$ and $g^*(z^*) = g^*(x_2^*)$, and Proposition 4.1.20 imply that $f^*(x_1^*) + g^*(x_2^*) = (f+g)^*(x^*) = \text{cl}(f^*\square g^*)(x^*) \leq (f^*\square g^*)(x^*)$; that is, (4.56) holds.

Exercise 57: Observe that $A + N_C(x) = \partial \sigma_A(\theta) + \partial I_{\mathbb{R}_+(C-x)}(\theta)$. Also, since the function $I_{\mathbb{R}_+(C-x)}$ is continuous at some point in $\text{dom}\,\sigma_A$, by (4.58), we obtain that $A + N_C(x) = \partial \sigma_A(\theta) + \partial I_{\mathbb{R}_+(C-x)}(\theta) = \partial(\sigma_A + I_{\mathbb{R}_+(C-x)})(\theta)$, which yields the closedness of $A + N_C(x)$.

Exercise 58: Fix $x := (x_n) \in \ell_1$ such that $x_n > 0$ for all $n \geq 1$. If $x^* \in \partial g(x)$, then taking $e_k := (0, \cdots, 0, 1^{(k)}, 0, \cdots) \in \ell_1$ we see that, for all $t \in \mathbb{R}$ sufficiently small,

$$tx_k^* = \langle x^*, (x + te_k) - x \rangle \leq g(x + te_k) - g(x)$$
$$= \|x + te_k\|_{\ell_1} - \|x\|_{\ell_1} = |x_k + t| - |x_k| = t.$$

Thus, $x_k^* = 1$ and we deduce that $\partial g(x) \subset \{(1, 1, \ldots)\}$. Therefore, since g is continuous, we have $\partial g(x) \neq \emptyset$ (Proposition 4.1.22), and so $\partial g(x) = \{(1, 1, \ldots)\}$.

Exercise 59: See the bibliographical notes of chapter 4.

Exercise 60: (i) We have $\text{int}(\text{dom } f) \subset \text{int}(\text{dom } g)$. Also, we have for all $x \in \text{int}(\text{dom } f)$

$$\emptyset \neq \partial f(x) \cap \partial g(x) \subset \partial f(x) \cap (\partial g(x) + N_{\overline{\text{dom } f}}(x)) \subset \partial f(x) \cap \partial(g + I_{\overline{\text{dom } f}})(x).$$

But $\tilde{g} := g + I_{\overline{\text{dom} f}} \in \Gamma_0(X)$ and so, since $\text{int}(\text{dom}\,\tilde{g}) = \text{int}(\text{dom}\,g) \cap \text{int}(\overline{\text{dom}\,f}) \subset \text{int}(\text{dom}\,f)$, the conclusion follows by applying Proposition 4.4.5 with f and $\tilde{g}$.

(*ii*) The current assumption implies that $\text{dom}\,f = \overline{\text{dom}\,f} \subset \text{dom}\,g$. Also, for all $x \in \text{dom}\,f$ and $0 < \varepsilon \leq \varepsilon_0$, we have

$$\emptyset \neq \partial_\varepsilon f(x) \cap \partial_\varepsilon g(x) \subset \partial_\varepsilon f(x) \cap (\partial_\varepsilon g(x) + N_{\text{dom}\,f}(x)) \subset \partial_\varepsilon f(x) \cap \partial_\varepsilon (g + I_{\text{dom}\,f})(x).$$

Hence, $\tilde{g} := g + I_{\text{dom}\,f} \in \Gamma_0(X)$ and $\text{dom}\,\tilde{g} = \text{dom}\,g \cap \text{dom}\,f \subset \text{dom}\,f$, and so the conclusion follows by applying Corollary 4.4.7 with f and $\tilde{g}$.

Exercise 61: We may assume that $\theta \in \text{dom}\,f$, so that $Y := \text{aff}(\text{dom}\,f) = \text{span}(\text{dom}\,f)$ is a finite-dimensional Banach space. We consider the functions $f_1 := f_{|Y}$ and $g_1 := (g + I_Y)_{|Y}$, so that $f_1, g_1 \in \Gamma_0(Y)$ and the current assumption together with Exercise 55(*i*) implies that, for all $y \in Y$,

$$\partial f_1(y) = \left\{ x^*_{|Y} : x^* \in \partial f(y) \right\} \subset \left\{ x^*_{|Y} : x^* \in \partial g(y) \right\}. \quad (9.11)$$

But, since $g \leq g + I_Y$ and the two functions coincide on Y, we have that

$$\{x^*_{|Y} : x^* \in \partial g(y)\} \subset \{x^*_{|Y} : x^* \in \partial(g + I_Y)(y)\} = \partial g_1(y) \text{ for all } y \in Y,$$

where the last equality comes again by Exercise 55(*i*), and so (9.11) reads $\partial f_1(y) \subset \partial g_1(y)$ for all $y \in Y$. Then, applying Proposition 4.4.8 in Y, there exists some $c \in \mathbb{R}$ such that $f_1(y) = g_1(y) + c$ for all $y \in Y$, entailing that $f(y) = g(y) + c$ for all $y \in Y$.

Exercise 62: Take $x^* \in \partial_\varepsilon f(x)$. Then $f(x) \in \mathbb{R}$ and $\langle x^*, y - x \rangle + f(x) \leq f(y) + \varepsilon$ for all $y \in X$. Thus, the function $\overline{\text{co}}f$ satisfies the same inequality, and so. $(\overline{\text{co}}f)(x) \geq f(x) - \varepsilon$ as the affine function $\langle x^*, \cdot \rangle - \langle x^*, x \rangle + f(x) - \varepsilon$ is an affine minorant of $f + \varepsilon$.

Exercise 63: If $x^* \in \partial_{(\overline{\text{co}}f)(x) - f(x) + \varepsilon}(\overline{\text{co}}f)(x)$, then we can write, for all $y \in X$,

$$f(y) \geq (\overline{\text{co}}f)(y) \geq (\overline{\text{co}}f)(x) + \langle x^*, y - x \rangle - ((\overline{\text{co}}f)(x) - f(x) + \varepsilon)$$
$$= f(x) + \langle x^*, y - x \rangle - \varepsilon,$$

9.3 EXERCISES OF CHAPTER 4

and $x^* \in \partial_\varepsilon f(x)$. Conversely, if $x^* \in \partial_\varepsilon f(x)$, for all $y \in X$, the continuous affine function $h := \langle x^*, \cdot \rangle - (\langle x^*, x \rangle - f(x) + \varepsilon)$ is a minorant of f and this entails $\overline{\mathrm{co}} f \geq h$, in other words, for all $y \in X$,

$$(\overline{\mathrm{co}} f)(y) \geq h(y) = \langle x^*, y - x \rangle - (-f(x) + \varepsilon)$$
$$= (\overline{\mathrm{co}} f)(x) + \langle x^*, y - x \rangle - ((\overline{\mathrm{co}} f)(x) - f(x) + \varepsilon)),$$

i.e., we are done. Additionally, since $\overline{\mathrm{co}} f \in \Gamma_0(X)$, one has $\partial_\eta (\overline{\mathrm{co}} f)(x) \neq \emptyset$ for all $\eta > 0$, and we deduce from the first part that f is ε-subdifferentiable for each $\varepsilon > f(x) - (\overline{\mathrm{co}} f)(x)$. Finally, $\partial_\varepsilon f(x) \neq \emptyset$ for all $\varepsilon > 0$ if and only if $(\overline{\mathrm{co}} f)(x) - f(x) + \varepsilon$ for all $\varepsilon > 0$, if and only if $(\overline{\mathrm{co}} f)(x) - f(x) = 0$.

Exercise 64: We have $h(x^*) = \inf_{t \in T}(f_t^*(x^*) - \langle x^*, x \rangle) + f(x) \geq \inf_{t \in T}(-f_t(x)) + f(x) = 0$ and

$$h^*(z) = \sup_{t \in T}(f_t^* - \langle \cdot, x \rangle + f(x))^*(z) = \sup_{t \in T} \sup_{x^* \in X^*} \{\langle x^*, z \rangle - f_t^*(x^*) + \langle x^*, x \rangle - f(x)\}$$
$$= \sup_{t \in T} \left\{ \sup_{x^* \in X^*} (\langle x^*, x + z \rangle - f_t^*(x^*)) - f(x) \right\} = \sup_{t \in T} \{f_t^{**}(x + z) - f(x)\}$$
$$= \sup_{t \in T} \{f_t(x + z) - f(x)\}.$$

Finally, if $h(x^*) < \varepsilon$, there must exist $t_0 \in T$ such that $f_{t_0}^*(x^*) - \langle x^*, x \rangle + f(x) < \varepsilon$, and from this we get $-f_{t_0}(x) + f(x) \leq (f_{t_0}^*(x^*) - \langle x^*, x \rangle) + f(x) < \varepsilon$ and $f_{t_0}^*(x^*) - \langle x^*, x \rangle + f_{t_0}(x) \leq f_{t_0}^*(x^*) - \langle x^*, x \rangle + f(x) < \varepsilon$; that is, $t_0 \in T_\varepsilon(x)$ and $x^* \in \partial_\varepsilon f_{t_0}(x)$.

Exercise 65: Take $f := I_{B_X}$. Then $f^* = \|\cdot\|_{X^*}$, $(f^*)^* = I_{B_{X^{**}}}$ and Theorem 4.3.3 gives rise to $I_{B_{X^{**}}} = (f^*)^* = \mathrm{cl}^{w^{**}}(\hat{f}) = \mathrm{cl}^{w^{**}}(I_{B_X}) = I_{\mathrm{cl}^{w^{**}}(B_X)}$, where $\hat{f}$ is defined in (4.103).

Exercise 66: By Proposition 4.1.14, we have for all $x \in X$ and $\varepsilon \in (0, \delta/2)$

$$\partial_\varepsilon(\mathrm{cl}\, f)(x) = \bigcap_{0 < \gamma < \delta/2,\, V \in \mathcal{N}_X} \bigcup_{y \in V} \partial_{\varepsilon + \gamma} f(x + y)$$
$$\subset \bigcap_{0 < \gamma < \delta/2,\, V \in \mathcal{N}_X} \bigcup_{y \in V} \partial_{\varepsilon + \gamma} g(x + y) = \partial_\varepsilon(\mathrm{cl}\, g)(x).$$

Thus, since $\mathrm{cl}\, f \in \Gamma_0(X)$ thanks to Proposition 3.1.4 (as $(\mathrm{cl}\, f)^* = f^* \in \Gamma_0(X)$), Proposition 4.4.6 yields the existence of some c such that $\mathrm{cl}\, f = \mathrm{cl}\, g + c$.

9.4 Exercises of chapter 5

Exercise 67: (i) It is clear that $\sup_i g_i = f$, so that $\sup_i \operatorname{cl} g_i \leq \sup_i g_i = f$, and hence $\sup_i(\operatorname{cl} g_i) \leq \operatorname{cl} f = \operatorname{cl}(\sup_i g_i)$. Conversely, since $\operatorname{cl} g_i \geq \sup_{t \in T_i}(\operatorname{cl} f_t)$, we obtain the other inequality, $\sup_i(\operatorname{cl} g_i) \geq \sup_i \sup_{t \in T_i}(\operatorname{cl} f_t) = \sup_{t \in T}(\operatorname{cl} f_t) = \operatorname{cl} f = \operatorname{cl}(\sup_i g_i)$. Now, concerning the validity of (5.26) under the following condition $\operatorname{cl} f = \sup \left\{\operatorname{cl} f_t,\ t \in T_{\varepsilon_0}(x);\ \operatorname{cl}(\sup_{t \in T \setminus T_{\varepsilon_0}(x)} f_t)\right\}$, for some given $\varepsilon_0 > 0$, we only need to apply Theorem 5.2.2 to the family $g_t := f_t,\ t \in T_{\varepsilon_0}(x),\ g_1 := \sup_{t \in T \setminus T_{\varepsilon_0}(x)} f_t$. Indeed, setting $I := T_{\varepsilon_0}(x) \cup \{1\}$, with 1 being a new index, we have that

$$\partial f(x) = \partial(\sup_i g_i)(x) = \bigcap_{L \in \mathcal{F}(x),\, 0 < \varepsilon < \varepsilon_0} \overline{\operatorname{co}}\left\{\bigcup_{t \in I_\varepsilon(x)} \partial_\varepsilon g_i(x) + \mathrm{N}_{L \cap \operatorname{dom} f}(x)\right\},$$

where $I_\varepsilon(x) := \{i:\ g_i(x) \geq f(x) - \varepsilon\}$. Since for $0 < \varepsilon < \varepsilon_0$ we have $f(x) - \varepsilon > f(x) - \varepsilon_0 \geq \sup_{t \in T \setminus T_{\varepsilon_0}(x)} f_t(x) = g_1(x)$, it follows that $1 \notin I_\varepsilon(x)$; that is, $I_\varepsilon(x) = T_\varepsilon(x)$ and the last relation simplifies to

$$\partial f(x) = \bigcap_{L \in \mathcal{F}(x),\, \varepsilon > 0} \overline{\operatorname{co}}\left\{\bigcup_{t \in T_\varepsilon(x)} \partial_\varepsilon f_t(x) + \mathrm{N}_{L \cap \operatorname{dom} f}(x)\right\}.$$

(ii) Fix a convex set $A \supset \operatorname{dom} f$. Assume first that $f_t \in \Gamma_0(X)$ for all $t \in T$. Then we have

$$\sup_{t \in T,\, x^* \in \operatorname{dom} f_t^*} \operatorname{cl}(x^* - f_t^*(x^*) + \mathrm{I}_A) = \sup_{t \in T,\, x^* \in \operatorname{dom} f_t^*} \{x^* - f_t^*(x^*) + \mathrm{I}_{\overline{A}}\}$$
$$= \sup_{t \in T}\{f_t + \mathrm{I}_{\overline{A}}\} = f + \mathrm{I}_{\overline{A}} = f = \operatorname{cl} f = \operatorname{cl}(f + \mathrm{I}_A),$$

which shows that the family $\{x^* - f_t^*(x^*) + \mathrm{I}_A :\ t \in T,\ x^* \in \operatorname{dom} f_t^*\}$, whose supremum is $f + \mathrm{I}_A$, satisfies (5.10). Thus, by (i), the family

$$\{f_t + \mathrm{I}_A,\ t \in T\} = \left\{\sup_{x^* \in \operatorname{dom} f_t^*}\{x^* - f_t^*(x^*) + \mathrm{I}_A\},\ t \in T\right\}$$

satisfies (5.10). Now, we suppose that $\operatorname{cl} f_t \in \Gamma_0(X)$ for all $t \in T$, so that the last paragraph gives us

$$\sup_{t \in T} \operatorname{cl}((\operatorname{cl} f_t) + \mathrm{I}_A) = \operatorname{cl}(\sup_{t \in T}(\operatorname{cl} f_t) + \mathrm{I}_A) = \operatorname{cl}((\operatorname{cl} f) + \mathrm{I}_A).$$

Consequently, using the relations $\overline{A} \supset \overline{\operatorname{dom} f} \supset \operatorname{dom}(\operatorname{cl} f)$,

9.4 EXERCISES OF CHAPTER 5

$\sup_{t \in T} \operatorname{cl}(f_t + I_A) \geq \sup_{t \in T} \operatorname{cl}((\operatorname{cl} f_t) + I_A) = \operatorname{cl}((\operatorname{cl} f) + I_A) \geq \operatorname{cl}((\operatorname{cl} f) + I_{\overline{A}}) = (\operatorname{cl} f),$

and so

$$\sup_{t \in T} \operatorname{cl}(f_t + I_A) \geq \operatorname{cl} f = \operatorname{cl}(f + I_A) \geq \sup_{t \in T} \operatorname{cl}(f_t + I_A).$$

Thus, the family $\{f_t + I_A : t \in T\}$ satisfies (5.10) as required.

Exercise 68: Fix $x \in X$ and $\varepsilon \geq 0$. If $x^* \in \partial_\varepsilon \sigma_A(x)$ and $\delta > \varepsilon$, then (5.1) yields a net $(x_i^*)_i \subset \{z^* \in \operatorname{co} A : \langle z^*, x \rangle \geq \sigma_A(x) - \delta\}$ which w^*-converges to x^*. Hence, the inclusion "⊂" in (5.81) follows. The converse inclusion also holds since, by (4.4),

$$\{x^* \in \operatorname{co} A : \langle x^*, x \rangle \geq \sigma_A(x) - \delta\} \subset \partial_\delta \sigma_A(x) \text{ for all } \delta > \varepsilon.$$

Exercise 69: The same as the proof of Exercise 68, but with the use of Proposition 5.1.2 instead of (5.1).

Exercise 70: Combine (5.1) and Proposition 5.1.2.

Exercise 71: First, observe that only the inclusion "⊂" in (5.83) needs to be proved in the case where $\partial_\varepsilon f(x) \neq \emptyset$. Hence, $f(x) \leq (\operatorname{cl} f)(x) + \varepsilon \leq f(x) + \varepsilon < +\infty$, and so we may assume for simplicity that $x = \theta$ and $f(\theta) = 0$, so that $0 \geq \sup_{t \in T} \bar{f}_t(\theta) = \bar{f}(\theta) \geq f(\theta) - \varepsilon = -\varepsilon$.

(i) Since $\bar{f}_t(z) = -\infty$, for all $t \in T \setminus S_0$ and $z \in \operatorname{cl}(\operatorname{dom} f_t)$, the comment above implies that $S_0 \neq \emptyset$. Also, since

$$\sup_{t \in T} \bar{f}_t(z) = \begin{cases} \sup_{t \in S_0} \bar{f}_t(z), & \text{if } z \in \bigcap_{s \in T \setminus S_0} \operatorname{cl}(\operatorname{dom} f_s), \\ +\infty, & \text{if not}, \end{cases}$$

and $\operatorname{dom} f \subset D \subset \operatorname{cl}(D) \subset \operatorname{cl}(\operatorname{dom} f_s)$ for all $s \in T \setminus S_0$, (i) follows from (5.10) and

$$\bar{f} = \sup_{t \in T} \bar{f}_t = \sup_{t \in S_0, \, s \in T \setminus S_0} \bar{f}_t + I_{\operatorname{cl}(\operatorname{dom} f_s)} \leq \sup_{t \in S_0} \bar{f}_t + I_{\operatorname{cl}(D)} = \sup_{t \in S_0} g_t \leq \bar{f}.$$

(ii) It is clear that $\{g_t, t \in S_0\} \subset \Gamma_0(X)$. To verify that this family also satisfies the convex combinations closedness assumption we pick $\lambda \in \Delta(S_0)$, so that $\sum_{t \in \operatorname{supp} \lambda} \lambda_t g_t = \sum_{t \in \operatorname{supp} \lambda} \lambda_t f_t + I_{\operatorname{cl}(D)}$. But the family $\{f_t, t \in S_0\}$ is closed for convex combinations by the current assumption, and so there exists some $s \in S_0$ such that $\sum_{t \in \operatorname{supp} \lambda} \lambda_t f_t \leq$

f_s, showing that $\sum_{t\in\operatorname{supp}\lambda}\lambda_t \bar f_t \le \bar f_s$. Therefore, $\sum_{t\in\operatorname{supp}\lambda}\lambda_t g_t \le \bar f_s + \mathrm{I}_{\operatorname{cl}(D)} = g_s$ and $\{g_t, t \in S_0\}$ is closed for convex combinations.

(iii) Applying (5.82) to the family $\{g_t, S_0\} \subset \Gamma_0(X)$, we obtain

$$\partial_\varepsilon f(\theta) \subset \partial_{(\varepsilon+\bar f(\theta))}\bar f(\theta) = \operatorname{cl}\left\{\bigcup_{t\in S_0}\partial_{(\varepsilon+\bar f_t(\theta))}(\bar f_t + \mathrm{I}_{\bar D})(\theta)\right\}$$

$$\subset \operatorname{cl}\left\{\bigcup_{t\in S_0}\partial_{(\varepsilon+f_t(\theta))}(f_t + \mathrm{I}_D)(\theta)\right\},$$

which gives rise to the non-trivial inclusion in (5.83).

Exercise 72: Given $x \in \operatorname{dom} f$, we fix $\delta > \varepsilon$. By applying Theorem 5.1.4 and using the Mazur theorem, we obtain that

$$\partial_\varepsilon f(x) \subset \operatorname{cl}^{\|\cdot\|_*}\left(\operatorname{co}\left\{\bigcup_{\lambda\in\Delta_k,\ (t_i)_{1\le i\le k}\subset T,\ k\ge 1} A(\delta,\lambda,t_1,\ldots,t_k,k,1)\right\}\right),$$

where, for the sake of simplicity, we denoted

$$A(\eta,\lambda,t_1,\ldots,t_k,k,\gamma) := \partial_{\eta+\lambda_1 f_{t_1}(x)+\ldots+\lambda_k f_{t_k}(x)-\gamma f(x)}\left(\sum_{1\le i\le k}\lambda_i f_{t_i}\right)(x).$$

Take $x^* \in A(\delta,\lambda,t_1,\ldots,t_k,k,1)$ and $y^* \in A(\delta,\beta,s_1,\ldots,s_m,m,1)$, for $\lambda \in \Delta_k$, $\beta \in \Delta_m$, $(t_i)_{1\le i\le k}$, $(s_i)_{1\le i\le m} \subset T$ and $k,m \ge 1$. Then, for all $\alpha \in\,]0,1[$,

$$\alpha x^* + (1-\alpha)y^* \in A(\alpha\delta,\alpha\lambda,t_1,\ldots,t_k,k,\alpha) + A((1-\alpha)\delta,(1-\alpha)\beta,s_1,\ldots,s_m,m,(1-\alpha)),$$

and Proposition 4.1.6 gives us $\alpha x^* + (1-\alpha)y^* \in A(\delta,\gamma,t_1,\ldots,t_k,t_{k+1},\ldots,t_m,k+m,1)$, where $\gamma := (\alpha\lambda_1,\ldots,\alpha\lambda_k,(1-\alpha)\beta_1,\ldots,(1-\alpha)\beta_m) \in \Delta_{k+m}$. Thus, $\alpha x^* + (1-\alpha)y^* \in \bigcup_{\lambda\in\Delta_k,\ (t_i)_{1\le i\le k}\subset T,\ k\ge 1} A(\delta,\lambda,t_1,\ldots,t_k,k,1)$ and the inclusion "$\subset$" in (5.16) (with the norm closure) holds. The converse inclusion is immediate from (5.16) (with the w^*-closure).

Exercise 73: The arguments are similar to Exercise 72.

Exercise 74: Take $x^* \in \partial_\varepsilon f(x)$ and fix $\delta > 0$. Then $f(x) + f^*(x^*) \le \langle x^*, x\rangle + \varepsilon < \langle x^*, x\rangle + \varepsilon + \delta$. By Proposition 3.2.8$(i)$, we know that $f^* = \operatorname{cl}(\inf_i f_i^*)$, and so

$$f(x) + \operatorname{cl}(\inf_i f_i^*)(x^*) = f(x) + f^*(x^*) < \langle x^*, x\rangle + \varepsilon + \delta.$$

9.4 EXERCISES OF CHAPTER 5

Hence, there exists a net $(x_l^*)_l \subset X^*$ w^*-converging to x^* such that

$$f(x) + (\inf_i f_i^*)(x_l^*) < \langle x_l^*, x \rangle + \varepsilon + \delta \text{ for all } l.$$

In other words, for each l, there exists i_l such that $f_i(x) + f_i^*(x_i^*) \leq f(x) + f_{i_l}^*(x_l^*) < \langle x_l^*, x \rangle + \varepsilon + \delta$ for all $i_l \preccurlyeq i$; that is,

$$x_l^* \in \bigcap_{i_l \preccurlyeq i} \partial_{\varepsilon+\delta} f_i(x) \subset \bigcup_j \bigcap_{j \preccurlyeq i} \partial_{\varepsilon+\delta} f_i(x),$$

and by taking the limit on l, we get $x^* \in \mathrm{cl}\,(\bigcup_j \bigcap_{j \preccurlyeq i} \partial_{\varepsilon+\delta} f_j(x))$. Then the direct inclusion follows by intersecting over $\delta > 0$, while the opposite inclusion is straightforward.

Exercise 75: We consider the family $\{f_x := \langle \cdot, x \rangle - f(x),\ x \in \mathrm{dom}\, f\} \subset \Gamma_0(X)$, used in Example 5.1.8, so that

$$\sum_{x \in \mathrm{supp}\,\lambda} \lambda_x f_x = \left\langle \cdot, \sum_{x \in \mathrm{supp}\,\lambda} \lambda_x x \right\rangle - \sum_{x \in \mathrm{supp}\,\lambda} \lambda_x f(x) \text{ for all } \lambda \in \Delta(\mathrm{dom}\, f).$$

Then, by (5.16), for all $x^* \in X^*$ and $\varepsilon > 0$, we obtain

$$\partial_\varepsilon f^*(x^*) = \mathrm{cl} \left\{ \sum_{x \in \mathrm{supp}\,\lambda} \lambda_x x : \begin{array}{l} \alpha \geq 0,\ \lambda \in \Delta(\mathrm{dom}\, f), \\ \sum_{x \in \mathrm{supp}\,\lambda} \lambda_x f_x(x^*) \geq f^*(x^*) + \alpha - \varepsilon \end{array} \right\}$$

$$= \mathrm{cl} \left\{ \sum_{x \in \mathrm{dom}\, f} \lambda_x x : \begin{array}{l} \lambda \in \Delta(\mathrm{dom}\, f), \\ \sum_{x \in \mathrm{supp}\,\lambda} \lambda_x (\langle x^*, x \rangle - f(x)) - f^*(x^*) \geq -\varepsilon \end{array} \right\}.$$

Next, taking $\varepsilon_x := f(x) + f^*(x^*) - \langle x^*, x \rangle$, we get $\varepsilon_x \geq 0$, $\sum_{x \in \mathrm{dom}\, f} \lambda_x \varepsilon_x \leq \varepsilon$, and the last expression simplifies to (5.84).

Exercise 76: We denote $g_n := \max_{1 \leq k \leq n} f_k$, $n \geq 1$, so that $(g_n)_n$ is non-decreasing and $f = \sup_n g_n$. Then, by Example 5.1.6, for every given $x \in X$ and $\varepsilon \geq 0$, we have

$$\partial_\varepsilon f(x) = \bigcap_{\delta > 0} \mathrm{cl} \left(\bigcup_{k \geq 1} \bigcap_{n \geq k} \partial_{\varepsilon+\delta} g_n(x) \right).$$

Hence, given $\delta > 0$ and $U \in \mathcal{N}_{X^*}$ and denoting $h_\lambda := \sum_{1 \leq i \leq n} \lambda_i f_i$, for each $x^* \in \partial_\varepsilon f(x)$ there exists some $k_0 \geq 1$ such that

$$x^* \in \bigcap_{n \geq k_0} \partial_{\varepsilon+\frac{\delta}{2}} g_n(x) + U = \bigcap_{n \geq k_0} \bigcup_{\lambda \in \Delta_n} \partial_{\varepsilon+\frac{\delta}{2}+\Sigma_{1 \leq i \leq n} \lambda_i f_i(x) - g_n(x)} h_\lambda(x) + U,$$

where the last equality comes from Corollary 5.1.9. Hence, if $n_0 \geq k_0$ is such that $g_n(x) \geq f(x) - \frac{\delta}{2}$ for all $n \geq n_0$, we deduce that

$$x^* \in \bigcap_{n \geq n_0} \bigcup_{\lambda \in \Delta_n} \partial_{\varepsilon + \frac{\delta}{2} + \Sigma_{1 \leq i \leq n} \lambda_i f_i(x) - g_n(x)} h_\lambda(x) + U$$

$$\subset \bigcup_{n \geq n_0, \lambda \in \Delta_n} \partial_{\varepsilon + \delta + \Sigma_{1 \leq i \leq n} \lambda_i f_i(x) - f(x)} h_\lambda(x) + U,$$

and, a fortiori, $\partial_\varepsilon f(x) \subset \bigcup_{n \geq 1, \lambda \in \Delta_n} \partial_{\varepsilon + \delta + \Sigma_{1 \leq i \leq n} \lambda_i f_i(x) - f(x)} h_\lambda(x) + U$. Finally, the inclusion "$\subset$" follows by intersecting over $U \in \mathcal{N}_{X^*}$ and after over $\delta > 0$. To verify the converse inclusion, we take $x^* \in \partial_{\varepsilon + \delta + \Sigma_{1 \leq i \leq n} \lambda_i f_i(x) - f(x)} h_\lambda(x)$ for given $\delta > 0$, $n \geq 1$ and $\lambda \in \Delta_n$. Then, for every $y \in X$,

$$\langle x^*, y - x \rangle \leq h_\lambda(y) - h_\lambda(x) + \varepsilon + \delta + h_\lambda(x) - f(x) \leq f(y) - f(x) + \varepsilon + \delta,$$

so that $x^* \in \partial_{\varepsilon + \delta} f(x)$. Consequently, using (4.15),

$$\bigcap_{\delta > 0} \mathrm{cl} \left(\bigcup_{n \geq 1, \lambda \in \Delta_n} \partial_{\varepsilon + \delta + \Sigma_{1 \leq i \leq n} \lambda_i f_i(x) - f(x)} h_\lambda(x) \right) \subset \bigcap_{\delta > 0} \mathrm{cl}\, \partial_{\varepsilon + \delta} f(x) = \partial_\varepsilon f(x).$$

Exercise 77: (i) Fix $x \in \mathrm{dom}\, f$, $\varepsilon \geq 0$ and take $x^* \in \partial_\varepsilon f(x)$. Then, taking into account the Mazur theorem, (5.22) yields

$$\partial_\varepsilon f(x) \subset \bigcap_{\delta > \varepsilon} \overline{\mathrm{co}} \left\{ \bigcup_{J \in \mathcal{T},\, J \subset S_0} \partial_{(\delta + f_J(x) - f(x))} (f_J + \mathrm{I}_D)(x) \right\}$$

$$= \bigcap_{\delta > \varepsilon} \overline{\mathrm{co}}^{\|\cdot\|} \left\{ \bigcup_{J \in \mathcal{T},\, J \subset S_0} \partial_{(\delta + f_J(x) - f(x))} (f_J + \mathrm{I}_D)(x) \right\}. \quad (9.12)$$

Moreover, for every $\alpha \in \Delta_k$ and every selection of finite sets $J_1, \ldots, J_k \subset S_0$, $k \geq 2$, the set $J := \bigcup_{1 \leq i \leq k} J_i \in S_0$ is finite and satisfies, for all $\delta > \varepsilon$,

$$\Sigma_{1 \leq i \leq k} \alpha_i \partial_{(\delta + f_{J_i}(x) - f(x))} (f_{J_i} + \mathrm{I}_D)(x) \subset \partial_{(\delta + \Sigma_{1 \leq i \leq k} \alpha_i f_{J_i}(x) - f(x))} \left(\Sigma_{1 \leq i \leq k} \alpha_i f_{J_i} + \mathrm{I}_D \right)(x)$$

$$\subset \partial_{(\delta + f_J(x) - f(x))} (f_J + \mathrm{I}_D)(x),$$

and consequently, (9.12) implies that

$$\partial_\varepsilon f(x) \subset \bigcap_{\delta > \varepsilon} \mathrm{cl}^{\|\cdot\|} \left\{ \bigcup_{J \in \mathcal{T},\, J \subset S_0} \partial_{(\delta + f_J(x) - f(x))} (f_J + \mathrm{I}_D)(x) \right\}. \quad (9.13)$$

9.4 EXERCISES OF CHAPTER 5

Now, take $x^* \in \partial_\varepsilon f(x)$. Then, by the last relation, for each $n \geq 1$, there exists a (finite) set $J_n \subset T$ such that $J_n \subset S_0$, $f_{J_n}(x) \geq f(x) - \varepsilon - \frac{1}{n}$ ($f_{J_n}(x) \geq f(x) - \varepsilon$, if $\varepsilon > 0$), and

$$x^* \in \partial_{\left(\varepsilon + \frac{1}{n} + f_{J_n}(x) - f(x)\right)} (f_{J_n} + I_D)(x) + \frac{1}{n} B_{X^*}$$
$$\subset \partial_{\left(\varepsilon + \frac{1}{n} + f_J(x) - f(x)\right)} (f_J + I_D)(x) + \frac{1}{n} B_{X^*},$$

where $J := \cup_{n \geq 1} J_n$ and B_{X^*} is the closed unit ball in X^*. Therefore, by taking the limits as $n \to \infty$, we deduce that $x^* \in \partial_{\varepsilon + f_J(x) - f(x)} (f_J + I_D)(x)$, and since J is countable, (9.13) gives rise to the non-trivial inclusion "$\subset$" in (i)

(ii) The proof of this assertion is the same as (i), except that we use (5.23) instead of (5.22).

Exercise 78: (i) We have

$$\sup_{t \in T}(\operatorname{cl} \tilde{f}_t) = \sup_{t \in T}(\operatorname{cl}(f_t(\cdot + x) - f(x))) = \sup_{t \in T}(\operatorname{cl}(f_t(\cdot + x))) - f(x),$$

where $\operatorname{cl}(f_t(\cdot + x))(y) = \liminf_{z \to y} f_t(z + x) = (\operatorname{cl} f_t)(y + x)$ for all $y \in X$. Hence, since the family $\{f_t, t \in T\}$ satisfies (5.10),

$$\sup_{t \in T}(\operatorname{cl} \tilde{f}_t) = \sup_{t \in T}((\operatorname{cl} f_t)(\cdot + x)) - f(x) = (\operatorname{cl} f)(\cdot + x) - f(x) = \operatorname{cl}(f(\cdot + x) - f(x)) = \operatorname{cl} \tilde{f},$$

that is, $\{\tilde{f}_t, t \in T\}$ also satisfies (5.10).

(ii) It is obvious that $\tilde{f}(\theta) = \sup_{t \in T} \tilde{f}_t(\theta) = 0$, while $\partial \tilde{f}(\theta) = \partial f(x)$ and $\partial_\varepsilon \tilde{f}_t(\theta) = \partial_\varepsilon f_t(x)$ for all $t \in T$ and $\varepsilon > 0$. Hence, it suffices to prove formula (5.26) for the functions $\tilde{f}_t$, $t \in T$, as far as $\{t \in T : \tilde{f}_t(\theta) \geq -\varepsilon\} = \{t \in T : f_t(x) \geq f(x) - \varepsilon\} = T_\varepsilon(x)$.

Exercise 79: We denote by B the set in the left-hand side of (5.85). Then, taking into account the non-decreasing character of the A_ε's, we only need to prove that $B \subset \bigcap_{0 < \varepsilon < \varepsilon_0} \operatorname{cl}\{A_\varepsilon + N_C(x)\}$ in the non-trivial case when the sets A_ε's are assumed to be non-empty (otherwise, by (2.6) both sets in (5.85) are empty). Actually, we shall prove that, for every $u \in X$ and $0 < \varepsilon < \varepsilon_0$,

$$\sigma_B(u) \leq \sigma_{A_\varepsilon + N_C(x)}(u). \tag{9.14}$$

More precisely, since $\sigma_{A_\varepsilon + N_C(x)}(u) = \sigma_{A_\varepsilon}(u) + \sigma_{N_C(x)}(u) = \sigma_{A_\varepsilon}(u) + I_{(N_C(x))^\circ}(u)$, we deduce that $\sigma_{A_\varepsilon + N_C(x)}(u) = +\infty$ when $u \notin (N_C(x))^\circ =$

$\overline{\text{cone}}(C-x)$, and (9.14) trivially holds in this case. Assume now that $u \in \overline{\text{cone}}(C-x)$. If $u = y - x$, for some $y \in C$, then for all $0 < \delta < \varepsilon < \varepsilon_0$, we have

$$\sigma_B(u) \le \sigma_{A_\delta}(u) + \sigma_{N_C^\delta(x)}(y-x) \le \sigma_{A_\delta}(u) + \delta \le \sigma_{A_\varepsilon + N_C(x)}(u) + \delta, \tag{9.15}$$

which leads us to $\sigma_B(u) \le \sigma_{A_\varepsilon + N_C(x)}(u)$. Thus, taking into account the positive homogeneity of the support function, (9.14) holds for all $u \in \text{cone}(C-x)$. Thus, we are done under the closedness of $\mathbb{R}_+(C-x)$. Therefore, only the case $u \in \overline{\text{cone}}(C-x) \setminus \text{cone}(C-x)$ remains to be checked when there is some $u_0 \in \text{ri}(C-x) \cap \text{dom}\,\sigma_{A_{\varepsilon_0}}$ ($\subset \text{ri}(\text{cone}(C-x)) \cap \text{dom}\,\sigma_{A_{\varepsilon_0}}$). To this aim, for each $\lambda \in \,]0,1[$, we denote $u_\lambda := \lambda u + (1-\lambda)u_0$. Then, $u_\lambda \in \text{ri}(\text{cone}(C-x)) \subset \text{cone}(C-x)$ by (2.15), and so, by the paragraph above and the convexity of the support function, for every $\lambda \in \,]0,1[$ we obtain

$$\sigma_B(u_\lambda) \le \sigma_{A_\varepsilon + N_C(x)}(u_\lambda) \le \lambda \sigma_{A_\varepsilon + N_C(x)}(u) + (1-\lambda)\sigma_{A_\varepsilon + N_C(x)}(u_0).$$

Hence, using the lower semicontinuity and convexity of σ_B,

$$\sigma_B(u) \le \liminf_{\lambda \to 1} \sigma_B(u_\lambda) \le \sigma_{A_\varepsilon + N_C(x)}(u) + \limsup_{\lambda \uparrow 1}(1-\lambda)\sigma_{A_\varepsilon}(u_0)$$
$$\le \sigma_{A_\varepsilon + N_C(x)}(u) + \limsup_{\lambda \uparrow 1}(1-\lambda)\sigma_{A_{\varepsilon_0}}(u_0) = \sigma_{A_\varepsilon + N_C(x)}(u),$$

and the desired conclusion follows.

Exercise 80: As in the first part of the proof of Theorem 5.2.2, we are going to establish the following inclusion, when $\partial f(x) \ne \emptyset$ ($x = \theta$ and $f(\theta) = 0$ by Exercise 78)

$$\partial f(x) \subset A := \bigcap_{\varepsilon > 0, L \in \mathcal{F}(x)} \overline{\text{co}}\left(\bigcup_{t \in T_\varepsilon(x)} \partial_\varepsilon f_t(x) + N_{L \cap \text{dom}\,f}(x) \right). \tag{9.16}$$

Since some of the f_t's are lsc, but possibly some of them are not proper, we introduce the set of indices $I := \{t \in T : f_t \text{ is not proper}\}$. If $I = \emptyset$, then $f_t \in \Gamma_0(X)$ for all $t \in T$, and we are in the situation of the proof of Theorem 5.2.2. Otherwise, if $I \ne \emptyset$, then we define the functions $g_t : X \to \mathbb{R}_\infty$, $t \in T$, as

$$g_t(z) := \begin{cases} \max\{f_t(z), -1\}, & \text{for } t \in I, \\ f_t(z), & \text{for } t \in T \setminus I, \end{cases}$$

9.4 EXERCISES OF CHAPTER 5

together with the associated supremum function $g := \sup_{t \in T} g_t$; that is, $g = \max\{f, -1\}$, and so $\text{dom } f = \text{dom } g$. It is clear that $g_t \in \Gamma_0(X)$ for all $t \in T$, and that $T \setminus I \neq \emptyset$; if not, since $f_t(\theta) \leq f(\theta) = 0$ for all $t \in T$, then we would have $f_t(\theta) = -\infty$ for all $t \in T$, because these functions are lsc. This would lead to the contradiction $f(\theta) = \sup_{t \in T} f_t(\theta) = -\infty$. Moreover, since $f(\theta) = 0$ and f is lsc, there exists a θ-neighborhood $V \subset X$ such that $f(z) \geq -1$ for all $z \in V$, so that the functions f and g coincide on V. Consequently, since $g_t \in \Gamma_0(X)$ for all $t \in T$, as we have shown in the proof of Theorem 5.2.2 (for the case of functions in $\Gamma_0(X)$), we obtain that

$$\partial f(\theta) = \partial g(\theta) = \bigcap_{L \in \mathcal{F}(\theta),\, \varepsilon > 0} \overline{\text{co}}\left(\bigcup_{t \in T'_\varepsilon(\theta)} \partial_\varepsilon g_t(\theta) + \mathrm{N}_{L \cap \text{dom } g}(\theta)\right), \quad (9.17)$$

where $T'_\varepsilon(\theta) := \{t \in T : g_t(\theta) \geq -\varepsilon\}$, $\varepsilon > 0$. Moreover, since $\partial_\varepsilon g_t(\theta)$ and $T'_\varepsilon(\theta)$ do not increase as $\varepsilon \downarrow 0$, we may restrict ourselves to $\varepsilon \in \,]0,1[$. Take $t \in T'_\varepsilon(\theta)$, with $\varepsilon \in \,]0,1[$. If $t \in I$, then $f_t(\theta) = -\infty$ and so $g_t(\theta) = \max\{f_t(\theta), -1\} = -1 < -\varepsilon$, entailing a contradiction. Hence, $T'_\varepsilon(\theta) \subset T \setminus I$ and $g_t \equiv f_t$; that is, $T'_\varepsilon(\theta) \subset T_\varepsilon(\theta)$ and $\partial_\varepsilon g_t(\theta) = \partial_\varepsilon f_t(\theta)$, and so (9.16) follows as $\mathrm{N}_{L \cap \text{dom } g}(\theta) = \mathrm{N}_{L \cap \text{dom } f}(\theta)$. Thus, taking into account that the opposite inclusion $A \subset \partial f(x)$ always holds, Theorem 5.2.2 is proved in the current case.

Exercise 81: As in the proof of Exercise 80, we suppose that $x = \theta$, $\partial f(\theta) = \partial (\text{cl } f)(\theta) \neq \emptyset$ and $(\text{cl } f)(\theta) = f(\theta) = 0$. Then it suffices to prove that

$$\partial f(\theta) \subset A := \bigcap_{\varepsilon > 0,\, L \in \mathcal{F}(\theta)} \overline{\text{co}}\left(\bigcup_{t \in T_\varepsilon(\theta)} \partial_\varepsilon f_t(\theta) + \mathrm{N}_{L \cap \text{dom } f}(\theta)\right).$$

Let us fix $L \in \mathcal{F}(\theta)$ and define the convex and lsc functions $h_t := (\text{cl } f_t) + \mathrm{I}_{\overline{L \cap \text{dom } f}}$, $t \in T$, and $h := \sup_{t \in T} h_t$, so that

$$h(\theta) = \sup_{t \in T} h_t(\theta) = \sup_{t \in T}(\text{cl } f_t)(\theta) = (\text{cl } f)(\theta) = f(\theta) = 0, \quad (9.18)$$

$$h = \sup_{t \in T}(\text{cl } f_t) + \mathrm{I}_{\overline{L \cap \text{dom } f}} = (\text{cl } f) + \mathrm{I}_{\overline{L \cap \text{dom } f}}; \quad (9.19)$$

hence, $\text{dom } h = \text{dom}(\text{cl } f) \cap (\overline{L \cap \text{dom } f})$ and $L \cap \text{dom } f \subset \text{dom } h$. Now, by combining (9.19) and Remark 1, we obtain that $\partial f(\theta) = \partial (\text{cl } f)(\theta) \subset \partial((\text{cl } f) + \mathrm{I}_{\overline{L \cap \text{dom } f}})(\theta) = \partial h(\theta)$, and therefore (see Exercise 80),

$$\partial f(\theta) \subset \partial h(\theta) \subset \bigcap_{\varepsilon>0} \overline{\operatorname{co}}\left(\bigcup_{t\in \widetilde{T}_\varepsilon(\theta)} \partial_\varepsilon h_t(\theta) + \mathrm{N}_{L\cap \operatorname{dom} h}(\theta)\right), \qquad (9.20)$$

where $\widetilde{T}_\varepsilon(\theta) := \{t\in T:\ 0\geq h_t(\theta)\geq -\varepsilon\}$. Take $t\in \widetilde{T}_\varepsilon(\theta)$ with $\varepsilon>0$, so that $-\varepsilon\leq h_t(\theta) = (\operatorname{cl} f_t)(\theta)\leq f_t(\theta)\leq f(\theta) = 0$, $t\in T_\varepsilon(\theta)$, and $\operatorname{cl} f_t$ is proper. Moreover, since $\partial_\varepsilon(\operatorname{cl} f_t)(\theta)\subset \partial_{2\varepsilon} f_t(\theta)$, by (4.46), we obtain

$$\partial_\varepsilon h_t(\theta) \subset \operatorname{cl}(\partial_\varepsilon(\operatorname{cl} f_t)(\theta) + \mathrm{N}^\varepsilon_{\overline{L\cap \operatorname{dom} f}}(\theta)) \subset \operatorname{cl}(\partial_{2\varepsilon} f_t(\theta) + \mathrm{N}^\varepsilon_{L\cap \operatorname{dom} f}(\theta)).$$

Moreover, using (9.19),

$$\partial f(\theta) \subset \bigcap_{\varepsilon>0} \operatorname{cl}\left(E_\varepsilon + \mathrm{N}^\varepsilon_{L\cap \operatorname{dom} f}(\theta) + \mathrm{N}_{L\cap \operatorname{dom} h}(\theta)\right) \subset \bigcap_{\varepsilon>0} \operatorname{cl}\left(E_\varepsilon + \mathrm{N}^\varepsilon_{L\cap \operatorname{dom} f}(\theta)\right), \qquad (9.21)$$

where $E_\varepsilon := \operatorname{co}\left\{\bigcup_{t\in T_\varepsilon(\theta)} \partial_{2\varepsilon} f_t(\theta)\right\}$. But for every $z\in \operatorname{dom} f$ and $\varepsilon>0$, we have

$$\sigma_{E_\varepsilon}(z) = \sup_{t\in T_\varepsilon(\theta),\, z^*\in \partial_{2\varepsilon} f_t(\theta)} \langle z^*, z\rangle \leq \sup_{t\in T_\varepsilon(\theta)} (f_t(z) - f_t(\theta) + 2\varepsilon) \leq f(z) + 3\varepsilon < +\infty,$$

showing that $(\emptyset \neq \operatorname{ri}(L\cap \operatorname{dom} f)) \subset \operatorname{dom} f \subset \operatorname{dom} \sigma_{E_\varepsilon}$. So, (see Exercise 79)

$$\bigcap_{\varepsilon>0} \operatorname{cl}\left(E_\varepsilon + \mathrm{N}^\varepsilon_{L\cap \operatorname{dom} f}(\theta)\right) = \bigcap_{\varepsilon>0} \operatorname{cl}\left(E_\varepsilon + \mathrm{N}_{L\cap \operatorname{dom} f}(\theta)\right)$$
$$\subset \bigcap_{\varepsilon>0} \operatorname{cl}\left(\operatorname{co}\left\{\bigcup_{t\in T_{2\varepsilon}(\theta)} \partial_{2\varepsilon} f_t(\theta)\right\} + \mathrm{N}_{L\cap \operatorname{dom} f}(\theta)\right),$$

and the desired result follows from (9.21), as L was arbitrarily chosen in $\mathcal{F}(\theta)$.

Exercise 82: We only give the proof for formula (5.26). If

$$A := \bigcap_{\varepsilon>0,\, L\in \mathcal{F}(x)} \overline{\operatorname{co}}\left(\bigcup_{t\in T_\varepsilon(x)} \partial_\varepsilon f_t(x) + \mathrm{N}_{L\cap \operatorname{dom} f}(x)\right),$$

then we have

9.4 EXERCISES OF CHAPTER 5

$$A \subset \bigcap_{\substack{\varepsilon > 0 \\ L \in \mathcal{F}(x, A)}} \overline{\mathrm{co}} \left(\bigcup_{t \in T_\varepsilon(x)} \partial_\varepsilon f_t(x) + \mathrm{N}_{L \cap \mathrm{dom}\, f}(x) \right)$$

$$= \bigcap_{\substack{\varepsilon > 0 \\ L \in \mathcal{F}(x)}} \overline{\mathrm{co}} \left(\bigcup_{t \in T_\varepsilon(x)} \partial_\varepsilon f_t(x) + \mathrm{N}_{(L+\mathrm{span}\{A\}) \cap \mathrm{dom}\, f}(x) \right)$$

$$\subset \bigcap_{\substack{\varepsilon > 0 \\ L \in \mathcal{F}(x)}} \overline{\mathrm{co}} \left(\bigcup_{t \in T_\varepsilon(x)} \partial_\varepsilon f_t(x) + \mathrm{N}_{L \cap \mathrm{dom}\, f}(x) \right).$$

Exercise 83: Assertion (i) is a mere application of the Mazur theorem, while (ii) follows because the functions $f_t + \mathrm{I}_{L \cap \mathrm{dom}\, f}$ are defined in the finite-dimensional linear subspace L.

Exercise 84: See the bibliographical notes in chapter 5.

Exercise 85: Apply Proposition 5.3.1 and Exercise 31.

Exercise 86: Like in Exercise 85, apply Proposition 5.3.1 and Exercise 31.

Exercise 87: We may assume that $\partial f(x) \neq \emptyset$, so that $f(x) = (\mathrm{cl}\, f)(x)$ and $\partial f(x) = \partial(\mathrm{cl}\, f)(x)$ (Exercise 62). Then, by applying Theorem 5.2.12 to the family $\{\mathrm{cl}\, f_t,\ t \in T\}$, whose supremum is $\mathrm{cl}\, f$, we obtain

$$\partial f(x) = \partial(\mathrm{cl}\, f)(x)$$

$$\subset \bigcap_{\substack{\varepsilon > 0 \\ L \in \mathcal{F}(x), p \in \mathcal{P}}} \overline{\mathrm{co}} \left(\bigcup_{\substack{t \in \tilde{T}_\varepsilon(x) \\ y \in \hat{B}_{p,t}(x, \varepsilon)}} \partial((\mathrm{cl}\, f_t) + \mathrm{I}_{\overline{L \cap \mathrm{dom}(\mathrm{cl}\, f)}})(y) \cap S_\varepsilon(y - x) \right),$$

where $\hat{B}_{p,t}(x, \varepsilon) := \{y \in X : p(y - x) \leq \varepsilon,\ |(\mathrm{cl}\, f_t)(y) - (\mathrm{cl}\, f_t)(x)| \leq \varepsilon\}$ and $\tilde{T}_\varepsilon(x) := \{t \in T : (\mathrm{cl}\, f_t)(x) \geq f(x) - \varepsilon\}$. Observe that $f_t(x) - \varepsilon \leq f(x) - \varepsilon \leq (\mathrm{cl}\, f_t)(x) \leq f_t(x)$ for all $t \in \tilde{T}_\varepsilon(x)$, entailing that $|(\mathrm{cl}\, f_t)(x) - f_t(x)| \leq \varepsilon$. Consequently, $|(\mathrm{cl}\, f_t)(y) - f_t(x)| \leq |(\mathrm{cl}\, f_t)(y) - (\mathrm{cl}\, f_t)(x)| + |(\mathrm{cl}\, f_t)(x) - f_t(x)| \leq 2\varepsilon$ for all $t \in \tilde{T}_\varepsilon(x)$; that is, $\hat{B}_{p,t}(x, \varepsilon) \subset B_{p,t}(x, 2\varepsilon)$. Therefore, denoting $g_L := \mathrm{I}_{L \cap \mathrm{dom}\, f}$, the inclusion "$\subset$" follows as

$$\partial f(x) \subset \bigcap_{\substack{\varepsilon > 0 \\ L \in \mathcal{F}(x), p \in \mathcal{P}}} \overline{\mathrm{co}} \left(\bigcup_{\substack{t \in \tilde{T}_\varepsilon(x) \\ y \in B_{p,t}(x, 2\varepsilon)}} \partial((\mathrm{cl}\, f_t) + g_L)(y) \cap S_\varepsilon(y - x) \right).$$

To establish the opposite inclusion "⊃", we first verify that, for all $y \in B_{p,t}(x, \varepsilon)$ and $t \in \tilde{T}_\varepsilon(x)$, we have that $\partial((\mathrm{cl}\, f_t) + g_L)(y) \cap S_\varepsilon(y - x) \subset \partial_{3\varepsilon}(f + g_L)(x)$ (as in (5.59)). Indeed, if $z^* \in \partial((\mathrm{cl}\, f_t) + g_L)(y)$, so that $y \in L \cap \mathrm{dom}\, f$, then for all $z \in L \cap \mathrm{dom}\, f$, we get

$$\langle z^*, z - y \rangle \leq (\mathrm{cl}\, f_t)(z) + g_L(z) - (\mathrm{cl}\, f_t)(y) - g_L(y)$$
$$\leq (\mathrm{cl}\, f_t)(z) - f_t(x) + \varepsilon \leq f(z) - f(x) + 2\varepsilon.$$

If, in addition, $z^* \in S_\varepsilon(y - x)$, then we obtain $\langle z^*, z - x \rangle \leq \langle z^*, z - y \rangle + \langle z^*, y - x \rangle \leq f(z) - f(x) + 3\varepsilon$, proving that $z^* \in \partial_{3\varepsilon}(f + g_L)(x)$. Therefore, we deduce that

$$\bigcap_{\substack{\varepsilon > 0 \\ L \in \mathcal{F}(x), p \in \mathcal{P}}} \overline{\mathrm{co}} \left(\bigcup_{\substack{t \in \tilde{T}_\varepsilon(x) \\ y \in B_{p,t}(x, 2\varepsilon)}} \partial((\mathrm{cl}\, f_t) + g_L)(y) \cap S_\varepsilon(y - x) \right) \subset \bigcap_{\substack{\varepsilon > 0 \\ L \in \mathcal{F}(x)}} \partial_{3\varepsilon}(f + g_L)(x) = \partial f(x).$$

Exercise 88: Assertion (i) is a straightforward consequence of (5.42). Assertion (ii) comes from (5.60). Finally, (iii) is implied by (5.92) as $f_t + \mathrm{I}_{\overline{\mathrm{dom}\, f}} = f_t + \mathrm{I}_{\overline{\mathrm{dom}\, f_t}} = f_t + \mathrm{I}_{\mathrm{dom}\, f_t} + \mathrm{I}_{\overline{\mathrm{dom}\, f_t}} = f_t$, for all $t \in T$. Indeed, this observation together with formula (5.92) yields

$$\partial f(x) = \bigcap_{\varepsilon > 0} \overline{\mathrm{co}} \left(\bigcup_{t \in T_\varepsilon(x),\, y \in B_t(x, \varepsilon)} \partial f_t(y) \cap S_\varepsilon(y - x) \cap \partial_\varepsilon f(x) \right)$$
$$\subset \bigcap_{\varepsilon > 0} \overline{\mathrm{co}} \left(\bigcup_{t \in T_\varepsilon(x),\, y \in B_t(x, \varepsilon)} \partial f_t(y) \cap S_\varepsilon(y - x) \right) =: A.$$

Also, since $\partial f_t(y) \cap S_\varepsilon(y - x) \subset \partial_{3\varepsilon} f(x)$ by (5.45), we have that $A \subset \bigcap_{\varepsilon > 0} \partial_{3\varepsilon} f(x) = \partial f(x)$.

Exercise 89: Since condition (5.10) holds (as f is continuous at x_0, Proposition 5.2.4(i)), and all the lsc convex proper functions $f_{\varepsilon_0} := \sup_{t \in T_{\varepsilon_0}(x)} f_t$, $\varepsilon_0 > 0$, are finite and continuous at x_0 (again by the continuity of f at x_0), Corollary 5.3.3 yields $\partial f(x) = \mathrm{N}_{\mathrm{dom}\, f}(x) + \partial f_{\varepsilon_0}(x)$ for all $\varepsilon_0 > 0$. Hence, taking into account that the (lsc) functions f_t, $t \in T_{\varepsilon_0}(x)$, are proper, from the proof of (5.72) in Theorem 5.3.5, we obtain, for any given $\varepsilon_0 > 0$,

9.4 EXERCISES OF CHAPTER 5

$$\partial f(x) = \mathrm{N}_{\mathrm{dom}\, f}(x) + \mathrm{N}_{\mathrm{dom}\, f_{\varepsilon_0}}(x) + \bigcap_{\varepsilon>0, p\in\mathcal{P}} \overline{\mathrm{co}} \left\{ \bigcup_{t\in\hat{T}_\varepsilon(x), y\in\hat{B}_{p,t}(x,\varepsilon)} \partial f_t(y) \cap S_\varepsilon(y-x) \right\}$$

$$= \mathrm{N}_{\mathrm{dom}\, f}(x) + \mathrm{N}_{\mathrm{dom}\, f_{\varepsilon_0}}(x) + \bigcap_{0<\varepsilon<\varepsilon_0, p\in\mathcal{P}} \overline{\mathrm{co}} \left\{ \bigcup_{t\in\hat{T}_\varepsilon(x), y\in\hat{B}_{p,t}(x,\varepsilon)} \partial f_t(y) \cap S_\varepsilon(y-x) \right\},$$

where, for $0 < \varepsilon < \varepsilon_0$,

$$\hat{T}_\varepsilon(x) := \{t \in T_{\varepsilon_0}(x) : f_t(x) \geq f_{\varepsilon_0}(x) - \varepsilon\}$$
$$= \{t \in T_{\varepsilon_0}(x) : f_t(x) \geq f(x) - \varepsilon\} = T_\varepsilon(x).$$

Moreover, we have that $\mathrm{N}_{\mathrm{dom}\, f}(x) + \mathrm{N}_{\mathrm{dom}\, f_{\varepsilon_0}}(x) = \mathrm{N}_{\mathrm{dom}\, f \cap \mathrm{dom}\, f_{\varepsilon_0}}(x) = \mathrm{N}_{\mathrm{dom}\, f}(x)$, coming from Proposition 4.1.20, and the equality above entails

$$\partial f(x) = \mathrm{N}_{\mathrm{dom}\, f}(x) + \bigcap_{0<\varepsilon<\varepsilon_0, p\in\mathcal{P}} \overline{\mathrm{co}} \left\{ \bigcup_{t\in T_\varepsilon(x), y\in B_{p,t}(x,\varepsilon)} \partial f_t(y) \cap S_\varepsilon(y-x) \right\}$$

$$= \mathrm{N}_{\mathrm{dom}\, f}(x) + \bigcap_{\varepsilon>0, p\in\mathcal{P}} \overline{\mathrm{co}} \left\{ \bigcup_{t\in T_\varepsilon(x), y\in B_{p,t}(x,\varepsilon)} \partial f_t(y) \cap S_\varepsilon(y-x) \right\}.$$

Exercise 90: We denote

$$\tilde{A}_{\varepsilon,p} := \bigcup_{t\in T_\varepsilon(x), y\in \tilde{B}_{p,t}(x,\varepsilon)} \partial(\mathrm{cl}\, f_t)(y) \cap S_\varepsilon(y-x).$$

The reader will easily verify, like in the proof of Theorem 5.2.7, that the inclusion $\partial(\mathrm{cl}\, f_t)(y) \cap S_\varepsilon(y-x) \subset \partial_{3\varepsilon} f(x)$ holds for all $\varepsilon > 0$, $t \in T_\varepsilon(x)$ and $y \in \tilde{B}_{p,t}(x,\varepsilon)$. This entails

$$\mathrm{N}_{\mathrm{dom}\, f}(x) + \bigcap_{\varepsilon>0, p\in\mathcal{P}} \overline{\mathrm{co}}(\tilde{A}_{\varepsilon,p}) \subset \mathrm{N}_{\mathrm{dom}\, f}(x) + \bigcap_{\varepsilon>0} \partial_{3\varepsilon} f(x) = \mathrm{N}_{\mathrm{dom}\, f}(x) + \partial f(x) = \partial f(x),$$

and the inclusion "$\subset$" follows. Thus, we are done if $\partial f(x) = \emptyset$, and we only need to prove the statement when $\partial f(x) \neq \emptyset$; hence $\partial f(x) = \partial(\mathrm{cl}\, f)(x)$ and $f(x) = (\mathrm{cl}\, f)(x)$ (Exercise 62). Indeed, under the current continuity condition, (5.10) holds (Proposition 5.2.4(i)); that is, $\mathrm{cl}\, f = \sup_{t \in T}(\mathrm{cl}\, f_t)$. Thus, by applying formula (5.72) to the family $\{\mathrm{cl}\, f_t, t \in T\}$, we obtain

$$\partial f(x) = \partial(\mathrm{cl}\, f)(x) = \mathrm{N}_{\mathrm{dom}(\mathrm{cl}\, f)}(x) + \bigcap_{\varepsilon>0, p\in\mathcal{P}} \overline{\mathrm{co}} \left\{ \bigcup_{t\in\hat{T}_\varepsilon(x), y\in\hat{B}_{p,t}(x,\varepsilon)} \partial(\mathrm{cl}\, f_t)(y) \cap S_\varepsilon(y-x) \right\},$$

where $\hat{T}_\varepsilon(x) := \{t \in T : (\mathrm{cl}\, f_t)(x) \geq (\mathrm{cl}\, f)(x) - \varepsilon\} = \{t \in T : (\mathrm{cl}\, f_t)(x) \geq f(x) - \varepsilon\} \subset T_\varepsilon(x)$ and $\hat{B}_{p,t}(x,\varepsilon) := \{y \in X : p(y-x) \leq \varepsilon, |(\mathrm{cl}\, f_t)(y) - (\mathrm{cl}\, f_t)(x)| \leq \varepsilon\}$. Observe that for all $t \in \hat{T}_\varepsilon(x)$ and $y \in \hat{B}_{p,t}(x,\varepsilon)$, we have $f_t(x) \geq (\mathrm{cl}\, f_t)(x) \geq f(x) - \varepsilon \geq f_t(x) - \varepsilon$ and

$$|(\mathrm{cl}\, f_t)(y) - f_t(x)| \leq |(\mathrm{cl}\, f_t)(y) - (\mathrm{cl}\, f_t)(x)| + |(\mathrm{cl}\, f_t)(x) - f_t(x)| \leq 2\varepsilon;$$

that is, $\hat{B}_{p,t}(x,\varepsilon) \subset \tilde{B}_{p,t}(x, 2\varepsilon)$. So,

$$\partial f(x) \subset \mathrm{N}_{\mathrm{dom}\, f}(x) + \bigcap_{\varepsilon > 0, p \in \mathcal{P}} \overline{\mathrm{co}}\left\{\bigcup_{t \in T_\varepsilon(x), y \in \tilde{B}_{p,t}(x, 2\varepsilon)} \partial(\mathrm{cl}\, f_t)(y) \cap S_\varepsilon(y-x)\right\}$$

$$\subset \mathrm{N}_{\mathrm{dom}\, f}(x) + \bigcap_{\varepsilon > 0, p \in \mathcal{P}} \overline{\mathrm{co}}\left\{\bigcup_{t \in T_{2\varepsilon}(x), y \in \tilde{B}_{p,t}(x, 2\varepsilon)} \partial(\mathrm{cl}\, f_t)(y) \cap S_{2\varepsilon}(y-x)\right\},$$

and we are done.

Exercise 91: According to Corollary 5.2.3, we have $\partial f(x) = \mathrm{N}_{\mathrm{dom}\, f}(x) + \partial f_{\varepsilon_0}(x)$, and by applying Theorem 5.3.5 to the family $\{f_t, t \in T_{\varepsilon_0}(x)\}$, we get

$$\partial f_{\varepsilon_0}(x) = \mathrm{N}_{\mathrm{dom}\, f_{\varepsilon_0}}(x) + \bigcap_{\varepsilon > 0} \overline{\mathrm{co}}\left\{\bigcup_{t \in \tilde{T}_\varepsilon(x)} \partial_\varepsilon f_t(x)\right\}$$

$$= \mathrm{N}_{\mathrm{dom}\, f_{\varepsilon_0}}(x) + \bigcap_{0 < \varepsilon < \varepsilon_0} \overline{\mathrm{co}}\left\{\bigcup_{t \in \tilde{T}_\varepsilon(x)} \partial_\varepsilon f_t(x)\right\},$$

where $\tilde{T}_\varepsilon(x) := \{t \in T_{\varepsilon_0}(x) : f_t(x) \geq f_{\varepsilon_0}(x) - \varepsilon\} = \{t \in T_{\varepsilon_0}(x) : f_t(x) \geq f(x) - \varepsilon\}$; hence, $\tilde{T}_\varepsilon(x) = T_\varepsilon(x)$ for all $0 < \varepsilon < \varepsilon_0$. Therefore,

$$\partial f(x) = \mathrm{N}_{\mathrm{dom}\, f}(x) + \mathrm{N}_{\mathrm{dom}\, f_{\varepsilon_0}}(x) + \bigcap_{0 < \varepsilon < \varepsilon_0} \overline{\mathrm{co}}\left\{\bigcup_{t \in T_\varepsilon(x)} \partial_\varepsilon f_t(x)\right\},$$

and so, using the fact that $\mathrm{N}_{\mathrm{dom}\, f}(x) + \mathrm{N}_{\mathrm{dom}\, f_{\varepsilon_0}}(x) = \mathrm{N}_{\mathrm{dom}\, f \cap \mathrm{dom}\, f_{\varepsilon_0}}(x) = \mathrm{N}_{\mathrm{dom}\, f}(x)$, coming from Proposition 4.1.20, we obtain that

$$\partial f(x) = \mathrm{N}_{\mathrm{dom}\, f}(x) + \bigcap_{0 < \varepsilon < \varepsilon_0} \overline{\mathrm{co}}\left\{\bigcup_{t \in T_\varepsilon(x)} \partial_\varepsilon f_t(x)\right\}$$

$$= \mathrm{N}_{\mathrm{dom}\, f}(x) + \bigcap_{\varepsilon > 0} \overline{\mathrm{co}}\left\{\bigcup_{t \in T_\varepsilon(x)} \partial_\varepsilon f_t(x)\right\}.$$

9.5 Exercises of chapter 6

Exercise 92: The inclusion "⊃" in (6.16) comes from formula (5.26) as $T(x) \subset T_\varepsilon(x)$. To prove the inclusion "⊃" in (6.12), take $x^* \in \partial(f_t + I_{L \cap \text{dom} f})(x)$ with $L \in \mathcal{F}(x)$ and $t \in T(x)$. Then $\langle x^*, y - x \rangle \leq f_t(y) - f_t(x) \leq f(y) - f(x)$ for all $y \in L \cap \text{dom} f$, and $x^* \in \partial(f + I_{L \cap \text{dom} f})(x) = \partial(f + I_L)(x)$. Hence, by (4.16),

$$\bigcap_{L \in \mathcal{F}(x)} \text{co} \left\{ \bigcup_{t \in T(x)} \partial(f_t + I_{L \cap \text{dom} f})(x) \right\} \subset \bigcap_{L \in \mathcal{F}(x)} \partial(f + I_L)(x) = \partial f(x).$$

Exercise 93: We apply Theorem 6.1.4 to the (two-elements) family $\{f, f(x) - 1\}$. Indeed, since f is the unique active function of this family at x, we obtain

$$\partial g(x) = \bigcap_{L \in \mathcal{F}(x)} \partial(f + I_{L \cap \text{dom} g})(x) = \bigcap_{L \in \mathcal{F}(x)} \partial(f + I_L)(x) = \partial f(x).$$

Exercise 94: Take a sequence $(t_k)_k \subset T$ such that $f_{t_k}(x) \to f(x)$; for instance, choose $t_k \in T$ such that $f_{t_k}(x) > f(x) - 1/k$. Since T is compact, there exists a subnet $(t_{k_i})_i$ that converges to some $t \in T$. Hence, by the upper semicontinuity assumption, $f(x) = \lim_k f_{t_k}(x) = \lim_i f_{t_{k_i}}(x) \leq \limsup_{s \to t} f_s(x) \leq f_t(x)$ and $t \in T(x) \subset T_\varepsilon(x)$ for all $\varepsilon \geq 0$; that is, the sets $T_\varepsilon(x)$ are non-empty. Next, we proceed by showing that each $T_\varepsilon(x)$, $\varepsilon \geq 0$, is closed, and so compact. Indeed, given a net $(t_i)_i \subset T_\varepsilon(x)$ converging to $t \in T$, we have that $f(x) - \varepsilon \leq \limsup_i f_{t_i}(x) \leq \limsup_{s \to t} f_s(x) \leq f_t(x)$, and $t \in T_\varepsilon(x)$.

Exercise 95: (*i*) The set βT is a closed subset of the compact $\mathbb{S}$, and so it is compact.

(*ii*) If $(t_i)_i \subset T$ is a net converging to $t \in T$, then $\gamma_{t_i}(\varphi) = \varphi(t_i) \to \varphi(t) = \gamma_t(\varphi)$, for every $\varphi \in \mathcal{C}(T, [0,1])$; that is, $\mathfrak{w}(t_i) = \gamma_{t_i} \to \gamma_i = \mathfrak{w}(t)$ and $\mathfrak{w}$ is continuous. If T is compact, then its image by the continuous mapping $\mathfrak{w}$, $\beta T = \mathfrak{w}(T)$, is also compact in $\mathbb{S}$, and so closed as $\mathbb{S}$ is endowed with product topology which is Hausdorff.

(*iii*) If $t_i \to t$ in T, then $\gamma_{t_i} \to \gamma_t$ by the continuity of $\mathfrak{w}$. If $(t_i)_i \subset T$ and $t \in T$ are such that $\gamma_{t_i} \to \gamma_t$, we suppose by contradiction that $t_i \not\to t$. Then we find an open neighborhood V of t such that $t_i \notin V$, frequently. By the complete regularity of T, there exists $\varphi \in \mathcal{C}(T, [0,1])$ such that $\varphi(t) = 1$ and $\varphi(t_i) = 0$, frequently. But $\gamma_{t_i} \to \gamma_t$ implies that $\varphi(t_i) = 0 \to \varphi(t) = 1$, which is a contradiction.

Exercise 96: By proceeding as in the proof of Exercise 9, we see that condition (6.1) entails $U \subset \text{dom } f$; even more, we get $U \subset f^{-1}(\mathbb{R})$. Moreover, since the family $\{f_t,\ t \in T\}$ satisfies (5.10) (by Proposition 5.2.4(i)), for all $x \in U$ we have $(\text{cl } f)(x) = \sup_{t \in T}(\text{cl } f_t)(x) = \sup_{t \in T} f_t(x) = f(x)$, and f is lsc on U. Now, given $x \in U$, we choose $r > 0$ such that $\text{cl}(B_X(x, r)) \subset U$, and consider the function $\tilde{f} := f + \text{I}_{\text{cl}(B_X(x,r))}$. Then $\tilde{f} \in \Gamma_0(X)$ and $\text{int}(\text{dom } \tilde{f}) = B_X(x, r) \neq \emptyset$, so that Corollary 2.2.8 implies the continuity of $\tilde{f}$ on $B_X(x, r)$.

Exercise 97: It suffices to prove the inclusion "$\subset$" for x such that $\partial f(x) \neq \emptyset$; hence f^* is proper, $f(x) = f^{**}(x)$, and $\partial f(x) = \partial(\overline{\text{co}} f)(x) = \partial f^{**}(x)$. Thus, applying (6.73) to the family $\{f_t^{**},\ t \in T\}$,

$$\partial f(x) = \partial f^{**}(x) = \bigcap_{L \in \mathcal{F}(x)} \text{co}\left\{ \bigcap_{\varepsilon > 0} \text{cl}\left(\bigcup_{t \in T_\varepsilon^1(x)} \partial_\varepsilon f_t^{**}(x) + \text{N}_{L \cap \text{dom } f^{**}}(x) \right) \right\},$$

where $T_\varepsilon^1(x) := \{t \in T : f_t^{**}(x) \geq f(x) - \varepsilon\}$. Observe that every $t \in T_\varepsilon^1(x)$ satisfies $f_t(x) \geq f_t^{**}(x) \geq f(x) - \varepsilon \geq f_t(x) - \varepsilon$, so that $t \in T_\varepsilon(x)$ and $\partial_\varepsilon f_t^{**}(x) \subset \partial_{2\varepsilon} f_t(x)$. Additionally, the inequality $f^{**} \leq f$ implies that $\text{N}_{L \cap \text{dom } f^{**}}(x) \subset \text{N}_{L \cap \text{dom } f}(x)$, and the desired inclusion follows.

Exercise 98: It is clear that $\tilde{f} \leq f$, so that $\text{dom } f \subset \text{dom } \tilde{f}$ and $\text{N}_{\text{dom } \tilde{f}}(x) \subset \text{N}_{\text{dom } f}(x)$. Thus, since $\inf_{t \in T} \tilde{f}_t(x) > -\infty$, Proposition 6.4.1 implies that, for every fixed $\varepsilon > 0$,

$$\text{N}_{\text{dom } \tilde{f}}(x) = \left[\overline{\text{co}}\left(\cup_{t \in T} \partial_\varepsilon \tilde{f}_t(x) \right)\right]_\infty =: C.$$

So, we only need to verify that $\text{N}_{\text{dom } f}(x) \subset C$. Equivalently, due to (3.51) and the relation

$$\text{cl}\left(\text{dom}\left(\sigma_{\cup_{t \in T} \partial_\varepsilon \tilde{f}_t(x)}\right)\right) = C^-$$

coming from (3.52), we show that $\text{cl}\left(\text{dom}\left(\sigma_{\cup_{t \in T} \partial_\varepsilon \tilde{f}_t(x)}\right)\right) \subset \text{cl}(\mathbb{R}_+ (\text{dom } f - x))$. To this aim, taking into account (4.28) and (4.1), we pick

$$z \in \text{dom}\left(\sigma_{\cup_{t \in T} \partial_\varepsilon \tilde{f}_t(x)}\right) = \text{dom}\left(\sup_{t \in T} \sigma_{\partial_\varepsilon \tilde{f}_t(x)}\right);$$

that is,

9.5 EXERCISES OF CHAPTER 6

$$z \in \text{dom}\left(\sup_{t \in T}(\tilde{f}_t)'_\varepsilon(x;\cdot)\right) \subset \bigcap_{t \in T} \text{dom}(\tilde{f}_t)'_\varepsilon(x;\cdot) = \bigcap_{t \in T} \mathbb{R}_+(\text{dom}\,\tilde{f}_t - x). \tag{9.22}$$

Observe that, for every $t \in T \setminus T(x)$,

$$\text{dom}(\tilde{f}_t)'_\varepsilon(x;\cdot) = \mathbb{R}_+((\text{dom}(\mu_t f_t) \cap \text{dom}((1-\mu_t)f_{t_0})) - x)$$
$$= \mathbb{R}_+((\text{dom}\,f_t \cap \text{dom}\,f_{t_0}) - x) = (\mathbb{R}_+\,\text{dom}(f_t - x)) \cap (\mathbb{R}_+\,\text{dom}(f_{t_0} - x)).$$

Then, denoting $h_t := (f_t)'_\varepsilon(x;\cdot)$ if $t \in T(x)$, and $h_t := (\tilde{f}_t)'_\varepsilon(x;\cdot)$ if $t \in T \setminus T(x)$, relation (9.22) entails

$$z \in \bigcap_{t \in T(x)} \text{dom}\,h_t$$
$$= \left(\bigcap_{t \in T(x)} \mathbb{R}_+(\text{dom}\,f_t - x)\right) \cap \left(\bigcap_{t \in T \setminus T(x)} (\mathbb{R}_+\,\text{dom}(f_t - x)) \cap (\mathbb{R}_+\,\text{dom}(f_{t_0} - x))\right),$$

and so (Exercise 9),

$$z \in \bigcap_{t \in T} \mathbb{R}_+\,\text{dom}(f_t - x) = \mathbb{R}_+(\text{dom}\,f - x) \subset \text{cl}(\mathbb{R}_+(\text{dom}\,f - x)).$$

Exercise 99: Let us suppose, for simplicity, that $f(x) = 0$. Fix $\varepsilon > 0$. Then $0 < \rho_{t,\varepsilon} < 1$ and $\tilde{f}_{t,\varepsilon}(x) = \rho_{t,\varepsilon}f_t(x) > -\frac{\varepsilon}{2}$ for every $t \in T \setminus T(x)$. Moreover, since $\tilde{f}_{t,\varepsilon} \leq \max\{f_t, f_{t_0}\} \leq f$, we have $\cup_{t \in T \setminus T(x)} \partial_\varepsilon \tilde{f}_{t,\varepsilon}(x) \subset \partial_{\frac{3\varepsilon}{2}} f(x) \subset \partial_{2\varepsilon} f(x)$ and $\cup_{t \in T(x)} \partial_\varepsilon \tilde{f}_{t,\varepsilon}(x) \subset \partial_\varepsilon f(x) \subset \partial_{2\varepsilon} f(x)$. So, $\cup_{t \in T} \partial_\varepsilon \tilde{f}_{t,\varepsilon}(x) \subset \partial_{2\varepsilon} f(x)$ and the inclusion "$\supset$" follows by taking the closed convex hull and intersecting over $\varepsilon > 0$ (using (4.15)). To verify the inclusion "$\subset$", we fix $\varepsilon > 0$ and $L \in \mathcal{F}(x)$. Since the family $\{\tilde{f}_{t,\varepsilon}, t \in T; I_L\}$ satisfies $\inf_{t \in T} \tilde{f}_{t,\varepsilon}(x) \geq -\frac{\varepsilon}{2}$, we can show that (see Exercise 98)

$$N_{L \cap \text{dom}\,f}(x) = \left[\overline{\text{co}}\left(\bigcup_{t \in T} \partial_\varepsilon \tilde{f}_{t,\varepsilon}(x) \cup L^\perp\right)\right]_\infty = \left[\overline{\text{co}}\left(\bigcup_{t \in T} \partial_\varepsilon \tilde{f}_{t,\varepsilon}(x) + L^\perp\right)\right]_\infty,$$

where the last equality comes from (8.24). Therefore, by (6.16),

$$\partial f(x) \subset \overline{\operatorname{co}}\left(\bigcup_{t\in T(x)} \partial_\varepsilon f_t(x) + \mathrm{N}_{L\cap \operatorname{dom} f}(x)\right)$$

$$= \overline{\operatorname{co}}\left(\bigcup_{t\in T(x)} \partial_\varepsilon f_t(x) + \left[\overline{\operatorname{co}}\left(\bigcup_{t\in T} \partial_\varepsilon \tilde{f}_{t,\varepsilon}(x) + L^\perp\right)\right]_\infty\right)$$

$$\subset \overline{\operatorname{co}}\left(\bigcup_{t\in T} \partial_\varepsilon \tilde{f}_{t,\varepsilon}(x) + L^\perp\right) = \operatorname{cl}\left(\operatorname{co}\left(\bigcup_{t\in T} \partial_\varepsilon \tilde{f}_{t,\varepsilon}(x)\right) + L^\perp\right).$$

Intersecting over the L's in $\mathcal{F}(x)$, we get

$$\partial f(x) \subset \bigcap_{L\in\mathcal{F}(x)} \operatorname{cl}\left(\operatorname{co}\left(\bigcup_{t\in T} \partial_\varepsilon \tilde{f}_{t,\varepsilon}(x)\right) + L^\perp\right) = \overline{\operatorname{co}}\left(\bigcup_{t\in T} \partial_\varepsilon \tilde{f}_{t,\varepsilon}(x)\right),$$

where the last equality is found in Exercise 10(i).

Exercise 100: Fix positive integers m, n with $m > f(x)$ and take $\delta > 0$. Since $u^* \in \mathrm{N}_{\operatorname{dom} f}(x) \subset \mathrm{N}_{L\cap\operatorname{dom} f}(x)$, for every $y \in L$ we have $f(y) \leq m \Rightarrow \langle nu^*, y-x\rangle \leq 0 < \delta$; that is, $\langle nu^*, y-x\rangle \geq \delta$, $y \in L \Rightarrow f(y) > m$, and so $\langle nu^*, y-x\rangle \geq \delta$, $y \in L \Rightarrow \exists\, t \in T$ such that $f_t(y) > m$. In other words, $\langle nu^*, y-x\rangle \geq \delta$, $y \in L \Rightarrow y \in \bigcup_{t\in T}[f_t > m]$ and this shows that

$$\{y \in B_L(x,\delta) : \langle nu^*, y-x\rangle \geq \delta\} \subset \{y \in L : \langle nu^*, y-x\rangle \geq \delta\} \subset \bigcup_{t\in T}[f_t > m], \quad (9.23)$$

where $B_L(x,\delta)$ denotes the ball in L centered at x with radius δ (L endowed with the relative topology of X is isomorphic to an Euclidean space and, consequently, $B_L(x,\delta)$ is compact). Therefore, since the sets $[f_t > m]$ are open, by the lower semicontinuity of the f_t's, (9.23) gives rise to a finite set $\{t_1^{(n,m)}, \ldots, t_{k_{(n,m)}}^{(n,m)}\} \subset T$, $k_{(n,m)} \geq 1$, such that

$$\{y \in B_L(x,\delta) : \langle nu^*, y-x\rangle \geq \delta\} \subset \bigcup_{i=1,\ldots,k_{(n,m)}} \left[f_{t_i^{(n,m)}} > m\right].$$

Equivalently, if we define the functions $g_{(n,m)} := \max_{i=1,\ldots,k_{(n,m)}} f_{t_i^{(n,m)}}$, so that

$$[g_{(n,m)} \leq m] = \bigcap_{i=1,\ldots,k_{(n,m)}} \left[f_{t_i^{(n,m)}} \leq m\right] \subset (X \setminus B_L(x,\delta)) \cup \{y \in X : \langle nu^*, y-x\rangle < \delta\}.$$

Also, by denoting $g := \sup_{n,m\geq 1} g_{(n,m)}$, we have that $[g \leq m] \subset \bigcap_{n\geq 1}[g_{(n,m)} \leq m]$ and we obtain

9.5 EXERCISES OF CHAPTER 6

$$[g \leq m] \cap B_L(x,\delta) \subset \left(\bigcap_{n\geq 1} [g_{(n,m)} \leq m]\right) \cap B_L(x,\delta) \subset \{y \in X : \langle nu^*, y-x\rangle < \delta\}. \tag{9.24}$$

Hence, since $x \in [g \leq m] \cap B_L(x,\delta)$ (remember that $m > f(x)$), we get

$$nu^* \in \mathrm{N}^\delta_{[g\leq m]\cap B_L(x,\delta)}(x) \text{ for all } n \geq 1,$$

and by taking $n \uparrow +\infty$ we deduce that $u^* \in \mathrm{N}_{[g\leq m]\cap B_L(x,\delta)}(x) = \mathrm{N}_{[g\leq m]\cap L}(x)$ for all $m > f(x)$. Therefore, $u^* \in \cap_{m>f(x)}\mathrm{N}_{[g\leq m]\cap L}(x) \subset \mathrm{N}_{\cup_{m>f(x)}[g\leq m]\cap L}(x) = \mathrm{N}_{(\mathrm{dom}\, g)\cap L}(x)$ and we conclude the proof since

$$g = \sup_{n,m\geq 1} g_{(n,m)} = \sup_{n,m\geq 1, i=1,\ldots,k_{(n,m)}} f_{t_i^{(n,m)}}$$

is the supremum of a countable family.

Exercise 101: The relation that we aim to prove is obvious when $f(x) = -\infty$ (the sets in both sides are empty); then we suppose that $f(x) \in \mathbb{R}$. The inclusion

$$\partial f(x) \subset \bigcap_{0<\varepsilon\leq\varepsilon_0} \overline{\mathrm{co}}\left(\partial_\varepsilon f(x) + \{0,\varepsilon\}\partial_{\varepsilon+\delta} f(x)\right) \tag{9.25}$$

is obvious from $\partial f(x) = \cap_{\varepsilon>0}\partial_\varepsilon f(x) = \cap_{0<\varepsilon\leq\varepsilon_0}\partial_\varepsilon f(x)$. In order to prove the opposite inclusion "$\supset$" we suppose that $f(x) = 0$ (without loss of generality) and take x^* in the right-hand side set of (9.25). Hence, for each $\varepsilon \in]0,\varepsilon_0]$, we have

$$x^* \in \mathrm{cl}\left(\partial_\varepsilon f(x) + \mathrm{co}\left(\{0,\varepsilon\}\,\partial_{\varepsilon+\delta} f(x)\right)\right),$$

and so there are nets $(y_i^*)_i \subset \partial_\varepsilon f(x)$, $(\lambda_{i,k})_i \subset [0,1]$, $\Sigma_{k=1,\ldots,k_i}\lambda_{i,k} \leq 1$ and $(z_{i,k}^*)_i \subset \partial_{\varepsilon+\delta}f(x)$, $k=1,\ldots,k_i$, $k_i \geq 1$, such that $x^* = \lim_i \left(y_i^* + \sum_{k=1,\ldots,k_i}\varepsilon\lambda_{i,k}z_{i,k}^*\right)$. Thus, since $f(x) = 0$, for each $y \in \mathrm{dom}\, f$, we can write

$$\langle x^*, y-x\rangle = \lim_i \left\langle y_i^* + \varepsilon\sum_{k=1,\ldots,k_i}\lambda_{i,k}z_{i,k}^*, y-x\right\rangle$$

$$\leq \limsup_i \left((f(y)-f(x)+\varepsilon) + \varepsilon\left(\sum_{k=1,\ldots,k_i}\lambda_{i,k}(f(y)-f(x)+\varepsilon+\delta)\right)\right)$$

$$= \limsup_i \left(f(y)+\varepsilon+\varepsilon\sum_{k=1,\ldots,k_i}\lambda_{i,k}(f(y)+\varepsilon+\delta)\right)$$

$$\leq f(y)+\varepsilon+\varepsilon(f^+(y)+\varepsilon+\delta),$$

and $x^* \in \partial f(x)$, by taking $\varepsilon \downarrow 0$.

Exercise 102: The inclusion

$$\partial f(x) \subset \bigcap_{0<\varepsilon\leq\varepsilon_0} \overline{co}\left(\partial_\varepsilon f(x) \cup \varepsilon\partial_{\varepsilon+\delta}f(x)\right) \qquad (9.26)$$

follows from $\partial f(x) = \bigcap_{\varepsilon>0}\partial_\varepsilon f(x) = \bigcap_{0<\varepsilon\leq\varepsilon_0}\partial_\varepsilon f(x)$. To prove the opposite inclusion, take x^* in the right-hand side set in (9.26). Then, for each fixed $\varepsilon > 0$, there are nets $(\lambda_i)_i \subset [0,1]$, $(y_i^*)_i \subset \partial_\varepsilon f(x)$, and $(z_i^*)_i \subset \partial_{\varepsilon+\delta}f(x)$ such that $x^* = \lim_i(\lambda_i y_i^* + (1-\lambda_i)\varepsilon z_i^*)$. Thus, for each $y \in \text{dom } f$,

$$\begin{aligned}
\langle x^*, y-x \rangle &= \lim_i \langle \lambda_i y_i^* + (1-\lambda_i)\varepsilon z_i^*, y-x \rangle \\
&\leq \limsup_i (\lambda_i(f(y) - f(x) + \varepsilon) + (1-\lambda_i)\varepsilon(f(y) - f(x) + \varepsilon + \rho)) \\
&\leq f(y) - f(x) + \varepsilon + \varepsilon(f(y) - f(x) + \varepsilon + \rho),
\end{aligned}$$

as $f(y) \geq f(x)$. Now, taking $\varepsilon > 0$, we obtain $\langle x^*, y-x \rangle \leq f(y) - f(x)$ for all $y \in X$, showing that $x^* \in \partial f(x)$.

Exercise 103: (i) Fix $x \in \text{dom } f$ and assume, without loss of generality, that $f(x) = 0$. Fix $\varepsilon > 0$, $U \in \mathcal{N}_{X^*}$ and pick $L \in \mathcal{F}(x)$ such that $L^\perp \subset U$. Observe that the family $\{f_t, t \in T;\ I_L\} \subset \Gamma_0(X)$ also satisfies the compactness and upper semicontinuity assumptions as the family $\{f_t, t \in T\}$. Therefore, by applying Proposition 6.4.1 to the family $\{f_t, t \in T;\ I_L\}$, we obtain that (see Exercise 23)

$$\begin{aligned}
N_{L \cap \text{dom } f}(x) &= \left[\overline{co}\left(\left(\bigcup_{t \in T(x)} \partial_\varepsilon f_t(x) \cup L^\perp\right) \cup \left(\bigcup_{t \in T \setminus T(x)} \partial_\varepsilon f_t(x)\right)\right)\right]_\infty \\
&= \left[\overline{co}\left(\left(\bigcup_{t \in T(x)} \partial_\varepsilon f_t(x)\right) \cup \left(\bigcup_{t \in T \setminus T(x)} \varepsilon\partial_\varepsilon f_t(x)\right) \cup L^\perp\right)\right]_\infty,
\end{aligned}$$

Moreover, we have $N_{L \cap \text{dom } f}(x) = [\overline{co}A]_\infty$ (Exercise 22) where $A := \left(\bigcup_{t \in T(x)} \partial_\varepsilon f_t(x)\right) \cup \left(\bigcup_{t \in T \setminus T(x)} \varepsilon\partial_\varepsilon f_t(x) + L^\perp\right)$. Next, by combining this relation and (6.16) and denoting $C_t := \partial_\varepsilon f_t(x)$,

9.5 EXERCISES OF CHAPTER 6

$$\partial f(x) \subset \overline{\mathrm{co}}\left(\bigcup_{t\in T(x)} C_t + \mathrm{N}_{L\cap \mathrm{dom}\, f}(x)\right) = \overline{\mathrm{co}}\left(\bigcup_{t\in T(x)} C_t + [\overline{\mathrm{co}}A]_\infty\right)$$

$$\subset \overline{\mathrm{co}}\left((\overline{\mathrm{co}}A) + [\overline{\mathrm{co}}A]_\infty\right) = \overline{\mathrm{co}}A$$

$$\subset \overline{\mathrm{co}}\left(\left(\bigcup_{t\in T(x)} C_t + L^\perp\right) \cup \left(\bigcup_{t\in T\setminus T(x)} \varepsilon C_t + L^\perp\right)\right)$$

$$\subset \mathrm{co}\left(\left(\bigcup_{t\in T(x)} C_t\right) \cup \left(\bigcup_{t\in T\setminus T(x)} \varepsilon C_t\right)\right) + L^\perp + U$$

$$\subset \mathrm{co}\left(\left(\bigcup_{t\in T(x)} C_t\right) \cup \left(\bigcup_{t\in T\setminus T(x)} \varepsilon C_t\right)\right) + 2U.$$

Consequently, (6.116) follows by intersecting over $U \in \mathcal{N}_{X^*}$ and after over $\varepsilon > 0$. We proceed now by showing the opposite inclusion in (6.116), when $x \in X$ is such that $M := \inf_{t\in T} f_t(x) > -\infty$. Let $x^* \in X^*$ such that $x^* \in \overline{\mathrm{co}}\left(\left(\bigcup_{t\in T(x)} C_t\right) \cup \left(\bigcup_{t\in T\setminus T(x)} \varepsilon C_t\right)\right)$ for each $\varepsilon > 0$. Observe that, if $z^* \in C_t$ with $t \in T(x)$, then $\langle z^*, z - x\rangle \leq f_t(z) - f_t(x) + \varepsilon \leq f(z) + \varepsilon$ for every $z \in X$, and so $z^* \in \partial_\varepsilon f(x)$. Also, if $z^* \in C_t$ with $t \in T \setminus T(x)$, then $\langle z^*, z - x\rangle \leq f_t(z) - f_t(x) + \varepsilon \leq f(z) - M + \varepsilon$ for every $z \in X$, and so $z^* \in \partial_{\varepsilon - M} f_t(x)$ (observe that $M \leq f(x) = 0$). Thus, as $\varepsilon > 0$ was arbitrarily chosen, we obtain $x^* \in \bigcap_{\varepsilon>0} \overline{\mathrm{co}}\left(\partial_\varepsilon f(x) \cup \varepsilon \partial_{\varepsilon - M} f(x)\right) = \partial f(x)$ (see Exercise 102).

(ii) Denote $g := \sup\{f_t,\ t \in T;\ h\}$. Then, by the lower semicontinuity of the f_t's, the functions f and g coincide in a neighborhood of x, entailing $\partial f(x) = \partial g(x)$. If g satisfied (6.116) with equality, then taking into account that $\partial_\varepsilon h(x) = \{\theta\}$, we would have

$$\partial g(x) = \bigcap_{\varepsilon>0} \overline{\mathrm{co}}\left(\left(\bigcup_{t\in T(x)} \partial_\varepsilon f_t(x)\right) \cup \left(\bigcup_{t\in T\setminus T(x)} \varepsilon \partial_\varepsilon f_t(x) \bigcup \{\theta\}\right)\right).$$

But this implies that $\theta \in \partial g(x) = \partial f(x)$, which contradicts our assumption that x is not a minimum of f.

Exercise 104: First, as in the proof of Theorem 6.5.2, we may suppose that all the f_t's, $t \in T_c$, are continuous at some common point $x_0 \in \mathrm{dom}\, f$. By Corollary 6.1.6, we have

$$\partial f(x) = \partial\left(\sup_{t\in T(x)} f_t + \mathrm{I}_{\mathrm{dom}\, f}\right)(x) = \partial\left(\max\left\{\hat{f}, \tilde{f} + \mathrm{I}_{\mathrm{dom}\, f}\right\}\right)(x),$$

where $\hat{f} := \max_{t \in T_c} f_t$. Observe that $\hat{f}$ is continuous at x_0 as T_c is finite. Then Theorem 6.5.2 yields $\partial f(x) = \operatorname{co}\left\{\partial \hat{f}(x) \cup A\right\} + \operatorname{N}_{\operatorname{dom} f}(x)$, where $A := \partial(\tilde{f} + \operatorname{I}_{\operatorname{dom} f})(x)$, whereas Corollary 6.1.12 gives us $\partial \hat{f}(x) = \operatorname{N}_{\operatorname{dom} \hat{f}}(x) + \overline{\operatorname{co}}B \subset \operatorname{N}_{\operatorname{dom} f}(x) + \overline{\operatorname{co}}B$, where $B := \cup_{t \in T_c} \partial f_t(x)$. Hence, by combining these two relations,

$$\partial f(x) \subset \operatorname{co}\left\{(\operatorname{N}_{\operatorname{dom} f}(x) + \overline{\operatorname{co}}B) \cup A\right\} + \operatorname{N}_{\operatorname{dom} f}(x)$$
$$\subset \operatorname{co}\left\{(\operatorname{N}_{\operatorname{dom} f}(x) + \overline{\operatorname{co}}B) \cup (\operatorname{N}_{\operatorname{dom} f}(x) + A)\right\} + \operatorname{N}_{\operatorname{dom} f}(x)$$
$$= \operatorname{co}\left\{(\overline{\operatorname{co}}B) \cup A\right\} + \operatorname{N}_{\operatorname{dom} f}(x) \subset \overline{\operatorname{co}}\left\{B \cup A\right\} + \operatorname{N}_{\operatorname{dom} f}(x).$$

Exercise 105: The proof uses similar arguments to those composing Exercises 80 and 81, but here we take advantage of Theorem 6.1.4 (namely, one of its consequences given in Exercise 93), the whole thing proved after Theorem 5.2.2. To prove the direct inclusion "⊂" in Theorem 6.3.2, we assume without loss of generality that $x = \theta$, $f(\theta) = 0$ and $\partial f(\theta) \neq \emptyset$; hence, $\partial(\operatorname{cl} f)(\theta) = \partial f(\theta)$ and $f(\theta) = (\operatorname{cl} f)(\theta) = 0$. We consider the convex functions $g_t : X \to \mathbb{R}_\infty$, $t \in T$, defined by

$$g_t := \begin{cases} \operatorname{cl} f_t, & \text{if } \operatorname{cl} f_t \text{ is proper,} \\ \max\{\operatorname{cl} f_t, -1\}, & \text{otherwise.} \end{cases}$$

It is easy to see that $g_t \in \Gamma_0(X)$ and $g := \sup_{t \in T} g_t = \max\{\sup_{t \in T}(\operatorname{cl} f_t), -1\} = \max\{\operatorname{cl} f, -1\}$, due to (5.10). Hence, since $(\operatorname{cl} f)(\theta) = f(\theta) = 0$, we easily verily that f and g have the same value and the same subdifferential at θ. So, from the proof given in Theorem 6.3.2 (when $f_t \in \Gamma_0(X)$, for all $t \in T$), we have

$$\partial f(\theta) = \partial(\operatorname{cl} f)(\theta) = \partial g(\theta)$$
$$= \bigcap_{L \in \mathcal{F}(\theta)} \operatorname{co}\left\{\bigcap_{0 < \varepsilon < 1} \operatorname{cl}\left(\bigcup_{t \in T_\varepsilon^g(x)} \partial_\varepsilon g_t(\theta) + \operatorname{N}_{L \cap \operatorname{dom} g}(\theta)\right)\right\},$$
(9.27)

where, for all $\varepsilon \in \,]0, 1[$,

$$T_\varepsilon^g(x) := \{t \in T : g_t(\theta) \geq -\varepsilon\} = \{t \in T : (\operatorname{cl} f_t)(\theta) \geq -\varepsilon\}.$$

Take $t \in T_\varepsilon^g(x)$ with $\varepsilon \in \,]0, 1[$. Then the function $\operatorname{cl} f_t$ must be proper, because $-\varepsilon \leq (\operatorname{cl} f_t)(\theta) \leq f(\theta) = 0$, and so $g_t = \operatorname{cl} f_t$ by the definition of g_t. Thus, since $-\varepsilon \leq (\operatorname{cl} f_t)(\theta) \leq f_t(\theta)$, we have $t \in T_\varepsilon(\theta)$ and it can

9.5 EXERCISES OF CHAPTER 6

be easily proved that $\partial_\varepsilon g_t(\theta) = \partial_\varepsilon(\operatorname{cl} f_t)(\theta) \subset \partial_{2\varepsilon} f_t(\theta)$. Thus, taking into account that $\operatorname{dom} f \subset \operatorname{dom}(\operatorname{cl} f) = \operatorname{dom} g$, (9.27) yields

$$\partial f(\theta) \subset \bigcap_{L \in \mathcal{F}(\theta)} \operatorname{co} \left\{ \bigcap_{\varepsilon > 0} \operatorname{cl} \left(\bigcup_{t \in T_\varepsilon(\theta)} A_{t,\varepsilon} \right) \right\} \subset \bigcap_{L \in \mathcal{F}(\theta)} \operatorname{co} \left\{ \bigcap_{\varepsilon > 0} \operatorname{cl} \left(\bigcup_{t \in T_{2\varepsilon}(\theta)} A_{t,\varepsilon} \right) \right\},$$

where we denoted $A_{t,\varepsilon} := \partial_{2\varepsilon} f_t(\theta) + \mathrm{N}_{L \cap \operatorname{dom} f}(\theta)$.

Exercise 106: First, since $(z_i^*)_i \subset (U \cap E)^\circ$ and this last set is w^*-compact in E^*, by Theorem 2.1.9, we find a subnet $(z_{i_j|E}^*)_j$, where $z_{i_j|E}^*$ is the restriction to E of $z_{i_j}^*$, and $\tilde{z}^* \in E^*$ such that

$$\left\langle z_{i_j|E}^* - \tilde{z}^*, u \right\rangle \to_j 0 \text{ for all } u \in E.$$

Let $z^* \in X^*$ be an extension of $\tilde{z}^*$ to X^*, and fix $V \in \mathcal{N}_{X^*}$. Then we have (see section 2.1)

$$V_{|E} := \left\{ x_{|E}^* : x^* \in V \right\} \in \mathcal{N}_{E^*}.$$

Thus, by taking limits on j in the following inclusion

$$(z_{i_j|E}^*) \subset B := \left\{ u_{|E}^* \in E^* : u^* \in A \right\},$$

we infer that $z_{|E}^* = \tilde{z}^* \in \operatorname{cl}^{\sigma(E^*, E)} B \subset B + V_{|E}$. So, there are $u^* \in A$ and $v^* \in V$ such that $z_{|E}^* = u_{|E}^* + v_{|E}^*$; that is, $\langle z^* - (u^* + v^*), u \rangle = 0$ for all $u \in E$, and we obtain

$$z^* \in u^* + v^* + E^\perp \subset A + V + E^\perp.$$

Therefore, $z^* \in \operatorname{cl}(A + E^\perp)$, due to the arbitrariness of $V \in \mathcal{N}_{X^*}$.

Exercise 107: If $x \notin \operatorname{dom} f$, then $E_L = \emptyset$ for every $L \in \mathcal{F}(x)$, and we are obviously done. Thus, we fix $x \in \operatorname{dom} f$ and $L \in \mathcal{F}(x)$, so that $T_{\varepsilon_2}(x) \neq \emptyset$, by Exercise 94. We introduce the functions $\tilde{g}_t := f_t + \mathrm{I}_{L \cap \operatorname{dom} f}$, $t \in T_{\varepsilon_2}(x)$, entailing that $E_L = \operatorname{co} \left\{ \bigcup_{t \in T_{\varepsilon_2}(x)} \partial_{\varepsilon_1} \tilde{g}_t(x) \right\}$ and $\operatorname{dom} \tilde{g}_t = L \cap \operatorname{dom} f \cap \operatorname{dom} f_t = L \cap \operatorname{dom} f$. We also consider the associated restrictions to L, $g_t := \tilde{g}_t |_L$, $t \in T(x)$. Take a net $(u_i^*)_i \subset E_L$ such that $u_i^* \to^{w^*} u^* \in X^*$, and denote $z_i^* := u_i^* |_L$; hence $z_i^* \to^{w^*} u^* |_L =: z^*$ (the convergence in L^*) because $\langle z^*, y \rangle = \langle u^*, y \rangle = \lim_i \langle u_i^*, y \rangle = \lim_i \langle z_i^*, y \rangle$ for all $y \in L$. For each i, we write $u_i^* =$

$\mu_{i,1} u_{i,1}^* + \ldots + \mu_{i,k_i} u_{i,k_i}^*$, for some $u_{i,j}^* \in \partial_{\varepsilon_1} \tilde{g}_{t_{i,j}}(x)$, $t_{i,j} \in T_{\varepsilon_2}(x)$, $\mu_i \in \Delta_{k_i}$, $k_i \geq 1$. Thus, since $u_{i,j}^*|_L \in \partial_{\varepsilon_1} g_{t_{i,j}}(x)$ (Exercise 55(i)), we have

$$z_i^* = u_i^*|_L = \mu_{i,1} u_{i,1}^*|_L + \ldots + \mu_{i,k_i} u_{i,k_i}^*|_L$$

and so $(z_i^*)_i \subset \mathrm{co}\left\{\bigcup_{t \in T_{\varepsilon_2}(x)} \partial_{\varepsilon_1} g_t(x)\right\} \subset L^*$, where L^* is the dual of L (which is isomorphic to L, and hence it has the same dimension as L, say $m \geq 1$). Therefore, by the Carathéodory theorem, for each i, we find some $\lambda_i := (\lambda_{i,1}, \ldots, \lambda_{i,m+1}) \in \Delta_{m+1}$ together with $(t_{i,k})_i \subset T_{\varepsilon_2}(x)$ and $z_{i,k}^* \in \partial_{\varepsilon_1} g_{t_{i,k}}(x)$, $k \in K := \{1, \ldots, m+1\}$ such that $z_i^* = \lambda_{i,1} z_{i,1}^* + \ldots + \lambda_{i,m+1} z_{i,m+1}^*$. We may assume that $(\lambda_i)_i$ converges to some $\lambda \in \Delta_{m+1}$. Also, since $(t_{i,k})_i \subset T_{\varepsilon_2}(x)$ and $T_{\varepsilon_2}(x)$ is compact and closed, by Exercise 94, we may assume that $t_{i,k} \to t_k \in T_{\varepsilon_2}(x)$ for all $k \in K$. Consequently, for every $z \in L \cap \mathrm{dom}\, f$ ($= \mathrm{dom}\, g_{t_k}$, $k \in K$), we obtain

$$\begin{aligned}
\langle z^*, z - x \rangle &= \lim_i \langle \lambda_{i,1} z_{i,1}^* + \ldots + \lambda_{i,m+1} z_{i,m+1}^*, z - x \rangle \\
&\leq \lambda_1 \limsup_i (g_{t_{i,1}}(z) - g_{t_{i,1}}(x)) + \ldots \\
&\quad + \lambda_{m+1} \limsup_i (g_{t_{i,m+1}}(z) - g_{t_{i,m+1}}(x)) + \varepsilon_1 \\
&= \lambda_1 \limsup_i g_{t_{i,1}}(z) + \ldots + \lambda_{m+1} \limsup_i g_{t_{i,m+1}}(z) - f(x) + \varepsilon_1 \\
&\leq \lambda_1 \limsup_i g_{t_1}(z) + \ldots + \lambda_{m+1} \limsup_i g_{t_{m+1}}(z) - f(x) + \varepsilon_1 \\
&= \sum_{k \in K_+} \lambda_k g_{t_k}(z) - \sum_{k \in K_+} \lambda_k g_{t_k}(x) + \varepsilon_1,
\end{aligned}$$

where $K_+ := \{k \in K : \lambda_k > 0\}$; that is, $z^* \in \partial_{\varepsilon_1}\left(\sum_{k \in K_+} \lambda_k g_{t_k}\right)(x)$. Hence, using Proposition 4.1.26, as $g_{t_k} + \mathrm{I}_{L \cap \mathrm{dom}\, f} = g_{t_k}$ and $\mathrm{ri}(\mathrm{dom}\, g_{t_k}) = \mathrm{ri}(L \cap \mathrm{dom}\, f) \neq \emptyset$, for all $z \in L$ and $k \in K$ we obtain $z^* \in \partial_{\varepsilon_1}\left(\sum_{k \in K_+} \lambda_k g_{t_k}\right)(x) \subset \sum_{k \in K_+} \lambda_k \partial_{\varepsilon_1} g_{t_k}(x)$. Hence, there exists some $v_k^* \in \partial_{\varepsilon_1} g_{t_k}(x)$, $k \in K_+$, such that $z^* = \sum_{k \in K_+} \lambda_k v_k^*$. Since $\mathrm{dom}\, g_{t_k} = L \cap \mathrm{dom}\, f \subset L$, for every extension $\tilde{v}_k^*$ of v_k^*, $k \in K_+$, we have that $\tilde{v}_k^* \in \partial_{\varepsilon_1} \tilde{g}_{t_k}(x)$ (Exercise 55(i)), and so $\tilde{v}^* := \sum_{k \in K_+} \lambda_k \tilde{v}_k^* \in \sum_{k \in K_+} \lambda_k \partial_{\varepsilon_1} \tilde{g}_{t_k}(x)$ is an extension of z^* to X^*; that is, $\langle \tilde{v}^*, y \rangle = \langle z^*, y \rangle = \langle u^*, y \rangle$ for all $y \in L$ or, equivalently, $u^* - \tilde{v}^* \in L^\perp$. Therefore,

$$\begin{aligned}
u^* &\in \sum_{k \in K_+} \lambda_k \partial_{\varepsilon_1} \tilde{g}_{t_k}(x) + L^\perp \\
&\subset \sum_{k \in K_+} \lambda_k \partial_{\varepsilon_1}(f_t + \mathrm{I}_{L \cap \mathrm{dom}\, f} + \mathrm{I}_L)(x) = \sum_{k \in K_+} \lambda_k \partial_{\varepsilon_1} \tilde{g}_{t_k}(x) \subset E_L.
\end{aligned}$$

9.6 Exercises of chapter 7

Exercise 108: We proceed as in the beginning of the proof of Theorem 7.1.1; that is, we only need to verify the inclusion "$\subset$" when $\partial(g + f \circ A)(x) = \partial((\operatorname{cl} g) + (\operatorname{cl} f) \circ A)(x) \neq \emptyset$, $\operatorname{cl} g \in \Gamma_0(X)$, $\operatorname{cl} f \in \Gamma_0(Y)$, $(\operatorname{cl} g)(x) = g(x)$ and $(\operatorname{cl} f)(Ax) = f(Ax)$. Then, by (4.45),

$$\partial(g + f \circ A)(x) = \partial((\operatorname{cl} g) + (\operatorname{cl} f) \circ A)(x) = \bigcap_{\varepsilon > 0} \operatorname{cl}\left(\partial_\varepsilon (\operatorname{cl} g)(x) + A^* \partial_\varepsilon (\operatorname{cl} f)(Ax)\right)$$
$$= \bigcap_{\varepsilon > 0} \operatorname{cl}\left(\partial_\varepsilon g(x) + A^* \partial_\varepsilon f(Ax)\right).$$

Exercise 109: The proof is the same as the one of Exercise 108, but using the functions $\overline{\operatorname{co}} f$ and $\overline{\operatorname{co}} g$ instead of $\operatorname{cl} f$ and $\operatorname{cl} g$, respectively.

Exercise 110: Observe that f is proper and continuous on $\operatorname{int}(\operatorname{dom} f)$ so that $\operatorname{cl}(f + g)(x) = f(x) + (\operatorname{cl} g)(x)$ for all $x \in \operatorname{int}(\operatorname{dom} f)$. To prove the non-trivial inequality $\operatorname{cl}(f + g)(x) \leq (\operatorname{cl} f)(x) + (\operatorname{cl} g)(x)$ for $x \in \operatorname{dom}(\operatorname{cl} f) \cap \operatorname{dom}(\operatorname{cl} g)$, we choose $x_0 \in \operatorname{int}(\operatorname{dom} f) \cap \operatorname{dom}(\operatorname{cl} g)$ and denote $x_\lambda := \lambda x_0 + (1 - \lambda)x$, $0 < \lambda < 1$. Then $x_\lambda \in \operatorname{int}(\operatorname{dom} f) \cap \operatorname{dom}(\operatorname{cl} g)$ and so

$$\operatorname{cl}(f + g)(x_\lambda) = f(x_\lambda) + (\operatorname{cl} g)(x_\lambda) = (\operatorname{cl} f)(x_\lambda) + (\operatorname{cl} g)(x_\lambda).$$

But the functions $(\operatorname{cl} f)$ and $(\operatorname{cl} g)$ are convex and proper, and so

$$\operatorname{cl}(f + g)(x) \leq \liminf_{\lambda \downarrow 0} \operatorname{cl}(f + g)(x_\lambda) = \liminf_{\lambda \downarrow 0}((\operatorname{cl} f)(x_\lambda) + (\operatorname{cl} g)(x_\lambda))$$
$$\leq \liminf_{\lambda \downarrow 0}(\lambda((\operatorname{cl} f)(x_0) + (\operatorname{cl} g)(x_0)) + (1 - \lambda)((\operatorname{cl} f)(x) + (\operatorname{cl} g)(x)))$$
$$= (\operatorname{cl} f)(x) + (\operatorname{cl} g)(x),$$

yielding the desired inequality. To prove the second statement, take $x \in X$ such that $\partial(f + g)(x) \neq \emptyset$. Then, by taking into account the first statement (and Exercise 62), $(f + g)(x) = \operatorname{cl}(f + g)(x) = (\operatorname{cl} f)(x) + (\operatorname{cl} g)(x) \in \mathbb{R}$ and $\partial(f + g)(x) = \partial(\operatorname{cl}(f + g))(x) = \partial((\operatorname{cl} f) + (\operatorname{cl} g))(x)$. In particular, $f(x), g(x), (\operatorname{cl} f)(x), (\operatorname{cl} g)(x) \in \mathbb{R}$ and we obtain $f(x) - (\operatorname{cl} f)(x) + g(x) - (\operatorname{cl} g)(x) = 0$. This implies that $f(x) = (\operatorname{cl} f)(x)$ and $g(x) = (\operatorname{cl} g)(x)$. Moreover, since $\operatorname{cl} f \leq f$, the convex function $\operatorname{cl} f$ is continuous at some point in $\operatorname{dom}(\operatorname{cl} g)$, and Corollary 7.1.3 yields

$$\partial(f + g)(x) = \partial((\operatorname{cl} f) + (\operatorname{cl} g))(x) = \partial(\operatorname{cl} f)(x) + \partial(\operatorname{cl} g)(x) = \partial f(x) + \partial g(x).$$

Exercise 111: We start by proving that the set $\mathbb{R}_+(\text{epi}(f \circ A) - (x, f(Ax)))$ is closed. We proceed as in the proof of Theorem 7.2.5 and take nets $(\alpha_i)_i \subset \mathbb{R}_+$ and $(x_i, \lambda_i)_i \subset \text{epi}(f \circ A)$ such that $\alpha_i((x_i, \lambda_i) - (x, f(Ax))) \to (u, \mu) \in X \times \mathbb{R}$. Then $(Ax_i, \lambda_i) \in \text{epi } f$ and $\alpha_i((Ax_i, \lambda_i) - (Ax, f(Ax))) \to (Au, \mu)$, due to the continuity of A. But $\alpha_i((Ax_i, \lambda_i) - (Ax, f(Ax))) \in \mathbb{R}_+(\text{epi } f - (Ax, f(Ax)))$ and so, by the current hypothesis, $(Au, \mu) \in \mathbb{R}_+(\text{epi } f - (Ax, f(Ax)))$. Let $\alpha \in \mathbb{R}_+$ and $(y, \lambda) \in \text{epi } f$ such that $Au = \alpha(y - Ax)$ and $\mu = \alpha(\lambda - f(Ax))$. Next, if we take $\gamma > 0$ such that $\alpha\gamma < 1$, then the convexity of f yields

$$(f \circ A)(\gamma u + x) = f(\gamma\alpha(y - Ax) + Ax) \le \alpha\gamma f(y) + (1 - \gamma\alpha)f(Ax)$$
$$\le \alpha\gamma\lambda + (1 - \gamma\alpha)f(Ax) = f(Ax) + \gamma\mu,$$

and so $(u, \mu) \in \gamma^{-1}(\text{epi}(f \circ A) - (x, f(Ax))) \subset \mathbb{R}_+(\text{epi}(f \circ A) - (x, f(Ax)))$, showing that the set $\mathbb{R}_+(\text{epi}(f \circ A) - (x, f(Ax)))$ is closed. Hence, the functions $f \circ A$ and g satisfy condition (ii) of Theorem 7.2.2, and we get $\partial(g + f \circ A)(x) = \text{cl}(\partial g(x) + \partial(f \circ A)(x))$. Moreover, taking into account Theorem 7.2.3, the closedness of $\mathbb{R}_+(\text{epi } f - (Ax, f(Ax)))$ ensures that $\partial(f \circ A)(x) = \text{cl}(A^*\partial f(Ax))$. Consequently, the conclusion comes by combining these two identities,

$$\partial(g + f \circ A)(x) = \text{cl}(\partial g(x) + \text{cl}(A^*\partial f(Ax))) = \text{cl}(\partial g(x) + A^*\partial f(Ax)).$$

Exercise 112: The inclusion "$\supset$" is obvious. Let $\delta > 0$ be small enough to have $(\text{ri}(\text{dom } g) - x) \cap \text{dom } \sigma_{A_\delta} \ne \emptyset$. Observe that if we define the function $h : X \to \overline{\mathbb{R}}$ as $h(y) := \sigma_{A_\delta}(y - x)$, $y \in X$, then by (5.1), we have $A_\delta = \partial_\varepsilon \sigma_{A_\delta}(\theta) = \partial_\varepsilon h(x)$ for all $\varepsilon > 0$. Next, since the family $(A_\varepsilon)_{\varepsilon > 0}$ is non-decreasing, by applying consecutively Proposition 4.1.6(iii) and (4.15), we get

$$\bigcap_{\varepsilon > 0} \text{cl}(A_\varepsilon + \partial_\varepsilon g(x)) \subset \bigcap_{\varepsilon > 0} \text{cl}(A_\delta + \partial_\varepsilon g(x)) = \bigcap_{\varepsilon > 0} \text{cl}(\partial_\varepsilon h(x) + \partial_\varepsilon g(x))$$
$$\subset \bigcap_{\varepsilon > 0} \partial_{2\varepsilon}(h + g)(x) = \partial(h + g)(x).$$

Now, Theorem 7.2.1(ii) gives rise to

$$\bigcap_{\varepsilon > 0} \text{cl}(A_\varepsilon + \partial_\varepsilon g(x)) \subset \partial(h + g)(x) = \bigcap_{\varepsilon > 0} \text{cl}(\partial_\varepsilon h(x) + \partial g(x)) = \text{cl}(A_\delta + \partial g(x)).$$

Hence, since $\delta > 0$ is arbitrarily small, we deduce that

$$\bigcap_{\varepsilon > 0} \text{cl}(A_\varepsilon + \partial_\varepsilon g(x)) \subset \bigcap_{\delta > 0} \text{cl}(A_\delta + \partial g(x)),$$

9.6 EXERCISES OF CHAPTER 7

and we are done since the converse inclusion is obvious we are done.

Exercise 113: We represent by E the right-hand side set in (7.40), and take $x \in \mathrm{dom}\, f \cap \mathrm{dom}\, g$. For all $t \in T$, we have $\partial_\varepsilon f_t(x) + \partial_\varepsilon g(x) \subset \partial_{2\varepsilon}(f_t + g)(x)$. Moreover, for $t \in T_\varepsilon(x)$, we have $f_t \leq f$ and $f_t(x) \geq f(x) - \varepsilon$, and this implies $\partial_{2\varepsilon}(f_t + g)(x) \subset \partial_{3\varepsilon}(f + g)(x)$. Therefore, using (4.16) and (4.15),

$$E \subset \bigcap_{\varepsilon > 0,\ L \in \mathcal{F}(x)} \overline{\mathrm{co}}\,\{\partial_{3\varepsilon}(f+g)(x) + \mathrm{N}_{L \cap \mathrm{dom}\, f \cap \mathrm{dom}\, g}(x)\}$$

$$\subset \bigcap_{\varepsilon > 0,\ L \in \mathcal{F}(x)} \partial_{3\varepsilon}(f + g + \mathrm{I}_L)(x) = \partial(f+g)(x),$$

and the inclusion "$\supset$" follows. To show the opposite inclusion "$\subset$" we observe that $f + g = \sup_{t \in T}(f_t + g)$. Then, since condition (5.10) holds automatically in the current case, by Theorem 5.2.2 and making use of formula (4.46), we get

$$\partial_\varepsilon(f+g)(x) = \mathrm{cl}\left(\bigcup_{\substack{\varepsilon_1+\varepsilon_2=\varepsilon \\ \varepsilon_1,\varepsilon_2 \geq 0}} \partial_{\varepsilon_1} f(x) + \partial_{\varepsilon_2} g(x)\right) \subset \mathrm{cl}\,(A_\varepsilon + B_\varepsilon) \text{ for all } \varepsilon > 0,$$

where we denoted $A_\varepsilon := \partial_\varepsilon f(x)$ and $B_\varepsilon := \partial_\varepsilon g(x)$. So, denoting $C := \mathrm{N}_{L \cap \mathrm{dom}(f+g)}(x)$,

$$\partial(f+g)(x) = \bigcap_{\varepsilon > 0,\ L \in \mathcal{F}(x)} \overline{\mathrm{co}}\left\{\bigcup_{t \in T_\varepsilon(x)} \partial_\varepsilon(f_t + g)(x) + C\right\}$$

$$\subset \bigcap_{\varepsilon > 0,\ L \in \mathcal{F}(x)} \overline{\mathrm{co}}\left\{\bigcup_{t \in T_\varepsilon(x)} \mathrm{cl}\,(A_\varepsilon + B_\varepsilon) + C\right\}$$

$$\subset \bigcap_{\varepsilon > 0,\ L \in \mathcal{F}(x)} \overline{\mathrm{co}}\left(\bigcup_{t \in T_\varepsilon(x)} A_\varepsilon + B_\varepsilon + C\right).$$

Exercise 114: (i) Apply Theorem 7.3.2 with $g \equiv 0$.
(ii) Using (4.45) and the lower semicontinuity of the function $\mathrm{I}_{\mathbb{R}_+(\mathrm{dom}\, f - x)}$, for all $\varepsilon > 0$ and $L \in \mathcal{F}(x)$, we obtain

$$\mathrm{N}_{L \cap \mathrm{dom}\, f}(x) = \mathrm{N}_{L \cap (\mathbb{R}_+(\mathrm{dom}\, f - x))}(\theta)$$
$$\subset \mathrm{cl}\left(L^\perp + \mathrm{N}^\varepsilon_{\mathbb{R}_+(\mathrm{dom}\, f - x)}(\theta)\right) = \mathrm{cl}\left(L^\perp + \mathrm{N}_{\mathbb{R}_+(\mathrm{dom}\, f - x)}(\theta)\right).$$

Thus, according to Theorem 7.3.2 (taking $g \equiv 0$),

$$\partial f(x) \subset \bigcap_{\varepsilon > 0} \overline{\mathrm{co}} \left\{ \bigcup_{t \in T_\varepsilon(x)} \partial_\varepsilon f_t(x) + \mathrm{N}_{L \cap \mathrm{dom}\, f}(x) \right\}$$

$$\subset \bigcap_{\varepsilon > 0} \overline{\mathrm{co}} \left\{ \bigcup_{t \in T_\varepsilon(x)} \partial_\varepsilon f_t(x) + \mathrm{cl} \left(L^\perp + \mathrm{N}_{\mathbb{R}_+(\mathrm{dom}\, f - x)}(\theta) \right) \right\}$$

$$\subset \bigcap_{\varepsilon > 0} \overline{\mathrm{co}} \left\{ \bigcup_{t \in T_\varepsilon(x)} \partial_\varepsilon f_t(x) + \mathrm{N}_{\mathrm{dom}\, f}(x) + L^\perp \right\}.$$

Thus, the inclusion "⊂" follows by intersecting over the L's (Exercise 10(i)). The opposite inclusion in (7.41) is once again straightforward. Assume now that $\mathrm{ri}(\mathrm{cone}(\mathrm{dom}\, f - x)) \neq \emptyset$. Take $L \in \mathcal{F}(x)$ such that $L \cap \mathrm{ri}(\mathrm{cone}(\mathrm{dom}\, f - x)) \neq \emptyset$. Then, since $\mathrm{ri}\, L = L \neq \emptyset$, $\mathrm{ri}(\mathrm{cone}(\mathrm{dom}\, f - x)) \neq \emptyset$, and the convex function $(\mathrm{I}_L)_{|L}$ and $(\mathrm{I}_{\mathbb{R}_+(\mathrm{dom}\, f-x)})_{|\mathrm{aff}(\mathbb{R}_+(\mathrm{dom}\, f-x))}$ are continuous on L and $\mathrm{ri}(\mathrm{cone}(\mathrm{dom}\, f - x))$, respectively, Theorem 7.2.2(iii) yields

$$\mathrm{N}_{L \cap \mathrm{dom}\, f}(x) = \mathrm{N}_{L \cap (\mathbb{R}_+(\mathrm{dom}\, f - x))}(\theta)$$
$$= \partial (\mathrm{I}_L + \mathrm{I}_{\mathbb{R}_+(\mathrm{dom}\, f - x)})(\theta) = \mathrm{cl}\left(L^\perp + \mathrm{N}_{\mathbb{R}_+(\mathrm{dom}\, f - x)}(\theta) \right).$$

Thus, the conclusion follows similarly as in the paragraph above using Theorem 7.3.2.

(iii) Apply Corollary 7.3.5 with $g = 0$.

Exercise 115: Assume that $\mathrm{cl}\,(f + \mathrm{I}_D) = \sup_{t \in T}(\mathrm{cl}\, f_t)$ holds on D. Take x in $\mathrm{cl}(L \cap D)$ for certain $L \in \mathcal{F}$. Since $L \cap D \subset L$, we pick $x_0 \in \mathrm{ri}(L \cap D)$. Then $x_\lambda := \lambda x_0 + (1-\lambda)x \in L \cap D \subset D$ for $\lambda \in]0,1[$, and so

$$\mathrm{cl}\,(f + \mathrm{I}_D)(x) = \lim_{\lambda \to 0^+} \mathrm{cl}\,(f + \mathrm{I}_D)(x_\lambda) = \lim_{\lambda \to 0^+} \sup_{t \in T}(\mathrm{cl}\, f_t)(x_\lambda) = \sup_{t \in T}(\mathrm{cl}\, f_t)(x).$$

9.7 Exercises of chapter 8

Exercise 116: If $A(x) = \emptyset$, then $\bigcup_{t \in A(x)} \partial f_t(x) = \{\theta\}$ and the inclusion trivially holds as $F \subset C$. Take $x^* \in \mathrm{N}_C(x)$ and $t \in A(x)$. If $\partial f_t(x) = \emptyset$, then obviously $\mathrm{N}_C(x) + \partial f_t(x) = \emptyset \subset \mathrm{N}_F(x)$. Otherwise, take $x_t^* \in \partial f_t(x)$, $\lambda_t \geq 0$ and $z \in F$ ($\subset C$). Then

$$\langle x^* + \lambda_t x_t^*, z - x \rangle \leq \lambda_t \langle x_t^*, z - x \rangle \leq \lambda_t (f_t(z) - f_t(x)) = \lambda_t f_t(z) \leq 0,$$

that is, $\mathrm{N}_C(x) + \partial f_t(x) \subseteq \mathrm{N}_F(x)$. Hence,

9.7 EXERCISES OF CHAPTER 8

$$N_C(x) + \text{cone co}\left(\bigcup_{t\in A(x)} \partial f_t(x)\right) = \text{cone co}\left(\bigcup_{t\in A(x)} (N_C(x) + \partial f_t(x))\right)$$
$$\subseteq N_F(x).$$

Exercise 117: Take $x \in F$. If $\tilde{f}(x) < 0$, the continuity of $\tilde{f}$ together with condition $x \in \text{int } C$ entails x is an interior point of F. Then $N_F(x) = \{\theta\}$, and (8.18) and (8.21) are both trivially satisfied. Finally, if $\tilde{f}(x) = 0$, then $T(x) = A(x)$ and once again LFM and BCQ are equivalent.

Exercise 118: Observe that $F =]-\infty, 0]$, for $x = 0$ we have $T(0) = T$, and every finite subsystem has F as solution set. Despite this, we have

$$N_F(0) = [0, +\infty[\neq \{0\} = N_C(0) + \text{cone co}\left(\bigcup_{t\in T(0)} \partial f_t(0)\right),$$

because $\partial f_t(0) = \{0\}$ for all $t \in T$. Thus, $\mathcal{S}$ is not LFM.

Exercise 119: (a) Observe that

$$\text{cl}(\text{epi } g^* + \text{cl } K) = \text{cl}(\text{epi } g^* + K) = \text{epi } g^* + K \subset \text{epi } g^* + \text{cl } K,$$

and $\text{epi } g^* + \text{cl } K$ is w^*-closed.

(b) If g is linear (and continuous), $\text{epi } g^*$ is a vertical half-line and so, is locally compact. Moreover, $(-\text{epi } g^*)_\infty \cap \text{cl } K = \{(\theta, 0)\}$, as a consequence of (8.9), since (P) has feasible solutions. Then we apply Theorem 2.1.8 to deduce that $\text{epi } g^* + \text{cl } K$ is w^*-closed.

(c) If g is continuous at some point of F, the result comes from Proposition 4.1.20 and Lemma 8.1.2. In fact,

$$\text{epi}(g + I_F)^* = \text{epi } g^* + \text{epi } I_F^* = \text{epi } g^* + \text{epi } \sigma_F = \text{epi } g^* + \text{cl } K,$$

and $\text{epi } g^* + \text{cl } K$ is w^*-closed.

Exercise 120: See the bibliographical notes of chapter 8.

Exercise 121: According to Theorem 8.2.1, we have

$$\overline{x} \text{ is optimal for (P)} \Leftrightarrow \partial g(\overline{x}) \cap (-N_F(\overline{x})) \neq \emptyset \Leftrightarrow \theta \in \partial g(\overline{x}) + N_F(\overline{x}).$$

Since $\mathcal{S}$ is assumed to be LFM at $\overline{x}$, we obtain

$\bar{x}$ is optimal for (P) $\Leftrightarrow \theta \in \partial g(\bar{x}) + \text{cone co}\left(\bigcup_{t \in A(\bar{x})} \partial f_t(\bar{x})\right) + N_C(\bar{x}),$

and we are done.

Exercise 122: If $x^* \in N_F(\bar{x}) \setminus \{\theta\}$, where F is the feasible set of $\mathcal{S}$, then the point $\bar{x}$ turns out to be a minimum of the problem

$$\text{Min} - \langle x^*, x \rangle$$
$$\text{s.t. } f_t(x) \leq 0, \ t \in T, \ x \in C.$$

Hence, arguing as in the proof of Theorem 8.2.3, we conclude the existence of $\lambda \in \mathbb{R}_+^{(T)}$ such that $\theta \in -x^* + \sum_{t \in \text{supp}\,\lambda} \lambda_t \partial f_t(\bar{x}) + N_C(\bar{x})$ and $\lambda_t f_t(\bar{x}) = 0$ for all $t \in \text{supp}\,\lambda$. Thus, $\text{supp}\,\lambda \subset A(\bar{x})$ and

$$x^* \in N_C(\bar{x}) + \text{cone co}\left(\bigcup_{t \in A(\bar{x})} \partial f_t(\bar{x})\right),$$

so that $\mathcal{S}$ is LFM at $\bar{x}$.

Exercise 123: Let $\bar{x} \in F$ be a minimum of (P). Then $v(0) = g(\bar{x})$, and in the proof of Theorem 8.2.12, we established the existence of $\bar{\lambda} \in \mathbb{R}_+^{(T)}$ such that $g(x) + \sum_{t \in T} \bar{\lambda}_t f_t(x) \geq g(\bar{x})$ for all $x \in C$. It follows from this inequality (by letting $x = \bar{x}$) that $\sum_{t \in T} \bar{\lambda}_t f_t(\bar{x}) = 0$, and hence $L(x, \bar{\lambda}) \geq L(\bar{x}, \bar{\lambda}) = g(\bar{x})$ for all $x \in C$. In addition, for each $\lambda \in \mathbb{R}_+^{(T)}$, and since $f_t(\bar{x}) \leq 0$, we have $L(\bar{x}, \lambda) = g(\bar{x}) + \sum_{t \in T} \lambda_t f_t(\bar{x}) \leq g(\bar{x}) = L(\bar{x}, \bar{\lambda})$. Thus, (8.92) holds. Conversely, if there exists $\bar{\lambda} \in \mathbb{R}_+^{(T)}$ satisfying (8.92), by letting $\lambda = 0$ in (8.92), we get $g(\bar{x}) \leq g(x) + \sum_{t \in T} \bar{\lambda}_t f_t(x)$ for all $x \in C$. Thus, if $x \in F$, then $g(x) \geq g(\bar{x})$ as $\sum_{t \in T} \bar{\lambda}_t f_t(x) \leq 0$. This means that $\bar{x}$ is a minimum of (P).

Finally, from the paragraph above, we deduce that $v(\text{P}) = g(\bar{x}) \leq \inf_{x \in C} L(x, \bar{\lambda}) \leq -v(\mathcal{D})$, and we conclude that $\bar{\lambda}$ is a maximizer of $(\mathcal{D})$ by the weak duality.

Exercise 124: It is obvious that (8.93) holds trivially (all the sets are empty) when $f = +\infty$; i.e., $f^* = -\infty$ and $f^{**} = +\infty$. If, alternatively, $f^* \neq -\infty$, the assumption $\text{dom}\, f^* \neq \emptyset$ gives rise to the existence of $x_0^* \in Y$ such that $f^*(x_0^*) \in \mathbb{R}$, and $\langle \cdot, x_0^* \rangle - f^*(x_0^*)$ is a continuous affine minorant of f (which is proper). Now two possibilities arise. The first one corresponds to the case $m = \inf_X f = \inf_X f^{**} = -\infty$, where (8.93) holds due to the convention on $\varepsilon - \text{argmin}\, f$. Exploring the remaining case where f is bounded from below; i.e., $m \in \mathbb{R}$, we prove first the inclusion "$\subset$" in (8.93) under the assumption $m = \inf_X f =$

9.7 EXERCISES OF CHAPTER 8

$\inf_X f^{**} \in \mathbb{R}$. Then, for any $\varepsilon > 0$, $x^* \in \text{dom } f^*$, $u \in \varepsilon - \text{argmin } f$ and $z \in \{x^*\}^-$ one has

$$\langle x^*, u+z \rangle - f^*(x^*) \le \langle x^*, z \rangle + f(u) \le f(u) \le m + \varepsilon;$$

in other words, $\varepsilon - \text{argmin } f + \{x^*\}^- \subseteq [\langle x^*, \cdot \rangle - f^*(x^*) \le m + \varepsilon]$. Thus, since the set on the right hand side is closed and convex,

$$\overline{\text{co}}\left(\varepsilon - \text{argmin } f + \{x^*\}^-\right) \subseteq [\langle \cdot, x^* \rangle - f^*(x^*) \le m + \varepsilon]$$

and taking the intersection over $x^* \in \text{dom } f^*$, we obtain

$$\bigcap_{x^* \in \text{dom } f^*} \overline{\text{co}}\left(\varepsilon - \text{argmin } f + \{x^*\}^-\right) \subseteq \left[\sup_{x^* \in \text{dom } f^*} \{\langle \cdot, x^* \rangle - f^*(x^*)\} \le m + \varepsilon\right]$$
$$= [f^{**} \le m + \varepsilon] \equiv \varepsilon - \text{argmin } f^{**}.$$

We finish this part of the proof by taking the intersection over $\varepsilon > 0$. The converse inclusion "$\subseteq$" in (8.93) comes from (8.46) and the fact that $\varepsilon\text{-argmin } f + \text{N}_{L \cap \text{dom } f^*}(\theta) \subset \varepsilon\text{-argmin } f + \{x^*\}^-$, for every $x^* \in \text{dom } f^*$ and $L \in \mathcal{F}_{X^*}$ such that $x^* \in L$.

Exercise 125: Taking into account (3.2), inequality (8.94) implies that, for all $x^* \in X^*$,

$$f^*(x^*) \le (a \|\cdot\|)^*(x^*) - b = a(\sigma_{B_{X^*}})^*(\frac{x^*}{a}) - b$$
$$= \text{I}_{B_{X^*}}(\frac{x^*}{a}) - b = \text{I}_{aB_{X^*}}(x^*) - b,$$

showing that $aB_{X^*} \subset \text{dom } f^*$. Hence, f^* is *norm*-continuous in the interior of B_{X^*}. Then we apply Corollary 8.3.8.

Exercise 126: $(ii) \Rightarrow (i)$ Assume that (ii) holds. Since $\sigma_{\alpha B_{X^*}}(x) = \alpha \|x\|$ for all $x \in X$, inequality (8.95) implies that $f \ge \sigma_{x_0^* + \alpha B_{X^*}} + \mu$. So,

$$f^*(x^*) \le (\sigma_{x_0^* + \alpha B_{X^*}} + \mu)^* = \text{I}_{x_0^* + \alpha B_{X^*}}(x^*) - \mu$$
$$= -\mu \text{ for all } x^* \in x_0^* + \alpha B_{X^*},$$

and Proposition 2.2.6 implies that f^* is $\|\cdot\|_*$-continuous at x_0^*. In other words, (i) follows:

$(i) \Rightarrow (ii)$ If f^* is $\|\cdot\|_*$-continuous at some $x_0^* \in X^*$, then we find some $\mu \in \mathbb{R}$ and $\alpha > 0$ such that $\langle x^*, x \rangle - f(x) \le f^*(x^*) \le -\mu$ for all $x \in X$ and $x^* \in x_0^* + \alpha B_{X^*}$. Thus, for all $x \in X$,

$$\langle x_0^*, x \rangle - f(x) + \alpha \|x\| = \langle x_0^*, x \rangle - f(x) + \sup_{x^* \in \alpha B_{X^*}} \langle x^*, x \rangle \le -\mu,$$

and (8.95) follows. This yields (ii).

Exercise 127: Consider the function $\tilde{f} : X \to \mathbb{R}_\infty$ defined as $\tilde{f}(x) := f(x)$ if $x \in C$, and $\tilde{f}(x) := +\infty$ if not. We verify that $\tilde{f}$ is lsc and $\inf_X \tilde{f} = \inf_C f \in \mathbb{R}$. Moreover, if $M > 0$ is such that $C \subset MB_X$, we get

$$\tilde{f}^*(x^*) = \sup_{x \in C}\{\langle x^*, x\rangle - f(x)\} \leq \sup_{x \in C}\|x\|\,\|x^*\| - \inf_C f \leq M\|x^*\| - \inf_C f.$$

Hence, $\tilde{f}^*$ is *norm*-continuous on X^*, and we apply Corollary 8.3.9.

Exercise 128: Relation (8.81) reads

$$f^* = \inf_{L \in \mathcal{F}(x_0^*)} \left(g^* + \mathrm{I}_{\mathrm{cl}(L \cap \mathrm{dom}\, f^*)}\right) + c,$$

where $x_0^* \in \mathrm{dom}\, f^*$ ($\subset \mathrm{dom}\, g^*$, by (8.80)). Then, taking the conjugate with respect to the pair $((X, \mathfrak{T}_X), (X^*, \sigma(X^*, X)))$, the properness of f^* and Theorem 3.2.2 give rise to

$$\overline{\mathrm{co}}f = f^{**} = \sup_{L \in \mathcal{F}(x_0^*)} \left(g^* + \mathrm{I}_{\mathrm{cl}(L \cap \mathrm{dom}\, f^*)}\right)^* - c.$$

Moreover, since $x_0^* \in \mathrm{cl}(L \cap \mathrm{dom}\, f^*) \cap \mathrm{dom}\, g^*$, Proposition 4.1.16 implies that

$$\overline{\mathrm{co}}f = \sup_{L \in \mathcal{F}(x_0^*)} \mathrm{cl}\left((\overline{\mathrm{co}}g) \square \sigma_{L \cap \mathrm{dom}\, f^*}\right) + c = \sup_{L \in \mathcal{F}_{X^*}} \mathrm{cl}\left((\overline{\mathrm{co}}g) \square \sigma_{L \cap \mathrm{dom}\, f^*}\right) + c,$$

and we are done.

Exercise 129: We fix $\delta > 0$ and denote the function in the right-hand side by f_δ. Then we easily verify that $f_\delta \leq f$; hence $f_\delta \in \Gamma_0(X)$ and $f_\delta \leq \overline{\mathrm{co}}f$. Next, given $x \in X$ and $\varepsilon \in (0, \delta)$, we show that

$$\partial_\varepsilon f(x) \subset \partial_\varepsilon f_\delta(x). \tag{9.28}$$

Take $x^* \in \partial_\varepsilon f(x)$ and let $\alpha < f_\delta(x)$. Then there exist $n \in \mathbb{N}$, $(\varepsilon_i, x_i, x_i^*)$ $\in \mathbb{R} \times X \times X^*$ with $x_i^* \in \partial_{\varepsilon_i} f(x_i)$, for $i = 0, \ldots, n$, such that

$$f(x_0) + \sum_{i=0}^{n-1}\langle x_i^*, x_{i+1} - x_i\rangle + \langle x_n^*, x - x_n\rangle - \sum_{i=0}^{n}\varepsilon_i > \alpha. \tag{9.29}$$

Therefore, taking $(x_{n+1}, x_{n+1}^*) := (x, x^*)$ in the graph of $\partial_\varepsilon f$ and $\varepsilon_{n+1} := \varepsilon$, the definition of f_δ implies that, for each $y \in X$,

$$f_\delta(y) \geq f(x_0) + \sum_{i=0}^{n}\langle x_i^*, x_{i+1} - x_i\rangle + \langle x^*, y - x\rangle - \sum_{i=0}^{n+1}\varepsilon_i.$$

9.7 EXERCISES OF CHAPTER 8   425

So, (9.29) yields $f_\delta(y) > \alpha + \langle x^*, y - x \rangle - \varepsilon$ for all $\alpha < f_\delta(x)$, and we deduce $f_\delta(y) \geq f_\delta(x) + \langle x^*, y - x \rangle - \varepsilon$ for all $y \in X$. This proves that $x^* \in \partial_\varepsilon f_\delta(x)$, and so, (9.28) holds. Therefore, applying Corollary 8.4.8(iii), there exists $c \in \mathbb{R}$ such that $\overline{\operatorname{co}} f = f_\delta + c$, and we get the coincidence of $\overline{\operatorname{co}} f$ and f_δ because

$$f(x_0) = (\overline{\operatorname{co}} f)(x_0) \geq f_\delta(x_0) \geq f(x_0) + \sup_{\varepsilon_0 \geq 0}(-\varepsilon_0) = f(x_0).$$

Exercise 130: If g denotes the right-hand side in (8.96), then as in Exercise 129, we show that $g \in \Gamma_0(X)$, $g \leq f$, $g(x_0) = f(x_0)$, and $\partial f \subset \partial g$. Thus, by Theorem 8.4.3, there exists $c \in \mathbb{R}$ such that $f^{**} = g^{**} \Box \sigma_{\operatorname{dom} f^*} + c$. Moreover, since $g \leq f$, we have $f^* \leq g^*$ and $\operatorname{dom} g^* \subset \operatorname{dom} f^*$. Thus, taking the conjugates in the last relation above and using Theorem 3.2.2, we obtain $f^* = g^* + I_{\operatorname{dom} f^*} - c = g^* - c$. Consequently, again by taking the conjugates and using Theorem 3.2.2, we get $\overline{\operatorname{co}} f = g + c$, and the equality $\overline{\operatorname{co}} f = g$ follows because $c = (\overline{\operatorname{co}} f)(x_0) - g(x_0) = f(x_0) - g(x_0) = 0$.

Exercise 131: $(iii) \Rightarrow (ii)$ Assume (iii) and let $x \in X$ such that $\partial(f_1 + f_2)(x) \neq \emptyset$, implying that $\partial(f_1 + f_2)(x) = \partial(\overline{\operatorname{co}}(f_1 + f_2))(x) = \partial((\overline{\operatorname{co}} f_1) + (\overline{\operatorname{co}} f_2))(x)$ and $(\overline{\operatorname{co}} f_1)(x) + (\overline{\operatorname{co}} f_2)(x) \in \mathbb{R}$. Hence, $\overline{\operatorname{co}} f_1, \overline{\operatorname{co}} f_2 \in \Gamma_0(X)$, and Proposition 4.1.16 yields

$$\partial(f_1 + f_2)(x) = \partial((\overline{\operatorname{co}} f_1) + (\overline{\operatorname{co}} f_2))(x) = \bigcap_{\varepsilon > 0} \operatorname{cl}\left(\partial_\varepsilon (\overline{\operatorname{co}} f_1)(x) + \partial_\varepsilon (\overline{\operatorname{co}} f_2)(x)\right). \quad (9.30)$$

Also, by Exercise 62 and assumption (iii), we have $f_1(x) + f_2(x) = (\overline{\operatorname{co}}(f_1 + f_2))(x) = (\overline{\operatorname{co}} f_1)(x) + (\overline{\operatorname{co}} f_2)(x)$, and so $f_1(x) = (\overline{\operatorname{co}} f_1)(x)$, $f_2(x) = (\overline{\operatorname{co}} f_2)(x)$, $\partial(\overline{\operatorname{co}} f_1)(x) = \partial f_1(x)$, and $\partial(\overline{\operatorname{co}} f_2)(x) = \partial f_2(x)$. Thus, the conclusion of (ii) follows by (9.30).

$(i) \Rightarrow (iii)$ We fix $\varepsilon \in]0, \delta]$, $\alpha > 0$ and $x \in X$, and denote $\tilde{f}_1 := \overline{\operatorname{co}} f_1$, $\tilde{f}_2 := \overline{\operatorname{co}} f_2$. Then, by (i) and using Proposition 4.1.6,

$$\partial_\varepsilon (f_1 + f_2)(x) = \operatorname{cl}\left(\bigcup_{\substack{\varepsilon_1 + \varepsilon_2 \leq \varepsilon + \alpha \\ \varepsilon_1, \varepsilon_2 \geq 0}} \partial_{\varepsilon_1} f_1(x) + \partial_{\varepsilon_2} f_2(x)\right)$$

$$\subset \operatorname{cl}\left(\bigcup_{\substack{\varepsilon_1 + \varepsilon_2 \leq \varepsilon + \alpha \\ \varepsilon_1, \varepsilon_2 \geq 0}} \partial_{\varepsilon_1} \tilde{f}_1(x) + \partial_{\varepsilon_2} \tilde{f}_2(x)\right) \subset \partial_{\varepsilon + \alpha}(\tilde{f}_1 + \tilde{f}_2)(x),$$

and taking the intersection over $\alpha > 0$, we get $\partial_\varepsilon (f_1 + f_2)(x) \subset \partial_\varepsilon(\tilde{f}_1 + \tilde{f}_2)(x)$ for all $x \in X$ and $\varepsilon \in]0, \delta]$. Since $\tilde{f}_1 + \tilde{f}_2 \leq f_1 + f_2$, Corollary

8.4.8 (iii) applies and yields some $c \in \mathbb{R}$ such that $\overline{\text{co}}(f_1 + f_2) = \tilde{f}_1 + \tilde{f}_2 + c$. Observe that $c \geq 0$ because $\tilde{f}_1 + \tilde{f}_2 \leq \overline{\text{co}}(f_1 + f_2)$. Moreover, since $\text{dom}(f_1+f_2)^* \neq \emptyset$ by the current assumption, for each $\gamma \in \,]0, \frac{\delta}{2}[$ there is some $(x_\gamma, x^*) \in X \times X^*$ that satisfies $x^* \in \partial_\gamma (f_1+f_2)(x_\gamma)$. Hence, using (ii),

$$x^* \in \partial_\gamma (f_1+f_2)(x_\gamma) \subset \text{cl}\left(\bigcup_{\substack{\varepsilon_1+\varepsilon_2 \leq 2\gamma \\ \varepsilon_1, \varepsilon_2 \geq 0}} \partial_{\varepsilon_1} f_1(x_\gamma) + \partial_{\varepsilon_2} f_2(x_\gamma) \right)$$
$$\subset \text{cl}\left(\partial_{2\gamma} f_1(x_\gamma) + \partial_{2\gamma} f_2(x_\gamma) \right),$$

and we deduce that $\partial_{2\gamma} f_1(x_\gamma) \neq \emptyset$ and $\partial_{2\gamma} f_2(x_\gamma) \neq \emptyset$. Then, by Exercise 62, we get $f_1(x_\gamma) \geq (\overline{\text{co}} f_1)(x_\gamma) \geq f_1(x_\gamma) - 2\gamma$ and $f_2(x_\gamma) \geq (\overline{\text{co}} f_2)(x_\gamma) \geq f_2(x_\gamma) - 2\gamma$, so that

$$c = \overline{\text{co}}(f_1+f_2)(x_\gamma) - \tilde{f}_1(x_\gamma) - \tilde{f}_2(x_\gamma) \leq (f_1+f_2)(x_\gamma)$$
$$- f_1(x_\gamma) + 2\gamma - f_2(x_\gamma) + 2\gamma = 4\gamma.$$

Therefore, since $\gamma \in \,]0, \frac{\delta}{2}[$ is arbitrary, we infer that $c \leq 0$; that is, $c = 0$ and $\overline{\text{co}}(f_1+f_2) = \tilde{f}_1 + \tilde{f}_2$, as required in statement (iii).

$(ii) \Rightarrow (iii)$ when X is Banach with the RNP, f, g are lsc, and $\text{int}(\text{dom}(f+g)^*) \neq \emptyset$. Given $x \in X$, (ii) implies that

$$\partial (f_1+f_2)(x) = \bigcap_{\varepsilon>0} \text{cl}\left(\partial_\varepsilon f_1(x) + \partial_\varepsilon f_2(x) \right) \subset \bigcap_{\varepsilon>0} \text{cl}\left(\partial_\varepsilon \tilde{f}_1(x) + \partial_\varepsilon \tilde{f}_2(x) \right)$$
$$\subset \bigcap_{\varepsilon>0} \partial_\varepsilon (\tilde{f}_1+\tilde{f}_2)(x) = \partial(\tilde{f}_1+\tilde{f}_2)(x).$$

Thus, by Remark 20, there exists some $c \in \mathbb{R}$ such that $(f_1+f_2)^{**} = f_1^{**} + f_2^{**} + c$. Moreover, since $\text{int}(\text{dom}(f+g)^*) \neq \emptyset$ and X is Banach with the RNP, the w^*-lsc function $(f+g)^*$ is $(norm\text{-})$ continuous on $\text{int}(\text{dom}(f+g)^*)$, and so there exists a $(norm\text{-})$ dense set D of $\text{int}(\text{dom}(f+g)^*)$ such that $(f+g)^*$ is Fréchet-differentiable on D. Take $x_0^* \in D$. Then, by Theorem 8.3.7(i), we have that $\partial (f+g)^*(x_0^*) = \{((f+g)^*)'(x_0^*)\} = (\partial(f+g))^{-1}(x_0^*) =: x_0 \in X$. In other words, using (ii),

$$x_0^* \in \partial(f+g)(x_0) = \bigcap_{\varepsilon>0} \text{cl}\left(\partial_\varepsilon f_1(x_0) + \partial_\varepsilon f_2(x_0) \right).$$

In particular, $\partial(f+g)(x_0), \partial_\varepsilon f_1(x_0)$, and $\partial_\varepsilon f_2(x_0)$ are non-empty for all $\varepsilon > 0$, and we deduce by Exercise 62 that $(\overline{\text{co}}(f+g))(x_0) = (f+g)(x_0)$, $\tilde{f}_1(x_0) = f_1(x_0)$ and $\tilde{f}_2(x_0) = f_2(x_0)$. So, by Theorem 3.2.2, we infer that $c = (f_1+f_2)^{**}(x_0) - f_1^{**}(x_0) - f_2^{**}(x_0) = (f+g)(x_0) - f_1(x_0) - f_2(x_0) = 0$.

Glossary of notations

$\mathbb{R}_+^* := \,]0, +\infty[$

$\mathbb{R}_+ := [0, +\infty[$

$\overline{\mathbb{R}} := \mathbb{R} \cup \{\pm\infty\}$

$\mathbb{R}_\infty := \mathbb{R} \cup \{+\infty\}$

$\mathbb{R}^n$, n-dimensional Euclidean space

$\mathbb{R}_+^n$, nonnegative orthant in $\mathbb{R}^n$

0_n, zero-vector in $\mathbb{R}^n$

$\mathbb{R}^T := \{\lambda : T \to \mathbb{R}\}$

$\mathrm{supp}\,\lambda := \{t \in T : \lambda_t \neq 0\}$, $\lambda \in \mathbb{R}^T$

$\mathbb{R}^{(T)} := \{\lambda \in \mathbb{R}^T : \mathrm{supp}\,\lambda \text{ finite}\}$

$\mathbb{R}_+^{(T)} := \{\lambda \in \mathbb{R}^{(T)} : \lambda_t \geq 0 \,\forall t \in T\}$

$\mathbb{N}$, natural numbers

$|T|$, cardinality of T

θ, zero, or origin, in X (also in X^*)

$\mathfrak{T}_X$, initial topology in X

$\mathcal{N}_X$, $\begin{cases} \text{convex, closed, and balanced} \\ \text{neighborhoods of } \theta \text{ in } X \end{cases}$

ℓ_p, $p \geq 1$, sequences $(x_k)_{k=1}^\infty$ s.t. $\sum_{k=1}^\infty |x_k|^p < +\infty$

ℓ_∞, sequences $(x_k)_{k=1}^\infty$ s.t. $\sup_{k\geq 1} |x_k| < +\infty$

c_0, subspace of ℓ_∞ s.t. $x_k \to 0$

$c_{00} := \mathbb{R}^{(\mathbb{N})}$

$\mathcal{C}(T)$, continuous functions in $\mathbb{R}^T$

$\mathcal{C}(T, [0,1])$, continuous functions from T to $[0,1]$

$\mathcal{C}^1(T)$, continuously differentiable functions in $\mathbb{R}^T$

$\ell_\infty(T)$, bounded functions in $\mathbb{R}^T$

$B_X(x, r)$, closed ball of radius $r > 0$ centered at x

B_X, closed unit ball

X^* and X^{**}, (topological) dual and bidual spaces of X

$\Gamma_0(X)$, proper convex and lsc functions on X

$A + B := \{a + b : a \in A, b \in B\}$

$\Lambda A := \{\lambda a : \lambda \in \Lambda, a \in A\}, \Lambda \subset \mathbb{R}$

co$A$, convex hull of A

cone$A$, conic hull of A

aff$A$, affine hull of A

span$A$, linear hull of A

dim A, dimension of aff$A$

lin$A$, lineality space of A

$\mathcal{F}(x)$, finite-dimensional linear subspaces containing x

$\mathcal{F} := \mathcal{F}_X := \mathcal{F}(\theta)$

A°, polar set of A

A^-, (negative) dual cone of A

$A^\perp$, orthogonal subspace of A

$\mathrm{N}_A^\varepsilon(x)$, ε-normal set to A at x

$\mathrm{N}_A(x)$, normal cone to A at x

A_∞, recession cone of A

int$A$, interior of A

cl$A$ (or $\overline{A}$), closure of A

GLOSSARY OF NOTATIONS

bd$A$, boundary of A

$\overline{\text{co}}A := \text{cl}(\text{co}A)$

$\overline{\text{cone}}A := \text{cl}(\text{cone}A)$

ri$A$, relative interior of A

dom$f$, effective domain of f

epi$f$, epigraph of f

epi$_s f$, strict epigraph of f

gph$f$, graph of f

$[f \leq \alpha]$, level set of f

cl$f$, closed hull of f

co$f$, convex hull of f

$\overline{\text{co}}f$, closed convex hull of f

$f \square g$, inf-convolution of f and g

$f \circ g$, composition of f and g

f^*, Fenchel conjugate of f

f^{**}, biconjugate of f

σ_A, support function of A

$\rho_A := \sigma_{A \cup (-A)}$

I_A, indicator function of A

p_A, Minkowski gauge of A

d_A, distance to A

π_A, projection mapping on A

$\langle \cdot, \cdot \rangle$, duality pairing in $X^* \times X$

$\|\cdot\|$, norm

$A^* : Y^* \to X^*$, adjoint operator of $A : X \to Y$

$\partial_\varepsilon f$, ε-subdifferential of f

∂f, subdifferential of f

Bibliography

[1] Ch. Aliprantis and K. Border, *Infinite Dimensional Analysis. A Hitchhiker's Guide*, 3rd ed. Springer, Berlin, 2006.

[2] E.J. Anderson, M.A. Goberna, and M.A. López, Locally polyhedral linear inequality systems, Linear Algebra Appl. **270** (1998) 231–253.

[3] F.J. Aragón-Artacho, J.M. Borwein, V. Martín-Márquez, and L. Yao, Applications of convex analysis within mathematics, Math. Program. Ser. B **148** (2014) 49–88.

[4] E. Asplund, Differentiability of the metric projection in finite-dimensional Euclidean space, Proc. Amer. Math. Soc. **38** (1973) 218–219.

[5] E. Asplund and R. T. Rockafellar, Gradients of convex functions, Trans. Amer. Math. Soc. **139** (1969) 443–467.

[6] H. Attouch, J.-B. Baillon, and M. Théra, Variational sum of monotone operators, J. Convex Anal. 1 (1994) 1–29.

[7] H. Attouch and H. Brézis, Duality for the sum of convex functions in general Banach spaces. In "Aspects of Mathematics and Its Applications", North Holland, Amsterdam, 125–133, (1986).

[8] A. Auslender and M. Teboulle, *Asymptotic Cones and Functions in Optimization and Variational Inequalities*, Springer-Verlag, New York, 2003.

[9] D. Azé, *Élements d'Analyse Convexe et Variationnelle*, Ellipses, Math á l'Université/2E cycle, 1997.

[10] D. Azé, Duality for the sum of convex functions in general normed spaces, Arch. Math. (Basel) **62** (1994) 554–561.

[11] V. Barbu and Th. Precupanu, *Convexity and Optimization in Banach Spaces*, D. Reidel Publishing Company, Dordrecht, 1986.

[12] H.H. Bauschke and P.L. Combettes, *Convex Analysis and Monotone Operator Theory in Hilbert Spaces*, 2nd ed. Springer, New York, 2017.

[13] J. Benoist and A. Daniilidis, Integration of Fenchel subdifferentials of epi-pointed functions, SIAM J. Optim. **12** (2002) 575–582.

[14] J. Benoist and J.-B. Hiriart-Urruty, What is the subdifferential of the closed convex hull of a function? SIAM J. Math. Anal. **27** (1996) 1661–1679.

[15] D.P. Bertsekas, A. Nedić, and A.E. Ozdaglar, *Convex Analysis and Optimization*, Athena Scientific, Boston, MA, 2003.

[16] J.F. Bonnans and A. Shapiro, *Perturbation Analysis of Optimization Problems*, Springer-Verlag, New York, 2000.

[17] J.M. Borwein, Adjoint process duality, Math. Oper. Res. **8** (1983) 403–434.

[18] J.M. Borwein, Proximality and Chebyshev sets, Optim. Lett. **1** (2007) 21–32.

[19] J.M. Borwein and A.S. Lewis, Partially finite convex programming, part I: Quasi relative interiors and duality theory, Math. Program. **57** (1992) 15–48.

[20] J.M. Borwein and A.S. Lewis, *Convex Analysis and Nonlinear Optimization. Theory and Examples*, 2nd ed. CMS Books in Mathematics/Ouvrages de Mathématiques de la SMC, 3. Springer, New York, 2006.

[21] J.M. Borwein and D.W. Tingley, On supportless convex sets,Proc. Amer. Math. Soc. **94** (1985) 471–476.

[22] J.M. Borwein and J.D. Vanderwerff, Differentiability of conjugate functions and perturbed minimization principles, J. Convex Anal. **16** (2009) 707–711.

[23] J.M. Borwein and J.D. Vanderwerff, *Convex Functions: Constructions, Characterizations and Counterexamples*, Cambridge University Press, Cambridge (UK), 2010.

[24] J.M. Borwein and Q.J. Zhu, *Techniques of Variational Analysis*, CMS Books in Mathematics/Ouvrages de Mathématiques de la SMC, 20. Springer-Verlag, New York, 2005.

[25] J.M. Borwein and D. Zhuang, On Fan's minimax theorem. Math. Program. **34** (1986) 232–234.

[26] R.I. Bot, *Conjugate Duality in Convex Optimization*, Lecture Notes in Economics and Mathematical Systems, 637. Springer-Verlag, Berlin, 2010.

[27] R.I. Bot and G. Wanka, Farkas-type results with conjugate functions, SIAM J. Optim. **15** (2005) 540–554.

[28] B. Brighi, Sur l'enveloppe convexe d'une fonction de la variable réelle, Revue de Mathématiques Spéciales, **8** (1994) 547-550.

[29] A. Brøndsted, On the subdifferential of the supremum of two convex functions, Math. Scand.**31** (1972) 225–230.

[30] A. Brøndsted and R.T. Rockafellar, On the subdifferentiability of convex functions, Proc. Amer. Math. Soc. **16** (1965) 605–611.

[31] R.S. Burachik and V. Jeyakumar, Dual condition for the convex subdifferential sum formula with applications, J. Convex Anal. **12** (2005) 279–290.

[32] A. Cabot and L. Thibault, Sequential formulae for the normal cone to sublevel sets, Trans. Amer. Math. Soc. **366** (2014) 6591–6628.

[33] M.J. Cánovas, M.A. López, J. Parra, and M.I. Todorov, Stability and well-posedness in linear semi-infinite programming, SIAM J. Optim. **10** (1999) 82–98.

[34] M.J. Cánovas, M.A. López, J. Parra, and F.J. Toledo, Ill-posedness with respect to the solvability in linear optimization, Linear Algebra Appl. **416** (2006) 520–540.

[35] M.J. Cánovas, M.A. López, J. Parra, and F.J. Toledo, Sufficient Conditions for total Ill-posedness in linear optimization, Eur. J. Oper. Res. **181** (2006) 1126–1136.

[36] C. Castaing and M. Valadier, *Convex Analysis and Measurable Multifunctions*, Springer, Berlin, 1977.

[37] A. Charnes, W.W. Cooper, and K.O. Kortanek, On representations of semi-infinite programs which have no duality gaps, Manage. Sci. **12** (1965) 113–121.

[38] N.H. Chieu, V. Jeyakumar, G. Li, and H. Mohebi, Constraint qualifications for convex optimization without convexity of constraints: new connections and applications to best approximation, Eur. J. Oper. Res. **265** (2018) 19–25.

[39] F.H. Clarke, A new approach to Lagrange multipliers, Math. Oper. Res.**2** (1976) 165–174.

[40] F.H. Clarke, *Optimization and Nonsmooth Analysis*, John Wiley and Sons, New York, 1983.

[41] C. Combari, M. Laghdir, and L. Thibault, On subdifferential calculus for convex functions defined on locally convex spaces, Ann. Sci. Math. Qurbec **23** (1999) 23–36.

BIBLIOGRAPHY

[42] P.L. Combettes, Perspective functions: properties, constructions, and examples, Set-Valued Var. Anal. **26** (2018) 247–264.

[43] R. Correa, Y. García, and A. Hantoute, Nonconvex integration using ε-subdifferentials, Optimization **67** (2018) 2205–2227.

[44] R. Correa, Y. García, and A. Hantoute, Integration formulas via the Fenchel subdifferential of nonconvex functions, Nonlinear Anal. **75** (2012) 1188–1201.

[45] R. Correa and A. Hantoute, New formulas for the Fenchel subdifferential of the conjugate function, Set-Valued Var. Anal. **18** (2010) 405–422.

[46] R. Correa and A. Hantoute, Subdifferential of the conjugate function in general Banach spaces, TOP **20** (2012) 328–346.

[47] R. Correa and A. Hantoute, Lower semicontinuous convex relaxation in optimization, SIAM J. Optim. **23** (2013) 54–73.

[48] R. Correa, A. Hantoute, and A. Jourani, Characterizations of convex approximate subdifferential calculus in Banach spaces, Trans. Amer. Math. Soc. **368** (2016) 4831–4854.

[49] R. Correa, A. Hantoute, and M.A. López, Weaker conditions for subdifferential calculus of convex functions, J. Funct. Anal. **271** (2016) 1177–1212.

[50] R. Correa, A. Hantoute, and M.A. López, Towards supremum-sum subdifferential calculus free of qualification conditions, SIAM J. Optim. 26 (2016) 2219–2234.

[51] R. Correa, A. Hantoute, and M.A. López, Valadier-like formulas for the supremum function I, J. Convex Anal. **25** (2018) 1253–1278.

[52] R. Correa, A. Hantoute, and M.A. López, Moreau-Rockafellar type formulas for the subdifferential of the supremum function, SIAM J. Optim. **29** (2019) 1106–1130.

[53] R. Correa, A. Hantoute, and M.A. López, Valadier-like formulas for the supremum function II: the compactly indexed case, J. Convex Anal. **26** (2019) 299–324.

[54] R. Correa, A. Hantoute, and M.A. López, Subdifferential of the supremum via compactification of the index set, Vietnam J. Math. **48** (2020) 569–588.

[55] R. Correa, A. Hantoute, and M.A. López, Alternative representations of the normal cone to the domain of supremum functions and subdifferential calculus, Set-Valued Var. Anal. **29** (2021) 683–699.

[56] R. Correa, A. Hantoute, and M.A. López, Subdifferential of the supremum function: moving back and forth between continuous and non-continuous settings, Math. Program. Ser. B **189** (2021) 217–247.

[57] R. Correa, A. Hantoute, and M.A. López, Biconjugate Moreau theorem revisited: conjugacy analysis and minimax theorem, submitted, 2023.

[58] R. Correa, A. Hantoute, and M.A. López, Conjugacy based approach to subdifferential calculus, submitted, 2023.

[59] R. Correa, A. Hantoute, and A. Pérez-Aros, On the Klee–Saint Raymond's characterization of convexity, SIAM J. Optim. **26** (2016) 1312–1321.

[60] R. Correa, A. Hantoute, and A. Pérez-Aros, On Brøndsted–Rockafellar's Theorem for convex lower semicontinuous epi-pointed functions in locally convex spaces, Math. Program. Ser. B **168** (2018) 631–643.

[61] R. Correa, A. Hantoute, and A. Pérez-Aros, MVT, integration, and monotonicity of a generalized subdifferential in locally convex spaces, J. Convex Anal. **27** (2020) 1345–1362.

[62] R. Correa, A. Hantoute, and D. Salas, Integration of nonconvex epi-pointed functions in locally convex spaces, J. Convex Anal. **23** (2016) 511–530.

[63] R. Correa and A. Jofré, Some properties of semismooth and regular functions in nonsmooth analysis, Recent advances in system modeling and optimization (Santiago, 1984), 69–85, Lecture Notes in Control and Inform. Sci. 87, Springer, Berlin, 1986.

[64] R. Correa and A. Seeger, Directional derivative of a minmax function, Nonlinear Anal. **9** (1985) 14–22.

[65] A. Daniilidis, A. Hantoute, and C. Soto, Klee theorem and weak projections onto closed sets, submitted 2022.

[66] J.M. Danskin, The theory of max-min, with applications, SIAM J. Appl. Math. **14** (1966) 641–664.

[67] J.M. Danskin, *The Theory of Max-Min and its Applications to Weapons Allocations Problems*, Springer, New York, 1967.

[68] G. Debreu, *Theory of Value: An Axiomatic Analysis of Economic Equilibrium*, John Wiley & Sons, Inc., New York; Chapman & Hall, Ltd., London, 1959.

[69] J.-P. Dedieu, Cônes asymptotes d'ensembles non convexes, Bull. Soc. Math. France Mem. **60** (1979) 31–44.

[70] F. Deutsch, W. Li, and J. Ward, A dual approach to constrained interpolation from a convex subset of Hilbert space, J. Approx. Theory **90** (1997) 385–414.

[71] N. Dinh, M.A. Goberna, and M.A. López, From linear to convex systems: consistency, Farkas' lemma and applications, J. Convex Anal. **13** (2006) 113–133.

[72] N. Dinh, M.A. Goberna, M.A. López, and T.Q. Son, New Farkas-type constraint qualifications in convex infinite programming, ESAIM Control Optim. Calc. Var. **13** (2007) 580–597.

[73] N. Dinh, M.A. Goberna, M.A. López, and M. Volle, Functional inequalities in absence of convexity and lower semicontinuity, Applications to optimization, SIAM J. Optim. **20** (2010) 2540-2559.

[74] N. Dinh, V. Jeyakumar, and G.M. Lee, Sequential Lagrangian conditions for convex programs with applications to semidefinite programming, J. Optim. Theory Appl. **125** (2005) 85–112.

[75] A.L. Dontchev, *Lectures on Variational Analysis*, Applied Mathematical Sciences, 205. Springer, Cham, 2021.

[76] A.Y. Dubovitskii and A.A. Milyutin, Extremum problems in the presence of restrictions, Zh. Vychisl. Mat. Mat. Fiz. **5** (1965) 395–453 (in Russian).

[77] N. Dunford and J.T. Schwartz, *Linear Operators. Part I. General theory*, John Wiley & Sons, Inc., New York, 1988.

[78] J. Dutta and C.S. Lalitha, Optimality conditions in convex optimization revisited, Optim. Lett. **7** (2013) 221–229.

[79] I. Ekeland, On the variational principle, J. Math. Anal. Appl. **47** (1974) 324–353.

[80] I. Ekeland and R. Temam, *Convex Analysis and Variational Problems*, North-Holland & American Elsevier, Amsterdam, New York, 1976.

[81] M. Fabian, P. Habala, P. Hájek, V. Montesinos, and V. Zizler, *Banach Space Theory*, CMS Books in Mathematics/Ouvrages de Mathématiques de la SMC. Springer, New York, 2011.

[82] M.D. Fajardo and M.A. López, Locally Farkas-Minkowski systems in convex semi-infinite programming, J. Optim. Theory Appl. **103** (1999) 313–335.

[83] K. Fan, Minimax theorems, National Academy of Sciences, Washington, DC, Proceedings USA **39** (1953) 42–47.

[84] K. Fan, Existence theorems and extreme solutions for inequalities concerning convex functions or linear transformations, Math. Z. **68** (1957) 205–216.

[85] D.H. Fang, Ch. Li, and K.F. Ng, Constraint qualifications for optimality conditions and total Lagrange dualities in convex infinite programming, Nonlinear Anal. **73** (2010) 1143–1159.

[86] W. Fenchel, *Convex Cones, Sets and Functions*, Mimeographed lectures notes, Princeton University, 1951.

[87] J. Florence and M. Lassonde, Subdifferential estimate of the directional derivative, optimality criterion and separation principles, Optimization **62** (2013) 1267–1288.

[88] J.R. Giles, *Convex Analysis with Application in the Differentiation of Convex Functions*, Research Notes in Mathematics, 58. Pitman (Advanced Publishing Program), Boston, Mass.-London (1982).

[89] M.I.A. Ghitri and A. Hantoute, A non-convex relaxed version of the minimax theorem, submitted 2022.

[90] M.A. Goberna, V. Jeyakumar, and M.A. López, Necessary and sufficient constraint qualifications for solvability of systems of infinite convex inequalities, Nonlinear Anal. **68** (2008) 1184–1194.
[91] M.A. Goberna and M.A. López, Optimal value function in semi-infinite programming, J. Optim. Theory Appl. **59** (1988) 261–279.
[92] M.A. Goberna and M.A. López, *Linear Semi-infinite Optimization*, J. Wiley, Chichester, 1998.
[93] M.A. Goberna and M.A. López, *Post-Optimal Analysis in Linear Semi-Infinite Optimization*, Springer, New York, 2014.
[94] M.A. Goberna, M.A. López, Recent contributions to linear semi-infinite optimization: An update. Ann. Oper. Res. **271** (2018), 237–278.
[95] M.A. Goberna, M.A. López, and M.I. Todorov, Stability theory for linear inequality systems, SIAM J. Matrix Anal. Appl. **17** (1996) 730–743.
[96] M.A. Goberna, M.A. López, and M.I. Todorov, Stability theory for linear inequality systems II: Upper semicontinuity of the solution set mapping, SIAM J. Optim. **7** (1997) 1138–1151.
[97] A. Griewank and P.J. Rabier, On the smoothness of convex envelopes, Trans. Amer. Math. Soc. **322** (1991) 691–709.
[98] J. Gwinner, On results of Farkas type, Numer. Funct. Anal. Optim. **9** (1987) 471–520.
[99] A. Hantoute, Subdifferential set of the supremum of lower semi-continuous convex functions and the conical hull property, Top **14** (2006) 355–374.
[100] A. Hantoute and M.A. López, A complete characterization of the subdifferential set of the supremum of an arbitrary family of convex functions, J. Convex Anal. **15** (2008) 831–858.
[101] A. Hantoute and M.A. López, Characterization of total ill-posedness in linear semi-infinite optimization, J. Comput. Appl. Math.**217** (2008) 350–364.
[102] A. Hantoute and M.A. López, A new tour on the subdifferential of suprema. Highlighting the relationship between suprema and finite sums, J. Optim. Theory Appl. **193** (2022) 81–106.
[103] A. Hantoute, M.A. López, and C. Zălinescu, Subdifferential calculus rules in convex analysis: A unifying approach via pointwise supremum functions, SIAM J. Optim.**19** (2008) 863–882.
[104] A. Hantoute and J.E. Martínez-Legaz, Characterization of Lipschitz continuous difference of convex functions, J. Optim. Theory Appl. **159** (2013) 673–680.
[105] A. Hantoute and A. Svensson, General representation of δ-normal sets and normal cones to sublevel sets of convex functions, Set-Valued Var. Anal. **25** (2017) 651–678.
[106] A. Hantoute and T. Zakaryan, Subdifferentiation of the inf convolution and minimal time problems, J. Convex Anal. **27** (2020) 315–335.
[107] J.-B. Hiriart-Urruty, Convex analysis and optimization in the past 50 years: some snapshots. Constructive nonsmooth analysis and related topics, 245–253, Springer Optim. Appl. 87, Springer, New York, (2014).
[108] J.-B. Hiriart-Urruty and C. Lemaréchal, *Convex Analysis and Minimization Algorithms I, II*, Springer-Verlag, Berlin, 1993.
[109] J.-B. Hiriart-Urruty, M.A López, and M. Volle, The ε-strategy in variational analysis: illustration with the closed convexification of a function, Rev. Mat. Iberoam. **27** (2011) 449–474.
[110] J.-B. Hiriart-Urruty, M. Moussaoui, A. Seeger, and M. Volle, Subdifferential calculus without qualification conditions, using approximate subdifferentials: A survey, Nonlinear Anal. **24** (1995) 1727–1754.
[111] J.-B. Hiriart-Urruty and R.R. Phelps, Subdifferential calculus using ε-subdifferentials, J. Funct. Anal. **118** (1993) 154–166.
[112] A.D. Ioffe, *Variational Analysis of Regular Mappings. Theory and Applications*, Springer Monographs in Mathematics. Springer, Cham, 2017.
[113] A.D. Ioffe, A note on subdifferentials of pointwise suprema,Top **20** (2012) 456–466.

[114] A.D. Ioffe and V.L. Levin, Subdifferentials of convex functions, Trudy Moskov Mat. Obshch, **26** (1972) 3–73 (in Russian).

[115] A.D. Ioffe and V.H. Tikhomirov, *Theory of Extremal Problems*, Studies in Mathematics and its Applications, Vol. 6, North-Holland, Amsterdam, 1979.

[116] V. Jeyakumar, Asymptotic dual conditions characterizing optimality for infinite convex programs, J. Optim. Theory Appl. **93** (1997) 153–165.

[117] V. Jeyakumar, *Farkas' lemma: Generalizations*, in Encyclopedia of Optimization II, C.A. Floudas and P. Pardalos, edited by Kluwer, Dordrecht (2001) 87–91.

[118] V. Jeyakumar, Characterizing set containments involving infinite convex constraints and reverse-convex constraints, SIAM J. Optim. **13** (2003) 947–959.

[119] V. Jeyakumar, N. Dinh, and G.M. Lee, A new closed cone constraint qualification for convex Optimization, Applied Mathematics Research Report AMR04/8, UNSW, 2004. Unpublished manuscript. http://www.maths.unsw.edu.au/applied/reports/amr08.html

[120] V. Jeyakumar, G.M. Lee and N. Dinh, New sequential Lagrange multiplier conditions characterizing optimality without constraint qualifications for convex programs, SIAM J. Optim. **14** (2003) 534–547.

[121] V. Jeyakumar, A.M. Rubinov, B.M. Glover, and Y. Ishizuka, Inequality systems and global optimization, J. Math. Anal. Appl. **202** (1996) 900–919.

[122] A. Jofré and J.-P. Penot, A note on the directional derivative of a marginal function, Rev. Mat. Apl. **14** (1993) 37–54.

[123] F. Jules and M. Lassonde, Formulas for subdifferentials of sums of convex functions, J. Convex Anal. **9** (2002) 519–533.

[124] B. Kirchheim and J. Kristensen, Differentiability of convex envelopes, C. R. Acad. Sci. Paris Sér. I Math. **333** (2001) 725–728.

[125] V. Klee, Convexity of Chebyshev sets, Math. Ann. **142** (1961) 292–304.

[126] P. Kocourek, An elementary new proof of the determination of a convex function by its subdifferential, Optimization **59** (2010) 1231–1233.

[127] M. Lassonde, Asplund spaces, Stegall variational principle and the RNP, Set-Valued Var. Anal. **17** (2009) 183–193.

[128] S. László, Minimax results on dense sets and dense families of functionals, SIAM J. Optim. **27** (2017) 661–685.

[129] P.-J. Laurent, *Approximation et Optimization*, Hermann, Paris, 1972.

[130] G. Lebourg, Valeure moyenne pour gradient generalisé, C.R. Acad. Sci. Paris Sér. A **281** (1975) 795–797.

[131] V.L. Levin, An application of Helly's theorem in convex programming, problems of best approximation and related questions, Mat. Sb., Nov. Ser. **79** (121) (1969), 250–263. English transl.: Math. USSR, Sb. 8, 235–247.

[132] W. Li, C. Nahak, and I. Singer, Constraint qualification for semi-infinite systems of convex inequalities, SIAM J. Optim. **11** (2000) 31–52.

[133] Ch. Li and K.F. Ng, On constraint qualification for an infinite system of convex inequalities in a Banach space, SIAM J. Optim. **15** (2005) 488–512.

[134] Ch. Li and K.F. Ng, Subdifferential calculus rules for supremum functions in convex analysis, SIAM J. Optim. **21** (2011) 782–797.

[135] Ch. Li, K.F. Ng, and T.K. Pong, The SECQ, linear regularity, and the strong CHIP for an infinite system of closed convex sets in normed linear spaces, SIAM J. Optim. **18** (2007) 643–665.

[136] Ch. Li., K.F. Ng, and T.K. Pong, Constraint qualifications for convex inequality systems with applications in constrained optimization, SIAM J. Optim. **19** (2008) 163–187.

[137] Y. Liu and M.A. Goberna, Asymptotic optimality conditions for linear semi-infinite programming, Optimization **65** (2016) 387–414.

[138] M.A. López and G. Still, Semi-infinite programming, European J. Oper. Res. **180** (2007) 491–518.

BIBLIOGRAPHY

[139] O. Lopez and L. Thibault, Sequential formula for subdifferential of upper envelope of convex functions, J. Nonlinear Convex Anal. **14** (2013) 377–388.

[140] M.A. López and E. Vercher, Optimality conditions for nondifferentiable convex semi-infinite programming, Math. Program. **27** (1983) 307–319.

[141] M.A. López and M. Volle, A formula for the set of optimal solutions of a relaxed minimization problem. Applications to subdifferential calculus, J. Convex Anal. **17** (2010) 1057–1075.

[142] M.A. López and M. Volle, On the subdifferential of the supremum of an arbitrary family of extended real-valued functions, RACSAM **105** (2011) 3–21.

[143] M.A. López and M. Volle, Subdifferential of the closed convex hull of a function and integration with nonconvex data in general normed spaces, J. Math. Anal. Appl. **390** (2012) 307–312.

[144] M.A. López and M. Volle, The use of the relative interior in integration with nonconvex data, Set-Valued Var Anal. **22** (2014) 247–258.

[145] D.T. Luc, Recession cones and the domination property in vector optimization, Math. Program. **49** (1990) 113–122.

[146] R. Lucchetti, *Convexity and Well-Posed Problems*, CMS Books in Mathematics/Ouvrages de Mathématiques de la SMC, 22. Springer, New York, 2006.

[147] O.L. Mangasarian, Set containment characterization, J. Glob. Optim. **24** (2002) 473–480.

[148] S. Marcellin and L. Thibault, Integration of ε-Fenchel subdifferentials and maximal cyclic monotonicity, J. Glob. Optim. **32** (2005) 83–91.

[149] P. Maréchal, On a functional operation generating convex functions, Part 1: Duality, J. Optim. Theory Appl. **126** (2005) 175–189.

[150] P. Maréchal, On a functional operation generating convex functions, Part 2: Algebraic Properties, J. Optim. Theory Appl. **126** (2005) 357–366.

[151] J.E. Martínez-Legaz and M. Théra, ε-Subdifferentials in terms of subdifferentials, Set-Valued Anal. **4** (1996) 327–332.

[152] K. Meng, V. Roshchina, and X. Yang, On local coincidence of a convex set and its tangent cone, J. Optim. Theory Appl. **164** (2015) 123–137.

[153] G.J. Minty, On the monotonicity of the gradient of a convex function, Pacific J. Math. **14** (1964) 243–247.

[154] B.S. Mordukhovich, *Variational Analysis and Generalized Differentiation I, II*, Springer-Verlag, Berlin, 2006.

[155] B.S. Mordukhovich, *Variational Analysis and Applications*, Springer Monographs in Mathematics, Springer, 2018.

[156] B.S. Mordukhovich and N.M. Nam, *Convex Analysis and Beyond, Vol. 1. Basic Theory*, Springer Series in Operations Research and Financial Engineering. Springer, 2022.

[157] B.S. Mordukhovich and T.T.A. Nghia, Subdifferentials of nonconvex supremum functions and their applications to semi-infinite and infinite programs with Lipschitzian data, SIAM J. Optim. **23** (2013) 406–431.

[158] B.S. Mordukhovich and T.T.A. Nghia, Constraint qualifications and optimality conditions in nonlinear semi-infinite and infinite programming, Math. Program. **139** (2013) 271–300.

[159] B.S. Mordukhovich and T.T.A. Nghia, Nonconvex cone-constrained optimization with applications to semi-infinite programming, Math. Oper. Res. **39** (2014) 301–337.

[160] J. Moreau, Proximité et dualité dans un espace hilbertien, Bull. Soc. Math. France **93** (1965) 273–299.

[161] J.-J. Moreau, *Fonctionnelles Convexes*, Lectures Notes, Séminaire "Equations aux dérivées partielles", Collège de France, 1966, and Rome: Instituto Poligrafico e Zecca dello Stato, 2003.

[162] M. Moussaoui and V. Volle, Quasicontinuity and united functions in convex duality theory, Comm. Appl. Nonlinear Anal. **4** (1997) 73-89.
[163] J. Munkres, *Topology*, 2nd Ed. Prentice Hall, Upper Saddle River, 2000.
[164] D. Pallaschke and S. Rolewicz, *Foundations of Mathematical Optimization: Convex Analysis without Linearity*, Kluwer, Dordrecht, The Netherlands, 1998.
[165] J.-P. Penot, Subdifferential calculus without qualification assumptions, J. Convex Anal. **3** (1996) 207–219.
[166] J.-P. Penot, *Calculus Without Derivatives*, Graduate Texts in Mathematics, 266. Springer, New York, 2013.
[167] P. Pérez-Aros, Formulae for the conjugate and the subdifferential of the supremum function, J. Optim. Theory Appl. **180** (2019) 397–427.
[168] R.R. Phelps, *Convex Functions, Monotone Operators and Differentiability*, Lecture Notes in Mathematics. Springer-Verlag, Berlin, 1993.
[169] B.N. Pshenichnyi, Convex programming in a normalized space, Kibernetika, **5** (1965) 46–54 (in Russian); translated as Cybernetics **1** (1965) 46–57 (1966).
[170] B.N. Pshenichnyi, *Necessary Conditions for an Extremum*, Marcel Dekker, New York, 1971.
[171] R. Puente and V.N. Vera de Serio, Locally Farkas-Minkowski linear semi-infinite systems, TOP **7** (1999) 103–121.
[172] R.T. Rockafellar, On the subdifferentiability of convex functions, Proc. Amer. Math. Soc. **16** (1965) 605-611.
[173] R.T. Rockafellar, Characterization of the subdifferentials of convex functions, Pac. J. Math. **17** (1966) 497–510.
[174] R.T. Rockafellar, *Convex Analysis*, Princeton University Press, Princeton, N.J., 1970.
[175] R.T. Rockafellar, On the maximal monotonicity of subdifferential mappings, Pacific J. Math. **33** (1970) 209–216.
[176] R.T. Rockafellar, *Conjugate Duality and Optimization*, in: CBMS Regional Conference Series in Applied Mathematics **16**, SIAM VI, Philadelphia, Pa. (1974).
[177] R.T. Rockafellar, Directionally Lipschitzian functions and subdifferential calculus, Proc. London Math. Soc. **39** (1979) 331–355.
[178] R.T. Rockafellar and R. Wets, *Variational Analysis*, Grundlehren der mathematischen Wissenschaften [Fundamental Principles of Mathematical Sciences], 317. Springer-Verlag, Berlin, 1998.
[179] J. Saint Raymond, Characterizing convex functions by variational properties, J. Nonlinear Convex Anal. **14** (2013) 253–262.
[180] A. Shapiro, First and second order optimality conditions and perturbation analysis of semi-infinite programming problems, in: Semi-Infinite Programming, R. Reemtsen and J. Rückmann, edited by Kluwer, Dordrecht (1998) 103–133.
[181] S. Simons, The occasional distributivity of ∘ over □ and the change of variable formula for conjugate functions, Nonlinear Anal. **14** (1990) 1111–1120.
[182] S. Simons, *From Hahn-Banach to Monotonicity*, 2nd ed. Springer-Verlag, Berlin, 2008.
[183] M. Sion, On general minimax theorems, Pac. J. Math. **8** (1958) 171–176.
[184] V.N. Solov'ev, Duality for nonconvex optimization and its applications, Anal. Math. **19** (1993) 297–315.
[185] V.N. Solov'ev, The subdifferential and the directional derivatives of the maximum of a family of convex functions, Izvestiya RAN: Ser. Mat. **65** (2001) 107–132.
[186] J. Stoer and Ch. Witzgall, *Convexity and Optimization in Finite Dimensions, I*, Die Grundlehren der mathematischen Wissenschaften, Band 163 Springer-Verlag, New York-Berlin, 1970.
[187] L. Thibault, *Unilateral Variational Analysis in Banach Spaces, I, II*, World Scientific, 2022.

BIBLIOGRAPHY

[188] L. Thibault, Sequential convex subdifferential calculus and sequential Lagrange multipliers, SIAM J. Control Optim. **35** (1997) 1434–1444.

[189] J. van Tiel, *Convex Analysis. An Introductory Text*, John Wiley & Sons, Chichester, 1984.

[190] V.M. Tikhomirov, *Analysis II, Convex Analysis and Approximation Theory*, RX Gamkrelidze (Ed.), Encyclopedia of Mathematics Vol 14, 1990.

[191] M. Valadier, Sous-différentiels d'une borne supérieure et d'une somme continue de fonctions convexes, C. R. Acad. Sci. Paris Sér. A-B **268** (1969) A39–A42.

[192] M. Valadier, Sous-différentiel d'une enveloppe supérieure de fonctions convexes, C. R. Math. Acad. Sci. Paris **317** (1993) 845–849.

[193] A. Verona and M.E. Verona, Remarks on subgradients and ε-subgradients, Set-Valued Anal. **1** (1993) 261–272.

[194] L.P. Vlasov, The concept of approximative compactness and its variants, Mat. Zametki **16** (1974) 337–348 (in Russian).

[195] M. Volle, Sous-différentiel d'une enveloppe supérieure de fonctions convexes, C. R. Acad. Sci. Paris Sér. I Math. **317** (1993) 845–849.

[196] M. Volle, On the subdifferential of an upper envelope of convex functions, Acta Math. Vietnam. **19** (1994) 137–148.

[197] M. Volle, Calculus rules for approximate minima and applications to approximate subdifferential calculus, J. Global Optim. **5** (1994) 131–157.

[198] M. Volle, Complements on subdifferential calculus, Pac. J. Optim. **4** (2008) 621–628.

[199] C. Zălinescu, Stability for a class of nonlinear optimization problems and applications, in Nonsmooth optimization and related topics (Erice, 1988), 437–458, Ettore Majorana Internat. Sci. Ser. Phys. Sci., 43, Plenum, New York, 1989.

[200] C. Zălinescu, A comparison of constraint qualifications in infinite-dimensional convex programming revisited, J. Austral. Math. Soc. Ser. B **40** (1999) 353-378.

[201] C. Zălinescu, *Convex Analysis in General Vector Spaces*, World Scientific Publishing Co., NJ, 2002.

[202] C. Zălinescu, On several results about convex set functions, J. Math. Anal. Appl. **328** (2007) 1451–1470.

[203] C. Zălinescu, On the differentiability of support functions, J. Global Optim. **15** (2012) 719–731.

[204] C. Zălinescu, Relations between the convexity of a set and differentiability of its support function, J. Optim. **65** (2016) 651–670.

[205] X.Y. Zheng and K.F. Ng, Subsmooth semi-infinite and infinite optimization problems, Math. Program. **134**A (2012) 365–393.

Index

A
absorbing, 21
accesibility lemma, 24
active constraint, 314
adjoint mapping, 67
affine hull, 21
Alaoglu–Banach–Bourbaki theorem, 30
algebraic interior, 24
approximate KKT conditions, 317
approximate subdifferential, 98
approximately compact, 354
Asplund function, 355
Asplund space, 35

B
Baire category theorem, 17
Baire space, 17
balanced, 21
ball, 31
Banach space, 31
BCQ constraint qualification, 317
biconjugate, 63
bidual space, 32
bipolar theorem, 80
boundary, 17
bounded set, 25

C
canonical projections, 20
canonical simplex, 20
Carathéodory theorem, 22
Cartesian product, 17
Cauchy sequence, 17
Cauchy–Schwarz's inequality, 31
characteristic cone, 309
Chebychev set, 354
closed convex hull of a function, 39
closed for convex combinations, 177
closed hull of a function, 39
closed set, 16
closed unit ball, 31
closedness criterion, 177
closure, 17
cluster point, 18
compact, 17
compatibe dual pair, 27
compatible (or consistent) topologies, 27
complete metric space, 17
completely regular, 242
conic hull, 21
consequence, 310
consistent system, 308
constraint system, 308
continuous, 18
convergence, 18
convex, 21
convex function, 37
convex hull, 21
convex hull of a function, 39
convex optimization problem, 308
convex relaxed problem, 330
cyclically-monotone, 34

D

dc (difference of convex) optimization, 141
dense, 17
Dieudonné theorem, 26
dimension, 22
directed set, 15
directional derivative, 34
distance function, 50
dual cone, 27
dual norm, 31
dual pair, 22
dual paring, 22
dual problem, 138

E

e-cyclically and cyclically monotone, 99
Eberlein–Smulian, 32
effective domain (or domain), 36
Ekeland variational principle, 147
epi-pointed, 343
epigraph, 36
epigraphical sum, 49
epsilon-directional derivative, 96
epsilon-Fermat rule, 97
epsilon-minima, 97
epsilon-normal set, 27
epsilon-subdifferentiable, 98
epsilon-subgradient, 97
equivalence class, 16
equivalence relation, 16
evaluation function, 242
eventually, 15
exact inf-convolution, 49
exact post-composition, 50
exposed point, 35
extended active index set, 245
extended e-active index set, 245

F

Farkas theorem, 80
Farkas–Minkowski systems, 312
feasible set, 308
Fenchel conjugate, 60
Fenchel duality, 131
Fenchel inequality, 62
Fréchet-derivative, 35
Fréchet-differentiable, 34
frequently, 15
Fritz-John conditions, 324
fuzzy KKT conditions, 320

G

Gâteaux-derivative, 34
Gâteaux-differentiable, 34
G_delta set, 35
generalized Farkas lemma, 310
Goldstein theorem, 33
graph, 36

H

Hahn–Banach extension theorem, 25
Hausdorff, 17
Hilbert space, 32
homoemorphism, 18

I

indicator function, 46
inf-compact, 19
inf-convolution, 49
inner product, 31
integration criteria, 150
interior, 17

J

James theorem, 33
Jensen's inequality, 37

K

Kadec–Klee property, 32
KKT conditions, 318
Krein–Šmulian theorem, 32

L

Lagrange duality, 131
Lagrangian dual problem, 136
Lagrangian function, 136
Lebesgue integration, 150
left-derivative, 34
lineality space, 25
linear hull (or span), 21
linear isometry, 33
linear isomorphism, 31
linear semi-infinite problem, 308
Lipschitz continuous (or Lipschitz), 17
local Farkas–Minkowski property, 314
locally compact, 18
locally convex space, 26
locally Lipschitz, 17
lower semicontinuous (or lsc), 19

Index 443

M
Mackey theorem, 33
Mackey topology, 29
marginal value function, 52
maximally cyclically-monotone, 34
maximally monotone, 34
Mazur theorem, 32, 396
metric space, 17
minimax theorem, 83
Minkowski gauge, 30
Minkowski sum, 21
monotone operator, 33
Moreau theorem, 68
Moreau–Yosida approximation, 50

N
neighborhood, 16
neighborhood base, 16
net, 15
norm, 28
normal cone, 27
normed space, 31

O
objective function, 308
one-point compact extension, 250
open set, 16
orthogonal subspace (or annihilator), 27

P
partial order, 16
perspective function, 66
perturbed dual problem, 138
perturbed optimal value functions, 138
perturbed primal problem, 138
polar set, 27
positively homogeneous, 38
post-composition with linear mappings, 50
primal problem, 138
product space, 242
projection, 50, 52
proper function, 37
properly separated, 25
Pshenichnyi–Rockafellar theorem, 317

Q
quasi-convex functions, 38
quasi-relative interior, 128
quotient canonical projection, 16

quotient set, 16
quotient topological space, 17

R
Radon–Nikodym property, 35
recession cone, 25
recession function, 51
reflexive Banach space, 33
relative (or induced) topology, 16
relative interior, 23
Riesz representation theorem, 35
right-derivative, 34

S
saturated family, 30
scalar multiple, 21
scalar product, 21
selection, 103
semi-infinite convex optimization problem, 323
semiballs, 28
seminorm, 28
separable, 17
separated, 17
sequentially closed, 18
sequentially compact, 18
sequentially lsc, 19
sign function, 18
Singer–Toland duality, 140
Slater condition, 137
solvable problem, 308
star product, 66
Stegall variational principle, 150
Stone–Cech compactification, 242
strict epigraph, 36
strong CHIP, 324
strong duality, 139
strong minimum, 150
strong separation, 26
strong Slater point, 321
strong Slater qualification condition, 321
strong topology, 29
strongly exposed points, 35
subadditive, 38
subdifferential, 98
subgradient, 98
sublevel sets, 19
sublinear, 38
subnet, 15
support function, 47
support of a function, 20

T

topological dual, 22
topological space, 16
topological vector space, 22
topology, 16
Tychonoff, 242
Tychonoff theorem, 20

U

unit sphere, 31
upper and lower limits, 19

W

w*–Asplund, 35
weak duality, 138
weak projection, 354
weak topology, 28
weak*-topology, 28
Weierstrass theorem, 19

Z

zero duality gap, 139
zero-neighborhoods, 26

The manufacturer's authorised representative in the EU is Springer Nature Customer Service Centre GmbH, Europaplatz 3, 69115 Heidelberg, Germany. If you have any concerns regarding our products, please contact ProductSafety@springernature.com

Printed and bound by CPI Group (UK) Ltd, Croydon, CR0 4YY
25/03/2026
02078184-0001